To Mark with love
from a rather erratic
god father

Charles Darwin's Zoology Notes
&
Specimen Lists from H.M.S. Beagle

This transcription of notes made by Charles Darwin during the voyage of H.M.S. Beagle records his observations on the animals and plants that he encountered, and provides a valuable insight into the intellectual development of one of our most influential scientists. Darwin drew on many of these notes for his well known Journal of Researches (1839), but the great majority have remained unpublished. The volume provides numerous examples of his unimpeachable accuracy in describing the wide range of animals seen in the course of his travels, and of his closely analytical approach towards every one of his observations. Only at the very end of the voyage were his first doubts about the immutability of species consciously expressed, but here are to be found the initial seeds of his theory of evolution, and of the fields of behavioural and ecological study of which he was one of the founding fathers.

CD and another rifleman shooting guanacoes on the banks of the Rio Santa Cruz
on 2 May 1834. Watercolour painting by Conrad Martens.

CHARLES DARWIN'S ZOOLOGY NOTES & SPECIMEN LISTS FROM H.M.S. *BEAGLE*

EDITED BY

RICHARD KEYNES

Emeritus Professor of Physiology in the University of Cambridge, And Fellow of Churchill College

CAMBRIDGE
UNIVERSITY PRESS

PUBLISHED BY THE PRESS SYNDICATE OF THE UNIVERSITY OF CAMBRIDGE
The Pitt Building, Trumpington Street, Cambridge, United Kingdom

CAMBRIDGE UNIVERSITY PRESS
The Edinburgh Building, Cambridge CB2 2RU, UK http://www.cup.cam.ac.uk
40 West 20th Street, New York, NY 10011-4211, USA http://www.cup.org
10 Stamford Road, Oakleigh, Melbourne 3166, Australia
Ruiz de Alacón 13, 28014 Madrid, Spain

First published 2000

Printed in the United Kingdom at the University Press, Cambridge

Typeset by the editor

A catalogue record for this book is available from the British Library

ISBN 0 521 46569 9 hardback

To

CHARLES ROBERT DARWIN

whose dedication and skill as an observer
of Nature has set an example for all time

Contents

Introduction

Charles Darwin, referred to hereafter as CD, arrived in Plymouth on 24 October 1831 in order to accompany Captain Robert FitzRoy on H.M.S. *Beagle* as a scientist and companion. As he noted in his private journal[1], the ship was 'in a state of bustle and confusion'. The carpenters were hard at work fitting up the drawers in the poop cabin, but the corner assigned to him, where for the next five years he was destined to work at his microscope and write his notes, looked too small to accommodate all his possessions[2]. A month later he was able to carry his books and instruments on board, and soon his fears about lack of space had been dissipated. On 4 December he mastered the technique of getting into his hammock, and slept on board for the first time. There followed an endless succession of southwesterly gales that kept the *Beagle* at anchor, and forced the abandonment of two attempts to sail, until on 27 December the wind shifted to the east, and the ship at last got under way.

Although CD's most important achievements were ultimately in the realm of biology, it must not be forgotten that FitzRoy's original intention was that his scientist should examine the land while the officers of the *Beagle* looked after the hydrography[3]. Shortly after the return of the ship to England in 1836, the Captain duly reported[4] that 'Mr Charles Darwin will make known the results of his five years' voluntary seclusion and disinterested exertions in the cause of science. Geology has been his principal pursuit'. The total bulk of CD's *Geology Notes*[5] was nearly four times greater than that of the *Zoology Notes* transcribed here, and a very rough analysis of the scientific topics covered in his letters to Henslow[6] from the *Beagle* shows that about three times more space was devoted to geology and palaeontology than to natural history. CD's geological findings were duly reported to the Geological Society, of which he had just been elected one of the two secretaries, on 7 March 1838[7,59]. His contribution forming Volume III of the joint account of the voyage edited by FitzRoy that appeared in 1839[8] was entitled simply *Journal and Remarks. 1832-1836.*, and when it was reprinted on its own later that same year it became *Journal of researches into the geology and natural history of the various countries visited by H.M.S. Beagle...*[9] In the second and final edition published in 1845[10], the order in the title was changed to 'natural history and geology', and there it remained. Of his three geological books[11], *Coral reefs* was published in 1842, *Volcanic islands* in 1844, and *Geological observations on South America* in 1846, their writing having occupied four and a half years' steady work[12].

Depending on the opportunities offered to him at different periods, the strength of CD's relative liking for geology and natural history fluctuated, but generally geology came out on top. To his sister Catherine he wrote in April 1834[13] 'there is nothing like geology; the pleasure of the first days partridge shooting or first days hunting cannot be compared to finding a fine group of fossil bones, which tell their story of former times with almost a living tongue'; and to his cousin W.D. Fox he had admitted a year earlier[14] 'The pleasure of working with the Microscope ranks second to geology'. The reason was perhaps, as he put it in his *Autobiography*[15], that in comparison with natural history, 'the investigation of the geology of all the places visited was far more important, as reasoning here comes into play'. And it was indeed more in geology than in natural history that he was able to indulge his latent passion for theorising[16], as became apparent as soon as he landed at St Jago in the Cape Verde Islands on 16 January 1832[5,17].

Nevertheless, he was quickly acting in many of his *Zoology Notes* on the strongly felt principle often quoted later on by Emma Darwin[18]: 'it is a fatal fault to reason whilst observing, though so necessary beforehand and so useful afterwards'; while to Wallace he wrote in 1857[19] 'I am a firm believer that, without speculation there is no good & original observation'. Although at the end of his life he wrote somewhat misleadingly in his *Autobiography*[20] that 'My first note-book was opened in July 1837. I worked on true Baconian principles, and without any theory collected facts on a wholesale scale', the truth was otherwise. While dissecting specimens under his microscope, he was constantly questioning himself about the logical implications of his findings, and his interpretations of their complex internal anatomy were always very closely reasoned. Often he carried out little experiments to test the response of his specimens to mechanical stimulation, or exposure to water of the wrong salinity or to alcohol. Moreover, the entries describing the animals that he watched and sometimes captured in the field were models of critical observation, packed with well thought out comments on the possible reasons for their behaviour, distribution and relation with their environment. He was always ready to question the correctness of the conclusions of his predecessors if they conflicted with what he saw for himself, and his intensely analytical approach was from the beginning one of the characteristics that stands out most clearly in his scientific writing.

The first observation in his *Zoology Notes*, dated 6 January 1832, was concerned with luminous matter in the sea. His collecting began in earnest on 10 January, when having quickly constructed the plankton net of which he drew a sketch[21], 'it brought up a mass of small animals, & tomorrow I look forward to a greater harvest'. The captures described in his notes were some medusae, including a Portuguese man-of-war whose powerful toxin he inadvertently got on to his fingers and into his mouth; some salps; and 'a very simple animal' that was new to him, and remained unclassified until he returned to England. Specimen No. 1 in Spirits of Wine was listed as chiefly pteropods, *i.e.* shelled opisthobranch molluscs such as *Limacina* and *Creseis*. Specimen No. 1 not in Spirits was a cuttlefish extracted from the stomach of a black-backed gull on 6 January, followed by a locust (*Acrydium*) and other insects taken on board the ship during the next few days.

CD's note-taking was distinguished from the very start by its orderliness, and by the manner in which he adhered faithfully to his chosen layouts throughout the voyage. Both for his private journal and for the *Zoology Notes* he wrote in ink on gatherings of paper making pages 20 by 25 cm in size, faintly lined and with a red marginal line[22]. At the head of each page, its number and the month, year and location of the observations were entered. In the *Beagle Diary*, the margin was used only to record the day of the month, and occasionally the day of the week. In the *Zoology Notes* CD quickly settled down after the first few pages to writing in the margin an underlined generic or family name for the specimen under consideration, together with its number and sometimes a further brief description. He soon found himself needing often to add extra notes on the backs of the pages, identified by letters in brackets placed opposite the relevant part of the text. Sometimes these were immediate afterthoughts, and sometimes comments arising later from subsequent observations.

As he had begun, so he continued, and in the end well over half of the pages of the *Zoology Notes* were concerned with marine invertebrates. His concentration on this particular field may be attributed not so much to his admitted pleasure in working with his microscope, but to the fact that during the long periods when the *Beagle* was at sea few other activities were open to him. It should also be appreciated that the dissection of a single bryozoan or crab sometimes generated half a dozen pages of notes, whereas observations on a beetle or a frog or a bird seldom occupied more than a few sentences. Many years afterwards he wrote[23] that 'from not being able to draw and from not having sufficient anatomical knowledge a great pile of MS. which I made during the voyage has proved almost useless', a typically self-deprecatory judgement on the merits of his *Geology* and *Zoology Notes* that was quoted almost word for word by Thomas Huxley in his obituary of CD for the Royal Society[24]. There were, nevertheless, many splendid descriptive passages drawn from the *Zoology Notes* that provided the natural history in the *Journal of Researches*, and although the results of his anatomical studies on bryozoans, crustaceans and other invertebrates mostly remained unpublished, there were among them, as will be seen, many pioneering observations of considerable interest. CD's modesty about his skill as an artist was borne out, as Huxley confirmed, by the crudity of the sketches that he drew in the *Beagle Diary* and in his letters, so that it comes as a surprise to see the accuracy of the drawings in fine pencil or ink on separate sheets of unlined paper, of the specimens that he subjected to close examination under his single lens Bancks microscope[25], not infrequently showing new and previously unrecognised anatomical features. These formed the 20 Plates preserved in CUL MS DAR 29 and reproduced in this volume, which each comprise up to a dozen Figures. His cross referencing to further mention of an animal on another page of the *Notes*, or to the Plate and Figure illustrating a particular point in the text, was always impeccable. The efficiency with which he thus organised his written records under very cramped conditions in a ship at sea, often stricken by seasickness, was without doubt an extremely important factor in his success as an observer and a collector both in geology and in natural history.

Another striking feature of the *Zoology Notes* is their total professionalism, despite the fact that on the face of it CD had had little appropriate training. However, in company first with his brother Erasmus and then more importantly with Robert Grant[26], he had in 1827 explored the shores of the Firth of Forth as described in an early diary[27], illustrated with forerunners of his sketches in the *Zoology Notes*; and he had received valuable instruction from Grant on the marine invertebrates that were found there. When he then encountered in the open Atlantic a range of organisms with which he was unfamiliar, he at once began to make extremely effective use of the *Beagle*'s quite extensive library of reference books. They were chiefly the works of the notable French encyclopédistes, of which his favourite was what he called *Dic. Class.*, the 17 volumes of the 'Dictionnaire Classique d'histoire naturelle', but he also consulted Cuvier's 'Le règne animale', Lamarck's 'Histoire naturelle des animaux sans vertèbres', Lamouroux's 'Exposition méthodique des genres de l'ordre des polypiers', Rang's 'Manuel de l'histoire naturelle des mollusques et leurs coquilles', and others[28]. With their help he was able to give generic or family names to quite a number of the marine invertebrates that he collected, though not many of them are still in use today, and he ran into difficulties with organisms belonging to phyla whose existence had yet to be recognised. Thus it was ironic that the 'very simple animal' which he caught in his net on 11 January 1832, and of which his drawing in Figure 1 of Plate 1 (see p. 4 of this volume) is instantly recognisable today as a chaetognath or arrow worm[29], still had him 'at a loss where to rank it amongst other animals' when he found large numbers off the coast of Patagonia (see pp. 66-9), and was only identified after his return to England four years later. In 1832 he had

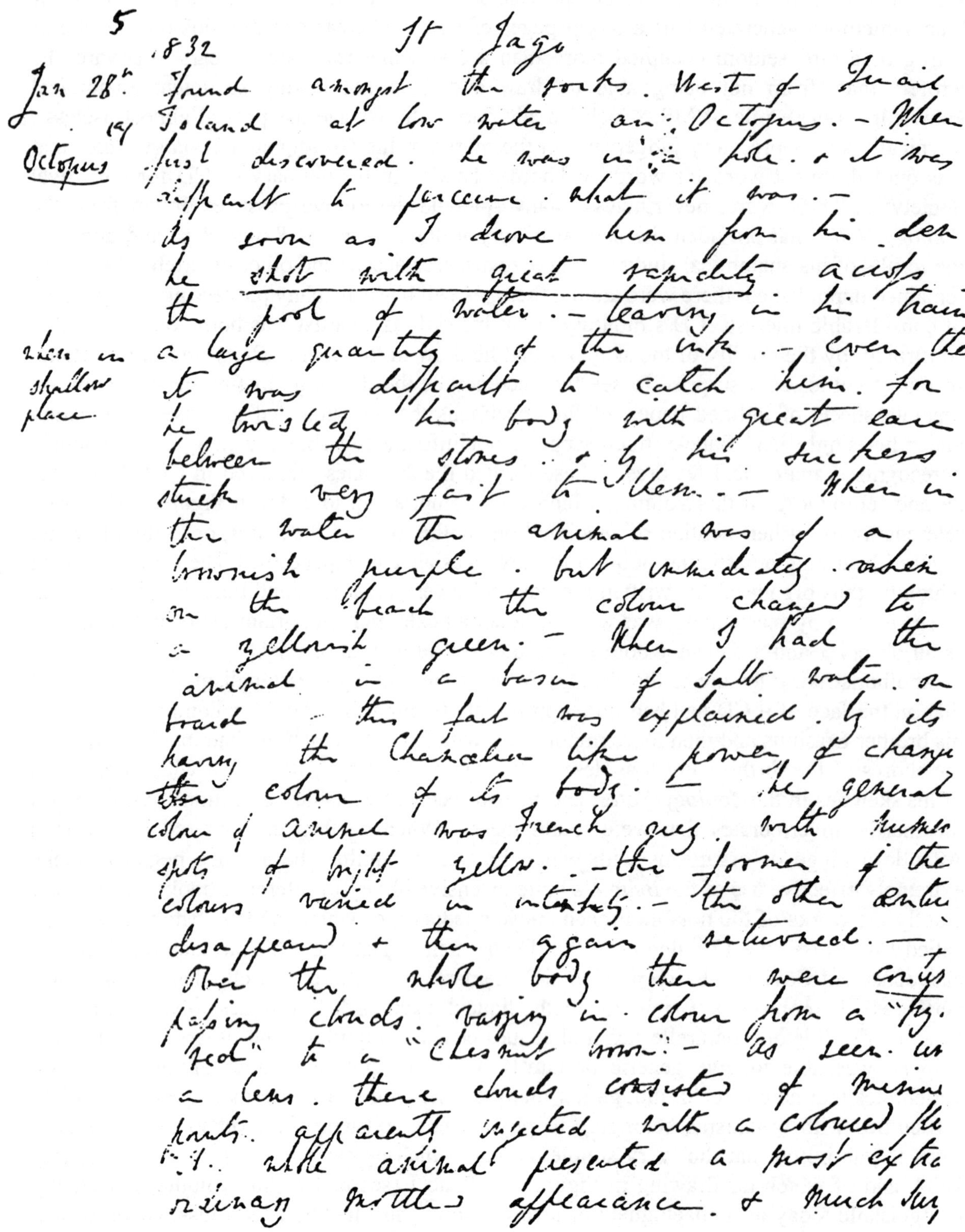
5 5
1832 St Jago
Jan 28th Found amongst the rocks. West of Quail
19 Island at low water an Octopus. — When
Octopus first discovered. he was in a hole. & it was
difficult to perceive what it was. —
As soon as I drove him from his den
he shot with great rapidity across
the pool of water. — leaving in his train
when in a large quantity of the ink. — even then
shallow it was difficult to catch him. for
place he twisted his body with great ease
between the stones. & by his suckers
stuck very fast to them. — When in
the water the animal was of a
brownish purple. but immediately when
on the beach the colour changed to
a yellowish green. — When I had the
animal in a basin of salt water on
board — this fact was explained by its
having the Chameleon like power of changing
the colour of its body. — The general
colour of animal was French grey with numerous
spots of bright yellow. — the former of these
colours varied in intensity. — the other entirely
disappeared & then again returned. —
Over the whole body there were continually
passing clouds. varying in colour from a "hyacinth
red" to a "Chesnut brown". — As seen under
a lens. these clouds consisted of minute
points. apparently injected with a coloured fluid.
The whole animal presented a most extraordinary
mottled appearance. & much surprise

A typical page from CD's *Zoology Notes*

been unaware of the foundation of the genus *Sagitta* by Quoy and Gaimard some five years previously, and he made up for it in a short note published in 1844[30] that he hoped would 'aid more competent judges than myself in ascertaining its true affinities.' But although the 100 species of Chaetognatha are common among marine plankton in tropical seas, even now the precise relationship of the phylum with the other pseudo-coelomates has not been finally settled.

It may also be noted that the barnacles collected by CD on the *Beagle* — eventually the subjects of his monograph on the Cirripedia[31] written at Down House — were always listed among the molluscs as shells, where they were still placed by Cuvier and Lamarck before J. Vaughan Thompson's discovery[32] in 1830 of their metamorphoses suggested their transfer to the crustaceans. But when CD was examining a shell that at first he had doubtfully entered as the marine snail *Conus* (see p. 135), he decided that because of its strikingly crustacean characteristics and possession of an external 'pied machoire' it was better identified as a barnacle common in the Falkland Islands. Hence independently of Thompson he had already observed the crustacean affinities of barnacles, and as he recognized later[23], a knowledge of crustaceans and of their larval stage the Zoea, was one of the most valuable outcomes of his dissections of marine invertebrates during the voyage.

The principal problem in classification encountered by CD in the 1830s lay in determining the true nature of some of the colonial plant-like invertebrates then still known colloquially as Zoophytes or Polypes, and nowadays separated into Cnidaria such as hydrozoa, anthozoa (including corals) and scyphozoa, Bryozoa and sponges. The smallest of these were the corallines, but thanks to the classical studies of John Ellis[33] it had been accepted in many quarters by the end of the 18th century that like some of the coelenterates closely similar to them in appearance, they belonged to the animal kingdom. Only Linnaeus was not wholly convinced, and coined the name Zoophyta — a group intermediate between plants and animals — for the corallines. In 1820 de Blainville discovered that the polyps of certain zoophytes possessed both mouth and anus, suggesting that they should be placed on a higher level than other coelenterates; and in 1827 Robert Grant[34] observed that some of them had ciliated tentacles and a recurved alimentary canal. In 1830, J. Vaughan Thompson, working independently on zoophytes off the southern shores of Ireland, also discovered that there were two anatomical forms of polyps, and added to his memoirs on crustaceans a fifth entitled *On Polyzoa, a new animal discovered as an inhabitant of some Zoophites*[35], in which he created a new animal class, the Polyzoa, to replace the Zoophyta. As has been explained by Ryland[36], the phylum concerned is now best known as the Bryozoa, while those animals in which the anus opens inside the circlet of tentacles belong to the phylum Entoprocta.

At the beginning of the voyage, CD referred to all such animals indiscriminately as corallines or coralls, although some of them were in fact hydrozoa or hydrocorals, some bryozoans, and some coralline algae. When in the end he had concluded[43] that his 'true corallinas' were indeed algae such as *Corallina* and *Amphiroa*, he listed this group as Nulliporae. The bryozoans were generally 'encrusting corallines' or Flustrae, and the reef-building hydrocorals were Madrepores. He had thus improved on the still prevailing confusion in the classification of the Zoophytes or Polypiferous Polypi in the accounts of Cuvier and Griffith[37] that he had with him on the *Beagle*.

The first corallines to be collected during the voyage were identified as *Sertularia*, a term applicable at that time both to bryozoans and hydrozoans, and a coralline alga *Amphiroa*.

Then in August 1832, off the coast of Patagonia at a depth of 14 fathoms, the first bryozoans were found, a 'corall' listed as 'Cellepora ?' and confirmed in 1901[38] as being *Cellepora eatonensis*, and a specimen which CD immediately and correctly recognised as related to *Flustra*, to whose leaf-like colonies in the Firth of Forth[27] he had been introduced by Robert Grant[26], and with which his first scientific paper, delivered to the Plinian Society in March 1827, had been concerned. But what, seen under his microscope, rendered the new genus 'singular' (see p. 73) was the occurrence of peculiar organs on the edges of the cells that resembled in shape the open beak of a vulture, and nodded continuously at a frequency of about five seconds. He asked himself what their function could possibly be, rejecting generation 'which is the last resource in all puzzling cases'. Later he found similar organs on other zoophytes, and speculated at some length on their role[39]. The organ in question was the type of anascan heterozooid now known as a pedunculate avicularium, and although a defensive role with adaptive value has been proposed for it, even today there is a shortage of firm evidence in support of this or any other hypothesis[40].

More 'coralls', identified by CD from Lamouroux's pictures[41] as *Celleporaria* and other bryozoans now placed in suborder Ascophora, were collected in Tierra del Fuego four months later, while in March 1833 a number of coralline algae were collected around East Falkland Island. Then at Port Desire in January 1834 considerable quantities of the 'Corallina' *Halimeda* were thrown up on the beach, and (see note (b) on p. 187) CD concluded from his examination of their articulation and mode of propagation that 'I do not believe Corallina to have any connection with the family of zoophites'. For as he wrote later to Henslow[42]: 'the 'gemmule' of a Halimeda contained several articulations united, & ready to burst their envelope & become attached to some basis [*i.e.* base]. I believe in Zoophites, universally the gemmule produces a single Polypus, which afterwards or at the same time grows with its cell or single articulation'. It followed that the zoophytes were definitely not plants, although this evidence was provided by the green alga *Halimeda* belonging to the Chlorophyta, and not by the coralline algae belonging to the Rhodophyta[43].

In March 1834, when the *Beagle* was once again in Tierra del Fuego, more specimens of *Flustra* were obtained that were bryozoans of several families of the order Cyclostomata as well as Cheilostomata. Pursuing a 'lately determined' intention, described in July to his sister Catherine[44], 'to work chiefly amongst the Zoophites or Coralls', CD engaged on an orgy of comparative anatomy, and anticipated a remarkable amount of bryozoan biology that had yet to be formally elucidated. He observed in Specimen 874 (see p. 195) the functioning of the autozooidal operculum 'like lower jaw of a bull-dog'. He correctly appreciated (see pp. 197 and 207) the phenomenon of degeneration and regeneration of the bryozoan polypide, and clearly saw the associated brown bodies[36]. He perceived (see p. 197) the relationship between the pedunculate vulture's-head avicularia of the erect species, and the adventitious sessile avicularia of an encrusting form. He observed (see pp. 198 and 206-207) ovicells brooding young, and (on p. 205) described the kenozooidal rootlets (rhizoids) that support and attach many erect forms. Bringing up in his net a specimen of a similar but new animal that he labelled 'polype?' (see pp. 199-201), he at once appreciated the different location of its anus that now distinguishes the phylum Entoprocta.

During the next months, off the coast of Patagonia and further to the south, he collected more bryozoans and accurately described (see pp. 222-3) the anatomy of the ascophoran *Eschara gigantea* with its calcified frontal wall. A few days later (see pp. 226-9) he found another specimen that he misidentified as the cyclostome *Crisia*, but which was in fact the anascan *Caberea minima*, belonging to the superfamily Cellularioidea. This species possessed the type of heterozooid now known as a vibraculum, a long tapering bristle-like seta mounted on a basal chamber containing musculature capable of swivelling and rotating the seta. CD's graphic description of the coordinated sweeping movements of the setae on each branch of the coralline was a triumph of accurate observation. Again he speculated on the function of the vibracula, and concluded that their role was not 'to drive away enemies and impurities', though the modern view[36] is rather that the species with well developed vibracula depend on their ability to discourage small organisms such as larvae, and particles of sand or silt, from settling on the surface of the colony.

Then at Port Famine in June 1834 he turned his attention (see p. 232) to specimen 983 in spirits of 'a very simple *Flustra*' — which was subsequently identified by S.F. Harmer[38] as belonging to the related species *Membranipora membranacea* — 'so that I might erect at some future day, my imperfect notions concerning the organization of the whole family of Dr Grant's Paper[34]'; and gave a good description of the organisation of its polypus. But this turned out to be the last bryozoan to be discussed in the *Zoology Notes*, and except for quite a number of specimens collected in January 1835 in the Chonos Archipelago and off Chiloe, and two hauls in the Galapagos Islands in September 1835, no more were collected during the second half of the voyage, though further specimens were taken of the coralline algae described by CD as corallinas, and of the true reef-building hydrocorals.

Nevertheless, CD's resolution to think further about the organization of the Flustrae was not wholly abandoned. It has survived, at least in part, as two loose pages of conclusions about the anatomy of corallines, probably written on board the *Beagle* early in 1836, to which attention was drawn by Sloan[45]. Here CD has in effect decided that the Flustraceae belong in a phylum of their own, although nowhere did he ever refer to Ellis[33] or Thompson[35], and is musing constructively on their biology. These two pages are listed as CUL MS DAR 5.98-9, the page numbering being that of the cataloguer, not CD's, and run as follows:

> [p. 98 commences]
> That the number of arms in Polypus of the Flustraceæ varies from 8 to 28 & is no more than a Specific character:
> That a proportion is kept up between simplicity of Polypier & number of arms.— that the same essential organ[s] are found in very varying forms of Polypier.—
> That the degree of stony nature in Corallines is entirely futile as a character[46].—
> That the ~~fo~~ orders of Lamouroux of Cellepora.— Cellaria & Flustra should be included in one family (probably also some Escharæ & Milleporæ[47]).—
> That one Sertularia ~~would~~ is also included.—
> That the structure of the Flustraceæ is most widely different from the Clytias[48]. not only in the Polypus, but in the generation in the former case each ovule & Polypus has some intimate connexion. in the latter it is a young Polypus altered.— (Manner of growth?)
> General Anatomical discussion.—
> [added later in pencil] (Study Hydra & Actinia & my Madrepore & Sigillina in Blainville) (Sigillina & Polypus) [pencil note ends]
> That the connexion of the cells although not apparent in the true Flustræ must exist:

from similarity in growth & chain of gradation in the Capsule Flustræ: & in the Flustra of **P 234** & true Flustræ & Cellariæ having same body.—

That the Polypier is the essential part in the Corallines, it produces the cells & ~~young~~ in young Polypi (& after death of Polypus consequent on generation reproduces them?)—

That the mere possession of arms has grouped very heterogenious animals.—

That Corallina is a plant.—

[in pencil] Does it not emit in Suns rays gaz[49].— [pencil note ends] [continued on verso]

In Virgularia[50] does the truncate extremity correspond to ~~extremity of branch~~ root in Corallium[51] ? Examine extremities & the bag to extremity of branch.

The relative position of Polypier, with living mass in the Lamelliform.—

The structure of transparent extremities of Corallina.— Regrowth of Corallines when separated.

In the capsule Flustræ, cells without Polypi have capsules (Moveable)? Yes? I believe strong proof of disconnection.—

[p. 99 commences]

A close connection & co-sensation between the Polypi of many Corallines is established by the co-movements of "Capsules Flustras" of the setæ in Crisia[52]: the flashes of light in Clytia[53]: strongly seen in Virgularia, & in Alcyonium an injury in the stem causing all to collapse: whilst one *[illeg.]* being injured did not affect the mass.— on the other hand, one point in a Synoicum Blainv: affected all round it for some distance[54].—

Have not the Escaræ in the growth of the Polypier an analogy with the Celleporariæ: where cells appear formed in a cellular tissue (or group of hoods, or angular tubes as in Favosites) of a stone?—

A cell reproduces its Polypus

The stony striæ, on outside of Lobularia[55], connecting link with stony Zoophites.—

The Lobated ~~form~~ *[illeg.]* position of tentacula in Chiloe Actinia perhaps is an analogy in change between a Caryophyllia & Gorgonia or Corallium?— it shows a passage of this arrangement, without material change in animal.—

It is important to see in Clytia[48], substance included in a young cell appearing equally ready to form Polypus or Ovules.— the Coralline must produce this matter; not the Polypus the gemmule.—

I am inclined to think in Corallines, such as Tubularia[56] & Flustra, the Polypier is as much a living ~~man~~ being as any Plant, (as a Lichen or Corallina) that it communicates with the circumambient fluid either simply as in Clytia, or in more complicated manner, as in Flustra.

[continued on verso]

How little organization can be seen in Corallina, yet even the basic articulations produce paps with gemmules.— In the Polypier of the Flustraceæ it seems to make little difference, whether a central living axis is clearly visible or whether it (probably) forms a thin fold at the base of cells, in the encrusting Flustræ.—

I imagine in the Lamelliform Coralls, the Polypier is only an ~~ex~~ internal secretions, (a bony axis to give support) the Polypier being then the mass of living Matter: we see it thus in Virgularia[50].—

There is an analogy between the corall-forming Polypi & turf-forming plants.— Hence here the soft matter ought to form the gemmules, as in the hard matter in the other cases.—
I think there is much analogy between Zoophites & Plants, the Polypi being buds: the gemmules the inflorescence which forms a bud & young plant.—
in Sertularia, the Capsules with gemmules appear to have no relation with any one Polypus; how could it form a totally different sort of capsule to its own, & in a place where it, the Polypus is never found.—
[p. 99 ends]

It has been suggested by Sloan[45] that CD's intellectual development as a biologist was strongly influenced by his early contacts with Robert Grant[26] in Edinburgh, which steered him to pursue on the *Beagle* a programme of research on marine invertebrates oriented from the start in the direction of transmutation. However, the validity of this proposition has to be questioned. In the first place, CD paid no special attention to corallines during the first eight months of the voyage, and when he found his first specimen of *Flustra* what at once excited him was not the issue of whether it was a plant or an animal, but the remarkable properties of its vulture's beak capsules, the possibility that these organs might have any role in generation being scornfully dismissed. Later on, he worked out correctly many of the details of the anatomy of bryozoans that subsequently served to differentiate between their several families, and when he came across one belonging to what is now recognised as the phylum Entoprocta, he immediately spotted the essential diagnostic feature. Hence his studies on bryozoans were primarily an exercise in comparative anatomy, very similar in nature, and in the end less useful to him[15], than the observations on the numerous crustaceans that he dissected. Although some mention is made of changes taking place between related animals in his final two pages of notes, and analogies are suggested between hydrocorals and turf-forming plants, it is difficult to read into them views on the transmutation of species that he had not yet begun to develop seriously in any other context.

Very soon after returning to England in 1836, CD was disconcerted to find[57] that 'the Zoologists seem to think a number of undescribed creatures rather a nuisance', and was unable to obtain expert assistance with the classification of his marine invertebrates. Although Robert Grant[26], who was by then Professor of Zoology at University College London, did express an interest in some of the corallines, it was not followed up. CD wrote of Grant later[58] that 'He did nothing more in science — a fact which has always been inexplicable to me'. CD's original intention to give an account of some invertebrates in *The Zoology of the voyage of H.M.S. Beagle* therefore fell by the wayside, although from the introduction to *Journal of Researches* **2** it would appear that as late as 1845 it had not been finally abandoned. At this time, CD was still referring to the bryozoans as zoophytes, and there is no record of his ever knowing of the successful naming of the group in 1830 by Thompson[35] as Polyzoa. When in March 1837 he was writing on page 130 of the Red Notebook[59] that 'if one species does change into another it must be per saltum — or species may perish', Zoophytes and Polypi must already have begun to fade from the picture as far as he was concerned. There are indeed fewer than a dozen brief references to them in the whole series of Transmutation Notebooks[60].

There were of course many other terrestrial invertebrates such as insects and spiders to which CD needed no introduction from Robert Grant, and which he collected avidly in a conventional way. He also took a great interest in the habits of some of the marine and terrestrial planarians that he found, which were free-living turbellarian flatworms now placed in orders Tricladida and Polycladida. In his paper published in 1844[61], a number of new species were described, though in the absence of further specimens from the areas of South America where he was working, they cannot always be assigned with certainty to a modern genus.

CD had less scope at the time of the voyage to theorise widely in discussing the animals that he collected than in his geological studies, but his *Zoology Notes* were nevertheless very much more than descriptions of the colouration and other details of his specimens that might be necessary for their taxonomic classification. At St Jago on 28 January 1832 he found an octopus among the rocks at low water, and recorded a splendid description of its changes in colour when he tried to grab hold of it, and of its responses on board the ship to the application of electric shocks and of being scratched. Seen under a lens he noted that the passing clouds of colour 'consisted of minute points apparently injected with a coloured fluid' — one of the earliest reports of the properties of their chromatophores. He was always interested in the locomotion of animals, and in the precise way in which they walked or ran or flew or swam, and soon we find him in Bahia (see p. 26) working out how the puffer fish *Diodon* takes up water by swallowing air in order to distend itself for regulation of its overall density and centre of gravity, and uses its pectoral fins after collapsing the caudals to enable itself, contrary to Cuvier's opinion, to swim while upside down. On the same day he caught a luminous click beetle, and critically examined the mechanism by which it bent its spine as a spring in order to jump suddenly into the air, this time finding grounds for disagreement with the account given in the *Dic. Class.* Next it was a migration of driver ants that attracted his attention, then the movements of some pulmonates, and a few weeks later (see p. 48) he came across 'the only butterfly I ever saw make use of its legs in running, this one will avoid being caught by shuffling to one side'. Many further examples could be quoted, among which one of the highlights would be his classical description (see p. 104) of the coating of the *Beagle*'s rigging off Monte Video by the gossamer web of spiders of family Linyphiidae that disperse by air. Others would be his accounts of the flights and feeding habits of *Rhynchops* (p. 159), humming birds (p. 235-6), condors (p. 254) and frigate birds (p. 300).

A field of biology of which CD was one of the founding fathers, together with Linnaeus, Buffon and Humboldt, was ecology, and many instances of his pioneering observations on the relations of animals with their environment are to be found in the *Zoology Notes*. Thus in May 1832 he wrote in Rio (see pp. 58-9):

> 'I could not help noticing how exactly the animals & plants in each region are adapted to each other.— Every one must have noticed how Lettuces & Cabbages suffer from the attacks of Caterpillars & Snails.— But when transplanted here in a foreign clime, the leaves remain as entire as if they contained poison.— Nature, when she formed these animals & these plants, knew they must reside together.—'

After the *Beagle*'s first visit to Tierra del Fuego in the southern summer of 1832-3, CD prefaced with an excellent account of the severity of the weather, backed up by some temperature records, some general observations correlating the climate with the growth of trees, the formation of peat, and the populations of particular species of mammals, birds and insects. He noted on p. 134 that although the thermometer often rose to about 60° [Fahrenheit]:

> 'Yet there were no Orthoptera, few diptera, still fewer butterflies & no bees, this together with [the] absence of flower feeding beetles throughily [*sic*] convinced me how poor a climate that of Tierra del F. is'.

Visiting the Falkland Islands for the second time in April 1834, he wrote in a memorable passage on p. 215 about the marine zoology:

> 'Its main striking feature is the immense quantity & number of kinds of organic beings which are intimately connected with the Kelp. . . I can only compare these great forests to terrestrial ones in the most teeming part of the Tropics; yet if the latter in any country were to be destroyed I do not believe nearly the same number of animals would perish in them as would happen in the case of Kelp: All the fishing quadrupeds & birds (& man) haunt the beds, attracted by the infinite number of small fish which live amongst the leaves: . . . On shaking the great entangled roots it is curious to see the heap of fish, shells, crabs, sea-eggs, Cuttle fish, star fish, Planariæ, Nereidæ, which fall out. . . One single plant form is an immense & most interesting menagerie.— If this Fucus was to cease living, with it would go many: the Seals, the Cormorants & certainly the small fish & then sooner or later the Fuegian man must follow.— the greater number of the invertebrates would likewise perish, but how many it is hard to conjecture.'

He commented frequently, and often tested his observations experimentally, on the adaptation of marine animals to fresh water and *vice versa*, as when near Rio having found a fresh water snail in a lake often made salty by the sea, he asked: 'Is not this fact curious, that fresh water shells should survive an inundation of salt water? In the neighbouring Lagoon, Balani were adhering to the rocks.' Sometimes his speculations were perhaps a little wide of the mark, as on finding fresh water beetles in a stream at the Cape Verde islands 'supposed to be part of Atlantis' (see p. 371 of Specimen List); or when (see pp. 109 and 137) he found barnacles in the Rio Plata and at the Falkland Islands that he thought might be especially adapted for brackish and even for fresh water, possibly by keeping their opercula more 'throughily' closed in fresh water. But when he found beetles alive and swimming actively in the sea seventeen miles off Cape Corrientes[62] he decided that they had survived being washed down from the Rio Plata, and that this was 'a very instrumental means in peopling Islands with insects'.

In the *Zoology Notes* themselves there is no direct evidence as to when CD's belief in the stability of species began to be shaken, for he was still thinking about 'centres of creation' when he arrived in the Galapagos in September 1835[63], and still speaking of a Creator when he was musing about the lion-ant in Australia four months later[64,65]. His doubts about the immutability of species were first expressed when he was reorganising his notes some time between mid-June and August 1836, and writing about the Galapagos mocking birds *Mimus thenca* in his *Ornithological Notes*[64,66], said:

'These birds are closely allied in appearance to the Thenca of Chile or Callandra of La Plata. In their habits I cannot point out a single difference. They are lively, inquisitive, active, run fast, frequent houses to pick the meat of the Tortoise which is hung up, sing tolerably well; are said to build a simple open nest; are very tame, a character in common with the other birds. I imagined however its note or cry was rather different from the Thenca of Chile? Are very abundant over the whole Island; are chiefly tempted up into the high & damp parts by the houses & cleared land. I have specimens from four of the larger Islands: the two above enumerated [males from Charles and Chatham Islands]; a female from Albemarle Isd. and a male from James Island. The specimens from Chatham & Albemarle Isd appear to be the same; but the other two are different. In each Isld each kind is exclusively found: habits of all are indistinguishable. When I recollect the fact that [from] the form of the body, shape of scales & general size, the Spaniards can at once pronounce from which Island any Tortoise may have been brought. When I see these Islands in sight of each other, & possessed of but a scanty stock of animals, tenanted by these birds but slightly differing in structure & filling the same place in Nature, I must suspect they are only varieties. The only fact of a similar kind of which I am aware is the constant asserted difference between the wolf-like Fox of East & West Falkland Islds. If there is the slightest foundation for these remarks the zoology of Archipelagoes will be well worth examining; for such facts would undermine the stability of Species.'

Nevertheless, several of the issues to which he often returned earlier may give some indication as to how, albeit subconsciously, his ideas about evolution were taking shape. Thus he always asked himself whether the rats and mice, and other domestic animals, were indigenous or introduced species, and how much variation they displayed. Finding a rat on Goriti Island near Maldonado, he thought (see p. 171) because of its huge size and habits that it was 'an aboriginal', but the final decision[67] was that it was an extra large variety of the European *Mus decumanus*. A similar problem arose in relation to the black rabbits and other animals found in the Falkland Islands (see p. 209), but the rabbits had been released by early settlers[68] and resembled 'the cattle & horses, which are of as varying color as a herd in England'. CD once more thought that the mice were indigenous, but his specimens were eventually identified[69] as a variety of the European *Mus musculus*. It was clear on the other hand that no foxes had been introduced, and like three mainland species *Canis magellicanus*, *C. fulvipes* and *C. azarae* that he collected in Chile and Argentina, the two varieties of the Falkland fox *C. antarcticus*[70] proved to be indigenous. They were, however, all too approachable, and CD concluded: 'very soon these confident animals must all be killed: How little evidence will then remain of what appears to me to be a centre of creation.' In Ynche Island in the Chonos Archipelago (see p. 281) he found 'very many wild goats' whose 'color was pretty uniform' and which were evidently 'retrograding into their original figure & kind'.

Again, he was always assiduous in collecting the parasites of his specimens of all kinds, and having collected the lice from the native guinea-pig known as Aperea in Maldonado, commented (see p. 340):

'it would be interesting to compare these parasites with those inhabiting an Europæan individual to observe whether they have been altered by transportation: It would be curious to make analogous observation with respect to various tribes of men.'

Later he collected a louse in Chiloe (see p. 283) that he considered to be identical with those carried by the Patagonians at Gregory Bay, and quoted evidence from a surgeon from an English whaler for the existence of differences with those of Europeans; but this has not been confirmed[71]. Two of the first specimens that he collected in the Galapagos (see p. 412) were Acari from a marine iguana and from the Pudenda of a tortoise. He did not confine himself to vertebrate parasites, but also (see p. 87) noted their presence in the body cavity of a ctenophore.

He also took a particular interest in coprophagous beetles. Noting in Maldonado (see p. 175) 'the ample repast afforded by the immense herds of horses & cattle almost untouched', he continued:

'This absence of Coprophagous beetles appears to me to be a very beautiful fact; as showing a connection in the creating between animals as widely apart as Mammalia & Insects. Coleoptera, which when one of them is removed out of its original Zone, can scarcely be produced by a length of time & the most favourable circumstances.— The same subject of investigation will recur in Australia: If proofs were wanting to show the Horse & Ox to be aboriginals of great Britain I think the very presence of so many species of insects feeding on their dung, would be a very strong one.'

And commenting much later on specimen 3819 (not in spirits) he said:

'Very common beetle beneath dung on higher parts of St Helena. This is the most extraordinary instance yet met with of transportal or change in habits of stercovorous insects.'

In Australia the native beetles turned out to be largely restricted to wooded rather than pastoral areas, so that as in Maldonado the dung of cattle and horses remained uneaten. However, the several species of Scarabaeidae that CD found in Tasmania under the dung of cows (see p. 234) were probably native to the island, and had no difficulty in adjusting themselves to a new and copious supply of food. Not until the 1960s were programmes set up by CSIRO for the introduction of dung beetles from Africa and Europe to Australia in order to control dung-breeding pests of cattle and man, and at the same time to bury more dung with consequent improvement of the pasture[72]. The dung beetles in St Helena were presumably of African origin, and able to make do with mouse dung.

A further theme with obvious implications for the species problem was the geographical distribution of different species, and their isolation on islands or by mountain ranges. Arriving at the Falkland Islands for the first time on 1 March 1833, just after the British flag had first been hoisted, he found it 'one of the quietest places we have ever been to', and with all the boats away had little to occupy him except for his thoughts. These he noted down in his pocketbook, and they include the following queries and comments[73]:

'March 2. Falkland—
To what animals did the dung beetles in S. America belong — Is not the closer connection of insects and plants as well as this fact point out closer connection than Migration.
Scarcity of Aphidians?
The peat not forming at present & but little of the Bog Plants of Tierra del F; no moss; perhaps decaying vegetables may slowly increase it. — beds ranging from 10 to one foot thick.
Great scarcity in Tierra del of Corallines, supplanted by Fuci: Clytra prevailing genus.
Tuesday 12th —
Examine Balanus in fresh water beneath high water mark.
Horses fond of catching cattle — aberration of instinct.
Examine pits for Peat. Specimen of do — Have there been any bones ever found &c or Timber.
Are there any reptiles? or Limestone?
21st
Saw a cormorant catch a fish & let it go 8 times successively like a cat does a mouse or otter a fish; & extreme wildness of shags.
22nd.
East of basin, peat above 12 feet thick resting on clay, & now eaten by the sea. Lower parts very compact, but not so good to burn as higher up; small bones are found in it like Rats — argument for original inhabitants: from big bones must be forming at present, but very slowly: Fossils in Slate: opposite points of dip: & mistake of stratification: What has become of lime?
It will be interesting to observe differences of species & proportionate Numbers: what also appear characters of different habitations.
Migration of geese in Falkland Islands as connected with Rio Negro?'

There are not many direct references in the *Zoology Notes* themselves to the geographical distribution of different species of mammals and birds except (see pp. 188-90) in the case of the ostrich that CD called the 'Avestruz Petise'. This was named *Rhea Darwinii* by Gould, when he mounted the specimen shot by Conrad Martens at Port Desire in January 1834[74], which was partly eaten before CD had realised that it belonged to a smaller and darker species than the *R. Americana*, that was common further to the north. The two birds came to provide the best known example of the manner in which closely related species with overlapping ranges replaced one another in proceeding southwards over the continent. There was next an essay written on board the *Beagle* in 1834 by CD[75] entitled 'Reflection on reading my Geological Notes', in which he developed a narrative framework for the history of life on the continent, and listed the mammals that could reasonably have migrated sequentially southwards from their northern original homes. And in two relevant notes on some of the birds of Chile[76] he wrote:

'These forms appear to our eyes singular to be the common birds throughout an extensive country. In T. del Fuego the Certhia & Troglodytes were the two most abundant kinds. In central Chile both are found, but extremely few in numbers. In that country (& in a like manner in a like case in other countries) one is apt to feel surprise that a species should have been created, which appears doomed to play so very insignificant a part in the great scheme of nature. One forgets that these same beings may be the most common in some other region, or might have been so in some anterior period, when circumstances were different. Remove the Southern extremity of America, & who would have supposed that Certhia, Troglodytes, Myothera, Furnarius had been the common birds over a great country.'

and

'It appears to me, that when the lists & collections of birds made in the different parts of S. Southern America are compared, a large number will be found to have surprisingly large geographic ranges. No doubt the similarity in physical constitution of the country; over T. del Fuego & the whole west coast as far north as Concepcion; & again between Patagonia, the lofty valleys of the Cordillera, & northern Chili; & lastly but in a much lesser degree between La Plata & central Chili, is the chief cause of this fact. I should observe, that in the few cases where I have spoken of Lima (Lat 12°) as the Northern Habitat of any species; it is probable that the real boundary lies ten degrees further north (near C. Blanco), where the arid open country of Peru is converted into the magnificent forest land of Guyaquil.'

It is probable, however, that these passages were added to the *Ornithological Notes* shortly after the return of the *Beagle* to England. For in a document now filed with his unpublished *Beagle* Animal (*i.e.* mammal) Notes[77], he drew up long lists of the closely related birds and mammals found on the east and west sides of the Andes, and considered possible reasons for their distribution. The *Animal Notes* were headed 'Gt Malbro' [St], where starting on 13 March 1837 he lived for 21 months in furnished rooms with his secretary and servant Syms Covington, so that such material belongs strictly to the period after the end of the voyage when he had already begun to develop his ideas on the transmutation of species. Nevertheless, the role of geographical distribution was clearly in his thoughts very early on.

The second field of biology to whose establishment CD made major contributions was the study of animal behaviour[78,79]. Most significantly, he appreciated from the start that behaviour was an important factor to be taken into account in identifying a species, as in the case mentioned on p. 50 of the butterflies which shuffled to one side, '& which from appearance & habits were I am sure the same species'. The following year (see p. 211) he noted that the carrion-feeding hawk caracara had a 'connection in habit as well as in structure with true Hawks'. Other examples could be quoted, and it was possibly the close similarity in habits of the various Geospizinae in the Galapagos except for the cactus finch (see p. 297), that deterred him from appreciating their significance when he saw them, though at the same time it was behavioural differences between the mainland species of mocking bird that had led him to distinguish *Mimus orpheus* in Monte Video from *M. patagonicus* on the Rio Santa Cruz.

There are many vivid descriptions of the behaviour of animals at all levels, from the ants in Bahia (see p. 29), through spiders spinning their webs and wasps preying on them (see p. 38), the 'monstrous' coconut crabs in the Cocos Keeling Islands (see p. 311), penguins and steamer-ducks in the Falklands (see p. 213), to the herds of guanaco on the pampas (see p. 181-2). CD's speculations on the underlying reasons, such as the attribution to an instinct 'to find new countries' that leads flocks of butterflies to fly out to sea (see p. 121), are not always successful. The motivation of the biscatche for collecting large piles of rubbish in front of their holes (see pp. 180-1) is described in more anthropomorphic terms than would be acceptable today, but this does not detract from the liveliness of his accounts, nor from his purposeful correlation of behaviour with details of structure and environment.

In this field, as in all else, CD was a superbly skilful and accurate observer who thanks to his intensely analytical approach invariably made a highly effective use of the opportunities offered to him, whether to conduct studies of the comparative anatomy of marine invertebrates, or to examine the distribution, ecology and behaviour of a wide range of terrestrial animals. He was thus enabled to examine the animals occupying many different environments, and had the very good fortune to be taken by the *Beagle* to the Galapagos, which turned out eventually to be an ideal place, rivalled only by Hawaii and Madagascar, for studying the evolution of new species in isolated islands. In addition, the *Beagle* landed him at places where exceptionally informative fossils were lodged in the cliffs, and enabled him to visit the Andes and the coastal plains on either side of the continent where there was much for a geologist to learn about the formation of a mountain range and the accompanying rise and fall of the land. It might not be an exaggeration to say that he was exposed in those five years to more new facts than any previous scientist, and such were his talent for observation and his genius afterwards to arrive by hard thinking at fundamentally new explanations for what he had seen, that *On the Origin of Species* was the inevitable outcome.

CD himself summed up the whole story rather nicely in a letter to his sister Catherine written from Maldonado on 22 May 1833[80]:

> 'I am quite delighted to find the hide of the Megatherium has given you all some little interest in my employments. These fragments are not however by any means the most valuable of the Geological relics. I trust & believe that the time spent in this voyage, if thrown away for all other respects, will produce its full worth in Nat: History. And it appears to me the doing what little one can to encrease the general stock of knowledge is as respectable an object of life as one can in any likelihood pursue. It is more the result of such reflections (as I have already said) than much immediate pleasure which now makes me continue the voyage. Together with the glorious prospect of the future, when passing the Straits of Magellan, we have in truth the world before us. Think of the Andes; the luxuriant forest of the Guayquil; the islands of the South Sea & new South Wale[s]. How many magnificent & characteristic views, how many & curious tribes of men we shall see. What fine opportunities for geology & for studying the infinite host of living beings: is not this a prospect to keep up the most flagging spirit? If I was to throw it away, I don't think I should ever rest quiet in my grave: I certainly should be a ghost & haunt the Brit. Museum.'

So now let his *Zoology Notes* speak for themselves.

Endnotes to Introduction

1 *Beagle Diary* p. 4.
2 In June 1833 the Captain gave him all the drawers in the poop cabin formerly belonging to John Lort Stokes, mate and surveyor, so that he had it to himself (see *Correspondence* **1**:313); and to accommodate his specimens, he had in addition a very small cabin under the forecastle. See Vol. 1, pp. 218-24 of *The life and letters of Charles Darwin*. Edited by Francis Darwin. John Murray, 1887.
3 *Narrative* **1**:385.
4 R. Fitz-Roy (1836) Sketch of the Surveying Voyages of his Majesty's Ships Adventure and Beagle, 1825-1836. *J. Royal Geog. Soc. Lond.* **6**:311-43.
5 Cambridge University Library MSS: DAR 32-3 *Diary of observations on the geology of the places visited during the voyage*. Parts I and II; DAR 34-8 *Notes on the geology of the places visited during the voyage: maps, etc.* Parts I-V.
6 Nora Barlow, ed. *Darwin and Henslow. The growth of an idea. Letters 1831-1860.* John Murray, 1967; and *Correspondence* **1**.
7 Charles Darwin (read 7 March 1838) On the connexion of certain volcanic phenomena in South America; and on the formation of mountain chains and volcanos, as the effect of the same power by which continents are elevated. *Transactions of the Geological Society of London*, 2nd ser. pt. 3, 5 (1840):601-31. Reprinted in *Collected Papers* **1**:53-86.
8 *Narrative* **3**.
9 *Journal of Researches* **1**.
10 *Journal of Researches* **2**.
11 *The structure and distribution of coral reefs* etc. Also *Geological observations on the volcanic islands visited during the voyage of H.M.S. Beagle* etc. And *Geological observations on South America* etc. London, Smith Elder and Co.
12 *Autobiography* p. 116.
13 *Correspondence* **1**:379-82.
14 *Correspondence* **1**:315-17.
15 *Autobiography* pp. 77-8.
16 Sandra Herbert (1991) Charles Darwin as a prospective geological author. *British Journal of the History of Science* **24**:159-92. And see also Sandra Herbert (1977) The place of man in the development of Darwin's Theory of Transmutation. Part II. *Journal of the History of Biology* **10**:155-227.
17 *Beagle Diary* pp. 22-7.
18 *Autobiography* p. 159.
19 *Correspondence* **6**:514.
20 *Autobiography* p. 119.
21 *Beagle Diary* p. 21, and letter from John Coldstream of 13 September 1831 in *Correspondence* **1**:151-3.
22 In the *Zoology Notes* the supply of paper with a red marginal line seems to have been exhausted at **CD P. 315**.
23 *Autobiography* pp. 77-8.
24 *Proceedings of the Royal Society of London* **44**:i-xxv (1888).
25 This instrument, manufactured by Bancks & Son of 119 New Bond Street, had been

recommended to him by Robert Brown. See letters to Susan Darwin of 6 September 1831, and to W.D. Fox of 23 May 1833, in *Correspondence* **1**:143-5 and 315-17.

26 Robert Edmond Grant (1793-1874) was a local physician and lecturer in comparative anatomy at Edinburgh University when CD was a student there in 1825-1827, and was Professor of Zoology and Comparative Anatomy at University College London 1827-1874. CD accompanied him on local expeditions around Edinburgh, and was closely associated with his researches on marine invertebrates.

27 Cambridge University Library MS DAR 118.

28 For a list of the books on board the *Beagle* see *Correspondence* **1**:553-66.

29 Q. Bone, H. Kapp and A.C. Pierrot-Bults (1991) *The Biology of Chaetognaths*. Oxford University Press. See also C. Nielsen (1995) *Animal Evolution. Inter-relationships of the Living Phyla*. Oxford University Press.

30 Charles Darwin (1844) Observations on the Structure and Propagation of the Genus *Sagitta*. *Annals and Magazine of Natural History, including Zoology, Botany, and Geology* **13**:1-6. Reprinted in *Collected Papers* **1**:177-82.

31 *A monograph on the sub-class Cirripedia, with figures of all the species. Vol. I. The Lepadidae, or pedunculated Cirripedes. Vol. II. The Balanidae, or sessile Cirripedes; the Verrucidae, etc., etc., etc.* The Ray Society, London, 1851 and 1854.

32 John V. Thompson (1830) Memoir IV. On the Cirripedes or Barnacles; demonstrating their deceptive character; the extraordinary Metamorphosis they undergo, and the Class of Animals to which they indisputably belong. *Zoological Researches and Illustrations . . .* King and Ridings, Cork.

33 John Ellis (1755) *An essay towards a natural history of the corallines, and other marine productions of the like kind, commonly found on the coasts of Great Britain and Ireland. To which is added the description of a large marine polype.* London, 1755.

34 R.E. Grant (1827) Observations on the Structure and Nature of Flustræ. *Edinburgh New Philosophical Journal* **3**:107-18; 337-42. The paper was read before the Wernerian Natural History Society on 24 March 1827, three days before CD presented the contribution of his own to the Plinian Society that is reproduced in *Collected Papers* **2**:285-91.

35 John V. Thompson (1830) Memoir V. On Polyzoa, a new animal discovered as an inhabitant of some Zoophites — with a description of the newly instituted Genera of Pedicellaria and Vesicularia, and their Species. *Zoological Researches and Illustrations . . .* King and Ridings, Cork.

36 John Ryland (1970) *Bryozoans*. Hutchinson, London.

37 Edward Griffith and others. *The animal kingdom arranged in conformity with its organization by the Baron Cuvier . . . with supplementary additions to each order.* 16 vols. Edinburgh, 1827-35. See also *Cuvier*, 2nd edition, vols. 4, 5.

38 S.F. Harmer (1862-1950), later Sir Sidney Harmer FRS, was in 1901 Superintendent of the University Museum of Zoology in Cambridge, when with the aid of CD's Specimen lists lent to him by Francis Darwin he identified a number of the specimens of marine invertebrates presented some years earlier to the Museum.

39 *Journal of Researches* **1**:258-62 and **2**:201-3.

40 Judith Winston (1984) Why bryozoans have avicularia — a review of the evidence. *Novitates No. 2789.* American Museum of Natural History, New York.

41 *Lamouroux* p. 66.

42 Letter from CD to Henslow of 24 July to 7 November 1834 in *Correspondence* **1**:397-403.
43 *Plant Notes* pp. 194-5.
44 Letter from CD to Catherine of 20-29 July 1834 in Correspondence **1**:391-4.
45 Phillip R. Sloan (1985) Darwin's invertebrate program, 1826-1836: Preconditions for transformism. Chapter 3, pp. 71-120 in *The Darwinian Heritage*, edited by David Kohn. Princeton University Press.
46 This has turned out not to be entirely true, since calcification of the zooids is characteristic of the Cheilostomata as opposed to the Ctenostomata.
47 The first four of these are indeed bryozoans, but Milleporae are hydrocorals.
48 *Clytia*, formerly included with bryozoans among the Sertularians, is a hydrozoan of order Leptothecata.
49 In a Memoir sent by CD to W.H. Harvey at the Herbarium of Trinity College Dublin on 7 April 1847 (*Correspondence* **4**:29) he said of observations made at Bahia on either the coralline alga *Melobesia* or on *Halimeda* in August 1836 that 'on several occasions having kept vigorous tufts of articulated Nulliporæ in sea-water in sun-light, it appeared as if a good deal of gas was exhaled; it w[d] be curious to ascertain what this is.' That bubbles of oxygen were released under such conditions had first been observed by Joseph Priestley in 1777, and was described more fully in 1779 by Jan Ingen-Housz in his book on *Experiments on Vegetables*.
50 *Virgularia* is a sea pen, a hydrozoan octocoral of order Pennatulacea.
51 *Corallium* is a brightly coloured octocoral of order Gorgonacea, but no specimen is recorded in the *Zoology Notes* or *Specimen Lists*.
52 CD's *Crisia* was not in fact this genus, but the anascan bryozoan *Caberea minima*, the coordinated movements of whose vibracula he described very nicely.
53 Bioluminescence is indeed common in cnidarians, and its propagation is controlled by their primitive nervous systems.
54 CD has here concluded perceptively that the coordinated movements of the vibracula in a bryozoan, the flashes of light in the thecate hydroid *Clytia* and the coral *Virgularia*, and the spread of injury in another coral and the tunicate *Synoicum*, indicate that all these 'heterogenious' animals must somehow be capable of internal communication between their individual polyps, and therefore heralds the first appearance of nervous systems in the eumetazoa. (See, for example, Chapter 9 by J.P. Thorpe on *Bryozoa* in *Electrical conduction and behaviour in "simple" invertebrates*, edited by G.A.B. Shelton. Clarendon Press, Oxford, 1982.) This crucial stage in the evolution of higher animals was reached in the cnidarians some 550 Ma ago (see Bertil Hille (1992) Evolution and diversity. Chapter 20 in *Ionic channels of excitable membranes*. 2nd edition. Sinauer Associates, Sunderland, Massachusetts.) It has also been pointed out recently by Richard Keynes & Fredrik Elinder (1999) *The screw-helical voltage gating of ion channels. Proc. R. Soc. Lond.* B **266**:843-52 that across the whole of the animal kingdom, voltage-gated ion channels of every type have genes in which several critical features have been perfectly conserved since that same era, though CD's addition of bryozoans to the list of animals that possess primitive nervous systems remains to be followed up by a detailed examination of the innervation of avicularia and vibracula, and by the cDNA sequencing of the ion channels in their nerve fibres.
55 *Lobularia* is a soft coral of order Alcyonacea, dead men's fingers, in which the

coenenchyme is sclerite-filled.
56 *Tubularia* is not a bryozoan, but a hydroid of suborder Anthoathecata.
57 Letter from CD to Caroline Darwin of 24 October 1836 in *Correspondence* **1**:509-10.
58 *Autobiography* p. 49.
59 Sandra Herbert (ed.) (1980) *The Red Notebook of Charles Darwin*. British Museum (Natural History) and Cornell University Press.
60 Paul H. Barrett, Peter J. Gautrey, Sandra Herbert, David Kohn and Sydney Smith (eds.) (1987) *Charles Darwin's Notebooks, 1836-1844*. British Museum (Natural History) and Cambridge University Press.
61 Charles Darwin (1844) Brief descriptions of several terrestrial *Planariae*, and of some remarkable marine species, with an account of their habits. *Annals and Magazine of Natural History, including Zoology, Botany and Geology* **14**:241-51. Reprinted in *Collected Papers* **1**:182-93.
62 See entry for Specimen 875 (not in spirits); and *Journal of Researches* **2**:158-9; also *Insect Notes* pp. 66-7.
63 *Beagle Diary* p. 356.
64 R.D. Keynes (1997) Steps on the path to the Origin of Species. *Journal of Theoretical Biology*. **187**:461-71.
65 *Beagle Diary* pp. 402-3.
66 *Ornithological Notes* p. 262.
67 *Zoology* **2**:31-4.
68 *Zoology* **2**:92.
69 *Zoology* **2**:38.
70 *Zoology* **2**:7-16.
71 *Insect Notes* pp. 43-4 and 88.
72 Information provided by Lindsay Barton Browne, formerly leader of the CSIRO program on 'Biological control of dung and dung breeding flies'. Dung beetles from southern Africa were introduced in northern Australia with limited success to control the buffalo fly, a blood sucking pest of cattle, and with greater success European beetles were introduced in south-eastern Australia to control another dung-breeding nuisance pest of man and cattle, the bushfly.
73 *Beagle Diary* pp. 144-9; and *CD and the voyage* pp. 177-9.
74 *Beagle Diary* p. 212; *Ornithological Notes* pp. 268-76; and *Zoology* **3**:123-5.
75 Sandra Herbert (1995) From Charles Darwin's portfolio: an early essay on South American geology and species. *Earth Sciences History* **14**:23-36.
76 *Ornithological Notes* pp. 259-60.
77 Cambridge University Library MS DAR 29.1.
78 Richard Burkhardt (1985) Darwin on animal behaviour and evolution. *Darwinian Heritage* Chapter 13, pp. 327-65.
79 Patrick Armstrong *Darwin's Desolate Islands: a Naturalist in the Falklands, 1833 and 1834*. Picton Publishing (Chippenham) Ltd., 1992. Also: An ethologist aboard HMS *Beagle*: the young Darwin's observations on animal behaviour. *Journal of the History of the Behavioural Sciences* **29**:339-44, 1993.
80 Letter from CD to Catherine Darwin of 22 May-14 July 1833 in *Correspondence* **1**:311-15.

Acknowledgements

I am grateful to George Pember Darwin for permission to publish Charles Darwin's Zoology Notes and the lists of Specimens collected by him during the voyage of HMS *Beagle*, 1831-1836, and to the National Maritime Museum for permission to reproduce the painting by Conrad Martens of CD shooting on the banks of the Rio Santa Cruz. I also thank the Syndics of the Cambridge University Library for making available MSS DAR 30 and 31 of the Zoology Notes and other papers, English Heritage for making available the *Beagle* Specimen Lists at Down House, the Cambridge University Zoology Museum for making available notes on CD's specimens by Leonard Jenyns and S.F. Harmer, and the Zoology Library of the Natural History Museum for making available MS 89FD containing Thomas Bell's notes on CD's amphibia and reptiles.

I once again wish to thank the editors of *The Correspondence of Charles Darwin* for setting such impeccably high standards for the transcription and publication of Darwin's manuscripts, and in their Volume 1 for Appendix II on the listing of Darwin's *Beagle* records, and Appendix IV on the books on board the *Beagle*. Frederick Burkhardt, Duncan Porter and Sandra Herbert gave me invaluable help and advice on a variety of editorial questions, and Duncan Porter was kind enough to check the proofs of the final text. Arieh Lew and Nigel Stevens advised me on computer programming problems and on the preparation of a camera ready text for the printers. Dee Hughes skilfully scanned photographs of CD's drawings made under his microscope to improve their reproduction in this volume. Godfrey Waller and other members of the staff of the Cambridge University Library were most helpful at all times in providing rapid access to the original manuscripts of the Notes, Specimen Lists and other Darwin papers, and to annotated books from CD's own library. Clare Osbourn did the same for books in the Balfour Library that I needed to consult. The cost of obtaining copyflow prints from microfilms of the Notes and of the Specimen Lists was met by a grant from the Darwin Fund of the Royal Society.

My deepest indebtedness is to the biologists, taxonomists and other specialists in various parts of the world who gave me so much of their time in advising on the probable identity of the many marine and terrestrial invertebrates and some cold-blooded vertebrates that were studied by CD during the voyage, but for whose identification he was unable to recruit any specialists when the *Beagle* returned to England. I thank also those now responsible for care of the birds and mammals collected by CD. They included Federico Achaval, Lindsay Barton Browne, John Bishop, Quentin Bone, Jean Bouillon, Geoffrey Boxshall, David Briggs, Lester Cannon, Paul Clark, Paul Cornelius, Greg Estes, Yves Finet, Adrian Friday, David George, Peter Grant, Eileen Harris, Paul Hilliard, Roger Lincoln, Colin McCarthy, Jenny Mallinson, Gillian Mapstone, John Parnell, Robert Prys-Jones, Brian Rosen, Frank Rowe, Richard Sabin, Roy Sawyer, Michael Schrödl, Jim Secord, Sharon Shute, Mary Spencer-Jones, Frank Steinheimer, John Topham, Kathie Way and Leigh Winsor. The responsibility is, however, mine alone for any errors in the final choices of species, genera, families and orders to which CD's specimens have been assigned. My last but not least acknowledgement is due to my wife for the forbearance and patience that she has exercised during the years that have been devoted to the transcription and editing of this volume.

Note on editorial policy

My aim has been to adopt the majority of the practices laid down and explained in full by the editors of *The Correspondence of Charles Darwin*, introducing a few changes only in the interests of making the text as easily readable as possible. One departure from convention has been to retain CD's underlining and double underlining as it stands in the manuscript, reserving italics to be used in the customary way in the footnotes for the Latin names of genera and species in former or current use. Liberties have been taken where necessary with CD's sometimes erratic punctuation, further complicated by the not infrequent dots, which have been omitted when they can reasonably be regarded as 'pen rests', but have otherwise been retained as commas or full stops according to the sense of the passage. CD's own idiosyncratic spelling of words such as *broard* and *throughily* is always preserved, but mistakes that are clearly a slip of the pen have been corrected, and missing letters have been inserted in square brackets. Where there is doubt, and there is no difficulty in deciding what his intention should have been, for example in the case of adding the final s to the plural of a noun, I have given him the benefit of it. Where it is hard to decide whether a word starts with a lower case or a capital letter, I have used a capital in the cases of proper names and places. His abbreviations appear as nearly as possible as they are written, with '&' almost invariably used in place of 'and'. Relatively few words have been crossed out by CD during the writing, and such corrections have been retained in the text rather than listing them separately, as has any later over-writing of a single letter. Round brackets used occasionally by CD are retained. Editorial interpolations are in square brackets. Italic square brackets enclose conjectured readings and descriptions of illegible passages. Material that is irrecoverable because the manuscript has been torn or damaged is indicated by angle brackets < >, and any text within them is the editor's. CD's paragraphing has in general been retained, with a fresh paragraph for each new entry, except that for entries running for more than a page, breaks have sometimes been introduced when the subject changes, in order to avoid overlong paragraphs.

A number of pages of the text have later been lined through vertically, not because CD wished to delete them, but to indicate that the material had been incorporated in a subsequent publication.

Many important footnotes, identified in the margin as (a), (b), (c) etc. placed opposite the passages to which they refer, were added later by CD, generally on the back of the page on which he had been writing. Those that were clearly almost immediate afterthoughts or corrections have been incorporated at the most appropriate point in the text itself. Those that were evidently written at a later, though not always recorded date, have been distinguished by their relegation to separate paragraphs.

The pages were numbered right and left at the top of each page, generally with the year in the margin beneath, and the month beside it, with the place in the centre of the page. The topic was always entered, underlined, in the margin at the head of each page. The year, month and place appear in the headings of each of the printed pages, as far as possible retaining CD's description of the place. CD's not infrequent cross referencing to his own pages is entered in heavy type as **'CD P. 00'**, as are editorial references to places in the manuscript where the text continues after the insertion of one of his notes, or a group of editorial footnotes. The pagination of the manuscript is shown by the numbers in heavy type between vertical lines, thus |**000**|.

Principal sources of references

Journal of Researches **1**
1st edition: Narrative of the surveying voyages of His Majesty's Ships Adventure and Beagle, between the years 1826 and 1836, describing their examination of the southern shores of South America, and the Beagle's circumnavigation of the globe. Volume III. Journal and remarks. 1832-1836. By Charles Darwin Esq., M.A. Sec. Geol. Soc. Henry Colburn, London, 1839.

Journal of Researches **2**
2nd edition: Journal of Researches into the Natural History and Geology of the Countries visited during the Voyage of H.M.S. Beagle under the Command of Capt. Fitz Roy, R.N. By Charles Darwin, M.A., F.R.S. John Murray, London, 1845.

Zoology **1**
The zoology of the voyage of H.M.S. Beagle under the command of Captain FitzRoy, R.N., during the years 1832 to 1836. Edited and superintended by Charles Darwin, Esq. M.A. F.R.S. Sec. G.S. Naturalist to the expedition. Part I. Fossil mammalia: by Richard Owen, Esq. F.R.S. Smith, Elder and Co., London, 1840.

Zoology **2**
The zoology of the voyage of H.M.S. Beagle . . . Part II. Mammalia by George R. Waterhouse, Esq. Smith, Elder and Co., London, 1839.

Zoology **3**
The zoology of the voyage of H.M.S. Beagle . . . Part III. Birds, by John Gould, Esq. F.L.S. Smith, Elder and Co., London, 1841.

Zoology **4**
The zoology of the voyage of H.M.S. Beagle . . . Part IV. Fish, by The Rev. Leonard Jenyns, M.A., F.L.S. Smith, Elder and Co., London, 1842.

Zoology **5**
The zoology of the voyage of H.M.S. Beagle . . . Part V. Reptiles, by Thomas Bell, Esq., F.R.S., F.L.S. Smith, Elder and Co., London, 1843.

Cirripedia
A monograph of the sub-class Cirripedia, with figures of all the species. The Balanidæ, (or sessile cirripedes); the Verrucidæ, etc., etc., etc. By Charles Darwin, F.R.S., F.G.S. The Ray Society, London, 1854.

Planaria
Brief descriptions of several terrestrial *Planariae,* and of some remarkable marine species, with an account of their habits. By Charles Darwin. *Annals and Magazine of Natural*

History, including Zoology, Botany, and Geology **14**:241-51 (1844). Reprinted in *Collected papers* **1**:182-93.

Origin of Species
On the Origin of Species by means of Natural Selection, or the Preservation of Favoured Races in the Struggle for Life. By Charles Darwin. John Murray, London, 1859.

Beagle Diary
Charles Darwin's *Beagle* Diary. Edited by Richard Darwin Keynes. Cambridge University Press, 1988.

Beagle Record
The *Beagle* Record. Selections from the original pictorial records and written accounts of the voyage of H.M.S. Beagle. Edited by Richard Darwin Keynes. Cambridge University Press, 1979.

Ornithological Notes
Darwin's Ornithological Notes. Edited by Nora Barlow. *Bulletin of the British Museum (Natural History)*. Historical Series. Vol. 2(7):201-78. 1963.

Insect Notes
Darwin's Insects. Edited by Kenneth G.V. Smith. *Bulletin of the British Museum (Natural History)*. Historical Series. Vol. 14(1):1-123. 1987.

Plant Notes
Darwin's notes on *Beagle* plants. Edited by Duncan M. Porter. *Bulletin of the British Museum (Natural History)*. Historical Series. Vol. 14(2):145-233. 1987.

Oxford Collections
Charles Darwin's *Beagle* collections in the Oxford University Museum. Edited by Gordon Chancellor, Angelo DiMauro, Ray Ingle and Gillian King. *Archives of Natural History* Vol. 15:197-231. 1988.

Autobiography
The Autobiography of Charles Darwin 1809-1882. Edited by Nora Barlow. Collins, London, 1958.

CD and the Voyage
Charles Darwin and the Voyage of the Beagle. Edited by Nora Barlow. Collins, London, 1945.

Collected papers **1** and **2**
The Collected Papers of Charles Darwin. Edited by Paul H. Barrett. 2 vols. University of Chicago Press, Chicago and London, 1977.

Narrative
Narrative of the surveying voyages of His Majesty's Ships Adventure and Beagle, between the years 1826 and 1836, describing their examination of the southern shores of South America, and the Beagle's circumnavigation of the Globe. 3 Vols and an Appendix. Henry Colburn, 1839.

Correspondence **1-6**
The Correspondence of Charles Darwin. Edited by Frederick Burkhardt and Sydney Smith. Vol. 1. 1821-1836. Vol. 6. 1856-1857. Cambridge University Press, 1985-90.

Cuvier
Le règne animale. By Georges Cuvier. 2nd edition. 5 vols. Paris, 1829-30.

Darwinian Heritage
The Darwinian Heritage. Edited by David Kohn. Princeton University Press, 1985.

Dic. Class.
Dictionnaire Classique d'histoire naturelle. Edited by Jean Baptiste Genevieve Marcellin Bory de Saint-Vincent. 17 vols. Paris, 1822-31.

Dic. Sciences Naturelles
Dictionnaire des sciences naturelles. Edited by Henri Marie Ducrotay de Blainville, Anselm-Gäetan Desmarest et plusieurs Professeurs du Jardin du Roi. ? vols. Paris, 1816-30.

Lamarck
Histoire naturelle des animaux sans vertèbres. By Jean Baptiste Pierre Antoine de Monet de Lamarck. 7 vols. Paris, 1815-22.

Lamouroux
Exposition méthodique des genres de l'ordre des polypiers. By Jean Vincent Félix Lamouroux. Paris, 1821.

Rang
Manuel de l'histoire naturelle des mollusques et leurs coquilles. By Sander Rang. Paris, 1829.

Charles Darwin's *Beagle*

Zoology Notes

1832-1836

[**CD P. 1** commences]

Jan. 6th (a)[1]
Santa Cruz
Luminous
Sea

The sea was luminous in specks & in <the> wake of the vessel of an uniform slight milky colour.— When the wa<ter> was put into a bottle it gave o<ut> sparks for some few minutes after having been drawn up.— When exa<mined> both at night & next morning, it wa<s> found full of numerous small (but ma<ny> bits visible to naked eye) irregular pieces of (a gelatinous?) matter.— The sea next morning was in the sa<me> place equally impure.—

Jan 10th (b)

Lat. 21. Sea very luminous, chiefly from a crustace<an> animal, which gave a very green ligh<t>, retaining [it] for some time after having been taken out of water.—[2]

Jan 11th. (c)
Velella[3]

V. A (3)

Lat 22°. A & B represent a beautiful little animal, magnified about 4 *[crossed out]* 5 times its size:— A is the animal expanded: B partially closed.— 1 is flat circul<ar> membrane: 2 a mantle, which the animal i<s> perpetually folding & unfolding: 3 retractile ten<tacula>.

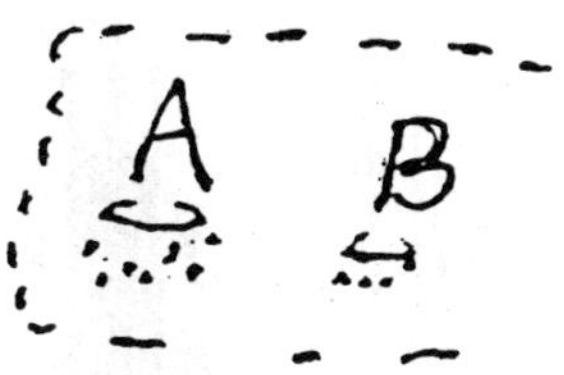

Do. (d)
V. A (4)
Medusa[4]

Allied to the Medusae (?). 1, a transparent membranous bag, with the lower margin sinuous: 2, hanging down in centre, coloured slightly red or purple: 3, four tentacula with adher<ing> cups at the ends.— Magnified about 10 times.

[drawing illegible]

Do. (e)
Physalia[5]

Caught a Portugeese Man of War, Physalia.— get<ting> some of the slime on my finger from the fila<ments> it gave considerable pain, & by accident putting my finger into my mouth, I experienced the |**2**| sensation that biting the root of the Arum produces.—

D.
Lat. 19 N

[note (D) added later at foot of P. 1] The animal is frequently seen with central depending part up & unfolded, like a*[n up]*right cork: tentacula & arm twisted bene<ath.> [note ends]

Jan 11th (a)
Limacina[6]

Limacina moving itself by the rapid motion of its expanded arm.—

Do. (b)

PL: 1, Fig: 1.— A very simple animal[7]: A. nat: size: B magnified:— E

Lat. 21°N

about 7 or 8 bristles on each side of the head with which the animal frequently clasped its head: C, the head with the bristles folden over it: D: a granular substance, ova (?).—

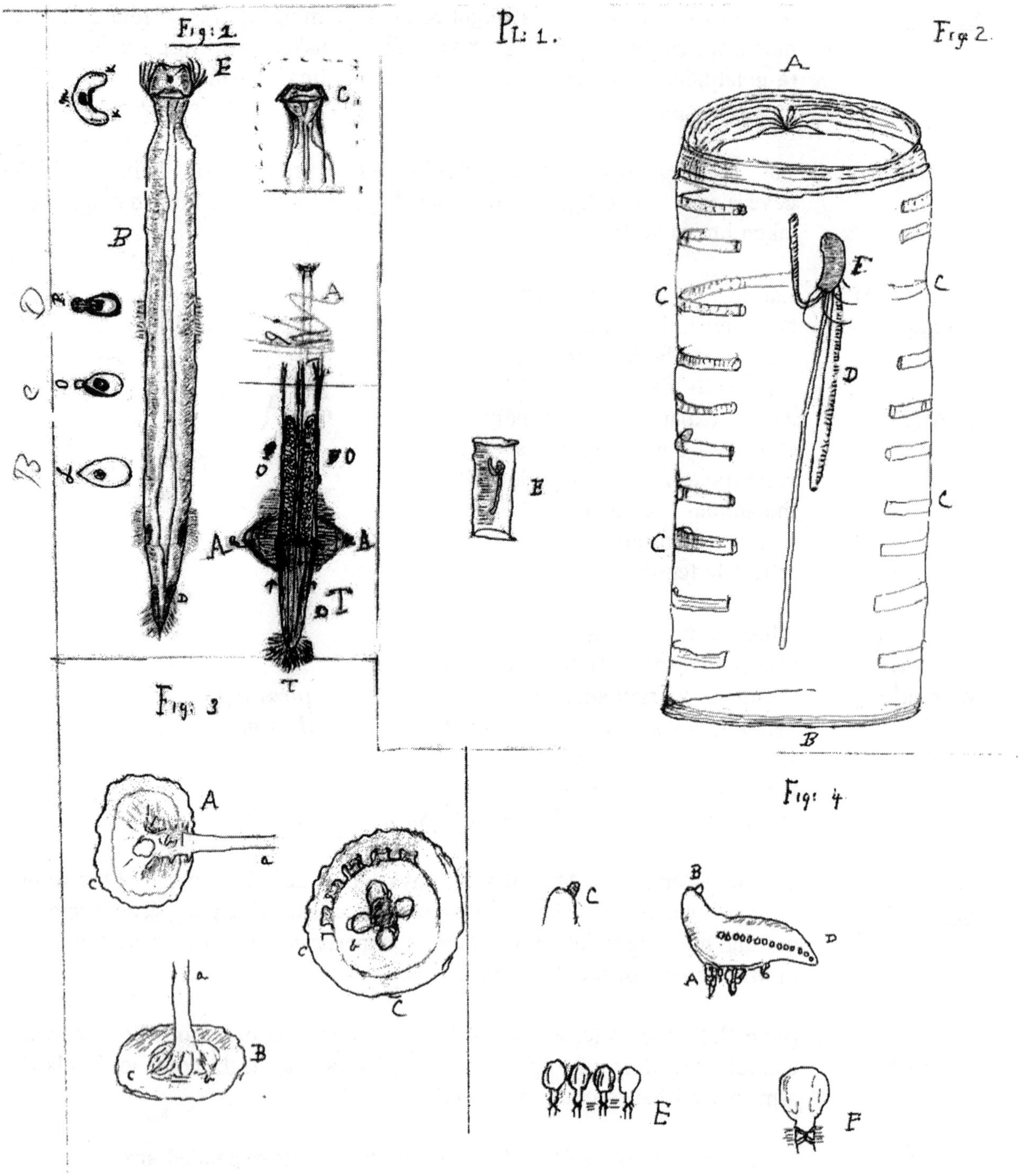

Plate 1, Figs. 1, 2, 3, 4

[note (b) added later on back of **P. 1**]

For more particulars Vide page (**73**) August 24th.—

A transverse section of the head gives an <*words lost*> the flat bunch of bristle are not placed at each end but rather nearer to each other.— In another specimen the granular mass (D) was absent.— But there was a much more transparent & less granular substance running up half way the animal from the tail.—

Description.— Animal transparent, membrane gelatinous: length .6 of inch: narrow: Head simple, rather wider than body: shape truncated cone with ~~terminal~~ orifice. on each side a flattened bunch of curved bristles about 8 in number, moveable & clasping mouth: Neck narrow. Body with thin vessel passing through centre.— Tapering towards the end *[illeg. word]* each side in some specim<ens> a small kidney shaped granular mass.— Extremity pointed, slightly downy.—

March 28th. few miles W of Abrolhos Island. 18°S Bottom at 20 fathoms! Caught ~~great~~ numbers of this animal In some, granular matter (D) was absent, in others it filled the whole tail or tapering extremity & from it were
No. 159[8] sent off 2 gut-shaped bags containing ~~small~~ grains or balls, larger than those at end.— There was an evident peristaltic motion in the internal tube or intestine: the animal could expan<d> this irregularly.— In the gut were curious small bodies, like beads strung to gether.— The animal moved through the water by starts, bending its body at the same time: could contract & shorten its self: has row of very fine hairs at tail sides of granular substance & middle <*word lost*>. [note ends]

[**CD P. 2** continues]

Do. (c)
Lat. 22°N
V A (2)
<u>Biphora</u>[9]

The net came up with a great number of Biphoræ: when placed in water it was quite wonderful with what perfect regularity the animal contracted itself: from five observations with a second watch there were precisely 19 pulsations in every 30 seconds.— PL: 1, Fig: 2. — represent[s] a very rough drawing of the animal: E nat: size: AB. the tunic: the upper end of which has its margin labiate: A represents exactly the appearance of a lip: the lower end B is simple:— Embracing ⅔ of cylinder there are ten flattened striated tubes (c), which are seen to contract during pulsation of the animal.— This uniform motion, together with the partial closing of the end or valve A, must drive the water through the animal: & its reaction accounts for the jumping motion by which it swims in the water:— D. is an appendage with marks on one side as represented: there are I should think tubes for there was an evident rapid circulation going on in them: F. bristles (?) in rapid & continual motion.— the heart, the membrane from transparency not visible, certainly the heart is not much clearer in Creseis. |**3**|

Jan 13th (a)
Lat 19°
(V (10))

Creseis[10] (?) Shell straight, conic, length .15, fragile extremity, contracted with oval ball at end. siphon striated lateral. A, magnified figure. B, extremity.— [text of entry crossed through]

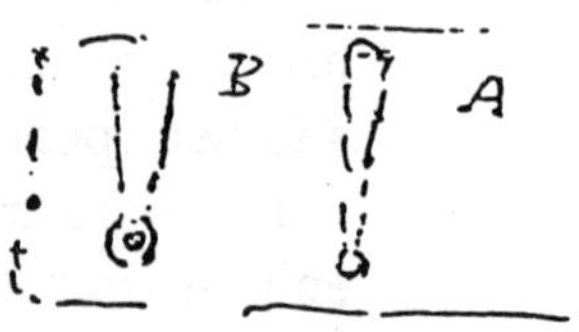

Jan 12th (b)
Lat 15°30'

Sea with numerous ova or rather balls of a brown granular substance in a gelatinous matter. [note (A) opposite] great number in a brown jelly invisible to the naked eye. [note ends]

[1] CD initially used marginal letters in brackets for cross references to his Catalogues of Specimens (see p. 317), but after the first three pages of his notes, the marking (a) etc. in the margin was always used to indicate that a further note correspondingly labelled had been inserted later on the back of the page, or sometimes opposite.
[2] It was on this day that CD used his plankton net for the first time, and drew a picture of it in his diary. See *Beagle Diary* p. 21.
[3] In list of Specimens preserved in spirits, No. 3 was identified as *Velella scaphidia*? *Velella* is a pelagic hydroid, the by-the-wind sailor. Unfortunately the drawing to accompany the picture was only partly completed.
[4] Another of the hydromedusae. The drawing was again not finished.
[5] *Physalia* is a siphonophore common in the warm North Atlantic.
[6] *Limacina* is a sea butterfly, a shelled pteropod of order Thecosomata.
[7] The 'very simple animal' was identified only after the end of the voyage as a chaetognath or arrow-worm, probably *Sagitta enflata*. Chaetognaths are predacious on other planktonic animals, which are seized by grasping spines located on either side of the head. Specimen 159 in spirits, later renumbered 1480, captured off the Abrolhos at the end of March 1832, was identified at the Zoology Museum in Cambridge in 1901 by S.F. Harmer as 'Sev. Chaetognaths'. CD described the anatomy of chaetognaths observed later off the coast of Patagonia in the entry that appears on p. 70.
[8] Numbers thus entered in the margin refer to the list in this case of Animals in Spirits of Wine.
[9] *Biphora* is a name used by Cuvier in 1804, later replaced by *Salpa*, for thaliacian tunicates that include the chainlike salps. For later discussions of this animal see pp. 59-63.
[10] Specimen No. 10 is identified in the list of Specimens not in Spirits as *Creseis agrice rotundo*, another sea butterfly. The modern name of the species is *C. acicula*.

[**CD P. 3** continues]

Jan 16th (c)
20 miles NW
of St. Jago
(V A(10))

Vide PL: 1 Fig. 3 — a delicate Medusaria[1] of a dirty orange colour: gelatinous, delicate, about .4 in diameter. A & B. represent a view from above: (a) is a long irregular narrow membrane, orange colour, terminating at (b) with four hole<s> on the umbrella. (c) an outside transparent membrane: C a view of bottom much magnified. at centre there are vermiform appendages.— beneath which is membrane (a).

Jan 16th (e)
Lat: 15°30'
82

PL: 1 Fig. 4: Physalia[2] length .8.— (D). crest on the side. (E) part of it magnified.— F much magnified.— (A) tentacula about mouth. of two sorts. one small & bright blue. the other longer. reddish brown with dark spots.— (B) small process. (C) magnified.—

[the further entries for 16th Jan. concerned with volcanic dust have later been crossed through[3]]

16th Jan^{y} (f)
V. (11)
(B)

At 8 oclock this morning the vane was taken down from the mast head & found on the under side to be covered with a very impalpable soft yellow-brown dust.—[3] It is probable it has been deposited lately as the ship has been on a tack for a day or two & this is the only way of accounting for the appearance of the dust on the lower side.— The dust under the blow pipe cakes & melts into black enamel: with soda gives a yellow one:— has a slight aluminous smell: under the microscope it is still quite impalpable.— It is probably of Volcanic origin:

(a)
Does not
Horsburg[4]
refer to
this

We are at present & were most part of yesterday |**4**| to the East of St Jago.— There was scarcely any wind this morning, but since noon of yesterday it has come from the East.— before which it was for 24 hour[s] E N E.— At noon of the 15th the Barom: stood at 30.16, by four oclock it had fallen .06.— it then rose gradually till this morning it was 30.2.— The weather generally has been light & fine, but very hazy. occasionally visible horizon. distant only one mile. There has been a long swell on the sea.— as if there had been not far off a heavy gale.— The dust might possibly have come from Mayo or Bonavista, but most probably owing to the wind from coast of Africa about Cape Verd.— I at first thought it might have been brought by the upper Equatorial current from some active Volcano.

(B)
April
1833

[note added later on back of **P. 2**] All the time we were at St Jago, this dust continued to fall so as to be a serious injury to astronomical instruments. Horsburgh[4] in E India Directory P 11 mentions the misty state of the atmosphere between the Cape Verd islands & mainland, & gives it as a reason for Ships avoiding this passage.— This shows to how great an extent it happens.— Although the amount deposited in the ocean during a short period may be small, yet when we consider the extreme constancy of the trade winds, in the course of centuries it must be great.— The dust would seem to be formed from the abrasion of Volcanic rocks & in

Geology of Quail island I show how hard a conglomerate is forming probably from the union of such decomposed rock with Lime.— May not this dust then be helping to consolidate (if mixed with other sediment) beds of mud at the bottom of the Atlantic. Aerial currents would not at first [be] supposed to be instrumental in geological changes.— (I see I have written this note twice) [notes end]

[further notes labelled (a) added later on back of **P. 3**]

(a) This fact of such quantities of Volcanic dust (& the wind in the island of St Jago constantly carried it to seaward to the great injury of fine astronomical instruments) must be in a great length of time of importance in a Geological point of view.— especially as it appears from the conglomerate at Quail Island is now forming from the union of Volcanic matter & lime ~~from~~ making so hard a matrix: perhaps at the bottom of the Atlantic it may form a hard rock.— The dust is formed at St Jago from the abrasion of the various Volcanic rocks:—

Mr Forbes when two miles from the coast of Africa found his sails covered with a brownish sand The wind had blown all night NE. The nearest land to the wind was the coast of Africa between C. Verd & the river Gambia.— Turners Sacred History P 149.[5] (Note): This brown sand doubtless is Volcanic dust: the great distance is very curious, as showing over what an extent this Geological phenomenon is acting.—

Lieut. Arlett (Geograph Journ Vol ?)[6] when surveying coast of Africa talks of quantity of dust: thinks water discoloured by it — Consult.— Charlottes statement about dust at Madeira.— Measure particles of dust transport of seeds of Cryptogams.— [notes end]

[1] Hydromedusa, a jellyfish.

[2] Portuguese man-of-war.

[3] This entry and the accompanying notes have later been crossed through vertically, which was a practice adopted by CD for particular topics on which he eventually wrote papers such as 'An account of the fine dust which often falls on vessels in the Atlantic Ocean', read to the Geological Society on 4 June 1845 (*Collected Papers* **1**:199-203). See also *Journal of Researches* **1**:4, and letter from Robert Bastard James to Charles Lyell (*Correspondence* **2**:77-8) about similar dust collected in 1838 on board H.M. Packet *Brig Spey*.

[4] See James Horsburg. *Directions for sailing to and from the East Indies, China, New Holland, Cape of Good Hope and the interjacent ports.* 2 parts. London, 1809-11. In *Beagle* library.

[5] See Sharon Turner. *The sacred history of the world . . .* Vol. 1. London, 1832. In *Beagle* library.

[6] See W. Arlett 'Survey of some of the Canary Islands and of part of the Western coast of Africa' *J. Roy. Geog. Soc. Lond.* **6**:296-310, 1836.

[**CD P. 4** continues, with next entries not crossed out]

St. Jago Jan. 19th (1) I had occasion to climb a sand bank this morning, which if it had been much steeper I should not have succeded in doing.— It was inclined at an angle of 30°.— The sand was very fine & the ~~greatest~~ slightest motion set it rolling.— I have often observed on flat sea-coast the sand furrowed into small regular ridges: as if it was mocking the waves that daily washed it.— The same appearance was presented by this bank of sand, only that in this case the furrows were longitudinal, in stead of being as on the coast transvers to the line of inclination.— [note (1) on back of P. 3] The dirt collected in the bottom of a basin groups itself in same manner in a direction transverse to the motion of the fluid.— [note ends] |**5**|

Jan 28th (a) Octopus Found amongst the rocks West of Quail Island at low water an Octopus.— When first discovered he was in a hole & it was difficult to perceive what it was.— As soon as I drove him from his den he shot with great rapidity across the pool of water.— leaving in his train a large quantity of the ink.— even then when [added in margin] in shallow place it was difficult to catch him, for he twisted his body with great ease between the stones & by his suckers stuck very fast to them.— When in the water the animal was of a brownish purple, but immediately when on the beach the colour changed to a yellowish green.— When I had the animal in a basin of salt water on board this fact was explained by its having the Chamælion like power of changing the colour of its body.— The general colour of animal was French grey with numerous spots of bright yellow.— the former of these colours varied in intensity.— the other entirely disappeared & then again returned.— Over the whole body there were continually passing clouds, varying in colour from a "hyacinth red" to a "Chesnut brown"[1].— As seen under a lens these clouds consisted of minute points apparently injected with a coloured fluid. The whole animal presented a most extraordinary mottled appearance, & much surprised |**6**| every body who saw it.— The edges of the sheath were orange.— this likewise ~~he~~ varied its tint.— The animal seemed susceptible to small shocks of galvanism: contracting itself & the parts between the point of contact of wires, became almost black.— this in a lesser degree followed from scratching the animal with a needle.— The cups were in double rows on the arms & coloured reddish.— The eye could be entirely closed by a circular eyelid.— the pupil was of a dark blue.— The animal was slightly phosphorescent at night.— [note (1)] Preserved in spirits No. (50). [note ends]

Jan 28th

(1)

[note (a) added later] Jan 30th. Found another. changed its colour in the same manner when first taken. Caught another: I first discovered him by his spouting water into my face when I certainly was 2 feet above him. When seen in water was of dark colour with rings: being with difficulty removed from a deep hole & placed in a puddle of water swam well & emitted a dark Chesnut brown ink.— he continued likewise to spout water, evidently being able to direct his siphon.— When on land did not walk well having difficulty in carrying its head which it continued filling with air as before with water.— From same cause the animal often made a noise when squirting out water. They are so strong & slippery that one hand is insufficient to hold them.— Whilst swimming generally changed colour & seemed to imitate colour of the rocks.—

Feb 3rd. Another upon merely seeing me instantly changed its colour, when in a deep hole being of a dark, but in shallow of a much paler colour.— From this cause & the stealthy way in which it creeps along occasionally darting forward had much difficulty in watching it.—

Cuvier[2] in introductory remarck to the Cephalopodous animals mentions the fact of changing colour. [notes end]

[1] Colours throughout the *Zoology Notes* that are quoted in inverted commas are taken from Patrick Syme, *Werner's nomenclature of colours with additions, arranged so as to render it highly useful to the arts and sciences. . .* 2nd edn. Edinburgh, 1821. There was a copy in the *Beagle*'s library, probably supplied by FitzRoy. The condition of the one now preserved among the books from Down House is spotless, so that the original must later have been replaced by CD. The spelling 'Chesnut' is not one of CD's idiosyncrasies, but is the form in use at the beginning of the 19th century, copied from Syme.
[2] See *Cuvier* Vol. 3, p. 10.

[**CD P. 6** continues]

(a)
52 & 92 Doris[1]. body oval. length 3.5 of inch. indigo blue slightly caudate. with surrounding membrane. [note (a)] feelers white: Branchiæ[2] short. conical. 8 in numb<er>. [note ends]

(b)
53 Doris length .4 slightly caudate. above light rose red with narrow orange rim: beneath with white marks: feelers & branchiæ white. [note (b)] Jan 30th. Doris. surrounding membrane large.— the pink colour in rays: Branchiæ 12 conical situated in semicircle, with points bent in [sketch in margin]. the branchiæ small at extremities the last one with small projection on it: perhaps may be considered as another:— each one with 2 opposite sets of transverse semilunar fringes.— No 79 [note ends]

(c)
55 & 56 & 54 Doris. 1 & ½ inch long, oblong, smooth flattened beneath, above convex.— colour Dutch orange, Mottled with chesnut brown.— feelers orange. broard membrane extending round body.— Branchiæ much
106 plumose, a tube leading from right side near anus.— [note (c)] Feb 5th. Branchiæ plumose. 8 united at their bas<e>. each arm much branched.— Feelers with tops obliquely lined on a tuberculated footstalk.— [note ends]

(d)
56 & 85
& 104

Cavolina[3] ~~(?)~~ ~~(has not the long feelers figured by Blainville[4])~~ Jan. 30th mistake [added above this erasure] Length .6. light flesh coloured, branchiæ dirty brown: feelers 4 white.— Generative organs (?). Much developed on right side: [note (d)] Jan 31st. Cavolina. tail ~~tapering~~ extremely pointed: feelers long taperi<ng>, posterior conical tuberculated: head narrow projecting with foot beneath: Branchiæ in two sets with intermediate dorsal line; placed in curved diagonal ~~lines~~ rows. 9 in each row, interior longer. About 10 rows on each side of back; colour brown with white membranous covering: each branch<iæ> simple. curved tapering.— [note ends] |7|

Jan. 28th
(a)
No. 51

Doris length 1 inch. very narrow cylindrical terminated by a pointed tail — — Membrane round the foot very little extended.— Above white with dark olive brown indentations: 2 narrow lines of orange surrounding back: tail & side blue mottled with white. Beneath & under side of head a fine blue.— Head above dark mottled with white.— Feelers with lower parts blue.— Branchiæ ~~about 14 tufts in number~~ blue tipped with white:— The animal firmly adheres by its tail to the rocks.— When dead & placed in water stains it "China blue". [note (a)] Jan. 30th found some more. Branchiæ straight conic<al> tuberculated.— Mouth whilst dying protruded .1, No. 79. [note ends]

(c)
57 & 79

Bulla[5]. like nitidula: shell with 2 reddish narrow lines following the whorls & sending out on each side alternate waving lines.— Animal transparent. edges of *[illeg. word]* membranes with narrow border of yellow, then emerald green.— Membrane itself marked with white opake spots.— [note (c)] 3d Feb.— took another Bulla, with three lines & the ~~intermediate~~ transverse ones waving, therefore the first must have been a variety. [note ends]

[1] Doridacean nudibranchs, sea slugs, probably Chromodorididae.
[2] The branchiæ, or in French 'branchies', are the gills of such animals.
[3] An aeolidacean nudibranch, family not readily identifiable.
[4] See Henri Marie Ducrotay de Blainville. *Conchyliologie et malachologie* in *Dictionnaire des Sciences Naturelles. Planches.* 2e Partie, *Zoologie.* Paris, 1816-30. In *Beagle* Library.
[5] Cephalaspidea, a bubble snail.

[CD P. 7 continues]

(d)
58 & 79

Worm[1].— about 7 inches long, body highly contractile, flattened, tail tapering. light flesh coloured with about 20 reddish lines, runni[n]g longitudinally *[illeg. deletion]* but not quite continuously.— [note (d)] Jan. 30th .— head flattened, with semicircular projection beneath mouth. Longitudinal edges folded. No. 79: Feb 5th under stones, about 11 inches long. [note ends]

60 (e) 62 63 64 — Fistularia[2]. length, .5-7 inches — Cylindrical: lower part with 4 irregular rows of yellowish ~~papillæ~~ suckers.— back "umber brown". With few papillæ. Tentacula white, surrounding mouth, about 25 or 26 in number.— Tentacula, with round foot stalk. bush shape at top: when expanded .3 in length. top .2 broard.— Body very *[word missing]*. |**8**|

Jan. 28th — Muscular[e].— with bony irregular shaped ring round throat.— They are common ~~amongst~~ beneath the rocks & appear to live on Tere~~bratulæ~~bellæ.— the sandy coats being in their stomachs.

61 (a) — Fistularia. body shorter. thicker flattened "deep reddish brown" sides with black tipped conaceous paps: tentaculæ more apart. larger. 20 in number.— only one specimen. [note (a)] Jan. 30th Fistularia. found several more.— when seized they squirted from Anus.— a considerable quantity of milky fluid.— which consisted of numerous fine white threads & most remarkably viscid.— even sticking fingers fast together.— Often has several largish pale coloured rings on the upper surface of the body. [note ends]

All the animals from page 5 were found amongst rocks to the West of Quail Island.—

69 Jan 30th <W> of Quail Is. — Fistularia. length .9. cylindrical soft transparent "primrose yellow": above covered with paps, beneath with suckers in 4 irregular rows: about mouth, about 15 "gamboge yellow" bush-like tentacula.

71 — Aplysia[3] length 1 & ½ inches, body lengthened: back convex: foot narrow: tail pointed: posterior feelers small, approximate, near to dorsal cavity; anterior feelers, dilated; edges simple, larger, covering mouth; may be considered as a folding membrane, with division near mouth: sides dirty flesh colour: beneath darker: membrane from operculum spotted with purple.— Branchiæ protruding, flesh colour: emitted purple liquor when taken: the folds of mantle seem to be used to aid respiration, or to cause water to flow over Branchiæ. |**9**|

[1] Identified by S.F. Harmer in 1901 as *Gephyrea*, a now obsolete term covering nonsegmented coelomate worms in the phyla Sipuncula, Echiura and Priapulida.
[2] An echinoderm of order Apodida. See p. 125.
[3] The sea hare *Aplysia* is a gastropod mollusc of order Anaspidea.

[**CD P. 9** commences]

No. 70 — <u>Actinia</u>[1]. Short, height ¾, breadth ¾.— Tentacula numerous. lengthened, pointed. "wood brown" bottom do: sides smooth dark greenish black with on overlapping edges about 10 bright blue spots.

(b) No 106 in Spirits *[illeg. word]* 80

Peronia[2]. (Blain[3]) .5, long oval flat; membrane contracted by anus, covering body, not broa~~r~~d, edges irregular.— upper surface blackish green covered with paps: beneath pale: Feelers short with black tips:— mouth divided longitudinally: over it a projecting bilabiate membrane (not very unlike anterior feelers of Aplysia).— Found in clusters under large stones at low water: when kept in a basin Crawled up sides.— Opening for lungs large, cylindrical cartilaginous.—

Peronia

[notes labelled (b) added later: 1st note] No^s^. 80 & 106. This animal according to Blain[3] has only been found in S Hemisphere! [2nd note] Peronia —— Onchidrium, Cuvier[4], who says 2 long retract<ile> tentacula? [3rd note] March 29th At the Abrolhos found a nest of the Onchidium[4] on a Coronula; which was adhering to a rock at high water mark: It looked different from those I caught at St Jago. Animal oval. Mantle fleshy, feelers very short tipped with black.— The length of specim<ens> varied from .2 to exceeding minute ones.— beneath white; above slightly tuberculated blackish green. a dorsal mark darker: pale rings on back giving a tortoise like appearance to the animal: pale lines from the centre to the circumference; these are best seen when by suction the animal firmly fixes itself to a flat surface.— Crawls very slowly.— V. No 174 Spirits [notes end]

[next two entries on **CD P. 9** are crossed through vertically]

N°. 81 Actinia[5]. .2 in heighth, globular, bare grey fibrous sides, "smoke grey" streaks longitudinally, overlapping edges darker.— tentacula greenish grey dappled.

N°. 83 Alcyonium[6]. spherical with short footstalk, base flattened. wrinkled.— colour light "Auricular purple". polypi darker.

No. 86 Doris[7]. oval, length .3, foot narrow: mantle fleshy little projecting.— Branchiæ short, upright, fimbriated, 10 in number.— Back slightly tuberculated of dirty light flesh colour.— with numerous rings of a darker tint —

No^r^. 87 Doris[8]. length .8, oblong: broad posteriorly: foot narrow mantle much projecting, with few brown spots. Branchiæ. large membranous, 6 in number, edges much divided: back light "liver brown". Slightly tuberculated, with darker patches. |**10**|

Actinia 94

Actinia[9]. cylindrical, length 1 & ½ inches, breadth ½: base contracted: sides longitudinally streaked with white point on the line.— these are bigger & more numerous in the upper folding edges & with small intermediate ones.— disk large flat: tentacula not numerous.— Body pale flesh colour, tentacula darker with paler bases .— The animal contract[s] body into a ring in any part of cylinder.—

Alcyonium Alcyonium[10]. growing in clusters: body spherical on a footstalk .2 high:
96 fine purple, semi transparent.

Doris Doris[11]: Length .4. breadth .25. fine orange: foot narrow: mantle much projecting, broardly oval. Posterior Feelers short, conical, with slanting
103 lines, tipped with white: Branchiæ 6 much ~~divided~~ branched, divided into two groups: tipped with white.— Feelers & Branchiæ darker reddish orange.— Feb 5th.—

Caryophillia Found growing on the ~~lower~~ surface of rocks at low tide, 2 Caryophillia[12] differing chiefly in colour. The stony part in both is of an "Aurora red",
99.....102 but in the one the back & part of animal is of an brilliant "orpiment orange", in the other of a bright "Gamboge yellow": in no part was the difference of colour so striking as in the internal tube or lip: perhaps also the orange coloured one was more sluggish in its motion & its lip was more fleshy .— I found them twice |**11**| united so close together that the
Caryophillia internal stony parts were joined or grafted: Are they different species? — The following observations were made on the yellow sort.— but they equally seem to apply to the orange one .— Height varying from one to two inches: diameter at extremity .3.— When thin covering of ~~fleshy~~ soft part is removed the coral is longitudinally striated & with fainter & more irregular & transverse ones:— At Extremity the points project.— Vide PL. 2. Fig 1. (the extremity here represents ~~poly~~ tentacula retracted).

Coral interiorly consists in the broarder branches of longitudinal plates; in the older & lower parts transverse divisions which being placed one below the other give a step like appearance, Fig. 7.— A transverse section gives a star ~~like a~~ with from a few in the younger to 20 in the older branches, Fig. 6.— The coral when dead becomes white & the centre part dies first, the dotted line Fig 7 represent[s] this.— The animal situated in the cup at extremity consists of an exterior row of slightly tuberculated conical tentacula about 30 in number, with aperture at extremity & growing on a fleshy highly retractile ring.— Fig. 2, magnified: Fig 3, extremity: Fig 4, outer circle tentacula retracted.— Within the tentacula is a short projectin[g] oral bilabiate tube, with 20 longitudinal ribs.— in smaller animals there are fewer.— ~~This largely~~ On upper part between each rib ~~are~~ is a minute dot.— This tube or lip can widely expand & ~~fall~~ fold back, & through it is seen large cavity: in Fig 4 it is seen in centre |**12**| [new page
Caryophillia headed Feb 3[d]] nearly closed: Fig. 5. magnified nearly closed & folding back: The cavity as seen through lip rests on the longitudinal division of coral & is lined with apparently fleshy ribs crossing each other.— When the animal is left perfectly at rest & the lip is expanded there is protruded
(a) a delicate membrane, with thick edges, folded up like bud of plant, Fig 8:

[note (a)] This membrane is continually in motion & highly sensative. [note ends] It is this which causes rib like appearance in cavity.— In the older branches one of these membranes is seen projecting from the end of each longitudinal division, when the outer ring of tentacula & lip are dissected away:— in the younger they (as well as transverse division in corals) are less numerous but extend much deeper down.—

The eggs are slowly sent out of animals mouth, are oval, orange colour, in diameter .04 of inch, they contain numerous irregular shaped grains, varying from .001 to .0001 in size.

[1] The sea anemones of order Actiniaria are solitary anthozoans.
[2] *Peronia* is the former name of *Onchidium*, an intertidal slug of order Systellommatophora, family Onchidiidae.
[3] See article by Henri de Blainville on Malacostracés in *Dic. Sciences Naturelles* **28**:138-425.
[4] See *Cuvier* Vol. 3, p. 46.
[5] Another sea anemone.
[6] A soft coral of order Alcyonacea.
[7] Doridacean nudibranch, a sea slug. Several of CD's specimens probably belonged to the family Chromodorididae, but cannot be more closely identified.
[8] ditto.
[9] Another species of sea anemone.
[10] Another soft coral.
[11] Another sea slug.
[12] Scleractinia, a small solitary stony coral. Identified by S.F. Harmer in 1901 as specimens of *Coenopsammia* together with some *Pyrgoma*.

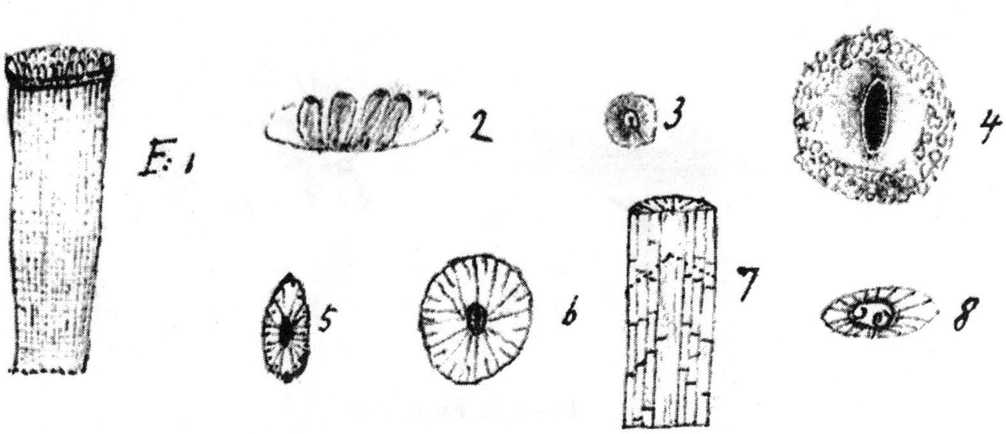

Plate 2, Figs. 1-8

[there follows an extra page inserted later]

to follow p. 12

1835 Octob. Appendix to **P. 12**

Caryophillia Ova (3261)

Having placed a living Specimen of this Corall in Basin of water whilst at James I^{d} in the Galapagos.— soon observed several orange coloured ovules *[illeg. mark]* swimming in the water. When the eye was four feet from the basin a progressive motion might be very distinctly seen.— Ova generally elongated oval, the narrower end slightly truncated.— length about ⅓ of inch.— body contractile as to alter form.— The motion is progressive, steady & quick. the obtuse end being the head.— Frequently there is also a random motion on the longer axis, but likewise on every possible axis.— A vibratory motion, with higher power might be seen on the surface, & a quick motion in the particles in the closely surrounding fluid.— When dead is surrounded by a halo of gelatinous matter, which I believe but am not sure is formed by the vibratory organs.— These probably coat the whole surface. I judge from the revolving motion on such varying axises.— [continued on back of page] Can fix itself temporarily to side of Watch Glass with sufficient force to resist the motion communicated to the Water.— Amongst some Ovules, one differed in form — perhaps being more developed — this was flask shaped [sketch in margin].— Power of attach<ment> lay in broard basis & which end always mount<ed> first — Apex colored more reddish orange. Here there is a most obscure trace of orific<e> & diverging rays.— Is not this a young Polypus, within which the stony plates will be produced? I may remark that on the Corall, near its base, there were several minu<te> living Polypi attached.— The motion of Ovules noticed in the Sertulariæ & Flustraceæ is now known to exist in the Lamelliform Coralls.

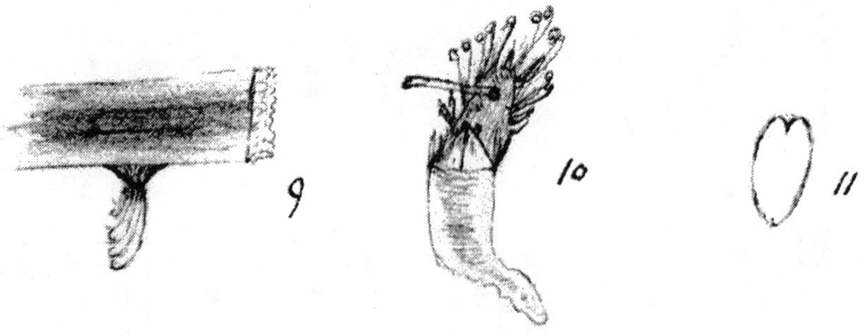

Plate 2, Figs. 9-11

[**CD P. 12** continues, the whole entry on *Pyrgoma* having later been crossed through vertically]

98, 99, & 127 & 128

Pyrgoma[1]. On both Caryophillia the shell is fixed.— Shell subglobular, conical; aperture small oval subcentral; calcareous smooth plate within,

(b) descending half way : externally an external crenulated ring, at which shell divides easily.— [note (b) also crossed through] Valve is fixed half way down the shell & is transversely lined. [note ends] Valve oval, in two pieces folding at one end. PL 2. Fig 11.— It is by curling in this end & joining the sides that the animal protects itself.— Through this valve the animal alternately protrudes & withdraws its ciliæ & has the power [of]
Pyrgoma expanding ~~them~~ & giving a rotatory motion to them: These ciliæ |**13**| surround the tube.— They are arranged in two rows obliquely on a wedge
3 pair shaped projection: The ciliæ are united at bases into pairs of which there are **4** 5 on each side. The centre ones are delicate articulate upright stalks with curled heads & hair from each side. the outside ones are merely curved at their extremities.— the other two pair[s] are much shorter & thicker & straight. Within these ciliæ is the trunk (or anus according to Cuvier[2]). it is as long as the ciliæ, contracted at the base, where it is united ~~itself~~ to the animal, & when seen under a high power appears to be made of rings.— Behind this & between 2 centre pair[s] of ciliæ is a sharp pointed projection.— The ciliæ are protruded at the folded end of valve.— At the other ~~end~~ is situated a ~~conical~~ triangular divided projection, surrounding by 6 small ciliæ, the two outside delicate, hairy, articulated & in continual perpendicular motion. 2 very small ones over division of the mouth (?) & 2 on the sides.— The 2 outermost ciliæ & tips of the other is the only part of animal that I have seen project beyond valve.— The animals body is terminated by an intestine shaped bag containing eggs.—
(a) this rests on a membranous cup which rest[s] on the Coral so that there is no calcareous bottom to the shell.—

[note (a)] I believe this is not correct. the membranous bag rest[s] on a cup-shaped base, which is as firmly imbedded in the Corals as easily to be mistaken for part of it.— The coral grows up around the base & half hides it.— & the soft back generally envelopes almost the whole shell.— In short the egg evidently fix[es] itself between the outside part & the central strong axis.— This animal differs from Pyrgoma of Blainville[3] in the shell not being thick & strength on each side.— & from that of Cuvier[2] in not being much depressed. All authors say animal unknown. [note ends]

[**CD P. 13** continues]

Pyrgoma Eggs are white, numerous, pointed, oval, with a darker substance in the interior: in some externally there |**14**| are a few small hairy ciliæ or arms which rapidly move.— I should undoubtedly have thought it a microscopic crustaceæ.— if I had not myself extracted it.— Vide Pl. 2, Fig. 9 & 10: 9 on Coral: 10 animal out of its shell & membranous valve, much too thick: Ciliæ too short: very badly drawn.—

(a) Jania Jania[4]. dichotomous, very much branched; short reddish: stems jointed, joints transparent, cylindrical, striated, diameter .002. Heads globular, with

199 neck transparent.— Neither Spirits of Wine or fresh water had any perceptible effect.— Feb 3^{d}. — Vide PL. 3, Fig. 1.— [note (a)] No. 199, not spirits. (Jania, Lamouroux) [note ends]

(b) Bacillarièes Bacillarièes[5] (Dic. Class:) growing on Jania. Vide PL 3, Fig 2.— drawn 200 times natural size.— Fig 3.— on a Fucus: Fig 4, in the sea, invisible to naked eye.— [note (b)] No. 200 not spirits. Fucus [note ends]

(c) Aplysia Aplysia[6]. narrow in front, rounder behind, with little tail: Mantle large, divided at each end. Anus surrounded with membrane: Shell transparent, oval, slightly beaked, with one shoulder scalloped out.— length about 5 inches. of a dirty "primrose yellow" traced with veins & rings of a purplish "umber brown" colour; about 10 ~~veins~~ rings in number on each side, 2 on head.— [margin: generally] Anterior feelers white.— Operculum purplish with purple descending fold, with a mark on centre. Foot of a darker yellow.— Stomach much contracted in centre, terminating in a sheath of muscles, round which are 7 to 10 pyramidal bits of semitransparent horn or teeth varying in size, one with another.— ~~Within~~ Stomach contains a quantity of a delicate pink Fucus & small pebbles, which I suppose are used like those in birds gizzards; in |**15**| the intestine, these appear to have been ground into sand.— [note (c)] 14, 18, 29, 30, 31 in spirits.— Shell in operculum & bones out of stomach 100. not in spirits.— [note ends] [margin: Aplysia]

These animals are very common, abounding amongst the stones at low water mark, especially where there is any mud.— I saw some small ones only one inch long.— When disturbed they emitted from under operculum a great quantity of a "Purplish red" fluid enough to stain the water for ~~over~~ a foot round; (b) [note (b)] Paper when stained with this beautiful colour, after a few days changed into a dirty red.— [note ends] When handled, the slime or purple caused a pricking sensation like the Physalia.— I never saw them use their mantle for swimming.— ~~If this animal is Aplysia depilans Linn: all authors badly describe the colour & zone of habitation: Blainville give<s> the animal too much tail.~~

[1] *Pyrgoma anglicum* belongs to a genus of barnacles in suborder Balanomorpha which are always found imbedded in corals. This specimen was described by CD in *A monograph on the sub-class Cirripedia. The Balanidae, or sessile cirripedes; the Verrucidae, etc.*, published by the Ray Society in 1854. See *Cirripedia* p. 360.
[2] See *Cuvier* Vol. 3, p. 178.
[3] Not found in *Dic. Sciences Naturelles.*
[4] 'Jania' is a genus of coralline algae, Rhodophyta. See *Plant Notes* pp. 156 and 187, and *Lamouroux* pp. 23-4.
[5] 'Bacillarièes' are diatoms, Chrysophyta. See *Plant Notes* pp. 155-6, and *Dic. Class.* **2**:127-9.
[6] A species of sea hare of order Anaspidea.

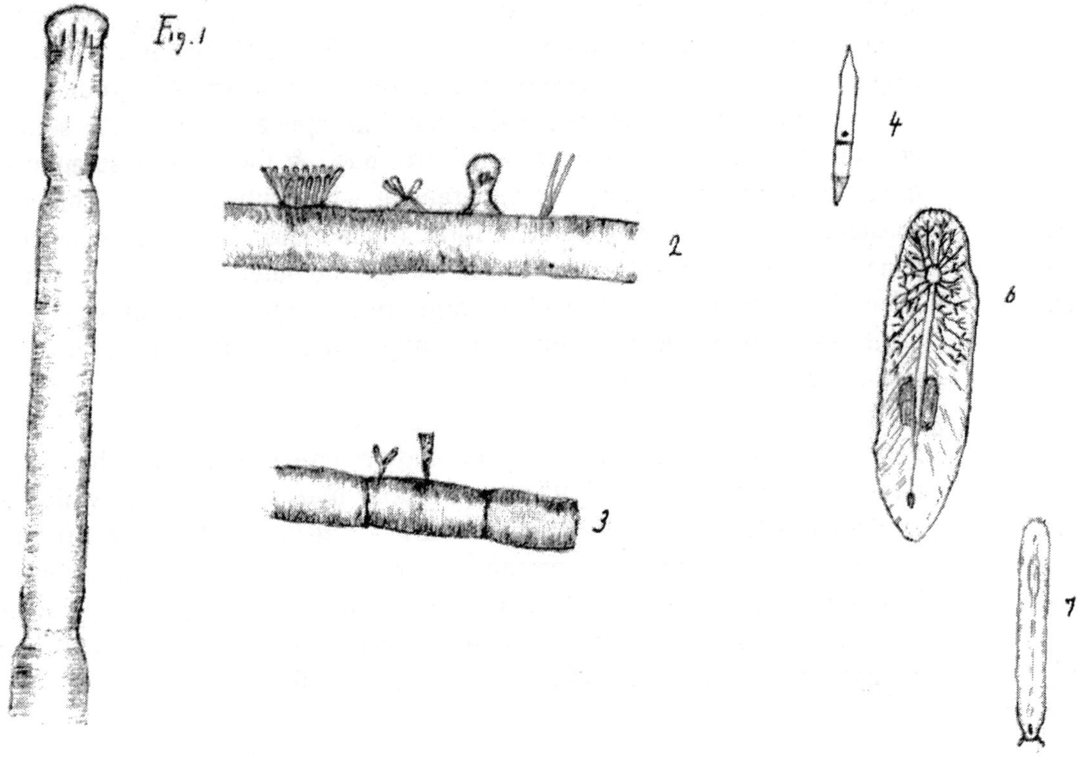

Plate 3, Figs. 1-4, 6-7

[**CD P. 15** continues with an account of some terrestrial Planarian worms, crossed through vertically to indicate its subsequent inclusion in a published paper]

Planaria — Planaria[1] (?) Jan 30th. W of Quail Island. Vide PL:3 : Fig. 5.— length one inch, breadth ¾ .1 — oval, creeping. highly contractile & active.—
84 & 105 — Body very flat. soft membranous.— divided anteriorly & posteriorly.— Pale. above finely reticulated with brownish purple.— At one extremity (A) ~~there~~ on the under side there are two paplike retractable orifices; the anterior one of which is largest.— From this point are sent off diverging rays — which nearly reach to the border: these act as muscles.— & when the animal contracts any part of body the rays to that part are raised.— A nearly continuous tube runs through the length of the animal, connecting the |**15bis**|

[CD continues on an extra page numbered in another hand **15bis**, and with 39 in a circle. The crossing through continues.]

Planaria — anterior orifices with one posterior one.— At middle of animal on each side of central tube is a mass of angular white grains.— & just above it a small orifice (B). This orifices is generally closed, & then invisible.— but the animal having been kept some time opened it, & through came out folds of highly transparent membrane continually contracting & dilating itself.—

(a) When first protruded it is folded up like bud of plant, but when expanded seems to be deeply divided into inverted wedge shaped portions, & extends as far as edge of body of animal.— [note (a)] V P.192 for some particulars respecting this organ observed in a terrestrial Planaria.— [note ends] When within, the membrane had a star like appearance.— As soon as animal died this membrane remained protruded & there likewise appeared to come another from between the granular white substance.— This latter substance likewise burst & sent forth numberless round balls, which I conceive to be the ova.— Under microscope the outer membrane consisted of numerous green grains & some larger brown egg shaped masses.— In death ~~the~~ its body became almost instantly soft & as [if] it were dissolved in the water.—

[illeg. note and brackets in pencil]

See better Specimen

Planaria

I could not preserve this specimen, but I afterwards procured another, which has kept well in Spirits of Wine.— Animal lives under stones which are imbedded in the shore at low-water mark.— It is very active & irritable, & has the power |**16**| of adhering most remarkably close to the stones.— This animal cannot be a true Planaria, although its external characters would show it to be such.— Its habits are more that of a Nereis: but as to its strange organization I am at a loss to what to refer it.—

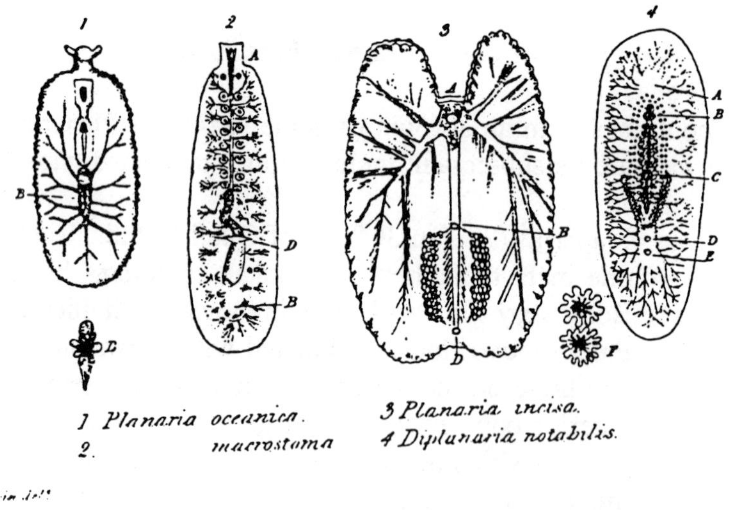

Plate V from Darwin (1844)

Planaria

105

Planaria[2]. length 1 & ½ inches. breadth .4: oblong: very flat, an elevated line running down the back, sending off lines on each side: Beneath the bands of a yellow substance bordering a central transparent space.— Signs of an aperture at each extremity.— Above light "Chesnut brown", beneath pale.— Habits similar to the first Planaria.— These animals evidently are closely allied, but differ in this latter one being narrower, of a different colour, & ~~not~~ being oval at each extremity.— But yet what a wide disparity between their Organizations!

Planaria

105

Planaria. length .7. breadth .2, very flat.— above pale brown.— sending off branched lines, especially on anterior parts.— Beneath pale, with anterior transparent spot (mouth?). Posterior spot likewise: Round anterior one, on each side are two rows of black specks, which contract with the animals skin.— Like former one crawls & sticks to stones: likewise can swim by a vertical motion of its body: often rolls itself into a ball.— Vide PL. 3, Fig. 6.— |**17**|

Planaria
105 (a)
Beneath with white opake mass

Planaria[3]. flat, linear, length (when fully extended) one inch, breadth .05: white, semi transparent, with a slightly elevated dorsal line: mouth retractile, with on each side short, curved feeler.— On these & mouth there is an irregular row of black specks.— Habits like the last: swims well, crawls with rapidity & occasionally walks on its extremities like a Leach.—

Vide PL 3, Fig. 7.— Feb 5th W of Quail Island.—

[1] Identified by CD as *Planaria (?) incisa* in a paper entitled 'Brief Descriptions of Several Terrestrial *Planariae*, and of Some Remarkable Marine Species, with an account of Their Habits' (*Annals and Magazine of Natural History, including Zoology, Botany, and Geology* **14**:241-51, 1844. Reprinted in *Collected papers* **1**:182-93). PL. 3, Fig. 5 has not been preserved, probably because it was redrawn by CD for the published paper. The animal is a turbellarian flatworm in order Polycladida, and was at one time classified as *Centrostomum incisum*, though recent authorities have left it as *incertae sedis*. CD has correctly described the extruding folds from B as pharynx, though in fact he may have been looking at eggs ready to be laid from the female pore behind it. The nearest polyclad group would be the Pseudocerotidae, though the precise position of the species is uncertain.
[2] This planarian and the next one are evidently those related to *P. incisa*, and both are probably polyclad turbellarians.
[3] By elimination this is the species from a new genus named *Diplanaria notabilis* by CD (see *Collected papers* **1**:191-3). Recent authorities have assigned it provisionally to the genus *Leptoplana*, but CD's description is inadequate by modern standards to place the species more certainly than among the polyclads.

[**CD P. 17** continues]

Cleodora
or
Creseis

107

Cleodora or Creseis[1], Rang:[2] Feb 14th. 2°30′ N[3].—
Shell extremely linear, pointed, length .4, diam .03.— straight.— Animal slight tinge of red: Membranous wing divided into three lobes.— 2 large, reticulated, orbicular, with pointed ears on each side.— The third is small: The animal easily propels itself by the 2 large ones.— keeping always that ~~the lowest~~ side of body the lowest in water.— Between them is a small linear opening with tube leading from it.— Surrounding this are excessively

minute ciliæ, which continue in such rapid motion that they are scarcely visible, & would not be perceived were it not for the motion communicated to all small particles near them.— These wings are situated on a footstalk or neck which leads into cavity of body.— [note (2)] Upon this, there are
(2) black spots, like eyes, fixed.— [note ends] This cavity consists of a membranous sheath or mantle, which terminates at the pointed extremity of the shell.— It is by the contraction of this that the animal is able to draw in his wings or head.— |**18**|

Creseis

February 14th. Upon upper edge of this mantle, which continually contracts & expands itself & which is rather irregular, are situated ciliæ such as described near mouth: likewise in centre they may be discovered rapidly driving about small grains of matter (& the ova?). The middle part of sheath is surrounded by lines or rings.— At upper extremity, near where the head joins, there are vermiform tubular appendages, which I am nearly sure can be protruded beyond the shell.— May they not be similar to those in Limacina?— .— Beneath this are sent forth two tubes.— one transparent & ending in globular ball (within another membrane?), which is always pulsating.— the heart?— the other is a strength gut which gradually tapers to the shells end.— At its upper end it continually contracts & dilates itself, close to which is [a] small dark organ, the liver?, & a mass of green small balls.— ova?—

Creseis
107

Tapering, extremity not much pointed, curved: animal same as former one.— only that perhaps vermiform appendages were more apparent: & necessarily from shortness of shell, the intestine beneath the liver & green granular substance much shorter.— As it is the mantle or sheath that surrounds this part, which chiefly aids in retracting the animal, it almost necessarily follows that this process would be slower when this part was shorter, & this is the case. |**19**|

~~Limacina~~
Atlanta (a)
Dic. Class.
107 & 155

Atlanta[4], ~~Lamacina Cuvier~~ Cuvier.— very small. fine violet. slightly carinate. whorls touching each other.— In one specimen, only small portion of whorl coloured.— 2 others uniformly.— I should think they were full grown & if so a new species.—

[note (a) added later] On 23^{d} of March: in about Lat 18°5′ & Long 36°W, the sea contained great numbers of this Atlanta.— Shells ~~varying~~ (largest specimens) 1/20 of inch in diameter.— Whorls four touching each other, the three internal ones purple, tapering suddenly: mouth of shell posteriorly cut out: not much carinated: [note ends]

[**CD P. 19** continues]

Porpita
(b)

Porpita[5]. Feb 14th.— 2°30′ N.— prussian blue. width .07. back rounded, slightly tuberculated, convex. slightly striated from centre, where there is a

108 brown mark. Surroundi<ng> membrane, narrow, stiff, scalloped.— Beneath depending, surrounded by numerous tentacula; extremities of
Fig. 8, B‴ which are divided into 4 papillæ, & being placed on one side give an hand like appearance.— Some of them are fully extended & are longer than diameter of the animal, the greatest number are much retracted.— Stalk of tentacula transparen<t>, with an interior tube which terminates in a bag at
Fig. 8, C the foot of the papillæ.— These papillæ are thin, delicate, transversely lined, with a globular much tuberculated head.— The occasional protrusion of some of their tentacula has given rise to the idea that there were 2 sorts of them.— Vide Cuvier.— Mouth white, membranous, tubercular, projecting, round which is a row of simple vermiform tentacula, of a China blue. |**20**|

[note (b) added later] PL. 3 Fig. 8A no side ridge such as in A. Peronia of Blainville[6]: Shell flat when seen from above (or edgewise) sides equal: whorls coiled obliquely & spiral.— so that on one side a slanting umbilicus can be seen ~~on the~~ & only a few of the whorls.— on the other all the whorls & no umbilicus: Only differs from No^{r} 8 (not spirits) in being of a purplish colour & generally smaller.— from 7 not spirits in whorls touching each other: If this latter, as I suppose, is A. Peronia, Blainville has drawn his figure with oblique ridges on the side which do not exist.— V No^{r} 385 (not spirits). [note ends]

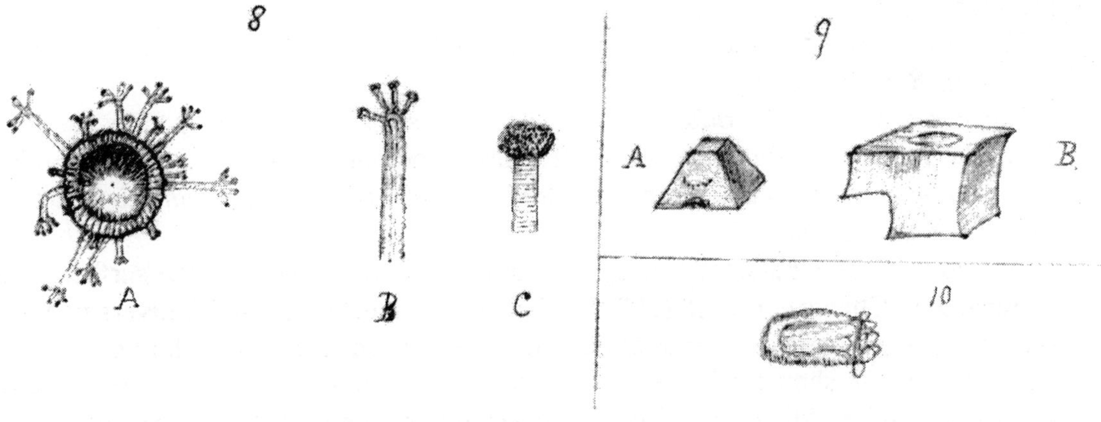

Plate 3, Figs. 8, 9, 10

[**CD P. 20** commences]

Mucor (a) No. 223 not in Spirits

<u>Mucor</u>[7]. Linn: growing on a lime from St. Jago, length .1 of inch. brown colour: pedical hollow, simple, transparent, diameter .0006.— At extremity ball containing sporules, diameter .007.— Sporules varying in size, very minute, about .00009 in diam: When the mould was placed in water, the balls burst longitudinally, & sent forth the Sporules.— at same time globules of air passed down the pedicel.— This took place with such

violence that the recoil on the ball gave it sufficient motion to be visible to the naked eye.— The same results occurred with greater force when Spirits of Wine was used instead of water.— Was it not a similar observation that first led Dutrochet[8] to the discovery of the Laws of Endosme?

[note (a) added later] Observed the same species growing on gum dissolved in vinegar.— (March 23d) found a sort very like this on old paste; the colour was yellow, & the stalks rather longer in proportion, were the only differences I could perceive. [note ends]

Dyphyes
No. 109

Dyphyes[9].— Plate 3 Fig 9.— Feb 17th. Lat 1°30′ S.—
(A) Square pyramidal, apex obliquely truncate. side slightly hollowed, with a projecting curved dotted rim on one side.

(B) Another species, an solid oblong placed on a square base & projecting over on one side.— ~~On~~ In upper parts is a net work bag. (the animal?) from which two appendages were sent off into the lower part.— I could see no signs of two animals joined.

Salpa
No. 109

Salpa[10]. Fig 10. Mantles rounded with four ridges or angles.— fringed beyond the mouth.— Mouth consists of a membrane stretched across the opening, with circular aperture.— |**21**|

[1] A shelled pteropod of order Thecosomata. Its measurements and those of other specimens are given in inches.
[2] See Sander Rang. *Manuel de l'histoire naturelle des mollusques et leurs coquilles*. Paris, 1829. In *Beagle* Library.
[3] The ship was now 150 miles from the Equator, see *Beagle Diary* p. 35.
[4] *Atlanta* is a pelagic snail of order Mesogastropoda and superfamily Heteropoda, whereas *Limacina* is another shelled pteropod. Only *A. inclinata* and *A. peronii* are found in the area. See *Dic. Class.* **2**:58.
[5] *Porpita* is a "blue button", a pelagic hyroid that floats on or near the surface. The animal identified by Chancellor *et al.* (1988) as No. 4/12904 in the Oxford University Museum, and classified as Amphipoda: Gammaridea, must have been in the same bottle.
[6] See Planche 58 showing Onchidie (Veronicelle), Planche 63 of Péronie de l'Isle de France, and Planche 68 of Atlanta de Péron in the section on *Conchyliologie et malacologie* by Henri Marie Ducrotay de Blainville in *Dic. Sciences Naturelles* Planches 2e partie, Zoologie.
[7] CD's specimen of the mold *Mucor* (Mucoraceae) was not well preserved, and Henslow wrote to CD in January 1833 'For goodness sake what is N°. 223 it looks like the remains of an electric explosion, a mere mass of soot—something very curious I dare say—' (see *Correspondence* **1**:294 and *Plant Notes* p. 153).
[8] Henri Dutrochet (1776-1847) was a French scientist who in the 1820s had published on the phenomenon of endosmosis, osmotic movement into cells.
[9] Diphyes is a pelagic hydroid of order Siphonophora.
[10] A thaliacean tunicate specialized for a free-swimming planktonic existence.

[CD P. 21 commences with an entry about a marine Planarian worm, crossed through vertically as before]

Planaria (a) [pencil notes written vertically] 80 miles from Fernando, 150 from America

Fig 1 Vide A

B

(Planoceros)

Planaria[1]. Plate 4, Fig 1.[2] — Feb^{r} 23^{d}. Lat 5° S, 33° W.— [note (a)] was unable to preserve it in spirits [note ends] length .2: Colour pale: membrane with edges jagged; anteriorly formed like a neck & head with 2 ear like processes.— Beneath near to the neck is an internal quadrangular membrane within which is a black spot, by the side of this is an opening which the animal can dilate & contract at its pleasure.— Joining to this part is an oval bag with an internal dark spot & delicate tube.— This bag terminates behind a central dark Mass formed by the union of eleven ~~veins~~ *[3 illeg. words]* or rather a congeries of grains.— In centre of the Mass there is a longitudinal opening through which the animal can protrude a dark coloured very delicate membrane.— This membrane is ~~close~~ when first seen, clewed up into eight divisions. The animal evidently bears a close relation to the Planaria (?) described in Page 15.— The two bags in this instance answering to the two pap-form orifices.— & the organization central mass is much the same in both cases: in this one however there is no mass of eggs.— One is tempted to consider the membrane as lungs, & the veins which branch off from the centre as a circulatory system.— Is it not extraordinary finding an animal adapted for creeping in such a situation; so many miles from shore.— |**22**|

[1] Identified by CD in his 1844 paper as *Planaria (?) oceanica.* The modern name of this polyclad turbellarian flatworm is *Planocera pellucida* Mertens.
[2] Plate 4 Fig. 1 is now missing, but it was redrawn in Plate V(1) of the 1844 paper without the label A (see p. 20).

[CD P. 22 commences with an entry about *Diodon*, much corrected with a different pen, probably because a shorter version was later included in *Journal of Researches* pp. 13-14]

Diodon No 132

(a)

March 10th a Diodon[1] was caught swimming in its unexpanded form near to the shore.— Length about an inch: above blackish brown, beneath spotted with yellow.— ~~Above~~ On head four soft projections; the upper ones longer like the feelers of a snail.— Eye with pupil dark blue; iris yellow mottled with black.— The dorsal caudal & anal fins are so close together that they act as one. [note (a)] These ~~fins~~ as well as the Pectorals which are placed just before branchial apertures, are in a continued state of tremulous motion even when the animal ~~is at~~ remains still.— [note ends] the animal propels its body by using these posterior fins in same manner as a boat is sculled, that is by moving them rapidly from side to side with an oblique surface exposed to the water.— The pectoral fins have great play, which is necessary to enable the animal to swim with ~~his~~ its back downwards.—

When handled a considerable quantity of a fine "Carmine red" fibrous secretion was emitted from the abdomen & stained paper, ivory &c of a high colour.— The fish has several means of defence it can bite hard & can squirt water to some distance from its Mouth, making at the same time a curious noise with its jaws.— After being taken out of water for a short time & then placed in again, it absorbed by the mouth (perhaps likewise by the branchial apertures) a considerable quantity of <u>water</u> & air, sufficient to distend its body into a perfect globe.— This process is effected by two

<u>Diodon</u> methods; ~~chiefly by gulping <&>~~ |**23**| ~~swallowing the air & water~~ & then forcing it into the ~~inside~~ cavity of the body, its return being prevented by

(a) Back of Page

a muscular contraction which is externally visible: ~~but also when the mouth was distended & motionless I observed a stream of water flowing in. this must have been caused~~ [and] by the dilatation of the animal producing suction.— [note (a)] The water however I observed entered in a stream through the mouth, which was ~~distended~~ wide open & motionless; hence this latter action must have been caused by some kind of suction [note ends] When the body ~~was~~ is thus distended, the papillæ with which it ~~was~~ is covered ~~with papillæ which by this action~~ become stiff, the above mentioned ~~ones~~ tentacula on the head ~~must~~ being excepted.— The animal ~~thus~~ being so much buoyed up, the branchial openings ~~was~~ are out of water, but a <u>stream</u> regularly flowed out of ~~it~~ them which was as constantly replenished by the mouth.—

After having remained in this state for a short time, ~~the body was emptied of~~ the air & water would be expelled with considerable force from the branchial apertures & the mouth.— The animal at its pleasure could emit a certain portion of the water & I think it is clear that this ~~water~~ is taken in ~~to~~ partly for the sake of regulating the specific gravity of its body.— The skin about the abdomen is much looser than that on the back & in consequence ~~is by far the~~ is most distended; ~~from same reason~~ hence the animal swims with its back downwards.— Cuvier doubts their being able to swim when in this position; but they clearly |**24**| can not only swim

<u>Diodon</u> forward, but also move round.— this they ~~do~~ effect, not like other fish by the action of their tails, but ~~by~~ collapsing the caudal fins, they move only by their pectorals.— When placed in fresh water seemed <u>singularly little inconvenienced.—</u>

<u>Vespertilio</u> No 134

Caught March 10th by flying into a room: it is an old female: This species would I think according to Dic. Class: be a new ~~Species~~ genus: but from Cuvier['s] sparing description is a 'Phyllosome sans queue'[2]. My specimen however by Dic Class does not agree with its teeth with this sub-division.— Head broard flattened.— 4 incisors in each jaw; of the superior two center

Copied

ones longest & bifid: Lower ones equal & slightly bifid.— Canine very sharp. superior ones nearly twice as long as the inferior.— 8 other teeth in the upper & 10 in lower jaw.— Nose with a flat semicircular membrane, retracted posteriorly & projecting upwards (or rt angles to the flat part) .3

of inch. lunar shaped with a fold or crease on each side.— No tail: membrane between thighs retracted. Ears oval with an interior denticulate pointed fold at base.—

Vespertilio

Copied

[Start of next paragraph marked by a deep square bracket] Body above darkish "clove brown". beneath much paler; wings (especially lower parts) dull "velvet black" with an irregular transparent colourless space at extremities. |**25**| Breadth from tip to tip 18 inches. Length from head to extremity of abdomen nearly 4.— Cuvier divides bats into those with three bony phalanges in middle finger & in 2 in all other & into those with one in index & 2 on all others.— I think this Species belongs to first division but I cannot perceive the 2 osseous joints in the index.—

Elater
No 352 (not
in spirits)

Caught March 10th Elater[3] (noctilucus) & took the opportunity of examining ~~their~~ its springing apparatus. It appears to me that this has not been well described in Dic. Class:.— When the insect prepares to jump it bends backwards the head & thorax.— by this process the spine is drawn out & the point rests on the edge of its tube.— a very little motion is sufficient for this, as in the usual position of animal the spine is only inserted a little way in the tube.— the muscles now having a fulcrum to act on the insect exerts its whole force & spine like a spring is bent.— The animal at this moment rests on its head & top of E[l]ytra & upon suddenly relaxing its efforts the head & thorax fly up & the spine suddenly is inserted in the tube.— by this ~~this~~ action the base of Elytra strikes |**26**| the supporting surface & by the reaction the insect is thrown up in the air: ~~It is precisely the same as when a spring curved at its extremities is foreibly held flat bowed in the contrary direction & this being loosed will spring upwards~~.—

Elater

(a)

The spine is notched at the end.— The points at base of thorax appear to serve as guys to steady it when ~~the animal~~ drawn backwards; ~~as likewise this does~~ the sheath of the spine during the spring seems to act in a similar manner.— In the account given in Dic: Class: stress is not sufficiently laid on the bowing of the spine; ~~&~~ it is this which explains the extraordinary manner in which the Elaters jump.— [note (a)] ~~Dic Class~~ The Author seems to think that the insect strikes the supporting surface with its head, thorax & tip of E[l]ytra, & that previous to the spring it bends its thorax inwards instead of in the contrary direction.— [note ends]

The light from the spots on thorax was brilliant & green.— it varied in intensity, being ~~most~~ brightest when the insect was annoyed.— There appeared to be a sort of internal pulsation within the bright spot.—

[1] Identified by Leonard Jenyns in *Zoology* **4**:151 as either a young example of *Diodon antenattus* Cuv.? or a new species. The *Beagle* had arrived at Bahia, Salvador on a modern map, on 28 February, when CD went into ecstasies at his first sight of a tropical forest (see

Beagle Diary pp. 41-2).
[2] Identified by George Waterhouse in *Zoology* **2**:4-5 as *Phyllostoma perspicillatum.*
[3] A click beetle of family Elateridae. Identified by CD in *Journal of Researches* p. 35 as *Pyrophorus luminosus* Illiger, and said by him to be 'the most common luminous insect'. See *Dic. Class.* **16**:70-6 and *Insect Notes* p. 48.

[**CD P. 26** continues]

General
Obser: Bahia

The sand on the beach is of a brilliant white colour & composed of minute grains of quartz: when walked over the friction of the particles caused a curious high note or chirp: The temperature of this sand a few inches beneath the surface was 108 in the open rays of the sun.— A person |**27**|

[**CD P. 27** is headed Bahia Feb 29th . . . March 19th, and commences]

General
Obser: Bahia

in a hot country might with closed eyes tell what colour the ground was on which he was walking.— The effects of reflection from a white surface preponderating those of radiation from a dark.—

Zoology

I was surprised at the scarceness of birds: the extreme thickness of the vegetation seems only to suit a few tribes.— Within the Tropic the insects take a ~~more~~ prominent part in the animal kingdom: the woods resound with their noise especially of the ~~Orthop~~ Hemipterous[1] tribes as Cicada & the eye is attracted by the gay and beautiful colours of the butterflies: these bespeak the Zone they inhabit far more plainly than the Coleoptera. The latter by their smallness, dark colours & European form much surprised me.— The genera that were most abunda<nt> were Haltica[2] & Galeruca[3]
(b) (or closely allied to it) & Curculio[4].— It was singular to find in the fresh water Berosus[5] & Hydroporus[6].

1832 (b)
March

[note (b) added later] ~~Do~~ Are not the Hydradephaga[7] remarkably constant in their forms in different stations & habitations. England: Patagonia: Tierra del Fuego: Cape Verds & Brazil:— [note ends]

[**CD P. 27** continues]

Carabidous[8] insects were rare. I only found three species, one Scarites and two Truncatipennis (~~Sebia & Odacantha?~~).— The wonderful number of Ants perhaps supply the place of these butchers of the colder climes.—

Formica
(a)
General
Obser: Bahia
No. 359...364

On first entering a Tropical forest one of the most striking things is the incessant Labour of the Ants.— [Note (a) No. 357 & 358 (not in spirits) [note ends] The paths in every |**28**| direction are traversed by hosts of them carrying parts of leaves larger than themselves & reminding one of the moving forest of Birnam in Macbeth: ~~Most of the trees contain large nests,~~

(not spirits) ~~which are 3 or 4 feet in length & 2 or 3 in breadth~~.—

[note (b), later struck out] August 20th. It is evident I have confounded the nest of the Termite with the real ants performing their marvellous labours.— [note ends]

[**CD P. 28** continues]

Some of the smaller species migrate in large bodies.— One day my
357, 358[9] attention was drawn by many spiders, Blattæ[10] & other insects rushing in
(not spirits) the greatest agitation across a bare bit of ground.— Behind this every stalk & leaf was blackened by a small ant: They crossed the open space till they arrived at a piece of old wall on the side of the road.— Here the swarm divided & descended on each side, by this many insects were fairly enclosed: & the efforts which the poor little creatures made to extricate themselves from such a death were ~~wonderful~~ surprising.— When the ants came to the road they changed their course & in narrow files reascended the wall & proceeding along one side in the course of a few hours (~~all~~ when I returned) they all had disappeared.—

When a small stone was placed in the track of one of their files, the whole of them first attacked it & then immediately ~~backed~~ retired: it would not on the open space have been one inch out of their way to have gone round the obstacle, & doubtless [continued at (a) on back of page] if it had previously been there, they would have done so.— In a few seconds another larger body returned to the attack, but they not succeeding in moving the stone, this line of direction was entirely given up.— [entry ends] |**29**|

[1] True bugs of order Hemiptera.
[2] Chrysomelidae, leaf beetles.
[3] Genera of Galerucinae and Chrysomelinae, leaf beetles.
[4] Curculionidae, weevils.
[5] Hydrophilidae, water beetles.
[6] More water beetles.
[7] Hydradephaga is an old term covering all water beetles.
[8] Carabidae, ground beetles.
[9] As reported in *Insect Notes* p. 48, these specimens have not survived, but from CD's description they were 'driver ants' of the subfamily Dorylinae, probably of the genus *Eciton*.
[10] *Blatta* is nowadays a cockroach, but was the name formerly applied to all insects of order Blattodea.

[**CD P. 29** commences]

March 23d Mucor[1] growing on green ginger: colour yellow, length from 1/20 to 1/15 of

Mucor

an inch.— Diameter of stalk .001, of ball at extremity .006.— Stalk transparent, cylindrical for about 1/10 of length, near to ball, it is flattened. angular & rather broarder: Terminal spherule full of grains, .0001 in diameter & sticking together in planes: When placed in water the ball partially burst & sent forth with granules large bubbles of air.— A rush of fluid was visible in the stalk or cylinder.— If merely breathed on, the spherule ~~was~~ expanded itself & three conical semitransparent projections were formed on surface.— (Much in the same manner as is seen in Pollen) These cones in a short time visibly were contracted & drawn within the spherule.—

Mantis (a)

Caught at Bahia on the 17th a Mantis[2] & as I thought killed it by holding for several minutes under water that was boiling, the head & thorax (to the insertion of the wings) & anterior legs.— These parts shortly were completely dead & became dry & brittle: but eight day[s] afterwards on the 25th the abdomen & hinder legs continued to possess a slight degree of irritability.— This appears a well marked instance of the ~~tenaciously~~ tenacity of life amongst insects.— |**30**|

Janthina (a)

The sea in Lat 18°6′ S & Long 36°6′ W. on the 26th contained numbers of specimens of Janthina[3].— Most of them were very small: the animal of rather a larger shell protruded itself & was of the same violet colour as the shell.— When touched emitted a fine purple colour. M. Rangs[4] states it to be de "couleur Laquesce" [note (a)] No. 158.— Cuvier mentions the fact about the colour. [note ends]

Fish (b)

In the above Lat & Long caught 2 specimens of a fish[5]; belly silvery white, mottled with brownish black. side blueish with dusky greenish markings. Iris yellow with dark blue pupil. Caudal fin with a pink tinge: these fish were 120 [miles] from the nearest land above water, namely Abbrolhos:— but the shoals are considerably nearer. [note (b)] Nos. 156 & 157. [note ends]

[1] See *Plant Notes* pp. 153-4.
[2] No specimen of Mantodea was found in CD's collection (see *Insect Notes* p. 48).
[3] Mesogastropoda, Ptenoglossa, the pelagic violet snails.
[4] See *Rang* pp. 196-7.
[5] Identified by Leonard Jenyns in *Zoology* **4**:73-4 as similar to *Psenes leucurus* Cuv.

[**CD P. 30** continues with observations on microscopic marine algae termed confervae found floating on the surface of the sea. The following 3 pages have later been crossed through, for the material was copied to appear in *Journal of Researches* pp. 14-20]

Colour of Sea

I had been struck by the beautiful colour of the sea when seen through the chinks of a straw hat.— To day 26th. Lat 18°6′ S: Long 36°6′ W. it was

according to Werner nomenclature "Indigo with a little Azure blue". The sky at the time was "Berlin with little Ultra marine [blue" & there were some cirro.cumili[1] scattered about.— No bottom could be found 230
(c) fathoms.— After running about 6 knots, soundings gave 30 fathoms & coral bottom, yet there was no change in the colour of the sea.—

Oily matter on surface
Oily matter on sea (d)

To day at noon I observed the sea covered with an oily matter.— The thin globules |31| displayed iridescent colours & were often time two inches in diameter.— A drop of water under a microscope showed on its surface minute globules of a transparent floating liquid, & which from its feel was of an oily nature.— it contained likewise irregularly shaped transparent minute fragments of matter: Three quarters of an hour after I first observed this appearance it was no longer visible. the ship in that [time] having sailed 2 & ½ knots.— I am at a loss to conjecture what could have been the origin of such a quantity of oily matter; it is stated that whales often produce this effect.— At night this water showed luminous particles.—

[note (d) added later] Octob 23[d]. South of Corrientes: I observe some of the Pelagic Amphipods contain in the intestinal vessels a considerable quantity of coloured oil:— Entomostraca The number of these Crustacea is often quite infinite [note ends]

Oily matter
(c)

27th at 10 AM the sea for yards was coated with the oil, having an iridescent appearance: It was in patches or streams & extended for a considerable distance.— [note (c)] In one hour & ½ afterwards having run 2 & ½ knots the water had its greasy covering. [note ends]

Tempera: of Sea

[note (c) from back of P. 30] ~~On approaching this bank at 4 P.M. no change of temperature was visible perceptible, the thermometer keeping at 82°. At 10 P M no bottom with 140 fathoms, & the thermometer instead of rising is at 81°.~~

The following is a table of thermometrical changes during crossing and recrossing the bank.

	[Hour]	[Depth]	[Temp]
26th	10 A M	230 Fathoms	82° Therm:
Lat. 18°6′ S	4 P M	30	82°
Long. 36°6′ W	10 P M	250	81°
27th	8½ A M	180	81⅔
Lat. 12°43′ S[2]	9 A M	150	81⅔
Long. 36°6′ W	10 A M	200	81¼
	11½ A M	250	81½
	1¼ P M	250	81⅔
	2¼ P M	30	81⅓

	3 P M	20	81⅔
	4 P M	22	81⅓
	5 P M	21	81½
	6 P M	21	81½
	7 P M	27	81
	8 P M	25	81½
	10 P M	27	81¼
	11 P M	27	81½
28th	8 A M	28	79⅔
	10 A M	10...30	79¼
	4 P M	do	78½
	9 P M	Anchored 20	76½

During this day (28th) the colour of sea varied, being sometimes black "Indigo blue", in evening very green.—

This table shows in some cases how little the Thermometer is affected. during the 26th & 27th, when not close to the Island, the mean of temp no bottom (at the lowest) at 150 fathom is within a very small fraction of degree of that when sounding<s> were at most 30.— On 28th the mean was 3 degrees lower than that of the 2 days previous: & we were then rapidly nearing the Islands. [note ends]

[**CD P. 31** continues]

Oscillaria[3] (a) 390:391 not spirits — At noon Lat 17°43′ S & Long 37°23′ W my attention was called by Mr Chaffers[4] observing that the sea was in places discoloured.— [notes (a)] No soundings with 250 fathoms. Nor not spirits 390, 391. An appearance similar to this one was seen between Canary & Cape de Verd at about ½ a mile distance from the ship. [notes end] Even from the Poop the cause was visible. it was owing to the presence of numberless minute whitish particles: These when examined under a lens whose focal distance was (e) ~~under~~ above 1/10 of inch appeared like bits of chopped rag, the ligneous fibres of which projected beyond the end.— [note (e)] Mr Brown[5] seems to have observed these Oscillariæ on the South shore of Australia. "particles 1/20 length, composed of cohering jointed fibres, of unequal length, so that the compound particle appeared as if torn" Flinders Voyage Vol 1 P 92[6]:— [note ends] |**32**|

Oscillaria x — These particles seen under a higher power consisted of about 20 fibrils adhering side by side & forming either a flat or a nearly cylindrical bit of mat.— These ~~cylin~~ fibrils or stalks were in length from .02 to .03 of inch; in diameter 1/2000: extremities round, rather broarder, transparent; internally a tube containing concentric layer of greenish brown granules. Hence appearing jointed: these layers are ~~close to~~ numerous. The external tube

(a) was marked by fine circular rings. (??) [note] (a) It required a 1/30″ focal lens in order to see the internal tube. [note ends]

No soundings at this spot.— But 6 knots before 36 h.

[note x] At noon on 31st of March, Lat 19°52′ S, Long 38°7′ W, the ship passed through a band of these Oscillariæ a mile in width. I reexamined them.— The bundle[s] were often cylindrical, containing from 20...to 60 fibrils.— a large one taking the extreme points was in length was .03 & in breadth .009.— Fibrils were perfectly straight: varied much in length; were I presume enveloped in a fluid.— as in ma<ny> of the bundles the fibrils did not touch each other.— Being kept till the following morning the particles became of a much brighter green & were partially decomposed: a considerab<le> quantity of brownish flocculent matter lying at the bottom of the cup.— The fresh Oscillaria placed in Alcohol uncoiled, moved the<se> & finally burst.— These appearances are called by the Sailors Spawn.— At 4 PM we passed through another irregular band running E & W.— about 10 yards wide & about 2 & ½ miles long.— The sea was the colour of thick reddish mud.— I believe each bundle of Oscillaria touched another.— I judge of this likewise by the thickness of the covering on some water brought up in a bucket.— (At this rate in this narrow band & at a *[illeg.]* moderate computation, in each square inch of surface there must have been 499950 fibrils or separate Oscillaria.— In the whole band: 323 967 600 00[0] 000: or 323 millions of millions &c.—?) Perhaps in square inch about 100 000. [note ends]

[**CD P. 32** continues]

(b) P 31

I once thought that I perceived a motion in these fibrils: from the description in Dic. Class.[7] I suppose it is an Oscillaria.— After being kept for an hour in water, most of them fell to the bottom of the Basin, & it appeared to me that in this state all the granules had been expelled: Figures are quite inadequate to give any idea of the numbers of these groups of Oscillaria which the sea contained.— A bucket which had been lowered for some water, had its interior sides (being left for short time at rest) literally coated with these minute particles.— I should think they extended for some distance; The sea 3 hours afterwards contained a few.—

Conferva (b) 392 not spirits

On 28th; 10 miles West of Abrolhos; there came up with the lead (17 fathoms) a piece of Fucus.— on which were growing numerous minute tufts of a Conferva[8].— Stems simple cylindrical white transparent jointed; end truncate; length 1/10 of inch, diameter 2/3000.— On this minute |**33**| plant & on a small coralline were crowded together a forest of numerous species of Bacillareès & Anthrodieès.—

[1] CD's spelling of the type of clouds on this day might not have met with FitzRoy's approval. See *Narrative,* Appendix to Vol. II, pp. 275-6.

2 According to CD, the *Beagle* had sailed over 5° northwards during the night! The table of compass variations during the voyage that appears in *Narrative*, Appendix to Vol. II, pp. 86-8, gives the Latitude on 27th March as 17°54′.

3 Identified by Porter (1987) in *Plant Notes* pp. 212-14 as a blue-green alga *Oscillatoria erythraea* (Ehrenberg) Kützing.

4 Edward Main Chaffers was Master of the *Beagle*.

5 Robert Brown (1773-1858) was a botanist and microscopist who discovered Brownian motion.

6 See Matthew Flinders. *A voyage to Terra Australis*. 2 vols., atlas. London, 1814. In *Beagle* Library.

7 See *Dic. Class.* **12**:457-85.

8 Although Specimens 390 and 391 are *O. erythraea*, Specimen 392 sent to J.D. Hooker in August 1844 with the comment 'Please throw away these specimens if of no use' (see *Correspondence* **3**:49-50 and *Plant Notes* p. 214) cannot now be identified.

[**CD P. 33** continues]

Phasianella (a) 169 Spirits	Animal with foot marked with black[1].— body blueish-lead colour; between feelers claret coloured.— feelers ringed with black.— these were nearly the only shell on coast of the Abrolhos.— they were however in the greatest profusion covering the rocks, & what appeared to me very singular, crawling up a bush which grew within high water mark.— The shells adhered to their leaves & bark far above the reach of the waves: From the habits when kept it is evidently an animal which passes much of its time out of the water.— Abrolhos.— March 29th.—
Tubiporèes 175 Spirits (Gorgonia?)	(Lamouroux ??) Abrolhos. 20 fathoms. March 30th.— Corallina[2], branched, stem rather flattened, horns, hollow.— Polypi when not expanded like buds scattered irregularly on sides & extremities of branches.— Stem slightly encrusted with red stony covering.— Polypi white, length .15, tentacula 8 in number, fimbriated. when partly collapsed having a leaf like appearance.— Tentacula situated on a fleshy tube proceeding from a slightly coriaceous one or cell.— Polypi highly irritable: but when fully expanded the Corallina had a beautiful flower-like appearance.— |**34**|
Abrolhos Islands Gen: Observ: (b)	The Abrolhos Islands seen from a short distance are of a bright green colour.— The vegetation consists of succulent plants & Gramina, interspersed with a few bushes & Cactuses.— [note (b)] Small as my collection of plants is from the Abrolhos I think it contains nearly every species then flowering[3].— [note ends] Birds of the family of Totipalmes are exceedingly abundant, such as ~~Sulas~~ Gannets, Tropic birds & Frigates.— The number of Saurians is perhaps the most surprising thing, almost every stone has its accompanying lizard: Spiders are in great numbers: likewise rats:— The bottom of the adjoining sea is thickly covered by enormous

brain stones[4]; many of them could not be less than a yard in diameter: Without being in the immediate presence of limestones how extraordinary it is that these Polypi should be able to obtain such an enormous stock of Carb of Lime[5].— [note (c)] This is an instance (perhaps not a strongly
(c) marked one) where there is a great formation of Coralls: & therefore the lime obtained without the neighbourhead of Volcanic action.— The currents in the ocean would however I think be sufficient for a ridge like this:— [note ends]

[1] Gastropoda, possibly superfamily Trochacea, a top snail.
[2] The Tubiporées listed by *Lamouroux* on pp. 65-7 among order XVII of the polypiers entièrement pierreux et non flexibles, are distinguished from the flexible Gorgonias in order IX on pp. 31-7. The organ-pipe coral *Tubipora* is a polyp of order Stolonifera. A specimen of *Idmonea milneana* Busk was identified by S.F. Harmer in No. 175 (in spirits).
[3] Although this claim was repeated in CD's letter to Henslow of 15 August 1832 (see *Correspondence* **1**:251), the only plant from the Abrolhos recorded in either of the Specimen Lists was a conferva (see *Plant Notes* p. 159).
[4] The brain corals are solitary stony corals of order Scleractinia, similar in appearance to a brain.
[5] Although the origin of limestone from the remains of marine organisms deposited in ancient seas was well established, CD was possibly unaware of the high concentration of calcium in sea water. The view that the growth of corals required a supply of lime brought up to the sea bed by volcanic action in the vicinity might have been derived from Lyell (see Charles Lyell. *Principles of Geology, being an attempt to explain the former changes of the earth's surface, by reference to causes now in operation*, Vol. 2, pp. 297-301. Murray, London, 1832), if note (c) was added after CD's copy reached him at Monte Video in November 1832.

[**CD P. 34** continues]

<u>Parmacella</u> Parmacella (Cuvier)[1] body lengthened; broardest across the mantle.—
198 Mouth ~~labiate~~ with upper lip bilabiate, inferior with a fold.— when closed it is folded into 5 irregular rays: Body beneath pale, above light dirty yellow; with few blueish lead coloured markings; colours more intense under the shell; 2 interrupted blueish lead coloured longitudinal bands on the back.— Shell transparent very brittle, oval, concave posteriorly beaked & slightly spiral; increases by concentric layers.— Lightly attached to the mantle, edges being overlapped by a membrane.— |**35**| Tentacula,
<u>Parmacella</u> superior lead coloured, inferior very short.— Length of large specimen 1.4 inch. breadth .4.— Habits, lives on aquatic plants & is partly amphibious.— When placed in water, turns its back downwards, draws in its tentacula & swims slowly till it finds some object to adhere to.— It moves in the same manner in the water as on a solid substance, viz by a wave-like motion in the foot; each wave is semicircular & travels upwards from the very extremity to the head.— It is not clear how the movement
(a) propels the animal.— Is it by a slight contraction after the formation of

(b)

each wave? [note (a)] The curved lines [sketch in margin] may represent the ridges or waves in the muscle of the foot; each one travelling onwards after the one before it; a new one of course continually commencing at the tail.— I fancied I perceived a slight contraction after the formation of each wave.— [note (b)] How naturally does the animal by its habits & organization connect through Succinea the terrestrial Pulmones without a shell, with those with one.— [notes end]

<u>Physa</u>
199

Physa[2].— shell sinestral.— Animal: foot thin. much separated from body. rounded in front, tail extremely pointed, lead coloured.— Tentacula long, tapering of same colour, with an attached membrane at the base.— Eyes within the base of feelers.— Over the mouth a large inverted wedge shaped bilabiate membrane.— Lives in grassy ditches, swims with its shell downward very rapidly by the aid of front projecting part of foot.— Steers itself by the head, perhaps membranes at base of feelers assist in this.— Body in nearly the same direction as opening of shell; when dead, not perfectly retractile: a fringed membrane projecting ~~from~~ around opening of shell. |**36**|

<u>Limnœa</u>
435 (not spirits)

Copied

Found <u>great numbers</u> of a species of Limnœa[3] adhering to aquatic plants in a lake situated between Mandetiba & Lagoa Araruama: The water was then fresh.— but the inhabitants affirmed that periodically once an year it became salt & sometimes oftener.— The period most probably in which the SW winds prevail; Is not this fact curious, that fresh water shells should survive an inundation of salt water? In the neighbouring Lagoon, Balani were adhering to the rocks.—

<u>Ceratophis</u>
185

My specimen[4] inhabited the dark & moist forest round Socêgo.— Its habits were those of an English toad, than a Frog. All its motions slow & feeble: proceeded by ~~slow~~ short jumps.— Colours in the Spirits have become rather fainter.— Iris bright copper colour.—

[1] Stylommatophora, a slug.
[2] A fresh water pulmonate of order Basommatophora.
[3] ditto.
[4] Not listed in *Zoology* **5**.

[**CD P. 36** continues]

<u>Aplysia</u>[1]
207

May 6th.— Animal with lateral crests unequal; right side nearly orbicular. very large.— measured internally to the back 2 & 3/4 inches wide.— left, posteriorly obliquely cut or slanted off & only 1 & 3/4 wide; the anterior basal parts of right one very thick & fleshy.— crests extend nearly length of whole body.— as the animal was dying when I found it, I am not sure of its shape — foot broa~~r~~d, length when contracted 4 & ½ inches, I have no doubt when crawling would be 6 inches — ~~Width~~ Depth with crest extended 4 & ½ (placed sideways on a plate).— |**37**| colour, purplish

Aplysia

dark brown with whitish marking, & in them minute snow white dots about 1/48th of inch in diameter.— on the edge of crests their markings are larger & more distinct.— Feelers same colour.— anterior fleshy, placed longitudinantly [*sic*], posterior small, near to anterior part of crests.— Mantle purplish, posteriorly forming simple tube; Branchiæ situated on a straight membrane on each side about ~~seven~~ eight corresponding tufts, primarily bifid.— A tube or line (?) running from between crests towards the head.— Connected with Generation? When first taken emitted a little purple.— If the Aplysia uses its lateral crests to swim.— Can this? Cuvier says Tectibranches have these Branchiæ not symmetrical.— Are not these?—

Hyla[2]
208

On the back, a band of "yellowish brown" width of head, sides copper yellow; abdomen silvery yellowish white slightly tuberculated: beneath the mouth, smooth dark yellow.— under sides of legs leaden flesh colour.— Can adhere to perpendicular surface of glass.— The fields resound with the noise which this little animal, as it sits on a blade of grass about an inch from the water, emits.— The note is very musical. I at first thought it must be a bird.— When several are together they chirp in harmony; each, beginning a lower note than the other, & then continuing upon two (I think these notes are thirds to each other).— |**38**|

Helix[3]
452 (a)
Copied

Body 1 & ½ inches long. Colour "Kings Yellow"; neck long, cylindrical, marked with longitudinal furrows which become reticulated on the sides; tentacula orange colour, bearing eyes at extremities, finely & regularly reticulated; anterior pair about 1/5 in length of posterior; beneath there are are angular projections forming sides of the mouth.— Mouth when protruded & closed, three folded (Y).— Foot & tail paler, the latter broard, rather pointed.— Inhabits thick woods on the hills.— [note (a)] 452 (not spirits, merely the shell) [note ends]

Spider[4]
212 (b)
in the Spider Bottle
213
taken in thick forest

Evidently by its four front strong equal legs being much longer than posterior; by its habits on a leaf of a tree, is a Laterigrade: It differs however most singularly from that tribe & is I think a new genus.— Eyes 10 in number, (!?) anterior ones red, situated on two curved longitudinal lines, thus the central triangular ones on an eminence: Machoires rounded inclined:

languettes bluntly arrow shaped: Cheliceres powerful with large aperture for poison.— Abdomen encrusted & with 5 conical peaks: Thorax with one small one: Crotchets to Tarsi, very strong (& with 2 small corresponding ones beneath?) Colour snow white, except tarsi & half of leg bright yellow.— also tops of abdominal points & line of eyes black.— It must

used
White

I think be new.— (Lithetron paradoxicus Darwin !!! [note (b)] 212 (in the Spider Bottle (213)) Taken in the fores<t> [note ends]

Helix[5]
481 (c)
Copied

Animal narrow, reticulated with lines all over body; colour brownish "Lavander purple" with snow white dorsal streak.— Superior feeler stout conical, terminated by a ball carrying the eye.— Eggs, white; .24 inch. diam: Shell effervesces with acids. Body when extended 3 inches long. [notes (c)] 481 (not spirits). Inhabits rocky wooded hills. [notes end] |**39**|

Leucauge[6]
(b) (a)
used
White

Spider, orbilates [orbitéles]; closely allied to Epeira (Leucauge. *[illeg.]*) Web, very regular nearly horizontal, animal rests in the centre on inferior surface: [note (a)] Vide No^{s} 235 & 214. [Notes (b)] Beneath the regular web with concentric circles, there is an <u>irregular</u> & then tissue of net work.— This irregular tissue work is sometimes above the concentric web.— [notes end] Machoires parallel, lengthened, thickening towards the end, square truncate: languette semicircular with central impression: Cheliceres cylindrical: eyes equal, thus placed [see sketch in margin]: thorax truncate, oval, depressed: 1st pair of legs *[illeg.]* longest. then 2^{d}, 4th & lastly 3^{d}: filieres little conical, projecting, distinct: Abdomen oblong, brilliant; the red like a ruby with a bright light behind.—

Bulimus[7] (?)
(c)
Copied

Animal with coarse reticulations, colour brownish yellow, becoming darker & forming a band on each side.— back white with central band.— tail broard flat pale.— Feelers yellowish, superiors long.— Mouth of the shell with anterior end flattened, animal protrudes itself in the same line as this.— Was found in the Botanic garden closely adhering to the species of firs which were originally brought from New S Wales.— [note (c)] V. n 240 [note ends]

Hymenop:[8]
(d)
(Rapaces)
Pompilus

I have frequently observed these insects [note (d): V No 535 Not spirits] carrying dead spiders, even the powerful genus Mygalus, & have found the clay (e) cells made for their larvæ [note (e): Vide (449) not spirits] filled with dying & dead small spiders: to day (June 2^{d}) I watched a contest between one of them & a large Lycosa.— The insect dashed against the spider & then flew away; it had evidently mortially [*sic*] wounded its enemy with its sting; for the spider crawled a little way & then rolled down the hill & scrambled into a tuft of grass.— The Hymenoptera most assuredly again found out the spider by the power of smell; regularly making small circuits

Hymenop:
(d)
(Rapaces)
Pompilus

|**40**| (like a dog) & rapidly vibrating its wings & antennæ: It was a most curious spectacle: the Spider had yet some life, & the Hymenop was most cautious to keep clear of the jaws; at last being stung twice more on under side of the thorax it became motionless.— The hymenop. apparently ascertained this by repeatedly putting its head close to the spider, & then dragged away the heavy Lycosa with its mandibles.— I then took them

both. (Hymenop. No 535)

[1] An opisthobranch of order Anaspidea, the sea hare.
[2] Not listed in *Zoology* **5**.
[3] Land snail.
[4] A crab spider, family Thomisidae, listed by Adam White in *Annals and Magazine of Natural History* 7:471-7 (1841) as *Eripus heterogaster*, but now known as *Epicadus heterogaster* (Guerin, 1831).
[5] Land snail.
[6] An orb-weaving orchard spider, family Tetragnathidae, named by Adam White (*loc. cit.*), using Darwin's name, as *Linyphia (Leucauge) argyrobapta*, this being the type species of the new genus *Leucauge*. Other species described by White from CD's collections in Brazil were *Linyphia (?) leucosternon* n.s., *Epeira (Singa) leucogramma* n.s., *Pholcus geniculatus* n.s.
[7] Rissoacea, a small snail with a conical shell.
[8] No specimen found, but could possibly be a *Trypoxylon* (Sphecidae). See *Insect Notes* p. 56.

[**CD P. 40** continues]

<u>Metereolog: Colour of sky &c</u>

(a)

In the course of to day (June 2^{d}) I have observed several trifling meterolog phenomena.— ~~The day~~ At noon it was very hot & calm: the sky dark blue & I remarked, what I have frequently before, that small Cumuli with defined edges float at ~~less~~ about 2000 feet elevation; they passed beneath the summit of the Caucovado.— These clouds to the eye had an appearance of great elevation.— For some hours the air, seen through for a short distance, had a prodigious transparency: but all colours at a greater were blended into a most beautiful tint.— giving to the landscape ~~an~~ serene appearance.— I have never observed this in England.— the colour was "French grey" with a very little prussian blue.— the sky in the Zenith was "Ultra marine" & "flax flower blue".— The Barometer had fallen .08 since the morning.— But from the same period, the dryness of atmosphere had much increased: the dew point was ~~64.5 & diff~~ 57°: diff 17°.— whilst in the morning the latter was only 7°.5.— |**41**|

[note (a) added later] Again the next ~~day~~ morning (June 3^{d}) a breeze set in from the NE. bearing with it a heavy bank of Cumuli.— This floated about 200 feet above the sea, & was not 600 thick, as the Sugar loaf peeped through its white covering, & looked like the peak of Teneriffe.— The rest of the sky was clear, with a few scattered Cirri.— As the white mass rolled inland, it rose in the atmosphere & was partially dissolved.— I never observed this phenomenon in any part of England. (The Barom was ~~not~~ but little affected)

June 8th.— From this fact of Cumili. with edge clearly defined against the blue sky, floating on a calm hot day. under 2000 feet of elevation. a landscape introducing it faithfully had to my eye, an unnatural appearance, although well aware of the truth of the fact.—

On May 5th & 17th there was a good instance of an appearance, which I had frequently witnessed with surprise on the Rio Macaè.— In ~~both~~ all cases for some hours the country had been drenched with rain; as soon [as] it ceased a most extraordinary evaporation commenced.— At 100 feet elevation the wooded hills were almost hidden in the clouds of vapour, which rising like column of smoke formed beds ~~of~~ not to be distinguished from the surrounding Cumili.— The most thickly wooded parts produced the greatest quantity.— I suppose this fact is owing to the great ~~extent~~ surface of heated foliage.— The atmosphere itself was not very damp DP 71. Temp 78. Diff: 7 [note ends]

[**CD P. 41** commences]

Meterolog: The thermometer (at same time) exposed on white cotton to the sun was at 2 PM 115°. The night was cloudless & a copious dew was falling. therm on the open turf fell to 61°.— So that the vegetation even in the winter season undergoes a range of 54 degrees.—

Mr Daniell[1] remarks that a cloud on a mountain sometimes is seen stationary, whilst a wind is blowing; the same phenomenon seen nearer on the Caucovado presented rather a different appearance.— Here the cloud ~~cloud~~ continued to curl over & pass by the summit & side of the peak & yet was not diminished. or increased in size.— The sun was setting & a gentle Southerly breeze came in.— this striking against the South side of the rock, which had not been exposed to the full rays of the sun & was open to the radiation of ~~an open~~ a clear sky, was cooled & the vapour condensed. but as it passed over the ridge it met the warmer air of the North sloping Bank & immediately the vapour was dissolved & cloud disappeared.—

[1] See John Frederic Daniell. *Meteorological essays and observations.* London, 1823. In *Beagle* Library.

[**CD P. 41** continues]

Lampyrus[1] (a) (d) (b) In the early part of the night of April & beginning of May. the marshy fields were illuminated by this beautiful insect; the light was green & more intense than the Elater noctelucis: it was visible at more than 200 yards.— [note (a)] 440, 441 not spirits [note (b)] It is remarkable how commonly that the light from animals is green.— Four Lampyr~~u~~is, Elater Noctelucis; ~~Marine~~ crustaceæ & other marine animals all partake of this tint.— [note (d)] Great numbers of this insect fall a prey to Epeirus [notes end]

the insect in its habits is very active & when most irritated emitted the most
Lampyrus brilliant |**42**| flashes: in the intervals, the two abdominal rings were completely obscure; the flash is almost instantaneous, but first appears in the upper ring.— The shining matter is fluid & very adhesive, & lies immediately under the skin: Carb of Soda added to it produced no immediate effect.— Places, where the skin was torn, in the interval remained bright & a scintillation was perceptible.— When the head of the insect was cut off, the rings continued interruptedly bright, but not so brilliant; pressing & pricking always increased the vividness; & it then appeared first of a bluer tint & in spots.— The abdomen remained luminous many hours more than 24 after the death of animal.— From all these facts it would appear that the vital action is more concerned in obscuring the light at intervals than in immediately producing it.—

Larva Larva of the above Lampyrus (I suppose) luminous not quite so strong as
(a) our glow worm.— [note (a)] No^s. not spirits 442, 443, 506, 507 [note (b)]
(b) I have no descriptions to recognise for certain the female from the larva of Lampyrus. I never however saw the winged ones near to where the apterous ones were crawling [notes end] Inhabits wet muddy places: when touched pretends death, & ceases to be luminous & irritation will not reproduce it.— Can swim well by a lateral serpentine motion of body; tibice rather *[illeg. word]* spinose.— Walks quickly by the aid of its tail.— This latter organ

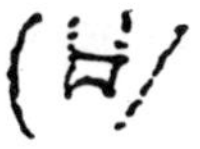

is curious: the last dorsal or tail plate is cut out [see sketch in margin] & the two inferior & posterior rings of abdomen with spines; beneath the
(c) penultimate is a cup, from which can be protruded an oval membranous tube, containing numerous approximate fillets, arranged |**43**| in a circle;
Larva of Lampyrus each of these is bifid & has the power of strongly adhering to any surface.— [note (c)] The cup rather rises at the junction of the last & penultimate joint.— the above mentioned spines in the penultimates are situated on the inferior surface, in the last joint at the very extremity.— The larger the specimen the more luminous it is. [note ends]

[**CD P. 43** continues]

The spines & tube being pointed posteriorly & the latter pulling in the same direction the animal can firmly attach itself by this means.— Mouth retractile.— Are strongly carnivorous, readily feeding on raw flesh.— Whilst so doing the tail is frequently applied to the mouth, which is partly drawn in; & a large drop of fluid is exuded from the terminal cup; this appears to act in both softening the mouth & the flesh.— The fluid neither affected Litmus or Turmeric: but like the gastric juice, the action of which Chemistry can so little explain[2], it doubtless aids digestion.— The tail was always guided to the mouth by first touching the neck.— These larvæ are in considerable number.— does not the fact of their being luminous render what has so often stated improbable, *[illeg.]* that the sexes shine in order to bring them together[3].— Amongst the specimens there is one of another species; the mouth protrudes further out & the dorsal plates are rounded.— I have likewise taken these species of the full grown Lampyrus. [note (a)] (a) No. 508 (not spirits) [note ends]

Hymen=-phallus (b) Copied — Growing in a very thick damp forest (June 4th) did not smell stronger than the Caninus: yet sufficient to be remarked by the inhabitants: the veil was inserted about ½ an inch beneath the cone at top.— top perforated: liquid on it yellowish brown: bag of jelly resembling impudicus. — the specimen is only in fragment<s>[4]. [note opposite] (b) No. 245.— A Leiodes[5] (550 not spirits) flew on it as I was carrying it.— |**44**|

[1] The glow worms and fireflies observed by CD were later identified by George Waterhouse as mostly *Lampyris occidentalis*, but none of the specimens have survived. See *Insect Notes* pp. 51 and 57.

[2] The first studies of the properties of organic catalysts were begun at about this time, but the name enzyme to describe them was introduced by Kühne only in 1878.

[3] In Vol. 1, p. 345 of *The descent of man, and selection in relation to sex* (John Murray, London, 1871), CD repeated his doubt whether the role of luminosity in the Lampyridae was truly the generation of mating signals, because this would not account for the particularly strong luminosity of the larvae.

[4] Caninus is the dog stinkhorn, and impudicus is *Phallus impudicus* L., but CD's specimen did not survive. See *Plant Notes* p. 219.

[5] The specimen has not been identified with certainty, but from other evidence cited in *Insect Notes* p. 56 is a Nitulid similar to British species but larger.

[CD P. 44 commences]

Bulimus ? 257 — Animal[1] crawling on the dry ground; shell destitute of an umbelicus.— (is it young Bulimus ??) — body 4 inches long .5 wide: superior feelers .9 long: inferior .2: foot very broard, thin at edges: back rugosely reticulated, colour dirty lead coloured; scales & tail more yellow.—

Vaginulus[2] specimen 256 — Veronicella Blainv: animal here described as in crawling.— ~~Mantle~~ Above rather pale "honey yellow". Mantle regularly rounded; smooth to the touch, but finely tuberculated; edges angular far projecting over foot, forming at anterior end a truncate hood; mouth & front part of foot retracted during inaction.— Mantle covering whole body length 5.5 inch, breadth .5, posterior end bluntly pointed.— foot of uniform breadth: thin, separated from mantle by an interval of sides: pointed at end & divided from extremity of mantle for .3 of an inches.— [note (c)] The side is I suppose (c) only the under edge of mantle; palish yellow [note ends] Between them fecal orifice; partly formed by groove in under surface of mantle.— it moves by wave-like motion of muscles as in Parmacella.— (an obscure hole on middle of rt side by edge of foot. generative?) Superior feelers, approximate, length .6. lead coloured terminated by ball, bearing an eye on superior surface: inferior with extremely blunt, length .2.— To the under surface for 2/3 of length is joined another organ, giving to the feeler a forked appearance.— it is pointed at extremity, & whilst the animal moves this part is perpetually retracting & protruding: it appears to have an aperture & to exude small quantities of fluid: is it to moisten path before the body?

Vaginulus |**45**| The slime (on body) is exuded through parallel pore on the foot.— Animal slow, torpid, generally with mouth retracted, lives & feeds on leaves of a tree in a dense forest on a hill; remote from any water.—

Vaginulus 256 (d) Specimen 291

A small specimen only .5 length differs from the former in the following respects.— Anterior & posterior ends of mantle black.— with 4 faint dorsal lines of same colour.— rather more tuberculated & with white dots: ~~edges~~ space between edges of foot & mantle white.— Lived in same forest: caught it in the sweeping entomological net.— Is it a different species or merely the young?— Are young snails generally darker coloured?—

[note (d)] June (23^{d}) Found an injured specimen of this animal; colour uniform yellowish green, tuberculated with white dots; sides & foot concolorés.— Number (291) [note ends]

[**CD P. 45** continues]

Comatula[3] 283 (c) Comatula

Botofogo Bay. 15th.— Ventral surface "deep reddish browne" arms & with their pinnæ banded with white.— dorsal plate & cirrhi pale.— Suckers on the pinnæ minute, numerous; on inferior surface of arms a fine canal, bending alternately to each pinnæ, meets on the ventral disk with the ~~other~~ canal from the brother arm: (proving that the number 5 is normal, although here apparently there are 10).— The junctions of these canals irregular; meeting in the irregular central mouth.— [note (c)] Lamarck seems to deny this mouth. Cuvier states there to be one.— it certainly is by no means so apparent as in Asterias. [note ends] Anus submarginal, tubular, ejecting fæces.— The pinnæ on the lower half of arms are at their base, fleshy & not banded with white. The animal was found adhering on the over |**46**| hanging ~~project~~ ledge of rocks.— its dorsal cirrhi were firmly fixed in an encrusting sponge.— & the arms widely extended, so as much to resemble an enormous Polypus.— irritable. Motion passing down the body as in a sensitive plant.— arms have considerable power of motion, can curl themselves into a perfect spire.— When placed in fresh water emitted a strong odour & stained the water with a brownish yellow ~~tint~~.— The animal had a most graceful appearance.—

Nudibranch[4] (allied to Scyllæa?

Branchiæ dorsal (resembling Doris). each arm conical with simple short cirrhi; 6 in number, 4 anterior longest; between posterior ones there is circular anal orifice.— foot narrow, doubled into a groove incapable of

264 adhering to flat surface, anterior end flat, enlarged into a natatory organ.— Mantle projecting ~~over~~ with longitudinal slit for mouth.— Feelers 6; 4 anterior simple, tapering thin.— of which the two first are more approximate. behind & within there are 2 oval strong ones, on a footstalk
(b) & with circular ridges.— [note (b)] These are semi-retractile.— [note ends] (in this specimen the rt one is only left, but I think I can perceive where the other was) The lines joining on each side the anterior feelers are raised into a sort of rudimentary membrane, which traverses the back, enclosing branchiæ, & meet at the tail.— this membrane is fringed with ~~projections~~
<u>Nudibranch</u> paps on its edges.— |47| Tail ~~eyl~~ round, pointed.— On right side between branchiæ & mouth, a closed orifice was visible.— Generative?— A strong pulsation was perceptible on the back before the Branchiæ.— Length of body .3 (probably young specimen) colours most beautiful; side blue & white with projecting white paps & with irregular transverse rows of bright orange spots.— Back with less blue.— Branchiæ & posterior pair of feelers coloured as the sides.— Animal was found crawling on the stalks of fine Corallines.— could swim well.— & had power of turning its head vertically back as far as Branchiæ.— It would seem to have some relation with Scyllæa & some with Polycera.—

<u>Coralline</u>[5] Cells oval, attached by one end in irregular scattered groups on irregular
(a) cylindrical jointed hollow transparent much branched stems. Polypus
(b) tubular, conical, lengthened with 8 long tapering arms.— Growing in large
Lardiman? tufts at low-water mark.— [note (a)] 265. Polypi hanging out.— [note (b)] Stems irregularly divided, interwoven, membrano-gelatinous.— [notes end]

[1] Rissoacea, a snail with a conical shell.

[2] Stylommatophora, a pulmonate land slug. See Planche 58 showing Veronicelle lisse as portrayed by Blainville in *Dic. Sciences Naturelles* Planches 2e partie, Zoologie.

[3] A feather star, a stalkless unattached crinoid of order Comatulida. See Lamarck *Animaux sans vertèbres*, Vol. 2, pp. 530-35, and Cuvier *Le règne animal*, 2nd edition, Paris 1829-30, Vol. 3, p. 228.

[4] Doridacean nudibranch, probably *Polycera cf. odhneri*, Marcus 1958.

[5] Specimen 265 in Spirits of Wine was further described as *Sertularia* Lamarck, a term formerly covering both bryozoans and hydrozoans, but it was listed among those thrown away by S.F. Harmer in 1901 as so much macerated that they could not be identified.

[CD P. 47 continues]

<u>Spider</u>[1] Ábdomen triangular, filières pointed inferiorly at rt angles to the body.—
? Machoires enlarged ~~into~~ & rounded at extremity, languette rounded.— 1st
266 pair of legs much longest.— 2d pair next.— Eyes like Epeira, but anterior & lateral on eminence.— Claw of cheliceres, small but little oblique, internal edge finely serrated.— This curious little spider inhabits with

Spider impunity the strong |**48**| web of the genus Epeira: & far most generally of a large one (specimens in No 252). division of Dic: Class: (I: ++:).— Indeed few of the webs can be found without these intruders.— There appears to be ~~one~~ more than one species: the more lengthened brown coloured one is the male of silvered abdomen sort.— as I think I observed them in copulation head to head.— When touched they either pretend death by shaking forward all front pair of legs.— or fall down, being attached by line to the Epeira web.— I know not where to place this genus.—

Epeira[2]
Nor 238
(b)
(c)

Div: in Dic: Class: (II.++.1), very common especially on coast amongst the Aloes.— the web is strengthened in a curious manner.— the rays from centre have of course the concentric circles, also on opposite sides of centre two adjoining spokes are connected by a Zig Zag band of web.— the case is sometimes double so as to be at right angles to each other, thus.

When the spider is touched it falls down instead of as is common in Epeira run to the corner.— Stands head lowermost in centre of web.— [note (b)] Some all<ied> species even have a regular piece of mat-work in the centre of their web.— [note ends] When an insect is caught (for instance I saw small wasp & grasshopper) the spider rushes on it & by rapidly revolving it ~~with~~ in a few seconds involves it in a thick mesh.— as this proceeds from the filieres, it looks like a silver ribbon.— The spider then examines its prey & (in case of wasp) bit it several times with cheliceres on the back or thorax.— & immediately retreats |**49**| to its usual place the centre of

Epeira web.— The insects in about ½ a minute being taken out of the mesh were quite dead & relaxed.— How much more powerful is this than any poison man knows of.— Prussic acid being rubbed into a Blaps seemed only to cause a slight paralysis, which in short time went off.—

[note (c)] June 25th. I again watched one of these spiders; it is chiefly when the web is over an aloe or thick bush that the insect suddenly falls to the ground.— If the space beneath is clear, the spider disturbed only moves with great quickness through a hole near the centre from one side to the other.— It also practises another most curious meeneuvre [*sic*] when still further disturbed; by rapidly contracting & expanding its legs & the meshes being attached to elastic twigs, it soon gives to the whole web such a vibratory motion, that even the outline of Spider is rendered indistinct.— I may mention that when animal perfectly stationary the ~~web~~ filières can lengthen the thread, which was attached to a point, previous to falling.— The spider being still further molested, instead of leaving a single line as a train, emitted the same mass of web as described in enveloping its prey.—

[**CD P. 49** continues]

Theridion[3] (a) (d) — In bottle (252) there are specimens of a small red spider. head, extremities of abdomen & half the legs black.— I believe it to be a Theridion.— They are exceedingly numerous: fresh turned up ground & short turf being in most places coated by its the small irregular web.— in the morning bespangled by Dew.— [note (a)] Specimens in bottle 252. [note (d)] Latreille referring to this appearance in Europe, refers it to young Lycosæ: in same manner as he does the Gossamer to grand [?] Arachnidæ.— (Vide P 117)[4] [note ends]

Myrmecia[5] (b) — This singular looking spider is not uncommon in the wooded hills, amongst the foliage;— it is the "rufrum", but the colours vary, especially the black marking.— the abdomen & posterior segments of thorax obscure; the general colour of legs & body is not "fauve" but a mixture of "Orpiment orange & Vermilion red". [note (b)] Inhabits a leaf curled up; is very active in running & looks singularly like an ant. Specimens in bottle (252) [note ends]

Tetragnatha[6] (c) — Common over water & may be seen in the evening forming its web.— When frightened, either remains stationary or runs to one corner, & stretches forward in a bundle its long legs.— Web horizontal, meshes large, points of attachment far apart.— it is generally attached to flags or rushes & is beautifully adapted to withstand being shaken by the wind.— I observed one, stretched across a very rapid brook, & joining to a central stone: how does the animal contrive to effect this?.— [note (c)] Specimens in bottle (252) [note ends] |**50**|

[1] A comb-footed spider of family Theridiidae, *Argyrodes* sp. or spp. This genus includes kleptoparasitic spiders that live in the webs of larger spiders, but it is not clear whether CD is referring to one or two species (male or female). It was of this period that CD wrote to Henslow 'I am at present red-hot with Spiders, they are very interesting, & if I am not mistaken, I have already taken some new genera'. See *Correspondence* **1**:238.

[2] Orb-weaving spider of family Araneidae, *Argiope* sp. *Epeira* is no longer a valid genus. See *Cuvier* Vol. 4, p. 247.

[3] Tangle-web weaver, family Theridiidae, *Theridion* sp. See *Cuvier* Vol. 4, p. 243.

[4] For CD's account of the invasion of the *Beagle* by gossamer spiders when sailing from Buenos Aires to Monte Video, see pp. 106-8.

[5] Ant-mimicking spider of family Corinnidae, *Myrmecium rufum* Latreille. According to *Cuvier* Vol. 4, p. 261 'La Myrmécie fauve . . . se trouve aux environs de Rio-Janeiro'. See also *Dic. Class.* **11**:587.

[6] Orb-weaving spider of family Tetragnathidae, *Tetragnatha* sp. See *Cuvier* Vol. 4, p. 247.

[**CD P. 50** commences with an entry on Planaria that as before has been crossed through, indicating that it has been used for publication]

Planaria[1] June 17th.— This very extraordinary animal was found, under the bark of a decaying tree, in the forest at a considerable elevation.— The place was

No. 278 quite dry & no water at all near.— Body soft, parenchymatous, covered with slime (like snails & leaving a track), not much flattened; when fully extended, 2 & ¼ inches long: in broardest parts only .13 wide.— Back arched, top rather flat; beneath, a level crawling surface (precisely resembles a gasteropode, only not separated from the body), with a slightly projecting membranous edge.— Anterior end extremely extensible, pointed lengthened; posterior half of body broardest, tail bluntly pointed.

Colours: back with glossy black stripe; on each side of this a primrose white one edged externally with black; these stripes reach to extremities, & become uniformly narrower.— sides & foot dirty "orpiment orange".— from the elegance of shape & great beauty of colours, the animal had a very striking appearance.—

The anterior extremity of foot rather grooved or arched.— on its edge is a regular row of round black dots (as in marine Planariæ) which are continued round the foot, but not regularly; foot thickly covered with very minute angular white marks or specks.— On the foot in centre, about ⅓ of length from the tail, is a[n] irregular circular white space, free from the specks.— Extending through the whole width of this, is a transverse slit, sides straight parallel, extremities rounded, 1/60th of inch long.— tolerably apparent.— (i.e. with my very weak lens) |**51**|

Planaria At the distance of .3 & nearer to the anterior extremity is another slit, resembling in every respect the former, but smaller & much more obscure (I did not perceive it till the animal was hurt by Salt Water).— Posteriorly trace of central dark vessel & I suspect anal orifice; I judge at this from the appearance on glass of something like fæces & diminution of dark coloured

(b) vessel.— [note (b)] This doubtless is an error, V. the Planaria P 53 [note ends]

The following[2] is the most remarkable phenomenon: I cannot doubt its accuracy as I observed it in several lights & with low powers chiefly 1/5 & 1/4 focal distances.— As the animal adheres to a plate of glass; in different parts of the foot, a slight contraction of the body includes & propels a coating or thin globule of air.— Instantly as the air comes in contact with surface of foot, a violent corpuscular motion is perceptible; in paroxysm & rather from centres; I cannot explain it, but by a simile which is most

(a) precise; it is a number of small eels in thick mud being disturbed by a stick.— [note (a)] I actually at first moment thought there were minute animalcules struggling in the slime.— it is like the motion of a linear animal (such as eel, tadpole, animalcule) struggling to release themselves.— [note ends] the motion was well seen by lens 1/5 focal distance, very rapid & serpentine.— I never observed it ~~except~~ on foot except where air was between it & the glass.— it was most singular to observe this motion as a globule of air was driven in, proceeding together with it.— A similar appearance was visible on rather smaller scale on the |**52**| dorsal surface;

Planaria I observed it once & most clearly on the very anterior extremity. I suppose

(b) this action is the absorbing or forcing air into minute cutaneous vessels.

[note (b)] I must have fallen into some error; to day 23[d] I saw same appearance on back of a Bulla, in places where the light was shining on the surface.— (the animal being out of water).— It remains however quite inexplicable to me what the cause of phenomenon is.— [note ends]

[**CD P. 52** continues]

The animal crawled like a Gasteropod, by wave like motion of foot; but differed in the anterior extremity being raised & stretched forward, & rather curved backward.— it appeared to use this part as a feeler.— could creep amongst moss.— appeared quite unused to water; salt water was highly destructive to it.— Motions slow; body irritable & irregularly contractile; quickly recovered from a cut, which I gave it in first taking it.— I should think from habits Phytovorous: kept it in tin box nearly 4 day[s]; could perceive no difference.— Was I think ~~perceptive~~ sensible to light.— From the above characters it is evident it is a Planaria of Cuvier[3].— It differs from those (marine) I have seen; in the narrowness of body & not being much flattened; in the well marked crawling surface or foot & in the beauty

(c) of colours & in manner of crawling.— [note (c)] Has not the rapid vivacious motion of the marine species.— [note ends] How much more wide is the difference in its habits.— who would ever suppose the soft pulpy body of a Planaria could withstand the action of the air.— When I first found it & before I had examined it.— I had no doubt it was a Vaginulus (Cuv). I feel sure from its general appearance, slime, &c most

(a) observers would at first fall into the same mistake.— [note (a)] Most certainly the real relation between a Planaria & Gasteropod (Pulmones) is very small; but ~~it appears~~ that ~~relation~~ of analogy is here well seen, as it often is in animals widely apart in the chain of Nature.— [note ends] |**53**|

Aplysia[4] Length of extended animal 1.^{in}; posterior feelers simple, conical, close at

279 bottom for 1/3 of length.— Colour pale green, with meshwork of brownish purple veins; circular spaces being left clear.— Head darkest coloured with the purple; from it a band leading to branchial covering.— The latter on edges with black dots.— Sides with few white dots.—

Bufo[5] (Bombinator). Back: "deep orange & chesnut brown". beneath pale, with

289 dark mark between front legs.— behind tympanum & under eye pale with

Copied — black marks.— legs banded slightly with black.— Iris yellow.— tongue large, fleshy .— Was found under piece of bark in forest, far from water.— Motions slow, jumps.— from the rich colours, the animal presented a curious appearance.

Planaria[6] 290 — This like the last (**Page 50**) was caught in the forest, crawling on soft decayed wood.— It is quite a different species.— Back, snow white, edged on each side by very fine ~~parallel~~ lines of reddish brown.— also within are two other approximate ones of same colour.— sides & foot white, nearer to the exterior red lines, thickly clouded by "pale blackish purple". animal beautifully coloured.— foot beneath with white specks.— but few black dots on edge & none on head.— length of body one inch, not so narrow in proportion as other species; & |**54**| anterior extremity not

Planaria — nearly so much lengthened.— the body in consequence of more uniform breadth.— like the former it rests on end of tail & bends out its head to

Colour — find object to crawl on.— In the colouring of the body three rings are left nearly of a pure white.— In the foot, & in ~~the line of~~ the two posterior rings; the two transverse slits or openings were clearly visible.— I examined very carefully by strongly concentrating the light, the posterior extremity & am convinced there is no anal ~~openin~~ orifice.— it appeared to consist of a uniform parenchymatous matter.— indeed every part of the body thus viewed had this appearance.— In all other respects this animal exactly resembles the Planaria of Page (50).— As the tree on which I found it was near to rapid brook, I again placed this specimen in water; far from being accustomed to it.— I think in short time it would have been drowned.— Having found this crawling slowly on the damp & rotten wood, & the other under the bark of a somewhat similar tree, in all probability they live on decayed vegetable matter.— Having found two species is fortunate as it more firmly establishes this new subdivision of the genus Planaria.— |**55**|

[1] Listed as *Planaria vaginuloides* by CD in his article on 'Brief Descriptions of Several Terrestrial *Planariae*, and of Some Remarkable Marine Species, with an Account of Their Habits' in *Annals and Magazine of Natural History, including Zoology, Botany, and Geology* **14**:241-51 (1844). Turbellarian flatworm in order Tricladida, now known as *Geoplana vaginuloides* Darwin.

[2] CD's 'remarkable phenomenon' was probably the action, made visible by the bubble of air, of the microscopic cilia beating. The same explanation would apply to his observations on *Bulla*.

[3] See *Cuvier* Vol. 3, p. 266.

[4] A sea hare, an opisthobranch gastropod of order Anaspidea.

[5] Not identified by Thomas Bell in *Zoology* **5**.

[6] Listed by CD in his article (*loc. cit.*) as *Planaria elegans*. In a letter to Henslow begun on 23 July 1832 (see *Correspondence* **1**:251), CD says 'Amongst the lower animals, nothing has so much interested me as finding 2 species of elegantly coloured true Planariæ, inhabiting the

dry forest! The false relation they bear to snails is the most extraordinary thing of the kind I have ever seen.— In the same genus (or more truly family) some of the marine species possess an organization so marvellous.—that I can scarcely credit my eyesight.—' Henslow was unconvinced, and on p. 5 of the edition of CD's letters to him printed for private distribution by the Cambridge Philosophical Society in 1835, the word 'true' was omitted, and (?) was added after 'Planariæ'. Nevertheless, CD's observations on the anatomy and behaviour of these flatworms were in the main accurate, and this species does indeed lack an anal aperture; but CD was wrong in thinking that they feed on decayed vegetable matter, for in fact they are carnivorous. This turbellarian is known today as *Geoplana (Barreirana) elegans* Darwin.

[**CD P. 55** commences]

Papilio[1]
Linnæus
615
not spirits

This insect is not uncommon & generally frequents the Orange groves; it is remarkable in several respects.— It flies high & continually settles on the trunks of trees; invariably with its head downwards & with its wings expanded ~~to further than~~ or opened to beyond the horizontal plane.— It is the only butterfly I ever saw make use of its legs in running, this one will avoid being caught by shuffling to one side.— Some time ago I saw (d) several pair[s], I presume males & females, of ~~these~~ butterflies chasing each other, & which from appearance & habits were I am sure the same species as this.— Strange as it may sound, they when fluttering about emitted a noise somewhat similar to cocking a small pistol; a sort of a click.— I observed it repeatedly.—

[note (d) added later] June 28th.— In same place I observed one of these butterflies resting as described on a trunk of tree; another happening to fly~~ing~~ past, immediately they chased each other, emitting (& there could be no mistake the space being open) the peculiar noise: this is continued for some time & is more like a small toothed wheel passing ~~over~~ under a spring pawl.— The noise would be heard ~~between~~ about 20 yards distant. This fact ~~(from Kirby)~~[2] would appear to be new. [note ends]

Cavolina?[3]
299

Anterior feelers very long, united at base projecting over the mouth; posterior ~~one~~ feelers conical with ~~transves~~ transverse ridges (like in many Doris): eyes situated posteriorly at the base of latter:— feelers orange coloured.— Branchiæ in longitudinal rows on each side (or rather in 2 sets of obliquely transverse ones).— Branchiæ simple, tapering, internally dark brown.— Tail pointed, enlarged near extremity, prehensile.— Found amongs[t] corallines at Botofogo Bay.— ~~(Examination very short!!)~~ |**56**|

Amphiroa[4]
282 & (595 not spirits) (3)

Branches very much flattened, formed of arched layers (a).— these are very brittle & stony, formed of parallel longitudinal fibres & appear in older branches solid.— Extreme layer white, semitransparent & so soft the least touch would injure it.— no trace of terminal aperture.— Joints (B) [see

sketch in margin] transparent horny & more generally at the bifurcation of branches.— ~~they~~ it would appear that these are formed rather by an alteration than continuation of central substance.— Without these joints the coralline would be rigid.— Branches irregular, generally dichotomous.— ~~From~~ The joints are formed by a crack in outer Calcareous coat & oval opening on each side: From the *[illeg. word]* & the terminal layer being soft, as they become dry they contract into hollows. V specimen (595).— I could by no means (fresh Water, Alcohol &c) perceive any signs of irritability.— On one side of this coralline there may be generally observed either irregularly or in double regular rows.— rounded projecting paps.— these have a distinct minute orifice: I am at a loss what to consider them, by no means could I make any animal protrude itself.— These cells are not fixed deeply into the branch.— Is it impossible to be a minute Pyrgoma; the recurrence in double rows *[illegible corrections in pencil]* on one side wars against this: yet it forcibly struck me to be the case.— The Coralline is in great quantity in Botofogo Bay.— |**58**|

[note (a) at the top of **CD P. 58**] I find I have pages **15** by mistake twice over[5], so that although late I have changed this page into **58** instead of **57**.— [note ends]

[note (3) added later] Corallina[6] growing abundantly on an mass of Ascidia throw<n> up on beach. June 1836. C. of Good Hope.— By accident
3686 nearly all the specimens were lost, the fragments preserve<d> showed on many of the cylindrical joints the small pap, formed bladders with little circular orifices. Being broken open beneath the microscope, there were seen 8-12 (about) small rather bright pink bodies, arranged in a sort of ring in a little flocculent matter; by a slight *[illeg.]* were easily detached & floated separately,— in form pear shaped, one side rather protuberant, ~~apex~~ one extremity pointed, the other rounded; the envelope was distinct, the central matter appeared granular & pink coloured. In size they could easily pass through the orifice of cell.— With 1/20" focal lens could perceive no particular organization in these ova.— I examined & opened several of the paps.— [note ends]

[**CD P. 58** commences]

Ctenus[7]
In tube (300)

All the specimens I have seen, have been on wooded hills; there appears to be 2 divisions in the genus.— The ones with body flattened, hairy & colours speckled, legs very long, line of four central eyes curved.— These live in decayed trees & may often be seen standing motionless with their legs stretched out near to some hole.— It is evident they can see to some distance; for the instant you draw back ~~they~~ out of sight, they dash into their holes.— The other division in their appearance & habits approaches closely to Lycosa: there are specimens of both in (300).

Oxyopes[8] (c) This genus was exceedingly numerous in May (during the wet season) & was universally found amongst the herbage, but more especially in damp places.— In its habits it is a Saltigrade; springing with all the activity of one of that tribe from leaf to leaf.— [note (c)] Numerous specimens in bottle (213). [note ends]

[1] A butterfly identified by CD in *Journal of Researches* **1**:38 as *Papilio feronia*. In a footnote he mentions that G.R. Waterhouse has examined the specimen, and cannot discover the source of the sound; but in the 1845 edition he refers to a paper about it by Doubleday in *Proc. ent. Soc. Lond.* p. 123 (1845). See *Insect Notes* p. 58.
[2] See William Kirby and William Spence. *An introduction to entomology*. 4 vols. London, 1815-26. Copy in *Beagle* Library.
[3] An aeolidacean nudibranch.
[4] The syntype specimens of the coralline alga *Amphiroa exilis* Harvey that were collected by CD at Botofogo are now in the Herbarium of Trinity College Dublin as described in *Plant Notes* pp. 186-90. Specimen 595 also included the bryozoan *Nichtina tuberculata* preserved in the Busk Collection at the Natural History Museum
[5] The second **CD P. 15** was later renumbered **15 bis.**
[6] Named *Amphiroa exilis* var. *crassiuscula* by Harvey. See *Plant Notes* pp. 199-200.
[7] A hunting spider of family Ctenidae, in CD's 2nd division. His 1st division would be flat hunting spiders, family Platoridae.
[8] A genus of lynx spiders, family Oxyopidae.

[**CD P. 58** continues]

General Observations The following remarks are grouped without order: The traveller in a country where every feature wears so totally a different aspect is liable to fall into errors from expecting contrasts & reversed order of things where they do not exist: from this cause a greater degree of caution is necessary in comparing the appearance of Nature in the two zones than would have at first have been expected. |**59**| After seeing a collection of Brazilian birds in a Museum; it would not easily be believed what little show they make in their native country.— Concealed in the universal mass of vegetation, the attention is not drawn to them by their notes.— The large swifts with pointed tail feathers, unlike to their congeners in England pursue in silence their airy circles.— Perhaps a bird allied to the Parrots (Krotophagus) possesses the most harmonious voice.— Nature in these Zones chooses her vocalists out of other tribes; in the evening some species of frogs make a concert no ways unpleasant. this as the night advances is accompanied by the endless cry of the Cicadas.— As far as regards insects, M. Lacordaire[1] states the months during which I have collected are by no means the most productive in insects.— This may account for the few numbers of large & brilliant beetles which I have seen.— Of the smaller species I certainly have succeeded in taking great numbers.—

General observations

[illeg. note in pencil]

Coleoptera[2]

(a)

Coleoptera.— Amongst the Carabidous beetles[3], the only ones I saw in plenty were Cicindela nivea[4], two Harpali[5] & a Lopha[6].— the other few chiefly belonged to Truncatipennis & ~~Scarce~~ Bipartis. [note (a)] Truncatipennis inhabiting the foliage in forest, the Bipartis sandy plains. [note ends] (I always allude more to number of individuals than of species) This family evidently more belongs to a higher latitude.— Amongst the Hydrocanthares[7] were several minute species of Hydroporus[7], Hygrotus[7] & Hyphidrus[7] & Noterus[8]. |**60**|

General Observations (a)

They are not however, so numerous as in England.— Gyrinis[9] frequent & might be seen dancing on the surface of a clear ditch; forcibly bringing to the recollection of an Entomologist his walks at home.—

(b) [marginal note in pencil] Lycodes[23] alighting upon Phallus. Stay at Barmouth[24].

Brachelytus[10] uncommon. chiefly on decaying vegetable matter.—
Elateridæ[11] most of species very small
Necrophagous insects very rare.— [note (b)] Animal matter putrifying too quickly for them.— [note ends]
Nitidulidæ[12] feeding on decayed fruits.—
Hydrophilidæ[13] very numerous, & many of species very minute.—
Scarabeidæ[14] not abundant (owing I suppose to season)
Heteromeræ[15] not abundant.
Tetramera[16] are by far the most numerous.
Rhyncophores[17] exceedingly numerous, both in number & species: as might have been expected from the abundance of Forest Land
Longicornis[18]. scarce (owing to Season?)
Crioceridees[19], Cassidanus[19], Clavipalpes & especially Galerucites[19] extraordinary & abundant & appear preeminently to characterise Tropical entomology.— The true Chrysomalines[20] scarce (excepting few Creptocephalis)
Trimera[3], Cricinella[21] & Pselaphus[22] not very common.—

Orthoptera

This order in every family is very numerous, both in species & individuals.— the latter ~~mat~~ is much increased in appearance by those in the Pupa state being active.— The order makes a prominent feature in the Entomology.

Hemiptera

Not so numerous as the last.— Cicadella[25] is preeminently numerous.— Many beautifully coloured.— |**61**|

[1] See Jean Théodore Lacordaire. Mémoire sur les habitudes des coléoptères de l'Amérique méridionale. *Annales des Sciences Naturelles* 20 (1830): 185-291; 21 (1830): 149-94. In *Beagle* Library.
[2] See *Insect Notes* pp. 49-59 for a full account of the insects collected by CD in Rio during April, May and June of 1832.
[3] Ground beetles of the family Carabidae.

[4] Tiger beetles.
[5] Ground beetles of the subfamily Harpalinae.
[6] Probably a beetle of the subfamily Bembidiini.
[7] Predaceous diving beetles, genera of Dytiscidae.
[8] Burrowing water beetle, genus of Noteridae.
[9] Whirligig beetles, genus of Gyrinidae.
[10] Possibly a term used to describe species of Staphylinidae.
[11] Click beetles.
[12] Sap beetles.
[13] Water beetles.
[14] Dung beetles.
[15] Darkling beetles, an old division of coleoptera now known as Tenebrionidae.
[16] Division of phytophagous beetles with 4-4-4 tarsal formula.
[17] Palm weevils of family Dryopthoridae.
[18] Wood boring beetles with long antennae of family Cerambycidae.
[19] Leaf beetles from subfamilies of Chrysomelidae.
[20] Species of the subfamily Chrysomelinae.
[21] Tiger beetle.
[22] Short winged mould beetle.
[23] Round fungus beetle, genus of Leodidae. See *Plant Notes* p. 56.
[24] CD had collected insects at Barmouth in the summers of 1828 and 1829. See *Correspondence* **1**.
[25] A species of bug.

[note (a) on **CD P. 60** follows]

(a) I will give a specimen of one days collecting[1].— June 23[d], after a continuance of dry weather (which is injurious) I went to the Forest. Where I did not pay particular attention to Coleoptera (for instance I took amongst other things 37 species of Arachnidæ) nor was ~~particularly~~ lucky.—

		Brought over	27
Truncatipennis	1	Scarabeides	2
Bembididous[2]	1	Curculionidæ[10]	15
Brachelyties	4	Lyctus[11]	1
Buprestis[3]	2	Corticaria[12]	1
Elater[4]	1	Criocerides[13]	1
Malacoderms[5]	4	Crysomela	1
Ptiniores[6]	1	Galeruca	3
Scaphidites[7]	2	Altica[14]	9
Nitidularus[8]	7	Phalacrus[15] Agathidicus	6
Byrrhidæ[9]	4	Cocanella[16] Poclaphs	2
	27		68

These were chiefly taken by sweeping on the borders of the forest.

~~Amongst the Carabidous~~ The Trenactipennis, like many of their congeners in England are found ~~amongst~~ upon the foliage.— [note ends]

[1] See *Insect Notes* p. 58.
[2] Bembidiini tribe of small ground beetles.
[3] Buprestidae, jewel beetles.
[4] Elateridae, click beetles.
[5] Old term for Lampyridae, fireflies and glow worms.
[6] ? Ptinidae, spider beetles.
[7] Scaphidiidae, shining fungus beetles.
[8] Nitidulidae, sap beetles.
[9] Pill beetles.
[10] Weevils.
[11] Genus of Lyctidae, powder post beetles.
[12] Genus of Lathridiidae, plaster beetles.
[13] Criocerinae, plaster beetles.
[14] Genus of Alticinae (Chrysomelidae), leaf beetles.
[15] Genus of Phalacridae, shining flower beetles.
[16] Genus of Coccinellidae, lady birds.

[**CD P. 61** commences]

Neuroptera General Observations — Libellula[1] very numerous: Many Agrions[2] in the forest.— I only saw one Hemiroti[3] & 2 Frigania[4].— Termites not so numerous as at Bahia & still less than at Fernando Noronha.

Hymenoptera — The division Rapaces[5] (Lamarck) in great number & characteristic of Entomol: especially Guepiariès[6].— Melliferes[7] are not at all abundant, & this strongly contrasts against England. Some of the Rapaces (solitary ones) prey on Spiders, & thus balance the very much increased number of latter.—

Lepidoptera — The Diurnes[8], perhaps by the brilliancy of colours, largeness of size, more than any tribe of animals show the region they inhabit.— they are very numerous.— Crepuscularis[9] scarce.— ~~Phalanee~~ Nocturus[10], (considering how well adapted the country appears for them) are wonderfully uncommon.—

Diptera — These became tolerably abundant during the time there was any rain.— but with the exception of Culicidæ[11] & some few Muscæ[12] at other times they are not abundant.—

These observations were made during the months of May June; part of which was wet & part dry.— I must again mention, that in these notes I very much refer to the abundance of individuals: that is the general & first appearance which the Entomology presents in the Brazils.— |**62**|

General Observations Arachnidæ

In this division of Articulated animals the number of species & individuals which they contain is very great: it appears to me no ~~no~~ other order, as compared to England is so very much increased.— Mygalus[13] is not uncommon in holes (chiefly rotten trees) on the wooded hills.— A small (a) red Theridion coats the turf with its web.— [note (a)] (**49**, Page in this journal) [note ends] & Pholcus[14] under rocks & in the corner of every room may be seen violently agitating with its long legs the web.—

Epeira

Amongst the next division Orbiteles.— Epeira[15] is most singularly numerous & interesting: it is a large & numerous family not a genus.— The paths in the forest are barricaded with the strong yellow web of (the (b) division Dic Class I++). [note (b)] **CD P. 48** [note ends] Also others of same division & of (II++1) are exceedingly abundant.— Number construct their webs over the water: especially one with a red coniceous covering to abdomen.— Many belonging to this latter section are singular by strange form & colour.— The species of Epeira with the tibiæ of 2nd pair of legs enlarged & spinose.— There is no end to the singularity & numbers of this genus.—

(c) Tetragnatha. Several species are common amongst the rushes over water. [note (c)] **CD P. 49** [note ends]

Vagabondes[16]

Amongst the spiders, the Vagabondes are here in exceeding plenty.— Every walk is crossed by Ctenus & Lycosa.— & upon the blades of grass Oxyopes (in its habits belonging to the next division) actively springs (d) about.— [note] (d)] **CD P. 58** [note ends] |**63**|

General Observat: Arachnidæ Vagabondes

In the Saltigrades the typical genus Salticus[17] is almost infinite in species.— In sweeping amongst herbage nearly as many spiders as Coleoptera are taken, especially of this last family.— And lastly under rotting wood Phalangium[18] is abundant: & still more the sub-genus Gonoleptes.— I found one strange species, at superior base of hinder legs was a claw, & also corresponding ones on the hips, which together formed a pair of posterior pincers with which the insect seized any object.— Living in same site as these latter were Cloporta[19], Tuli[20] & Polydermi[21].— together with few Scolopendiæ[22].—

[1] Libellulidae, dragonflies.
[2] Coenagriidae, damsel flies.
[3] Not identified.
[4] Not identified.
[5] Wasps of families Pompilidae or Sphecidae.

[6] Another wasp.
[7] Honey bees.
[8] Day flying moths.
[9] Not a modern name, but presumably twilight flying moths.
[10] Nocturnal moths.
[11] For diptera taken by CD in Rio see *Insect Notes* pp. 50-7.
[12] Muscidae, houseflies.
[13] Mygalomorphae, tarantula-like spiders such as the genus *Grammostola*.
[14] Daddy-long legs spider, White's *Pholcus geniculatus* (see p. 73).
[15] A golden orb-weaver, family Tetragnathidae, *Nephila clavipes*.
[16] Vagabondes are hunting spiders that do not spin webs. *Ctenus*, *Lycosa* and *Oxyopes* are still valid genera.
[17] *Salticus* is unlikely to be the correct genus, but the Salticidae are jumping spiders with a cosmopolitan distribution.
[18] *Phalangium* is unlikely to be the correct genus, but both it and *Gonoleptes* are harvestmen in order Opiliones, related to spiders.
[19] Not identified.
[20] Possibly *Julus*, a cylindrical millipede.
[21] Possibly *Polydesmus*, a flat-backed millipede.
[22] *Scolopendra*, a large pantropical genus of big centipedes.

[**CD P. 63** continues]

(a) Proceeding to the Coast: the rocks as at Bahia & other Tropical places are frequented by large bodies of Ligia[1].— Beneath the water are many species of Pilumnus[2].— On the Fuci are some Amphipodes & many Læmodipodes. Either from the exposed site or zone, there were no Stony Coralls: certainly the flexible such as Cellaria[3], Sertularia[3], Amphiroa[4] were more abundant than in lower Latitudes.— [note (a)] I observed, cast up on the beach, those waxy looking balls, formed of flattened cells, which contain the eggs of the Bucinum[5].— [note ends] In the fresh water, besides Coleoptera already mentioned are Leaches & Crustacean Entomostraca[6].— [note (b)] Monoclass Ostracordes, Blainville. [note ends] & numerous Molloscous animals such as Planorbis[7], Ampullaria[8] in most wonderful numbers & Physa[7], Cyclas & Chondras.— If Tertiary strata are formed in Tropical countries the numbers of fresh-water |**64**| shells is easily understood.— It would appear that these shells (& certainly Ampullaria), when the puddles of water dry up, bury themselves in the mud & thus like the Crocodiles mentioned by Humboldt undergo a sort [of] Hybernation or more properly Aestivation.— When the rain first fell I was astonished & could not explain the numbers which appeared of full size in every ditch & little pool[s] which had previously been dry.—

General Observ: (c)

[note (c) added later] June 1833.— Maldonado.— I accidentally kept an Ampullaria in a room for more than a month, at the end of which time there was much water within the shell & the animal was quite alive.— A lake having suddenly been drained by the breaking of an embankment, I noticed the manner in which the Ampullariæ buried themselves in the sand.— With the mouth of shell on the surface they revolved (I imagine by the slight motion of Operculum) excessively slowly in a direction towards outer edge of mouth of shell.— i.e. this edge would meet the sand.— By turning a shell in this direction, it acts something like a centre-bit, & by its own weight will bury itself.— [note ends]

1 Oniscoidea, a shore-dwelling isopod.
2 Xanthidae, a mud crab.
3 Anascan bryozoans.
4 Coralline alga.
5 Buccinidae, a whelk.
6 A term that formerly included all the crustaceans except Malacostraca.
7 Basommatophora, freshwater snails.
8 Mesogastropoda, Cyclophoridae, modern name *Pila*.

[**CD P. 64** continues]

In my geological notes I have mentioned the lagoons on the coast which contain either salt or fresh water.— The Lagoa near the Botanic Garden is one of this class.— the water is not so salt as the sea, for only once in the year a passage is cut for sake of the fishes.— The beach is composed of large grains of quartz & very clean. if cemented into a breccia or sandstone
(a) it would precisely resemble ~~the one~~ a rock at Bahia containing marine shells.— [note (a)] Page of Geology[1], 35 (2nd bed) [note ends] A small
No. 619 (not spirits) Turbo[2] appeared the only proper inhabitant, & thus differed from the lagoons on the Northern coast in the absence of those large bodies of Bivalves.— I was surprised on the borders to see a few Hydrophili inhabiting this salt water, & some Dolimedes running on the surface.

General <u>Observations</u>

Whilst I ascended the Caucovado.— I measured some of the trees; the circumference |**65**| of the greater number of trees, as in the interior, is not more than from 3 to 4 feet.— I only saw one 7^{ft} & another the largest 9^{ft} & 7 inches.— One of those remarkable trees which have plates running from the roots up the trunk had an apparent diameter of 7^{ft} 3^{inch}.— One of the plates projected at a mean distance of 3 feet & was not above 2 or 3 inches thick.— This fact has been noticed by all travellers.—

I could not help noticing how exactly the animals & plants in each region are adapted to each other.— Every one must have noticed how Lettuces & Cabbages suffer from the attacks of Caterpillars & Snails.— But when transplanted here in a foreign clime, the leaves remain as entire as if they contained poison.— Nature, when she formed these animals & these plants, knew they must reside together.—

Metereology
Botofogo
Temperature
Metereology

My observations in Metereology have been very scanty.— The Thermomometer taken at 9 AM & 9 PM from May 14th . . . to June 8th (with some exceptions altogether 43 observations) give as a mean result Temperature 71°.84.— The highest at which I saw it (at those times) was 75° & lowest 65°.— May 26th 1 PM. Therm: on white cotton exposed to rays of sun stood at 122°.— Running water at the elevation of some 2 or 3 hundred feet at Tijeuka & on Caucovado was 66°.— |**66**| Thermometer plunged into a spring on Caucovado (May 30th) stood at 73°.—

Barometer

The mean height from same number of observa: as Therm: & times of day & period is 30.333.— Attached Therm: 71.7.— ~~therefore~~ & corrected height 30.295.— The highest I ever observed it (uncorrected) was 30.545, & lowest 30.072.— Although the whole range of variation is small; yet the height of mercury even for few hours never remained constant.—

Hygrometer

From May 14th to June 12th with some exceptions, 23 observa: taken at 8 AM.— give mean results.—

Dew Point	63°.26	Force 0^{inch}.587
Temp:	69°.99	Weight of Cub. foot 6.335 grain
Diff:	6°.73	

(a) [note (a)] The Tem: is taken from Thermometrical observations as being more accurate.— [note ends] On May 17th the Diff: was 9°, which was the greatest: it is remarkable on this day the upper regions of atmosphere were surcharged with clouds & in one hour Therm fell 4° & Barom rose 0.021 & heavy rain commenced. Vide infrà.—

(b) On May 30th ascended Caucovado (elevation 2300 feet) & was in a thin cloud. [note (b)] Captain King from 5 observations with Barom: makes the height 2330 (I; 2225) [note ends] the diff between Dew P & Temp was scarcely perceptible, both being 60.5.— Observation made below 4 & ½ hours previously gave dew P. 61.7.— & Temp 68°.— So that in ascending the latter fell 7°.5, whilst Dew point only 1°.2.—

Winds were generally light & sky very frequently overcast. (V page 40 respecting the latter).—

Rain

From May 10th . . . to June 8th inches 3.75 fell.— On May 17 it rained very heavily, between 9 AM & 3 PM 1.60.— out of which 1.06 fell in three hours.— During 6 minutes 0.38.— |**67**|

[1] See CD's *Diary of observations on the geology of the places visited during the voyage. Part I.* CUL DAR32.1.

[2] Trochacea, a turban snail.

[**CD P. 67** commences with an entry written in Rio and dated May 4th]

Trichodes[1] Having placed a Murex[2] in fresh water, the fluid in the course of two days became rather putrid: & contained an infinite number of Trichodes invisible to naked eye. I think there were at least three species.

Plagiotricha Animalcule[3] flattened eggshaped, sides (not those flattened) not quite corresponding; white very transparent, containing in interior from about 5 to 15 minute balls.— largest specimens in length .002, the greater number half that.— Moved rapidly, with the broard flattened side uppermost, either end first, chiefly rotatory; & by starts.— Body slightly contractile. As their power became exhausted, on the upper side & near to one end might be seen a linear apparatus rapidly vibrating.— As the surrounding water dries up, death irrecoverable comes on suddenly.— Mixture of Spirits of Wine did not act so decisively as I expected.— I have this animal from Bory St Vincents article in Dic Class: the shapes does not agree with species figured in Plate B Genus 44.— Fig: 16 & 17

Plagiotricha Animalcule. Much flattened, elliptic, length .0005. swims not so universally on broard side.—

Oxitricha[4] Animalcule shaped like a partially opened muscle [*sic*] shell, division reaching to the base, has the power of extending itself almost into a straight line.— length .002.— Moves rapidly with ~~one~~ divided end first, generally with a rotatory motion on the long axis of body.— there were but few of these.— Differs from the one figured in Dic: Class: Plate C Genus 46, in the division reaching much further down than those drawn.— |**68**|

[1] Ciliated protozoa in class Polyhymenophora. See *Dic. Class.* **16**:556.
[2] Muricidae, a carnivorous prosobranch that drills into the shells of other molluscs.
[3] Hypotrichida, dorsoventrally flattened ciliates. See *Dic. Class.* **14**:8.
[4] Hypotrichida, another species. The spelling of 'mussel' as 'muscle' was one of CD's habitual idiosyncrasies.

[**CD P. 68** commences with the *Beagle* now at sea]

Moon Coloured rings At 11 oclock PM of the 14th of July (off St Catherines[1]) the moon was surrounded by beautifully coloured rings.— Around the disk there was a highly luminous circle edged with red.— The diameter of this (including the moon) was 1°.45′.— Then came one of greenish blue also edged with red, this as broard as to make the diameter of whole halo to be 2°.90′. The appearance only lasted a short time & disappeared gradually.— The sky was of a pale blue; & was traversed with some scattered Cumili driven

swiftly along by a Northerly breeze.—

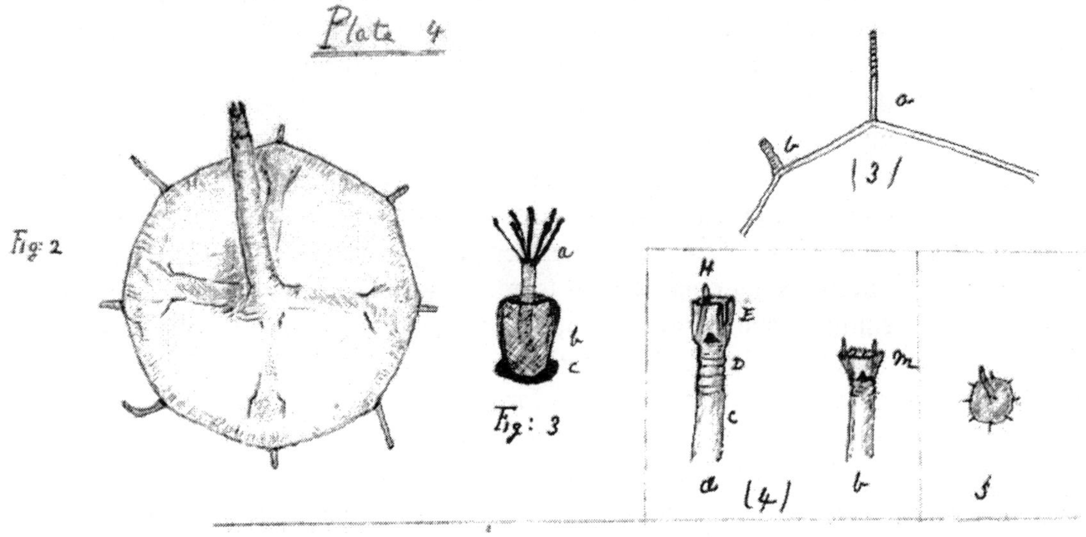

Plate 4, Figs. 2, 3, (3), (4), 5

[**CD P. 68** continues]

Dianœa[2] (Pelagia. Cuvier?) July 19th[3].— Lat 30° 31′.— Plate 4: Fig: 5 represents
Lamarck animal natural size, diameter .2.— Fig: 2 is the dorsal surface (as
(K) afterwards will be shown this probably is not the commonest form of
(c) animal). [note (c)] Nor 310 (in tube with Biphoræ) [note ends] back
convex, octagonal.— at each angle a projecting fibril, which is highly
flexible & contractile, & capable of seizing any object (?) — These are of
two sorts (Fig 3)[4], one shorter thicker & striated transversely; the other
long transparent within about seven little balls.— [note (a)] Are these
(a) minute balls Ova? & the shorter fibrils ovaria without the eggs.— these
shorter are exactly equal either in order (Vide Figure) or in size.— [note
ends] These fibrils are seated on a tube running round the edge.— which
also is contractile.— In centre is cylindrical hollow projecting tube,
(peduncle terminated by an organ capable of assuming various shapes.— Fig: 4[4] (a)
Lamarck) is end of simple tube: (D) is part rather narrower, with transverse folds &
capable of much contraction & expansion: (E) is ~~the~~ rather quadrilateral,
margin uneven.— within this are 2 lateral, fine, pointed transparent tubes,
(b) either capable of being protruded, & highly irritable.— [note (b)]
Occasionally the part (D) being much drawn in, the extremity E forms a cap
over tube (c) [note ends] The terminal organ (E) is capable of being |**69**|
Dianœa expanded into a funnel shaped cup.— in this case the pair of vermiform
tubes are more easily seen.—

In Fig: (2) on the convex surface there may be seen a faint cross of fibres:

it would appear to be the muscular organ of contraction.— From the octagonal margin (& not drawn in plate) there depends a delicate membrane which is slightly contractile at its inferior margin, forming a sort of bag.— In this shape I found the animal, but being kept it altered shape of body very remarkably & I think this latter the most natural.— The dorsal surface became much inflated, but was protruded through the octagonal margin on the other & inferior surface.— & the depending veil was turned upwards.— so that the central tube was now in the inside of body (In short the animal turned itself outside inward, every part except the tube.—[)] If now taken, it would be described as a transparent bag with ~~central~~ octagonal girth round the centre & ~~an~~ depending internal tube.— the basal aperture of tube being open (which formerly was interiorly) & now exterior.— The animal assumed another modification of this form. by much contracting the octagonal rim & the inferior margin of the veil, its shape was that of 2 spheres united, in the superior one of which is the internal tube.— How
(a) strange that the same body should have such shapes as the first & this latter.— This Animal as others, [continued at (a) opposite] Medusæ moving by sudden contractions.— Body highly transpa<rent> colourless.—

The sea contained Lat 33°.15′ S Long 50° 8′ W *[word 'contain' repeated]* vast numbers of ~~these~~ Radiata **|70|**

[note (K) added later opposite **CD P. 68**]

August 23^{d}. Lat 37°8 S & Long 56.46 W, found considerable numbers of this animal; having a better opportunity of more accurately examined it.— The peduncle was internal (as in the second & evidently most common case) & the depending veil within the marginal tentacula: (if the animal had been in state as Plate the depending veil would of course have been outside the tentacula):— The concave (convex in Plate) "ombrelle" (Fig 2) is of considerable thickness, but so very transparent, that I did not formerly perceive it.— Again I find the tail of peduncle opens within this thick part & not externally; also that the finer cross of striæ is not contractile or muscular but internal.— The mouth of peduncle is quadrangular & capable of much motion: the true vermiform arm (H) approximates at base & between them is a conical pap.— The margin of "ombrelle" was not so regular as drawn: the two sorts of tentacula (Fig 3) regularly alternate.— the shorter (b) is composed of concentric rings & is highly extensible; these 4 are situated at extremities of the cross. Behind each of them was another small one, internally connected with it.— The other tentacula (a) are curved & have a narrow footstalk, the little balls lie on one side & are from 7 to 9 in number.— During the time I kept them altered their appearance XX [continued at XX opposite **P. 69**] & one seemed to burst & sent forth its eggs. In all probability these correspond to the four ovaries in ~~G~~ Cyanœa.—

The animal moves by taking in water in the bag formed by concave surface of "ombrelle" & depending veil, & expelling it with violence.— I thought the Medusa used its powers of motion to avoid bei<ng> taken?—

The end of peduncle can fold its margin back over itself.— Some of these animals being kept in water till they were dead.— were luminous. |**70**|

[1] Isla Santa Catarina, off Florianopolis in the south of Brazil.
[2] Trachylina, Geryoniidae, the jellyfish *Liriope tetraphylla* (Chamisso & Eysenhardt, 1821), a primitive but abundant pelagic coelenterate which had been described by Quoy & Gaimard as *Dianœa exigua* in 1827.
[3] It was a calm day! See *Beagle Diary* p. 82.
[4] There was some confusion in the labelling of the drawings in Plate 4. The relevant Figs. for *Dianœa* are 2, 3, (3), 4 and 5.

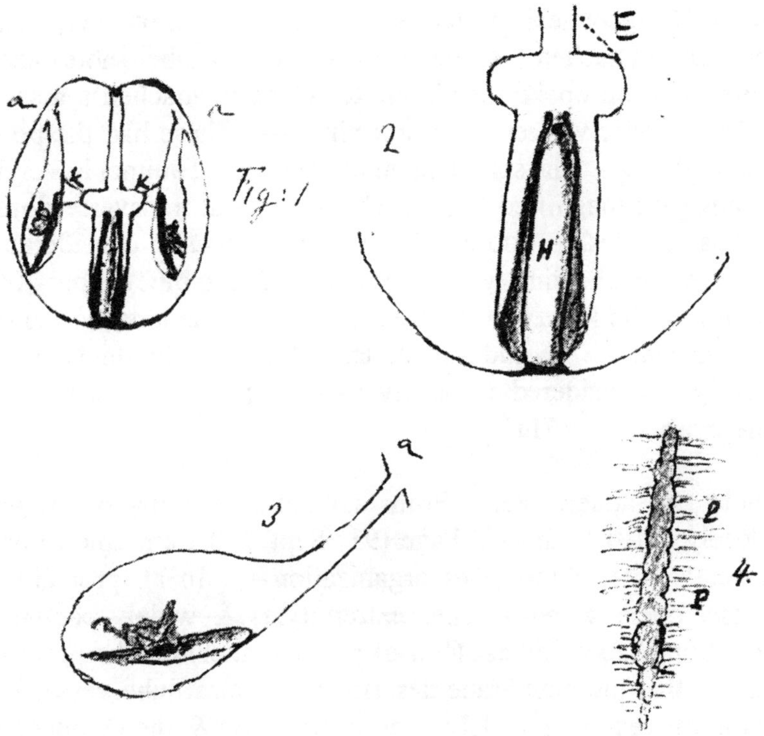

Plate 5, Figs. 1-4

[**CD P. 70** commences]

Biphora[1] allied to — Plate 5: — Body transparent, spherical, incurved at the poles.— Length about .4.— Fig 1: represents it as seen under microscope, from extreme

(e) No^{r} 310 transparency everything lies in one plane.— Through centre is a most thin
(X) tube, open at each extremity, enlarged in middle & one half much
(b) broarder.— [note (b)] The external aperture was not very distinct.— nor was the current of the water.— [note ends] This in Fig 2. is seen containing a membranous sack, much broarder & capacious posteriorly, & divided by longitudinal slit through its whole length. On the upper side the edge has power of expanding & contracting (NB this organ may lie above the central tube & not in it; the extreme transparency not allowing the ascertainment
(m) of this) This upper edge is thickened in its upper half & coloured pink.— [note (m) added later] Is it not possible that this thickened edge is a vessel as in animal described in **P 82**, Aug 30th?.— [note ends]

[**CD P. 70** continues]

At the point (a)[2] in Fig: 2 there was visible a pulsation, similar to what I have before seen in ~~this animals tribe~~.— In Fig 1: on each side is a highly
(a) delicate bag, not attached to outer coat of animal.— [note (a)] Both large & small specimens possessed this organ.— [note (k)] These bags were delicately attached (as drawn) to the central tube. [notes end] Within this (Fig: 3) is an opake membrane to which is attached a mass of vermiform
(k) tubes, precisely resembling intestines.— These had the power of moving themselves.— The spherical outer covering of animal has eight longitudinal bands (one magnified Fig 4), which send out transverse fibres & appear to
(d) act as muscles.— [note (d)] This animal was in considerable numbers:— We were in shoaling water (not coloured fine blue) about 100 miles off the mouth of the Plata.— [note ends] I cannot understand the organization of this animal.— I could not see Branchiæ:— The thickened pink edge can hardly be considered as the Liver.— Perhaps the pear-shaped bags may be the ovaries.— |**71**|

[note (X) added later] From a careful examination of an animal very closely allied to this (V Page (**91**) Septr 5th) I am able to pronounce upon several parts of this ones organization.— In all probability, the central vessel opens at ~~both~~ upper extremity~~ies~~ ~~& widely posteriorly~~: that the intestine shaped tubes (full of pulpy matter) can be protruded at orifices (a a): that the membrane described in central ~~tube~~ vessel is really a sack, lying on vessel, & widely open posteriorly; & the coloured rims, 2 folded up vessels in sack.— it is not impossible that I may have overlooked a circulation connecting central vessel with 8 external bands.— I forgot to say; that the intestine tubes are partly received in a receptacle here described as "an opake membrane".—

N.B. For more information about this animal V. note (b) **Page 96**. [note ends]

[1] Phylum Ctenophora, order Cydippida, probably *Pleurobrachia*. See p. 109.
[2] Missing from the drawing.

[**CD P. 71** commences with an entry[2] headed August 15th Monte Video]

Vaginulus[1]
330 & 471
(c)

Not uncommon under stones on the Mount[3]. Length varying in my specimens from three inches to ½; breadth (of largest) .8.— Measures taken when crawling; when at rest & its head retracted under mantle it was only 1.8 in length & .9 in breadth.— Mantle much flattened, oblong, of a uniform breadth; bluntly rounded at each extremity; rugosely punctured; projecting much laterally over foot:— foot narrow, caudal extremity appearing under mantle when animal crawls.— Mouth retractile under mantle; feelers short, superior rounded at end, bearing eyes, length .2; inferior appearing forked; the lower fork with extremities pointed; feelers coloured yellowish.— Foot & under side of mantle white.— Mantle pale dirty yellow, thickly mottled with purplish dark brown, so arranged that 2 pale irregular streaks are left tracing the form of foot.— The brown is sometimes so thick as [to] become of a uniform colour.— The youngest specimen was the darkest coloured.— Anal & Branchial orifice ~~? Mouth~~.— large (Novr 20th) [correction added later] These animals were found on summit of Mount (450 feet above the sea), where there is only herbage & no trees.—

[marginal note with different pen] Also Buenos Ayres found under stones.— This species differs from the Rio Janeiro species in its shorter & more depressed form [note ends]

[further note (c) added later] November 20th.— The summer is now far advanced & yet I find this animal under stones.— is it Nocturnal? I found it also at Buenos Ayres in same sit<es>. This species differs most strikingly from that of Rio de Janeiro in its shorter depressed body.— I may mention in this place, having found on an Agave a true Limax[4], but unfortunately lost it.— it would appear to have been hitherto not found in S America.—

[written with another pen] June. Maldonado.— I have found this latter animal & immense numbers of the Vaginulus. [note ends]

[1] Stylommatophora, land slug.
[2] The entries for the next few pages are all headed 15 August, but this was the date when they were written, for the *Beagle* had actually reached Monte Video on 26 July, in scenes of some confusion (see *Beagle Diary* p. 85).
[3] As described by CD in *Beagle Diary* pp. 85-6, the Mount was a hill 450 feet high overlooking the whole area which gave Monte Video its name.
[4] Stylommatophora, land slug.

[**CD P. 71** continues with two entries about Planaria also dated 15 August that as before have been crossed through vertically]

Planaria[1] (b) *[3 illeg. words]*

Inhabits same site as the last animal under dry stones on the Mount.— [note (b)] 331.— The situation being comparatively lofty & the stones large, the habitat must be very dry.— [note ends] The description of Planaria (Page 50) agrees with this in so many particulars, manner of walking &c &c that it may be considered as generic & the following as only specific. Body slightly flattened, length (when crawling) 1.9: breadth

Planaria (b)

.1 |**72**| Anterior extremity grooved beneath, much pointed, body gradually widening from this to the tail which is bluntly pointed.— [note (b)] The family Tremato*[des]* to which Planaria belongs is characterized by having beneath its body, "des organes en forme de ventouses".— perhaps the grooved surface at anterior extremity corresponds with this: although I never saw it used for any purpose, but as a sort of a feeler to direct its way.— [note ends]

[marginal note with different pen at bottom of **CD P. 71**] Eyes scattered at regular intervals on anterior part of body [note ends]

[**CD P. 72** continues]

Back coloured rich "umber brown" with a central dorsal narrow streak of "broccoli brown" reaching its whole length.— Beneath, of this latter colour.— On the under surface were two white spots, where (from the exact resemblance to the Planaria of Rio) I have no doubt there are apertures.— I believe I could perceive one.— I could perceive as formerly (**page 51**) the vibratory motion in the slimy surface of whole animal.— it ~~occurs~~ was seen wherever there was a gleam of light & it made no difference whether this was direct or reflected.— The animal seems to find presence of air to be necessary on the under surface.— Salt water (brine) killed & almost dissolved the body.— Animal not uncommon.—

Planaria[2] 332

Habitat &c same as last.— Body throughout of a more uniform narrowness.— (not tapering so much from head to tail) more cylindrical: length 1.3; breadth about .07.— Colour above pale dirty yellow with 2 dorsal stripes of "umber brown", which become narrower & unity at each extremity.— These Planariæ when first taken were rather inactive & were found on the earth beneath stones.— |**73**|

[notes scribbled roughly on back of **CD P. 72** concern Planaria No. 643 (in spirits) taken at Maldonado the following year]
rather less than 2/3 of length from anterior orifice, posterior 25/100 from anterior orifice.
Seen in *[illeg.]* Ocelli very numerous, minute & at regular intervals at anterior extremity, in groups of two or three at sides <of> body [notes end]

1 Listed by Darwin in his 1844 paper (see *Planaria* p. 186) as *Planaria pulla*, currently known as *Pseudogeoplana pulla* Darwin because of insufficient information about its internal features. CD notes correctly its use of chemosensory pits of the anterior tip. In a letter to Henslow dated 15 August 1832, CD says 'I have to day to my astonishment found 2 Planariæ living under dry stones. Ask L. Jenyns if he has ever heard of this fact.' See *Correspondence* **1**:252. Most terrestrial flatworms like a moist but not too wet microhabitat, but there are some species adapted for particularly arid situations, and others that occupy fully submerged habitats.

2 Listed by CD (*loc. cit.*) as *Planaria bilinearis*, currently classified as *Pseudogeoplana bilinearis* Darwin. Specimen 643 was listed as *Planaria nigro-fusca*, and is now *Pseudogeoplana nigro-fusca*.

[**CD P. 73** commences]

Cavia capybara[1] (c) Copied

Shot August 15th one of these animals, when I first saw it was on the rocks under the Mount. They do not appear to congregate in herds as described in other places.— perhaps the want of shelter may influence them.— The specimen[2] was a female & weighed 98 pounds.— Girth 3ft..2: Length from tip of snout to the tail 3ft..8½: Height from toes to top of shoulder 1ft..9.—

[note (c) added later] The dung in shape is rounded oval; when drie<d> & burnt smells like, but pleasanter, to Cedar wo<od>. This animal is very abundant in Rio St. Lucia: the hides are valuable being very tough: but the me<at> is very indifferent eating.— Cap. Paget of the Samarang[3] killed 45 of these animals.— For more particulars V **192**.— [note ends]

[**CD P. 73** continues]

Luminous Sea (a) (b)

August 22nd. between Points St Antonio & Corrientes: the sea was very luminous: light, pale, sparkling, but not as in Tropics either milky or in flashes.— The Luminous particles passed through fine gauze.— In the water were some minute Crustaceæ of the genus Cyclops[4]. I should not be surprised if these added to the effect.— During the day the sea has abounded with Dianœa[5].— & I find these when kept in water till they are dead render it luminous.— can this be the cause of the appearance in the ocean.—

[notes added later] (a) Sept: 6th.— Lat 40° S.— I observe that during this night, Crustaceæ of the Schiropodes & some other Macrouris, appear to abound on the surface, whilst during the day few can be taken. This applies to animal (**Page 73**):— as certainly many crustaceæ are luminous may this not ~~explain~~ help to explain the phenomenon of the luminous sea.— (b) Octob: 23d — Lat: *[not entered]* Sea wonderfully luminous; milky when seen in the mass; sparkling in numerous bright spots when seen in a tumbler; but I could not succeed in making by agitation, water in a watch glass show luminous particles, although certainly abundant in it.— The breakers & bows & wake of ship, i.e. when air acts on water, is luminous: this was after a heavy sea — Can this by destroying numbers of small animals be the cause:— [notes end]

[1] Listed by George Waterhouse in *Zoology* **2**:91 as *Hydrochœrus Capybara* Auct.
[2] No capybara was added to the collection, but Specimen No. 672 was an Acarus from *Cavia capybara* (see *Insect Notes* p. 60).
[3] H.M.S. *Samarang* was at Bahia, Rio de Janeiro and Monte Video at the same time as the *Beagle*, and CD dined with Captain Paget in Monte Video on 29 October 1832 (see *Beagle Diary* p. 112).
[4] Cyclopoida, copepod.
[5] The modern name of this jellyfish is *Liriope tetraphylla*.

[**CD P. 73** continues with a long entry crossed through vertically up to the end of P. 76, indicating its subsequent publication in a paper[1]]

Polype ? <u>undescribed</u>

At page (2) this animal[1] is described, but having opportunity throughly to examine one, I found some curious facts.—

August 24th. Lati: 37°.26′ S Long: 56°.58′ W: Sounding 10 Fath: This specimen agreed with those found at the Abrolhos.— PL 1. Fig 1.[2] I have drawn the posterior half of animal.— The tail, or that part which the central intestinal tube does not penetrate is filled with a fine granular pulpy matter. With <u>.3</u> focal distance lens, a longitudinal division & one on each side of this might be seen, so as to divide the |**74**| pulpy mass into ~~three~~ four columns. Within these I <u>clearly</u> saw a circulation somewhat like that in the Chara[3]: it was double the matter flowing upwards on the 2 outsides & then returning by the central divisions.— The circulation was strongest on the outside in the outer & inside of inner columns.— it was also much more rapid at the base of tail than at its extremity.— I ~~frequ~~ could see the grains turn round & pursue an opposite course at each extremity of tail. With 1/20 focal distance lens the matter (as nearly as I could judge) passed over 2 divisions of 1/500 micrometer in 5″.— but about the tail in double the time.— at the 5″ rate the progress is one inch in 20′..8″. And the tail being .15 of inch long, any grain would ~~pass~~ perform whole [see sketch in margin] circuit in 6′..2″.— this I daresay is accurate as the greater & lesser rates at base & end of tail would counterbalance each other.— I cannot even guess ~~what~~ what this is analogous to in other animals: as mentioned at Page (2) the granular matter is sometimes confined to small kidney shaped masses.— I could not clearly see that there was any communication with the intestinal tube; perhaps there was with the two gut-shaped bags at their inner ~~edges~~ crosses.—

<u>Polype ?</u> (a)

[marginal note added later] July 1834: Found some 4 feet beneath surface:

off Valparaiso.— [note ends]

Septemb 4th [note (a) on back of **CD P. 74** added later] Lat: 40°S.— The sea contained an incredible number of these animals.— I am enabled to add some new & verify former facts.— Within the body, in same plane as the mouth, there is <u>flattened</u> tube or cavity, which I have called the stomach. now this itself contains a delicate vessel (best seen posteriorly), & which pretty clearly terminates in an anus ~~of~~ on one side [of] the body, just at commencement of the tail.— Examining many specimens (V Pt 1 Fig 1.) I find both some of gut-shaped bags (FF)[4] & included globules vary; also that size & quantity of globules in (D) varies.— The globules in F are always much larger than in D: when there are but few in D the circulation is languid, & the 2 points of greatest intensity were at the bases of gut-shaped bags or the point of returning.— When (D) is in this case, F is small.— but when globules in F were highly perfected, D was full of regularly circulating granules.— I have no doubt at the internal base of (FF) there is a communication with D, although the included matter is distinct.— When globules were large in FF, I perceived on the external base a conical pap (V Fig. n n), which even projected slightly beyond line of body.— It is probable that the ova are excluded through this when
XXX ready.— In this specimen globules were very easily detached.—

[note continues at XXX on back of **CD P. 75**] I have formerly mentioned that in some specimens FF is almost evanescent.— From these facts showing connection in the two parts, I imagine that the ova (are first formed in D & then pass on into ?) F where they are perfected & then excluded or burst forth by the pap (n). If (D) had no connection with ova, why should the quantity & size of small globules or grains vary.— Again if it was a vascular *[illeg.]* the communication with rest of body would be more evident.—

I watched one of the ova after being removed from ovary.— (never taking my eyes from it).— the process as described ~~when~~ went on till the ova appeared made up of two equal balls.— they then separated; a capsule remaining; the other composed of globular mass of pulpy-granular matter, in which was ~~the~~ a small transparent ball.— This is the real ovum & is the same which is seen in balls (L,O).— I had imagined that the whole of excluded mass consisted of granular matter.— The process of separation took about 10 minutes.— Before they parted a line of division appeared which gradually widened on each side.— [note in margin] For particulars about Ova resembling these, V **104**. [notes end]

[**CD P. 74** continues]

At extremity of tail a fan of ciliæ is visible almost with naked eye: but they are so fine as not to be individually visible with 1/10 focal distance, with 1/20 Polype? they appeared delicate transparent hairs, arranged very close in one plane.— (a) they would seem to be locomotive organs |**75**| or rather to act on the water when the animal propels itself by starts.— [note (a)] Sept 4th These ciliæ adhere laterally, so as almost to form membrane.— in same manner as happens in a birds feather.— The animal uses its tail in another manner; when placed in a basin, it adheres firmly to the smooth sides, so as to prevent the water washing its body away.— [note ends]

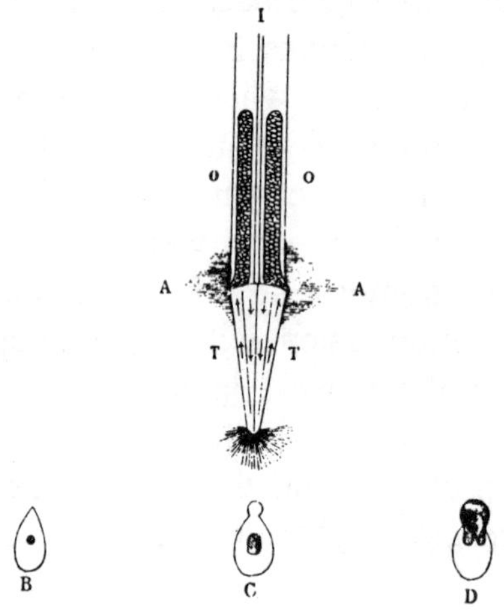

Pl. I. Structure of Sagitta. I. intestinal tube; *oo*, ovaries; AA, apertures of the ovaries, and lateral fins; TT, tail divided into four columns of circulating granular matter, the course of which is shown by the arrows; B, egg just liberated from the ovary; C, egg in first state of change; D, egg in a succeeding state.

Plate 1 from CD's 1844 paper on *Sagitta*

[**CD P. 75** continues]

On each side of the intestinal tube is a gut shaped bag (F)[4] filled with large grains, & if connected at all with the tail it is at the base by the side of intestinal tube.— The grains or globules are transparent, vary in length from 1/100 to 1/50 of inch, in shape are pointed oval & attached by the sharpest end in rows to the receptacle:— (L) represents a large one when first liberated, with high power a small internal ball may be seen not quite eggs so transparent: (I saw following phenomenon take place in two good instances) in a few minute<s> (L) altered its shape & became like (O) with a small globule at its apex: in short time afterwards a greater change took place, the little globule (as in P) increased in size & the internal matter in both became opake & granular.— This went on till all the granular matter was expelled out of the larger into the smaller: the former being left an empty capsule, the latter separating as a small ball of granules.— After the change of transparent fluid into the granular mass, the expulsion (as represented at P) wore the appearance of an internal case or membrane contracting & thus expelling it into the globule.— I must suppose the gut-shaped bag to be ovaries.— & the granules eggs collected in capsules (L: O).—

The construction of the head is beautiful & simple, but not easily described.— When not in action |**76**| the shape is a truncate cone (as before p. 2 described) & a transverse section of base would be an oval.— But when in action (mkk) is a transverse section of base; the dots are places of bristles, seated on moveable arm or jaw kk.— [note (e)] Fig E badly represents the head or mouth in action, the arm (kk) partly expanded.— [note ends] These when closed in, form the oval.— The semicircular part (m) is continued upwards rather higher than the bristles when erect; near its summit are 2 rows of very minute bristle which project out transversely; that is, cross the summit of the larger upright bristles.— [note (a)] The smaller bristles only cross the others when the latter are clasped together. I did not perceive these, till I had a high power in microscope. [note ends] The animal having seized any prey with the larger one, these smaller ones like a comb would effectually prevent its escape between their extremities.— The mouth is within (m).— [note (F)] Sept 4th. The orifice of mouth is longitudinal, & situated on oblique surface formed by the back part of head.— [note ends] The bristles are 16 in number; 8 on each side, curved, slightly hooked at extremities & strong; besides the power of clamping together on the head, each bristle can separate itself from the next, so as to take in greater span.— [note (b)] the central bristles are longest: teeth would be a more appropriate term. [note ends] When we consider this together with the power of motion in base (k), it makes a formidable instrument to seize any object, & when once within, the comb of small transverse bristles would effectually prevent its egress.— The substance of body is very sticky & gelatinous.— The range of Latitude is great of this animal: The more I understand of its organization, the more I am at a loss where to rank it amongst other animals.— |**77**|

Margin notes: Polype? (e) (a) (F) (b)

[1] The animal was eventually identified as belonging to the genus *Sagitta*, a carnivorous chaetognath or arrow worm, and was described by CD in a paper entitled 'Observations on the Structure and Propagation of the Genus *Sagitta*' in *Ann. Mag. Nat. Hist.* **13**:1-6 (1844). (see *Collected Papers* **1**:177-82.)

[2] CD's Plate 1 Fig. 1 (see p. 4), showing one of the first specimens caught in his plankton net in Lat. 21° on 11 January 1832, was redrawn to illustrate his paper on *Sagitta*, and is reproduced here.

[3] Chara is an aquatic alga with giant cells inside which rapid streaming of the cytoplasm may be observed.

[4] The labels F F were later altered to o o in Plate 1 Fig. 1, and the gut-shaped bags were the ovaries labelled oo in the published illustration.

[CD P. 77 commences]

Fish[1]
347

Above pale, regularly or symmetrically marked with "brownish red" (by the tip of each scale being so coloured).— Beneath silvery white: side with faint coppery tinge: Ventral fins yellowish.— Pupil of eye intense black.—

When cooked was good eating.—

Fish[2]
348

Many specimens exceeded a foot in length.— Above aureous-coppery; with wavelike lines of dark brown, then often collect into 4 or 5 transverse bands.— fins leaden colour.— beneath obscure: pupil dark blue.— When caught vomited up small fish & a Pilumnus.— Mr Earl[3] states these fish are plentiful at Tristan d Acunha, where it is called the Devil fish, from the bands being supposed the marks of the Devils fingers.— Was tough for eating, but good.— This sort was taken in very great numbers.—

Fish[4]
354

Above pale "Chesnut brown" so arranged as to form transverse bands on sides: Sides, head, fins, with a black tinge: beneath irregularly white: under lip pink: Eyes, with pupil black, with yellow ~~sclerotica~~ iris.—

Cellepora[5] ?
N^r 356

August 26th — Lat 38°..20′ Sounding 14 fathoms.— Coral, stony; brittle; branched; orange coloured, white at tips of branches ~~white~~; stems composed of numerous irregular circular small tubes, the former cells of polype.— Surface rough with little transparent cones, obliquely truncate, open.— I never saw polype protrude from these.— but from regular minute circular apertures with no external rim.— Polype very numerous.— Tentacula 12 round the mouth seated on a tube; **|78|** This is contained

Cellepora

in a case: tubular ~~with~~ rather wider at mouth protrudable.— Vide Pl 4: Fig: 3.— (a) Tentacula on tube, (b) the case: drawn as fully protruded from coral (c).—

[1] Listed by Leonard Jenyns in *Zoology* **4**:23-4 as *Percophis Brasilianus* Cuv.
[2] Listed in *Zoology* **4**:11-12 as *Plectropoma Patachonica* Jen.
[3] Augustus Earle was the first official artist on board the *Beagle*.
[4] Listed in *Zoology* **4**:20-1 as *Pinguipes fasciatus* Jen.
[5] Identified by S.F. Harmer as *Cellepora eatonensis* Busk.

[**CD P. 78** continues]

Flustra[1]
355
(a)

Habitat same as last: Coralline is closely allied to Flustra, but is a distinct & new genus.— Stem much & irregularly branched, flexible, about 2 inches high, coloured reddish.— Cells in 2, 3 or 4 rows according to breadth of branch, opening on one side.— Cells applied rather obliquely so as not to form distinct lines. On the ~~face~~ surface, when the cells open they overlap each other.— The other & back side, smooth, channelled by as many lines as rows of cells: thus seen (Pt 4, Fig 4) the cells appear of the shape drawn at (k), each anteriorly ending in point: widest in middle. Seen on upper surface quadrangular & oblong: the anterior opening with a spine at each corner.— Polype with 16 approximate, long (length 1/40 of inch), curved tentacula, seated within a lip on the extensible tube or mouth.— When in inaction, this is withdrawn to nearly the base of cell.—

I clearly saw at a spot where the tube & red intestine joined a sort of pulsation or rather a rapid revolution of small ~~grains~~ particles.— at the very base of cell, I saw in many a small mass of collected granules, which I suppose to be Ovules.—

[note (a) added later] For some particulars of Coralline somewhat resembling this (V **P 219**) [note ends]

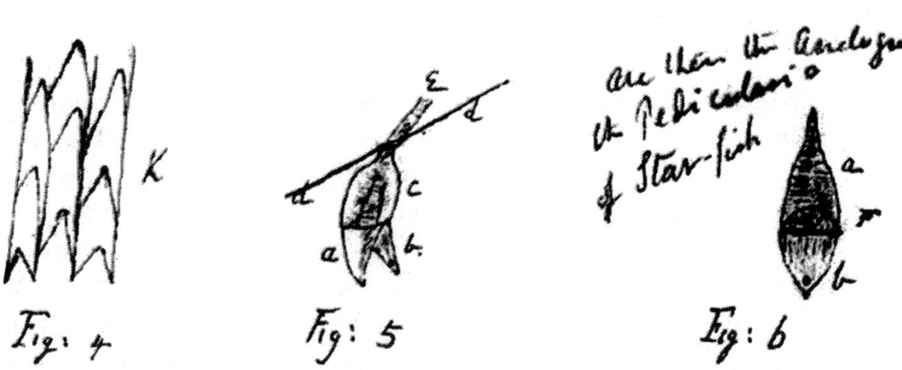

Plate 4, Figs. 4-6

Flustra

But what renders this coralline singular is the occurrence on the |**79**| edge of the cells of a peculiar organ[2].— In shape it curiously resembles the beak & head of a Vulture: is transparent, colourless, 1/75 of inch in length: is attached to the superior external edge of cell at its middle, by a short peduncle.— This peduncle appeared to communicate by a delicate tube to base of cell.— The head or capsule is connected to the peduncle at its superior base (above situation of neck in Vultures head).— The peduncle has great power of motion in a vertical direction (vertical being applied as to birds head).— Head empty oblong: upper mandible curved & much hooked at extremity; grooved within:— lower mandible closely fitting to superior with sharp projecting tooth at extremity, which fits into superior mandible; has the power of being opened so far as to make straight line with the other: at the joint is semicircular opening, which appears to lead
(a) by delicate tube to the peduncle.— The capsule (or head) lies close to the cell laterally & rather obliquely in direction: [note (a)] I mean by laterally that the cheek of the head is applied to the side of cell: ~~but that at either it is~~ & that the mouth or lower mandible opens in opposi<te> direction in the

case mentioned below.— [note ends] its ~~point is~~ base is towards base of cell: with respect to the surface in which ~~cells~~ aperture of cells are, the beak opens in different ways.— generally towards the under or back surface; but I saw a branch in which on one side the upper mandible was upwards, on the other, downwards.— Each cell has a capsule, but with this remarkable difference that when there are more than two rows, the central |**80**| ones

Flustra

have a capsule not more than 1/4th the size of the external ones.— Moreover the terminal cells in which the Polype are colourless have not them?— Pla: 4. Fig 5: represents one seen obliquely from above. (a) upper mandible: (b) lower with dot representing tooth: (c) head: (dd) ~~sides~~ edge of cell: e the delicate tube within:— Fig 6 represents the mouth wide open so that the peduncle is not seen.— F is the semicircular opening or gullet at base of upper mandible.—

When the Coralline is in water, whether the Polype is within or out of cell, the capsule generally is wide open (as in Fig: 6), & the whole head <u>on peduncle</u> turns backwards & forwards, vertically going through at least 90°. — They perform the whole motion in about 5" seconds.— Most of the Capsules perform it isochronously.— Occassionally they close for an instant the lower mandible.— In a small branch so many capsules moving caused in it a trembling.— A point of needle being inserted within the jaws was always seized so fast, as to be able to drag small branch.— The motion in these became fainter, as the Polype lost strength.— Polype, although so irritable of motion, took no notice of the motion of Capsule.— What office does this organ perform? It would appear superfluous for same

(a) animal to possess tentacula & another organ for seizing its prey.— [note (a)] And the absence of communication with intestinal tube.— [note ends]

Flustra

|**81**| Although its movements with the needle would indicate this.— In all probability by its motion a stream of water might be forced into base

(a) of cell. Can it have any relation with respiration & the revolution of particles (above mentioned) with circulation. [note (a)] the regularity of movement, & independence of the position of polype favors this idea.— [note ends] It is difficult to believe in so complicated [an] organization.— As far as regards generation (which is the last resource in all puzzling cases) what utility can so complicated an organ [have]? How different from the simple vesicles in other Zoophites.— Assuredly at base of cell there was an appearance of ball of ova.— I am quite at a loss from the want of all analogy.— But in any of these cases, how can it be explained that the old central cells have such small & comparatively speaking inefficient ones.—

<u>Squalus</u>[2]
Linn.
359

August 28th. Lat 38°.25′ S. Soundings 14 fathoms. Caught by a hook a specimen of genus Squalus: Body "blueish grey"; above, with rather blacker tinge; beneath much white:— Its eye was the most beautiful thing I ever saw.— pupil pale "Verdegris green", but with lustre of a jewel, appearing like a Sapphire or Beryl.— Iris pearly edge dark.— Sclerotica pearly:— In stomach was remains of large fish.— In the uterus the young ones for a long time after the viscera were opened continued to move: good specimen for dissecting:— |**82**|

[1] Cheilostomata, an anascan bryozoan. CD had a particular interest in suborder Anasca, of which he had previously collected Scottish specimens in the Firth of Forth, and having observed that its ova were motile had read a paper to the Plinian Society in Edinburgh on 27 March 1827. See MS notes in CUL DAR 118, and *Collected Papers* **2**:285-91. Plate 4, Fig. 4 suggests that specimen 355 was a *Bugula*, or a close relative, as was confirmed when S.F. Harmer examined the specimen in 1901.
[2] This was the first occasion on which CD observed the type of anascan heterozooid now termed a pedunculate avicularium, constantly moving and resembling a vulture's beak with jaws wide open. He discussed these organs at some length in *Journal of Researches* **1**:258-62.
[3] Squaloidei, an angel-fish. But not listed in *Zoology* **4**.

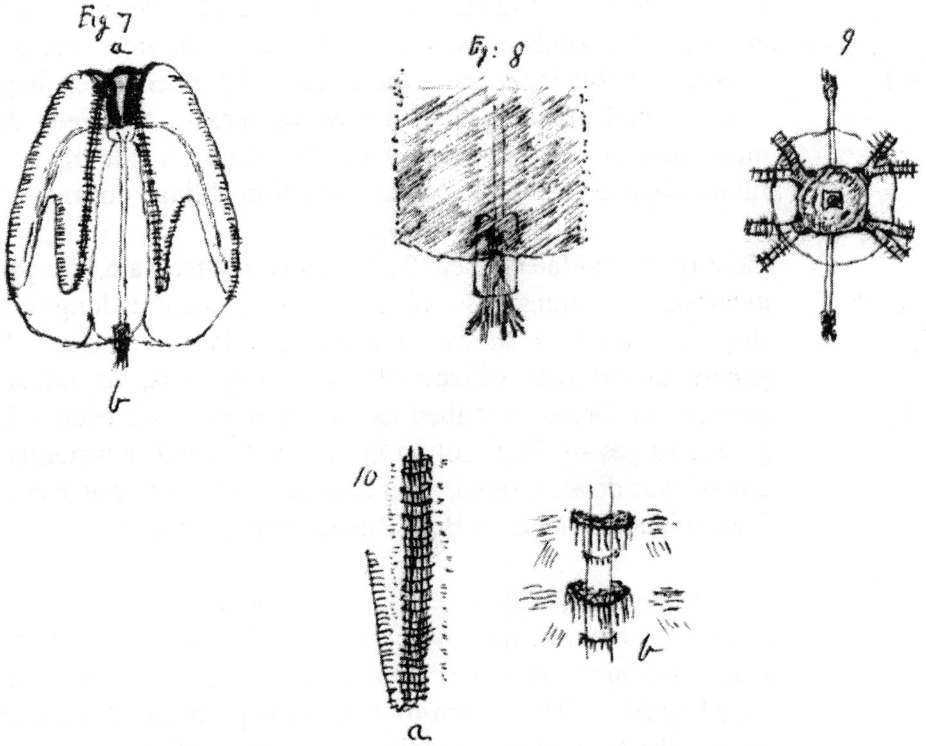

Plate 4, Figs. 7-10

[**CD P. 82** commences]

Moll:Tunicata allied to

August 30th. In Lat 38°.39′ S: sea contained great numbers of an animal of this division. On calm days floating near the surface, but in other

Biphora[1] weather they were brought up in a dredge.— Varied in length from one &
(d) ½ inch to a few 1/10^{ths}:— [note (b)] Sept: 3^{d}. Having procured a small &
N^{r} 360 (b) very perfect specimen (Lat. 39°9′) I am fortunately enabled to correct some errors & to certify the rest.— [note ends] perfectly transparent: colourless: shape a little flattened ~~oval~~ egg-shape; ~~at base~~ apex reflected inwards at the pole for 1/4th of the total length: Plate 4. Fig: 7: on the external surface ~~are~~ 8 bands, possessing vibratory organs, are clearly visible, they rise near to the base, pass over the apex & approximate in central depression: at apex
(a) they ~~give~~ cause ridges in the outline [note (a)] by depressing the soft substance of body [note ends]: round the mouth, in central depression, the bands are united in pairs; 2 pairs ~~being~~ approximating on one side & 2 on the other: so as to enable in describing to divide the animal in two halves: the plane of division being at right angles to the broarder or flattened one.—

[**CD P. 82** continues]

Pl: 4: (a b Fig: 7): (Fig: 9 is a view of central depression & mouth from above:) The bands consist in a tube on which are numerous semicircular
(c) rims of membrane; & from these, curved pointed fillets depend: these are in very rapid motion, directed towards apex.— between these are seated much smaller ones (V Fig: 10 a & b): on each side of the membrane are fibres which appear to act as muscles: also oblique ones.— (V Fig 10 b).—

[note (c) added later] Sept 29th This is not accurate, the part described as membrane is a transverse ridge or developement of longitudinal vessel; its shape is thus [see sketch in margin] it is not external, but within the gelatinous external surface; the vibrating ciliæ, or rather fillets solely project; the fibres described as muscular arise on each side between the greater ridges.— The motion in the fillets is either instantaneous in whole line or runs down it rapidly but regularly: high nervous communication[2].— This animal abounds in Baia Blanca, being 2 inches long.— [note ends]

[note (d) added later] Decemb: 7th. Lat 43°S.— on calm day float in great numbers, from near the surface to some feet deep: when then in water their shape is conical, & power of motion seems to be confined to expanding their bodies.— They seem to supply the place of Medusæ in this Zone:— These vibratory ciliæ are really transparent fillets [see sketch in margin], ragged at extremities: about 5 on each disc: motion lies in base of each one separately: when alive showed most beautiful prismatic colours: I should think only locomotive.— Those fillets, which are placed in the simple festoons.— have a vessel running near their bases but I could not see any actual connection, any more than than in those of the discs with the longitudinal vessels: The animal floats generally some way beneath surface & is continually revolving: one specimen in basin, being torn, had only fillets on one side at extremity, but these were sufficient to make it steadily

revolve:— [note in margin] April 17th St Josephs bay[3] abundant: There were also many Medusæ Lat. 42°, 30′ [notes end]

[**CD P. 82** concludes]

Biphora allied to

The motion in the ciliæ or |**83**| fillets sometimes commences at apex & thus runs down the vessel, but more generally is irregularly continued through its whole length.— When not in motion they lie close down (as 10 a). I should conceive when all in motion they would propel the animal with its base first.— Only the vessels are ~~continued~~ reflected within central depression.— At the base, the bands of each division are united, but in different manners: the 2 central ones of the four are united at the base by a simple curved ridge or membrane on which is seated a single row of vibrating ciliæ: but the lateral ones have a ridge running up ½ length of body forming acute angle, on the external half of which is seated row of
(a) ciliæ.— these vibrate in a direction at rt angles to the main ones on the longitudinal band, & towards these: they would have a tendency to move body round its axis.— [note (a)] Between each long cilia there is a minute one, in same manner as between each semicircular ridge is a small one.— [note ends] At Fig 7. the bases of the bands of the two divisions are seen disunited: but the lateral ones in each [are] joined by the acute angled ridge.— Within the vessels is a rapid circulation, the globules moved to & fro at the base of the vibrating fillets: so that I suppose they are connected with respiration.—

Within the body, beneath the mouth, the corresponding & opposite pairs from each division unite & form 2 central vessels in body.— For sake of simplicity, I will describe the organization of the one set (I now in describing divide the body directly oppositely to what I did before).— |**84**|
Allied to Biphora
(Plate 6 Fig: 1). At the point within body, ~~beneath~~ where the two pair of vessels unite, there is a semilunar shaped organ which performs the
(a) functions of heart: the two vessels after uniting form one central one: At the heart the circulation is exceedingly vigorous, but not very regular; as far as I was able to judge (from great motion in the ship) the fluid passes down central tube & is returned by the branches.— but at the same time it is
(a) certain that this was by no means universal, the same globules travelling some short ~~way~~ distance in one direction & then return.— Near the heart there were numbers of globules, slightly coloured, answering to blood; these appeared to be propelled in every direction, so that they entered different vessels, but by some power were driven back till they found their right
(a) course.— the heart lies within the main vessel, & it is difficult to understand to understand how it acts.—

[note (a)] Sept: 2d. I could not exactly in this specimen see the heart; but most clearly the centres of the double circulation lie at the upper extremity or junction of central vessels with the external ones.— Neither could I perceive the order of circulation; in junction of external vessels I saw ~~a~~ globules rapidly move backwards & forwards, till at last having entered the external vessel were carried onwards with great celerity.— In the same external vessel I saw circulation proceeding in opposite directions.— [note ends]

[CD P. 84 continues]

I cannot help imagining that the heart in some of these animals acts more in the manner of a fan, than of a pumping receptacle.— There was nothing like a systole & diastole: the ~~particles~~ globules only revolving with rapidity
(b) round a centre.— Just beneath the heart a narrow vessel arises which is continued in an arch close under the external surface to the base of the body. [note (b)] Or more accurately just beneath junction of two external vessels.— [note ends] (In Pl 6 Fig: 1, This is drawn on one side, its real course; it is in same line but above the central vessel): At the extremity
Allied to Biphora |**85**| the tube is widened into an oblong cavity. the posterior half projects beyond body. (Pl 4. Fig 8) Within this receptacle is a bundle of darker coloured parallel threads or filaments, viscous & extensible & capable of
(b) slight motion.— I at first thought these organs (of course there is a corresponding one on opposite side of body) connected with generation: but finding them as perfect in specimen only $3/10^{th}$ long it does not appear
(b) probable: if they were connected with respiration, there would be a circulation in the connecting tubes: from their opening just above neck of stomach (vide infrà) & from darker colour I conjecture they perform
(b) function of liver; the gall tube is certainly very long & it is most strange its being exposed to the open water:— If another system of vessels, precisely the same as above described, be placed directly beneath (as far as I was able
(a) to perceive) it will be a correct representation of internal organization.

[note (a)] Pl 4. Fig 11.— Here the drawing represents a plane at right angles to the one mentioned, so that both central tubes & both livers are seen.— the greater part of this drawing is incorrect: it only serves to show the relative position of the organs. [note ends]

Sept. 2^{d} [note (b)] The intestine-shaped threads seen under $1/10^{th}$ lens is composed of numbers of globules, united in irregular lines in a pulpy mass.— The globules resembled those in the circulating medium & were about 1/6000 in diameter.— A circulatio<n> is visible in the vessel which connects this organ to the central vessels: as mentioned, they do not open into stomach: my supposing the organ bears an analogy to liver is I think absurd. Is it generative? [notes end]

[CD P. 85 continues]

Pl 6. Fig 1:— What I am now going to describe is common to both

(c) systems.— Within centre of body there is a tube or bag formed of soft pulpy membrane.— at its superior extremity it receives, just beneath the heart, both central vessels & opening from the mouth.— at its base it widens & is united to the external covering of body.— ~~The central vessels~~
(c) This I imagine to be the stomach.— From the superior half of central
Allied to vessels, there are delicate |**86**| tubes sent off, which become gradually
Biphora finer; these I suppose to be absorbents.— The central vessels ~~having~~ being continued to the extreme base of the body turn off at right angles, & gradually become obscure; I could however pretty clearly trace the fluid into the lower branches of the external bands on vessels: (Pl 6. Fig 2. this turning off is represented; in Fig 1 it is not seen because the branch is in same plane as central vessel):— Thus it would seem generally to exist; but I saw two instances where instead of a single rectangular branch, there were two: this appearance is shown Pl 4 Fig 12: In the specimen from which this was drawn.— the central vessel appeared likewise to open at base of body by a projecting tube, as shown in Fig^rs 12 & 11.— This must remain in uncertainty.—

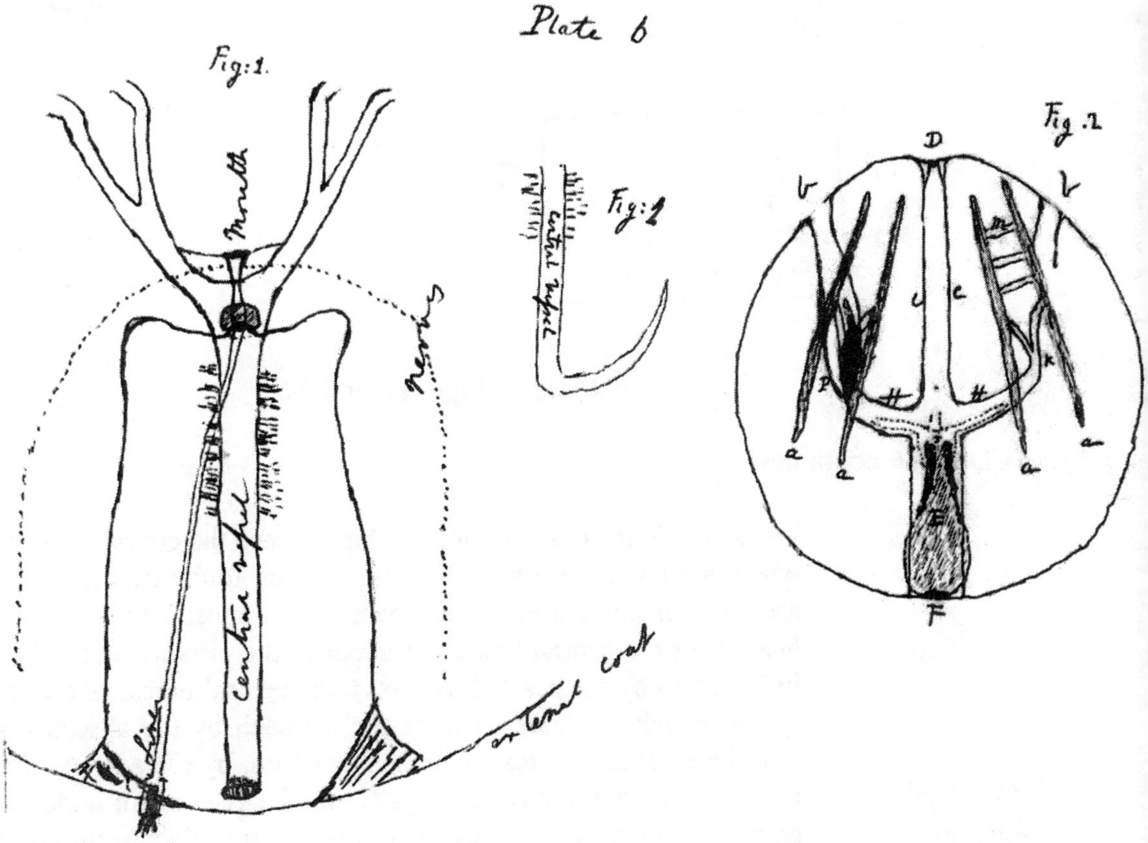

Plate 6, Figs. 1 and 2

Sept^{r}: 2nd [note (c) for **CD P. 85**] The stomach is capable of much motion, expand<ing> itself & contracting itself, irregularly.— much flattened; The central vessels do <u>not</u> pass within it, but lie close on the outside (I am not surpri<sed> at my mistake):— In this case I did not see absorbing tubes: The central vessels, having reach<ed> base of body, turn off (as described) vertically at rt angles; after which I see it obscurely branches into two which communicate with the external vessels, one on each side the <u>Liver</u>. This explains case in Pl 4 Fig: 12; where I did not perceive the first rectangular turn, or perhaps from transparency, the depending part (K) might be this.— [note ends]

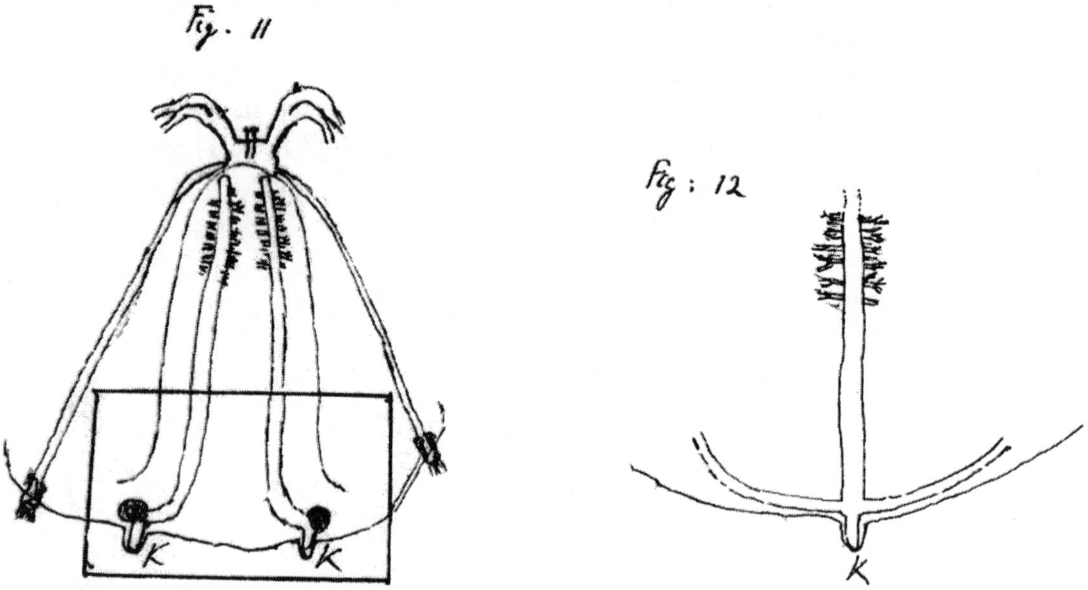

Plate 4, Figs. 11 and 12

[**CD P. 86** continues]

I was unable from the motion in ship to trace the course of any globule; the whole system of vessels is thus united, the four external ones (with ciliæ) are on each side united at their bases; but opposite pairs ~~of each~~ join at the heart with the central vessels: I suppose the circulating medium being put into motion by the heart flows down the central tubes, where it is joined by lymph which is separated from the stomach by the absorbents; passes on into the external vessels, & is then acted on, by the agency of the ciliæ, by the water, is then returned to |**87**| the heart & again undergoes the same course.— of course I cannot say whether any globule in the blood always goes through one heart or otherwise; I have shown that there is a complete communication between all parts:— The internal bag, or stomach, is joined by the gullet between the two hearts.— ~~& as far as I was able to judge from excessive transparency also by the bile ducts~~. [note (a)] I can hardly

Allied to <u>Biphora</u>

(a)

say that I could actually trace the gullet into the stomach; but just over it. [note ends]

[CD P. 87 continues]

(b) Pl: 6 Fig: 1 The mouth is situated in centre of square funnel shaped projection, which becoming narrower forms the gullet.— [note (b)] Sept: 2^{d} The situation of mouth is strongly marked by a black dot: it always appears closed.— [note ends] The situation of the mouth, as before mentioned, is in rather a deep depression;— the edges of this contract very suddenly if touched; & I suppose by this manner any minute object is caught, which may afford support to the animal.—

(c) I was totally unable to find any anus[4] ~~& I cannot easily believe that the mouth in so highly organized an animal performs this office:—~~ [note (c)] Sept: 2^{d} I am not much surprised at overlooking the ~~anus~~ basal orifice: the body is so very soft & tender & transparent, that without a small specimen can be placed under microscope it would be difficult to find it.— The stomach at base opens by a long slit (in direction of flattened side, i.e. at rt angles to the plane in which central vessels & (Livers!) are).— This orifice can be very accurately closed & widely expanded; so as rather to form a passage (as in Biphora) than an anus. The orifice was very sensitive & would instantly close.— When open I could fairly see into the stomach or internal tube: [note ends]

[CD P. 87 continues]

When I saw specimen figured PL 4: Fig 11 & 12, I thought the projecting paps (kk) were connected with this organ.— Round the gullet, beneath the funnel shaped mouth, is a collar of most delicate filaments; from each side a bundle is sent off & floats in the body between external coat & stomach: their direction is between central vessels & therefore at right angles to the bile ducts.— The bundle [of] filaments reaches to the base of body, in its
Allied to course |**88**| sending off some threads, it becomes both fewer in numbers
<u>Biphora</u> & finer:— This clearly is the nervous system:— The animal is highly
(a) sensitive & irritable & in a manner quite different from the Medusæ, to which in outward appearance it bears a great resemblance: [note (a)] The nervous system is represented in Pl: 6 Fig 1 by the arcs of dotted lines.— Sept: 2^{d} The nervous system was very plain in this specimen, following the course of lateral edge of stomach (as described). [note ends]

[CD P. 88 continues]

I could find no Generative organs: Animal is slimy: body very luminous, chiefly in the bands of ciliæ, to such an extent that the form of animal
(c) might be traced by the green light.— [note (c)] I do not think I ever saw

any animal more beautifully so:— Sept: 6th.— [note ends] I suspect from what I saw that the Petrels feed on them.— This animal from its organization belongs evidently to the "Clerphales sans coquilles" of Cuvier;
(b) & although so widely different comes nearest to Biphora.— if my observation is accurate, ~~the not having the two open perforations, or the mouth & anus is the most wide difference~~.— [note (b)] Sept: 2d If the organ which I have described as the stomach is considered as the inner tunic as in Biphora; the most wide anomaly in this animal is the absence of stomach, intestine, anus & Liver.— I never perceived any signs of water flowing through the body.— [note ends] The organ which I have described as Liver (??) bears some analogy to an organ in animal (Page 70).—

[1] The animal was a planktonic ctenophore of order Cydippida, probably *Pleurobrachia*, a comb jelly or sea gooseberry. The ctenophores were first placed in a separate phylum by Eschscholtz in 1839. CD's 8 bands of vibrating fillets are the rows of combs or ctenes controlled by an apical statocyst that serve for locomotion.
[2] Although ctenophores are regarded as among the most primitive living metazoa, they have more specialized nervous systems than cnidarians, as CD has not failed to note.
[3] St Joseph's Bay is situated on the north side of the Valdes Peninsula, and opens into the Gulf of St Mathias.
[4] The anal aperture is obscured by having the structure drawn in Plate 4 Fig. 9 at its centre, which CD must be forgiven for not recognising as the statocyst.

[**CD P. 88** continues]

<u>Erichthus</u>[1]
361

Septem 2d. Lat39°.9′. Sounding 15 F, 4 miles from shore: This species comes near to 'àrmé' of Desmarets[2]. length .2. (Organs of locomotion named from analogy from Squilla): 1st pair of "pieds machoire" long, cylindrical, terminated by ciliæ: 2nd strong with "griffe", penultimate joint broad, receiving griffe in a grove protected on each side by recurved spines: 3rd & 4th pairs, with claw, & penultimate joint enlarged, globular; vesicles at base: 5th rudimentary without claw.— True feet 6 in number, mere stumps: 5 pair of circular |**89**| ciliated caudal swimmers, when at rest they are applied indifferently either towards head or tail.— Terminal plate excised, finely dentated, with spine on each side, also others at base.— On the under surface there is a longitudinal slit, which is the anus.— this ~~intestine~~ opens into an enlargement of intestine. Lateral antennæ shorter than plate.— Frontal spine very long: ~~also~~ so is likewise the posterior Lateral.— Within these latter, there is a vessel in which I could perceive a circulation.— Dorsal spine long recurved.— on each segment of tail there is a small spine bent in same manner as the last.— Respiratory organs in form of plates, situated under edge of shell at base of "pieds machoires".— Body transparent, colourless, excepting the eyes which are dark green; all that was to be seen, when animal was in the water, were two

<u>Erichthus</u>

black spots, the eyes.— In its motions not active; swims in oblique direction; & frequently rolls from side to side:— Has the power of withdrawing large part of body from beneath shell.—

Mysis[3] 361 (a) (b) Mysis

Habitat &c same as last:— Species allied to "integer".— [note (a)] Sept: 4th.— Lat 40° S.— The sea contained vast numbers of this species.— [note ends] Body coloured slightly red: especially 2nd pair of "pieds machoires", inner part of: Females had attached near to base of last pair of legs, a curved circular ciliated membrane, when folded in, forming prominent pouches; in each of these were two young animals, length about 1/15 of inch; differed from old |**90**| Specimens by the greater proportional largeness of eyes; also by the less distinct separation of thorax & tail.— [note (b)] In the membrane were dark coloured vessels, much branched.— & I suppose by these pouches convey nutrition to the young animal.— [note ends] They possessed but very little irritability.— The females with young were larger & darker coloured than the others.—

Amph: Hetero =podes new genus[4] 361

Habitat &c same as last:— lateral antennæ & their peduncles very long: internal short: Thorax divided into many segments: 4 anterior legs, with very strong claw; the next 6 with claw less so: next 2 simply natatory, very long: last 2 simple natatory shorter:— Extremity of tail, with 2 jointed sitaceous appendages; beneath it 4 double stylets; on dorsal surface there is a short cylindrical fleshy projection: Body flattened, narrow, long; colour orange:

Loligo[5] 363 (a)

Calmar (Cuv). Lat 40° S. Sept: 4th: caught in open sea, together with great numbers of Mysis.— Arms 8 unequal; 2d pair rather longer than first; & 3d pair finer, but equal to 1st.— the 5th very delicate, half the length of others.— The 2 feelers (or long arm, making 10 in number) are .4 long, & about twice length of other arms: suckers at *[illeg.]* terminal half.— Suckers small, in double rows, alternate, circular, pedunculated.— Anal tube short, in line between eyes: body bluntly pointed, with 2 irregular rhomboidal membranes at apex. Body .6 long: pure white with angular ~~obliterated~~ scattered red markings. Eyes large, pupil black, iris pearly; ~~base~~ inferior base of sclerotica coppery red:— [note (a)] Emitted small quantity of ink [note ends] |**91**|

1 Erichthus was the term formerly applied to a larval mantis shrimp of order Stomatopoda.
2 See A.-G. Desmarest. In *Dictionnaire des sciences naturelles*. Paris, 1816-30.
3 Mysidacea, opossum shrimp.
4 Amphipod of suborder Hyperiidea.
5 It was concluded by S.F. Harmer on examination of Specimens 304, 363 and 368, labelled by CD as *Loligo* Lamarck, that 304 was *Sepiola*, but that CD's written description did not fit well with 363 or 368.

[**CD P. 91** commences]

Pelagia[1] Body transparent, shape half an oval ~~spheroid~~; internal cavity flatly arched;
364 membranous sides not so transparent: surrounding its edge on the inside
(a) there are about 40 tubular tentacula; extremities dark coloured, tuberculated,
(b) adhæsive.— These open into the space between internal cavity & exterior surface. In this, 4 delicate vessels, rising at base unite at summit of interior cavity forming a cross at their junction.— On each side of these vessels for their whole length, there are short transverse fibres which act as muscles & are capable of contracting so much as to give body a four lobed appearance.— [note (b)] Outside of the tentacula there is a short depending membrane.— [note ends] Depending within cavity is a short peduncle; terminal part coloured dark red.— Surrounding this there are four, small irregular shaped oval, membranous, flat semi-transparent sacks, placed cross wise (in centre in base of peduncle). The four delicate vessels run along (at the apex) the edges of these sacks, if they do not empty themselves into them.— Diameter of body .2: Habitat &c same as last animal (Loligo).— [note (a)] Dianœa (Lamarck).— All the species mentioned by him are North of Equator. I have found two species south of the Tropic Capricorn.— [note ends]

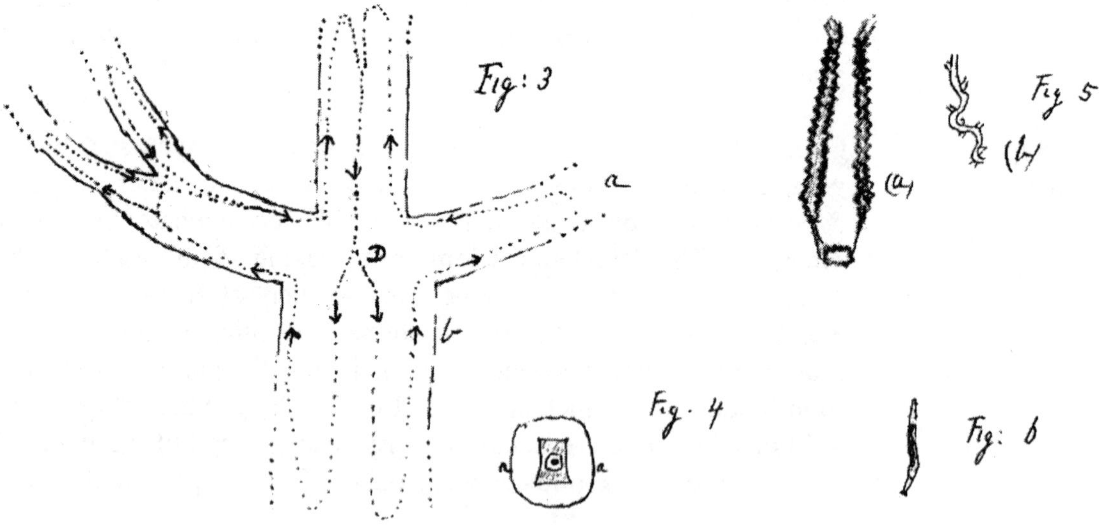

Plate 6, Figs. 3-6

[**CD P. 91** continues]

Molluscous Septemb: 5th.— Lat 40°.— Soundings 10 fathoms: Body, nearly spherical;
Tunicata[2] transparent; diameter .3. [note (c)] I have seen ~~much~~ larger specimens.—
365 (c) [note ends] On the surface there are 8 opaker lines; arranged in two sets:

they reach from near the apex longitudinally for 2/3 of whole length:— Each line is a shallow sack filled with granules, at centre of under surface arises a tube, which uniting with one from the next, forms a pair.— |**92**|
Molluscs: Tunicata (b) Two of these pair lead (thus connecting four of the external lines) on each side to a main transverse vessel.— Pl: 6. Fig 2^3 (a a a a) represent two pairs belonging to the opposite sides: (H H).

[note (b)] September 6th.— in a small specimen I observed the important fact of vibrating ciliæ, placed in numerous transverse rows on these lines or sack.— In direction, manner & appearance of motion, they precisely resembled those in animal (described **page 82**). And to which animal I show there is a relation in other respects: in this case the sacks were very empty.— I forgot to mention that the ciliæ or fillets are easily separated, & that they then possess much irritability.— It is to me inexplicable the occurrence or absence of such important organs in the same animal:— [note ends]

[**CD P. 92** continues with a description referring to Plate 6 Fig. 2 on p. 79, and Figs. 3-6 on p. 84]

Main transverse vessel: at K. on right side, the brother pair would join if drawn:— Through the centre or pole of sphere a tube runs (c c). A little below the middle it unites with the main transverse one.— Beneath this it increases in ~~diameter~~ breadth, but is very flat.— it terminates (F) on outside membrane, but I could not see aperture.— At the upper extremity or mouth (D) there is an appearance of an internal tube: mouth square, with central black spot which perhaps is the orifice closed:— in [Plate 6] Fig 4.— the mouth is seen from directly above, (a a) is the central vessel:— In this system of vessels there is a very powerful circulation.— The fluid is composed of variously sized globules, very faintly coloured.— The circulation varies much [in] intensity.— I only saw it once in full play.— Fig 3 will then show its course, generally it returns on inner side of the smaller branches & flows out on the outer: In central vessel the circulation reaches to both extremities; in the transverse ones to the point where the
(a) smaller vessels unite with ~~external~~ (near the surface) superficial sacks.— it even enters within these, but does not extend far, by degrees however the whole of the included grains or globules I have no doubt pass into circulation.— [note (a)] Do these outer longitudinal sacks perform any office similar to respiration?— [note ends]

Molluscous Tunicate As I have said, the |**93**| circulation suddenly becomes languid; in this case the order (Fig 3) ceases: when <u>most</u> ~~so~~ languid the globules may be seen moving in the last bifurcation, & especially at the point where the membranous stomach (below T) unites with central vessel.— Also in this case each part of vessel becomes a centre of a circulation; a globule may be seen for some time performing a small circuit & then pass on.— The blood likewise takes ~~of~~ different courses; for instance I saw stream (b) (Fig 3) instead of passing to right hand, flow round (D); ~~thus~~ proving that separate vessels do not conduct the complicated circulation in Fig: 3:— I frequently observed one pair of vessels with their blood in rapid circulation, whilst the others were nearly quiescent.— From these facts I do not believe there is a heart[4]; but that the parts of the different vessels by some unknown power act on the contained fluid.—

On the posterior & broard part of central vessels there lies a delicate very flat membranous gradually widening sack.— it is ~~highly~~ expansible & contractile.— till I saw [it] project beyond the line of central vessel, I thought it was contained within it.— The sack can be largely opened at its base, but is generally kept closed.— at its apex I do not know whether it communicates with central vessel or whether it has a separate tube leading to the mouth; at the mouth, there is an appearance (as already mentioned) of this:.— The sack has on each side two serpentine approximate vessels, which send off minute branches: Fig 5.(b) |**94**| These unite & from each go to surround basal opening.— These tubes are situated on the internal surface of sack.— When the latter (as is generally the case) is contracted, these tubes present a very different appearance; they are so much doubled up as to look like lobes in ~~some organ~~ a membrane.— this I have represented [in] Fig 5 (a).— In one instance there were two small oval organs attached to them; what were they?.— I was much surprised by seeing a rounded opake mass, slowly revolving at base of sack.— at last it was protruded through basal opening; it appeared to be the fæces, it was pulpy & adhæsive.— I presume the object of the revolving was to form into a properly shaped pellet.—

Molluscous
Tunicata

From extreme transparency I am not certain of what follows.— the lateral serpentine vessels at the summit unite & send off a delicate tube into the lateral circulating system.— [note (a)] I fancied that just above (T) there was a collar of nerves.— [note ends] I have represented these uncertain vessels by dotted lines in Fig 2.— I presume the sack is the stomach & the serpentine vessels the absorbents; the food is taken in by mouth, but I am ignorant of its course to the stomach.—

(a)

I have mentioned that when ~~the~~ two primary branches on one side in the circulatory system unite & form a pair, Fig 2. behind (k), another similar joins & so forms main vessel.— Within the segment of body contained by the latter & greater bifurcation, there is a curious organ.— It consists in cavity of form of bag with neck |**95**| which ~~rise~~ has its orifice not far from the mouth & reach[es] half way down the body.— Of course there is a corresponding one on opposite side of body; they lie in same plane as broard side of stomach: At the base of this bag, on the interior side, there is a flat opake irregular receptacle, ~~this~~ from this protrudes & is partly contained a mass of intestine shaped cylindrical tubes, full of granular

Molluscous
Tunicata

(a)

matter.— [note (a)] Sept 6th: Found small specimen where the interior receptacle or capsule was empty, having apparently ejected all the intestine-shaped granular cylinders.— this only occurred on one side of body:— [note ends] This is capable of motion; & so extensible as when unwound to project beyond external orifice.— Behind the receptacle, this organ communicates with the main transverse vessel at its great bifurcation.—

I may mention that I saw a small body moving with great rapidity in this cavity: was it an Infusoria? In Pl 6. Fig 2. the sack is drawn only on the left hand; it is beneath the two external lines: on the other side it would lie at (k):— Between the external lines or cavities there are narrow bands about 5 in number; they are so fine as scarcely to be visible, & act I suppose as muscles. in Fig: 2 I have shown a few (m):— In the cavity of the body there was a very minute Intestinal worm (Fig: 6).— body capable of much contractility.— tail with minute terminal sucker.— internally there appears to be an irregular cavity & intestine.— This is a low animal to be infested with parasites[5].— |**96**|

Molluscous Tunicata (a) (b)

This animal is closely connected with that described (**Page 70**, Pl 5).— it differs chiefly in the form of vessel where the central & transverse [structures] meet; in the external bands & their muscular arrangement; it is not impossible I might have overlooked the circulation, if so it must have been very obscure.— [note (a)] The cavity (Fig 3 Pl 5) will almost do for either animal.— [note ends]

With animal (**P^{e} 82**) it is related by its complicated circulation; by its internal sack or stomach widely open posteriorly; & especially by its lateral organ (described as Liver! I am yet unable to guess what its real nature is), in both cases they are united to central circulation & are open to the water, although by different means; are composed of extensible moveable ~~tubes~~ threads or strings in a receptacle: (~~the most marked difference is the absence of the vibratory ciliæ.~~ Vide Suprà)

Decemb 2^{d} Lat: S.40° Coast of Patagon:

493

[note (b) added later] Caught several specimens of animal; still more closely proving the identity of that described at **P 70** & this one:— Length of body from from .3 to .4: Vide Plate 5[6]: the bags (Fig 3) opened externally: also the bag or stomach (H) does: In these respects, I have no doubt the animal of **P 70** agrees.— & that I did not before observe it: at E the vessel is not so suddenly rounded, as shown by dotted lines: but the most important fact is that at k a pair of vessels were given off, which were themselves divided precisely in same manner as in the animal of text.— on the other hand, on the superficies there were no sacks corresponding to the 8 vessels: nor were there bands, such as in Pl 5 Fig 4; but merely lines as at P: there was a strong circulation at (a) which extended a little way within main transverse vessel.— The stomach (H) is closely attached to the longitudinal vessel: Upon the whole, considering animal of **Page 70** & of text; the real essential difference consists in the superficial sacks or bands.— these organs we have now seen in four states; as simple sacks with included granules; as sacks with transverse plates, with vibratory fillets; as bands with numerous transverse lines; & as simple lines or scratches on the surface.— What can their office be ?? (I may mention in this case, some of the external lines were half & finely spiral in places?) In this case the 4 convoluted vessels at stomach were coloured red.— [note ends]

[1] Again *Liriope tetraphylla.*
[2] The use of Plate 6 to describe both the tunicate allied to *Biphora* of **P. 82** and this one shows that CD has decided that they are closely related. He later comes clean about this, and adds the animal of **P. 70** to his list. They were not in fact tunicates, but were comb jellies, ctenophores of order Cydippida.
[3] See p. 79.
[4] According to J.A. Colin Nicol in *The Biology of Marine Animals* (Pitman, London, 1960), changes of the direction of beat of locomotory cilia are characteristic of ctenophores.
[5] The example given by CD does indeed appear to be true parasitism, and it is now recognised that there are worms of several classes that are endoparasites of molluscs and other marine invertebrates.
[6] See p. 63.

[**CD P. 96** continues]

Crustaceæ Schizopod[1] 366

Sept: 6th.— Lat 39 : Long 61 W : new genus allied to Mysis: 8 pair of locomotive organs; the exterior branch of all these simply natatory; of the internals the 1st is short, rudimentary, 2 longer, with terminal joint flattened circular; both these help to close the mouth, & are capable of curling themselves up: the 3rd, 4th, 5th, 6th, 7th are long, & have on internal side a double row of fine straight ciliæ, inclined to each other at an obtuse angle: the last & 8th pair natatory: When the animal swims, the ~~10~~ 5 pair of ciliated internal branches directed anteriorly almost form a complete circle round the mouth: any small object caught by these might easily be |**97**| carried into the mouth, by the involving movement of the two upper pairs of internal branch:— Before the mouth there were two fine arms terminated by a curved claw.— I once imagined there was a small internal branch from this; if so there are 9 pair of legs.— now I observed in Macrourus (**Page 98**) that between ventral swimmers & legs there were jointed setæ or rudimentary legs; is it possible that the last pair in this animal, both branches of which are natatory, may correspond with this:— Tail formed of 5 pieces; central one excised, finely dentated; ventral swimming plates, narrow: peduncles of eyes rather long.— Superior antennæ with two long divisions: inferior with protecting plate.— Body nearly transparent; except stomach & intestines, which are like quicksilver; This animal differs from Mysis principally in only having 2 divisions in antennæ & in form of legs.— They could swim well & jump a little: were

Crust: Schizopod

(a)

taken at night in vast numbers.—

Decemb. 2^{d}
Lat, 40°S.

[note (a) added later] The swimmers on the tail or abdomen are very small with a little jointed branch with internal ciliæ.— Mandible corners formed of a curved plate, square & smooth, with one of its corner[s] raised & toothed. this portion resembles the mandibles of Apus figured by Desmarets[2]; there were two sets: there also was an organ connected with the mouth in this shape [see sketch in margin] a tuft on a peduncle: the organs with claw are seated before the mouth & doubtless are palpi: This specimen was found dead & is female, from the capsular membrane at base of posterior legs: the central piece of the tail is not excised: strongly toothed: is this a different species, or is it sexual $diff^{e}$: or is my former description inaccurate. I do not think the latter probable: (I presume by 2 sets of mandibles, maxillæ a<re> meant). [note ends]

[**CD P. 97** continues]

Crustac:
Macrourus[3]
(b) 366
369
(c)

Habitat &c same as last.— Characters will not apply to any of Cuvier families, but most approximates to Salicoques.— [note (b)] The specimen (366) is with other crustaceæ at the top of tube; it is a perfect specimen: those in (369) are imperfect wanting lateral antennæ [note ends] Body one inch long; colourless or of a faint red: peduncle of eyes long.— External antennæ situated beneath the central ones & protected by large ciliated plate: these are of the extraordinary length of 2 & ½ inches, coloured red.— [note (c)] The external division of pieds machoires resembled Palpi?— [note ends] Superior antennæ with peduncle very long, basal joint thick, hollow, |**98**| carrying 2 very unequal branches, the longer one very fine; total length .3:— None of the legs are terminated "en pince" [note (a)] Have vesicles at base [note ends] 1st pair are shortest, & when in rest form a circle; the 4 other, long, slender, with double row of setæ, forming obtuse angle.— These precisely resemble interior branch in the last Schizopod animal: Ventral swimmers 5 pair; the 4 posterior approximate; each one divided into two ciliated plate[s]; the 1st pair are distant from the other, & single, & more formed for walking.— Between these & the true legs; there are 4 articulated setæ or arms, in line of legs; of these the anterior pair are much the longest:— The external division of caudal swimmer largest, central stylet pointed: Thorax with anterior sides, bi-dented.— This animal would in some respects connect the Salicoques & Schizopodes.—

Crustace:
Macrourus
(a)
(b)

Decemb: 4th
491

[note (b) added later] At Bay of San Blas took some specimens of a crab.— same genus as this, but ½ the length & I should think differing in other respects: anyhow it is sufficient to show that the description in text is most inaccurate.— the 3rd, 4, 5th pair of legs are terminated by an almost invisible (yet certain) "pince": the first pair of swimmers, which are single, have a small branch at base, which expands into a foliaceous organ & again contracts into articulate limb.— this fold covers eggs.— these are opake in transparent envelope, much oval: the mandible & palpi are ~~distinct~~ large, the former has anterior tooth, & very much longer than any of others: I could not understand the pied machoires.— they are evidently of a very simple structure.— the 1st pair of legs I almost suspect are the external pied-machoire.—

The organs which I saw are these 1st with simple palpi, oblong concave plate.— 2^{d} more rounded: 3^{d} & 4th united but at rt angles to each other: I thought at first (3) was ½ the labium.— there was also an obscure rounded ~~organ~~ plate behind all these: the pharynx was remarkable.—

Palpi very transparent

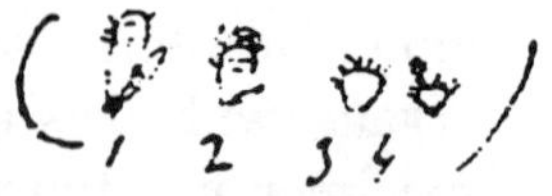

[**CD P. 98** continues]

Isopod <u>Cymothoudes</u>[4]

370

Habitat same as last &c.— taken in the sea.— is I think a new genus, comes nearest to Livoneca (Leach): Differs in having the eyes large, circular, black colour, faces very distinct: Mouth protected by shield, beneath which are 4 equal antennæ.— superior ones of same thickness even to terminal joint.— inferior ones pointed finer: Claws on feet strong, equal. received in ~~penultimate~~ ultimate joint by double row of short teeth:— Tail composed of 5 pieces; central one oval; lateral foleaceous ones equal; external plate pointed, oval — internal obliquely truncate:— |**99**|

Isopod <u>Cymothoudes</u>

Colour pale, with minute stars of reddish brown colour; these are thickly scattered on the back, so as to give it a dingy tint; there are a few on the lower surface:— [note (a)] The stars were only visible with a lens.— [note ends] Animal could swim very swiftly, & when at rest always turned its stomach upwards: could adhere even to a needle with great force:—

(a)

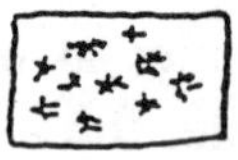

[1] Mysidacea, opossum shrimp.

[2] See Plate 52 of Apus cancriphorme by A.-G. Desmarest in *Dic. Sciences Naturelles* Planches de Zoologie 2e partie, Crustacés (Entomostracés).

[3] Macrourus was the term formerly used as a suborder for the long-tailed decapod crustaceans, including the shrimp-like and lobster-like forms. It is now replaced by the suborder Dendrobranchiata and part of the Pleocyemata. For further discussion of specimen 491 see *Oxford Collections* p. 206, and *Journal of Researches* **1**:189. See *Cuvier* Vol. 4, p. 91.

[4] Flabellifera, Cymothoidae, an ectoparasite of fish.

[**CD P. 99** continues]

Bufo[1]
377

Mr Bynoe has another specimen (For more particulars V **P 191**.)
Copied

Appears to approach nearest to Breviceps (Cuv.).— No tympanum or Parotid:— Mouth pointed: but the colours are the most extraordinary I have ever seen.— Body "ink black". under surface of feet, & base of abdomen & scattered patches of an intense "vermilion red" (the animal looked as if it had crawled over a newly painted surface).— back with scattered spots of "buff orange".— Inhabits the dry sandy pampas; there was no trace of water.— Sept: 11th.— Baia Blanca

Coluber[2]
383
(433 (b))
Copied

Heterodon (Cuv:). Above cream-coloured with symmetrical marks of dark brown; beneath with black & irregularly bright red.— The first of the maxillary teeth much developed & distinct.— Mouth dilatable & tongue very extensible, by these characters & shortness of tails, approximates to the Venimous serpents.— Was caught whilst eating a Lizard: Sandy plains: Sept 15th. Baia Blanca.

[note (b) added later] Mr Bynoe has ~~another & distinct species~~: (Trigonocephalus) Octob 4th. Monte Hermoso. B. Blanca.— Found this latter ~~species~~ on sandy hillocks near the sea.— Above marked with a chain of "umber brown", the intervals being "wood brown".— Aspect most hideous.— ~~I think finding these two species will establish the sub-genus "Heterodon"~~.—

Octob 8th: The triangular nose quite deceived me: this snake has no connection with the ~~one~~ Heterodon described. I caught a much larger one, coloured as above.— It is a Trigonocephalus, but does not exactly agree with any of Cuviers subdivisions.— Habits slow, strong, courageous. as long as it had life it would open its mouth very wide & protruding its fangs struck any object with great violence: ~~Iris~~ Pup<il> a vertical slit; iris mottled coppery: Tail with a pointed hard button at extremity.— When irritated the animal vibrated the last inch of tail with great rapidity, & this as it struck the blades of grass, & still more any sticks, made a distinctly audible noise.— As often as the snake was touched, its tail vibrated.— How beautifully does this snake both in structure & habits connect Crotalus & Vipera. As far as habits go Cuvier is right in ranking Trigonocephalus with Crotalus, contrary to Dic Class.— Inhabits the sandy hillocks & cannot be uncommon:—

Octob: 12

Found two more; the noise from tail audible at about 6 feet distance: live in holes: lizard in stomach: The orifice of the fang is very elliptic & placed on the anterior surface near extremity.— at the base the canal enters the fang at interior or concave surface.— [note ends]

[**CD P. 99** continues]

Perdrix-Scolopax[3] Vaginalis? 710 (not spirits) 388 more particulars V. **192**	This very singular bird was shot near the Fort.— In its first appearance partly resembles a lark & partly a Snipe.— In its flight & cry the former; inhabits dry sandy plains occasionally overflowed by sea. In small flocks.— Covering for the nostrils, soft: Baia Blanca Sept. 14th — Feeds on vegetable matter: Mr Bynoes has a good specimen. |**100**|
Fish[4] 390	Caught on a sand bank in the net:— body silvery: dorsal scales iridescent with green & copper; head greenish: tail yellow.
Fish[5] 391 (a)	Body pale, darker above; broard silvery band on sides; common:— [note (a)] This is probably the old fish of the small ones (367) taken at sea. [note ends]
Fish[6] 392	Body mottled with silver & green; dorsal & caudal fins lead colour: common
Fish[7] 393	Back coloured like Labrador feldspar; iris coppery: plentiful
Fish[8] 394	Above dirty reddish brown; beneath faint blue; iris yellow: plentiful
Fish[9] 395	Above pale purplish brown, with rounded darker markings:—
Agama?[10] 397 Copied	This is the most beautiful lizard I have ever seen: back with three rows of regular oblong marks of a rich brown: the other scales symetrically coloured either ash or light brown.— many also irregularly bright emerald green.— beneath pearly with semilunar marks of brilliant orange on throat.—
Buccinum[11] 412 Copied (b) Buccinum	Crawling in rushes on the sand banks & living on dead fish.— foot oblong, rounded anteriorly, the yellow operculum is placed obliquely on the upper part of extremity.— siphon lead colour, not closed; tentacula same colour pointed; mouth projecting over foot & between tentacula, when closed with small longitudinal division; from this there can be protruded a very long red coloured proboscis. terminal orifice with cartilaginous rim.— Very commonly on the whorls there are several ovules.— these are about 1/12" in diameter, rounded, conical, with broarder base, semitransparent, on the summit is a circular lid, which falls |**101**| off when the little shell is ready to obtain independent life.— [note (b)] The situation of the Ovules or eggs on the shell must be almost necessary, as the animal inhabits extensive sand banks, where there is no hard substance to fix them on.— [note ends] At first the capsules only contain a pulpy yellow matter.— but when further advanced the minute animal: the outline of the shell is rounded

oval, whorls not produced, the siphon not developed; but at the superior right corner, where the row of spines in old specimen commences, the edge of shell projects & is tranchant: animal after few minutes could crawl well; foot very large, thin; folding over the shell, fleshy siphon small; mouth & tentacula forming a triangle.—

[1] Identified by Thomas Bell from CD's specimens (see *Zoology* **5**:49-50) as *Phryniscus nigricans* Weigm. In *Journal of Researches* **1** p. 115, CD wrote 'If it is an unnamed species, surely it ought to be called *diabolicus*, for it is a fit toad to preach in the ear of Eve.'
[2] Most of the European snakes with which CD would have been familiar were non-venomous and oviparous species belonging to the family Colubridae. However, *Trigonocephalus* was probably a highly venomous pit viper, the Patagonian lancehead *Bothrops ammodytoides*, belonging to the Viperidae. CD considered this snake to be quite exceptionally ugly (see *Journal of Researches* **1** p. 114). *Crotalus* is another viper.
[3] Listed in *Zoology* **3**:117-18 as *Tinochorus rumicivorus* Eschsch.
[4] Listed in *Zoology* **4**:135-6 as *Alosa pectinata* Jen.
[5] Not listed in *Zoology* **4**, nor in MS list of Fishes in Spirits of Wine in CUL DAR 29(i).
[6] Listed in *Zoology* **4**:44 as *Umbrina arenata* Cuv. et Val.
[7] Listed in *Zoology* **4**:80-1 as *Mugil liza* Cuv. et Val.?
[8] Listed in *Zoology* **4**:137-8 as *Platessa orbignyana*.
[9] Listed in *Zoology* **4**:139 as *RHOMBUS*————?
[10] Identified by Thomas Bell in *Zoology* **5**:18-19 as *Proctotretus pectinatus*.
[11] Neogastropod, Buccinacea, probably a mud snail of family Nassariidae.

[**CD P. 101** continues]

Actinia[1]

413

Exteriorly dirty clouded yellow.— On the exterior rim are several rows, placed without order, of bluntly pointed tentacula; they have a minute orifice at extremity.— The inner ones are the largest.— They are coloured pale lead-colour:— Central orifice projecting.— Polype most widely expansible, fixed on stones.— Within the mouth is a collar with longitudinal ridges or plaits.— The whole sack or stomach is lined by delicate membranes ~~or rather bags~~ (which being double form thin bags) these project upwards & much folded in same manner as bud of plant.— the superior margin is thicker: The sides of polype are composed, first (exteriorly) of a thin covering of soft matter, this does not seem to extend to the adhæring surface: 2^{d} a strong white tough case, which must act as muscular; this on the interior surface is blueish & forms numerous longitudinal narrow plates.— between these bunches of the delicate |**102**|

Actinia

membrane is attached.— It is probable by these plates the tentacula communicate with the body.— I may mention these hasty observations as they show how singularly close the Actiniæ are in their organization to the Caryophillia as described at **Page (10)**.

Crepidula[2]

Adhering to the anchor, soundings 10 fathoms: shell with ~~concave~~ curved

(a) grooved spines: animal with foot rounded, posterior half lying on the
429 diaphragms of shell.— [note (a)] The young shells adhere to the old one. in these places the spines are absent.— [note ends] Tentacula pointed with minute black eyes situated near the base & on them: ~~mouth~~ between them the mouth opens on each side there being a rounded lobe, having a forked like appearance. Within mouth is <u>very</u> short proboscis.— Neck long.— On each side there is a membrane which when animal contracts itself closes
Copied the respiratory orifice: Branchiæ long, delicate, most regular, parallel, forming together apparently a rounded membrane — this adheres to the superior mantle by a longitudinal line.— The opening extends whole width of the shell.— From the appearance of fæces the anus must be on the right side:—

(429) There is another smaller & smooth species.— in this the foot anteriorly is crescent shaped with a horn at each corner:— also in some there on the right side ~~near~~ behind tentaculum was a long vermiform, tapering, generative organ.— |**103**|

[**CD P. 103** commences]

<u>Dipus:</u>[3] (Gme:) (777 not spirits) (a) Copied

This little animal does not appear to agree exactly with any of the subgenera of Cuvier.— It was caught Octob. 3^{d} at Monte Hermoso in B. Blanca.— In bringing at night a bush for fire wood, it ran out with its tail singed.— So that probably it inhabits bushes:— [note (a)] In sandy hillocks near the sea.— [note ends] it could not run very fast: it is a male: after skinning the head it has a much more elongated appearance than it had in Nature.—

<u>Clytia</u>[4] 437: 438 (b)

Coralline, with branches long, fine, colourless: bipinnate; polype either terminal or at the bifurcations, scattered; Polype in cups, which are ~~of~~ regular funnel-shaped enlargements of the tube or branch.— [note (b)] If, as I afterwards give reasons, the peduncles & branches may be considered as the same, then the Coralline will be both bi & tripinnate.— [note ends] each cup has a peduncle formed of elongated globular joints.— Those which arise at the bifurcations have 5 of these, of which the three basal ones are the largest: as the Coralline grows, the peduncle becomes a branch, being lengthened between 3 & 4th joint so that the terminal cups
(c) have but two articulations, but at the base of the branch there are three. [note (c)] This is by far the most general, but not universal case.— [note ends] These are ~~rather~~ from the thickening of branch ~~are~~ compressed, & may be considered as resulting from the form impressed on the branch when a peduncle.— Hence the Coralline ~~appears~~ is jointed, & at every bifurcation there are the three compressed globular articulations.— From this it would appear that the peduncle of the Clytia is really only the first form of the branch.— The peduncle is rather longer than the |**104**|
<u>Clytia</u> cup.— The central organized matter much developed included in a thin tube within the branches.— The polype unite at their bases with this.—

polype when retracted have a narrow base, like footstalk; ~~tentacula~~ arms short, 16 (?) in number situated round a central protruding mouth.—

Plate 7, Fig 1 is a drawing of a polype retracted in its cup, with the peduncle rising at a joint in a branch.— This coralline ought to form a distinct subgenus from Clytia of (Lamouroux), the latter having the peduncle twisted, & branches not jointed, & generally short creeping.— This would appear from structure of Coralline to be more closely allied (as Cuvier ranks it) to the Tubularia than to the Sertulariæ.— I never saw anything more beautifully luminous than this Coralline was; when rubbed in the dark every fibre might be traced by the blue light.— What was remarkable <was> that the light came in flashes, which appeared regularly to proceed up the branches: The coralline emitted a strong disagreeable odour.— Was brought from the bottom in abundance in 10 fathom water.— October 1st.—

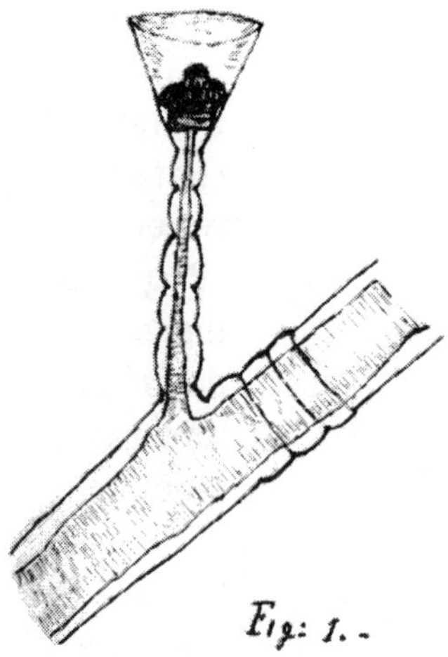

Plate 7, Fig. 1

[1] Actiniaria, a sea anemone.
[2] Mesogastropoda, Calyptraeacea, *Crepidula aculeata*, a slipper limpet.
[3] This mouse was identified by George Waterhouse as *Mus elegans* (see *Zoology* **2**:41-3.)
[4] Leptothecata, a thecate hydroid. See *Lamouroux* p. 13 and Plates 4*e,f,E,F* and 67, fig. 1 and 2.

[**CD P. 104** continues, the entries up to and including **P. 111** being crossed through with vertical lines and extensively corrected]

Ova[1] Sept 29th.— The sea contained great numbers of these: as the included animal bore a great resemblance to that described at **Page (2 & 73)**, I keep these notes: it only differed from the ova in that animal by the much greater size & having a pulsating |**105**| organ at the anterior extremity: Ova spherical about 1/14^{in} in diameter: they were in different states according to their ages. When least developed (& in this state they all were two days previously) they consist in a smaller sphere containing granular matter

(a) included in a larger: this matter gradually collects in a linear direction, & the next appearance presented is a projecting rim extending round 3/4 of the inner sphere, in which is a delicate vessel & one extremity, the anterior,

~~much~~ largest:— [note (a)] The part of sphere which contained the most developed granular matter projected beyond the outline of the rest:— [note ends] When the age is much more advanced, the inner sphere is pushed on one side & a small animal lies in curved position within the outer one.— (b) it possessed the same jumping motion as animal (of **Page 2**). At the anterior extremity. near the head, a heart might be seen pulsating.— a central vessel was very distinct, & ~~round~~ on side of the tail was a fine membrane which I imagine to be the ciliæ.— [note (b)] The tail is first liberated from the contact of the inner sphere:— [note ends] All the ova contained an air globule & in consequence floated on the surface of the water:— The largest ova which I extracted or saw in the body of animal (**Page 75**) were only $1/50^{in}$ in diameter; if I had any reason to suppose the<y> would increase so much in size, I should not have the slightest doubt of this being the animal; indeed I have not much now: It is very remarkable that in the extricated ova I mention a small transparent included globule.— Is not this the air globule of the larger ones?.— |**106**|

Virgularia[2] Octob 17th.— Bahia Blanca; N. Patagonia
~~This animal is~~ Found in the greatest numbers buried vertically in a flat of
401 muddy sand which is left uncovered at low water.— Their superior extremities projected ~~upwards~~ from one to 6 inches above the sand; The whole ~~ground is~~ surface was scattered over with them. *[several illegible words]* In length they vary (& in diameter in proportion) from about 8 inches to 2 feet. Colour yellow-orange.— When touched & especially if pulled they suddenly retract ~~their bodies~~ so entirely or nearly to disappear in the sand. This they do with so much force that the stony axis will break, before they can by force be dragged ~~far~~ out of the sand.— The superior extremity is truncate, with the axis uncovered, the other extremity terminates by a soft fleshy vermiform process of ~~a~~ greater thickness than rest of the ~~body~~ stem. This lowest part lies in a curved position buried in the sand.—

A section of the ~~axis~~ stem is rounded, oblong, surrounding it are ~~double~~ two
(a) rows of oblique fleshy folds bearing the polypi.— ~~These rise at one of the narrow sides of axis, & are then opposite & apart. From this point each fold winds obliquely downwards half round the stem, where, at the opposite edge to its origin it alternately crosses with the one from the other side.—~~ [note (a)] Not by any means universally so: sometimes the folds are for their whole length placed alternately with respect to each other.— [note ends] Hence at one edge there is a clear channel running down the stem.— These fillets or folds are exceedingly numerous. as they descend they become narrower & finer in proportion, at last they run into a point. here of course the opposite folds instead of meeting & interweaving on one side
Virgularia & nearly touching on the |**107**| other they are widely apart.— The place of termination is some way above the vermiform process & is generally marked by the stem being rather enlarged.—

These folds are composed by the junction of <u>numerous</u> polypi, side by side.— When animals alive & in the water, the folds are fringed by the widely expanded & plumar arms of the polype, & stand up at large angle (a) with the stem. On being touched they fold *[several illeg. words]* & the arms are folded together into a cone. [note (a)] The <u>arms</u> when collapsed form a cone, they are never withdrawn (like in Sertularia), but the papillæ on the surface are.— when fully expanded the arms are nearly horizontal like spokes of a wheel.— [note ends] ~~The~~ Each fold as it laps obliquely downwards becomes narrower.— Polype with elongated oval body united laterally, from the base of each a vessel runs through the supporting fillet.— These & not body of polype vary in length as the fold winds round the B stem.— Arms 8 in number, not tapering at extremities, with central vessel, [in pencil] covered irregularly & mostly at ends with delicate short <u>retractile</u> tentacula or papillæ; these arms surround a mouth with lips.— The polype vary in number; sometimes being as many perhaps as 40 on one fold.— The folds (c) are supported by hard pointed transparent spines ~~which passing through the base are free at their upper half:— They look like~~ which may be compared to the Calyx to a flower: [note (c)] It is only in the dried specimens that these project outwards.— [note ends] they are not attached to stony axis: (b) they are often 7 in number; but this is not constant, sometimes there being 10.— [note (b)] In this case the Polype are fewer in number. Can there be two species? Those in the bottle (401) were taken first, & were generally much longer than those floating loose in the great ~~bottle~~ jar ( ). [note ends] They can be applied close to the stem; but ~~will~~ can not diverge ~~form~~ at a <u>Virgularia</u> greater angle ~~with~~ from it than the |**108**| Polype do when fully expanded [inserted in pencil] C :—

The stem ~~is~~ terminates *[illeg. words]* bluntly pointed in vermiform process.— [It] is of larger diameter than the rest of the stem.— Within this process there are two large triangular cavities, separated by a division or septum.— These cavities or channels run up the whole stem, but in the upper parts are obscure & small.— Just above the vermiform process *[illeg. words]* they are larger & filled with ~~a~~ pulpy yellow matter.— Within one of the 2 cavities, as will presently be mentioned, the stony axis floats:— At the point ~~the~~ where the polypiferous folds commence, the ~~cavities~~ channels become smaller & the stony axis soon becomes attached to the septum & hence is central.— ~~the line joining the 2 channels or cavities is at right angles to that of the opposite folds:~~ [see sketch in margin] In the cylindrical soft part of stem beneath the polypi, section ~~gives first a covering of tough substance; within this & filling up the whole excepting the cavities, the substance is striated from the centre, & is composed of longitudinal~~ of the stem shows a number of plates radiating from the centre: [sketch in margin] in the parts ~~of stem~~ where the ~~folds~~ polypi are fully developed this structure nearly disappears, ~~& it is through this they first~~

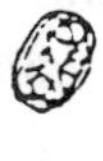

~~appear~~.— The calcareous ~~stony~~ axis is highly elastic & will break sooner than retain a new form; central parts brown, striated from centre. [note (a)] The rays have quite a crystalline fracture. [note ends] The *[illeg. words]* consists of a white softer substance.— (like to the marrow in bones!). The external part *[illeg. words]* white semi-opake; ~~superior extremity abruptly truncate, figure rounded oblong; at the narrower end there is a slight~~ |**109**|
Virgularia [inserted in pencil] D ~~depression or channell, & it is along these that the cavities within the stem run~~.—

The axis gradually tapers from its upper end to the finest point at the lower end.— ~~at this &~~ At the lower end, the extremity is suddenly curved backwards. Here the axis consists of a dark ~~line~~ centre part enveloped in a transparent covering, ~~afterward doubtless forming~~ which no doubt is connected with the exterior white stony ~~part~~ layer.— This recurved part of axis is included within a capacious membranous transparent elastic (irritable?) bag, which some way above the bend contracts round the axis & is probably continued all up the stem ~~together with it~~ in close apposition to the stony axis.— The lower part of the stony axis lies loose in one of the cavities, but in the higher parts where the polypi ~~are~~ stand, it is imbedded in or chiefly forms the septum between the two channels.— ~~it is attached at the corners of the polypiferous folds to the enveloping fleshy parts, & I think it probable that the vessels from the polypi lead into the membranous case of axis (?).~~

[**CD P. 109** continues]

In the elastic terminal bag, which encloses the terminal & recurved parts of the stony axis, there was a most distinct but irregular circulation of a fluid containing particles; this was even visible when the axis was entirely
(a) removed out of the body.— [note (a)] The circulation was strongest at the very bend; it was irregular [note ends] [marginal note in ink in very small handwriting, later crossed out in pencil, is here inserted] The circulation of the particles was strongest & most *[3 words illeg.]* at the point where the axis was most bent [insertion ends]

The axis evidently performs a very essential part in the economy of the animal; it is by this that the whole stem is kept in a vertical position & that the upper part stands upright in the water, & so allowing many of the polype to have free access to the surrounding ~~medium~~ fluid.— When a bit of the stem is cut off, the axis projects at ~~each~~ both extremities; & this
Virgularia shows the high contractility of the softer parts of |**110**| the stem.— ~~Hence~~ By this power the ~~animal~~ whole stem ~~its body~~ is easily withdrawn into the sand; but at first sight its manner of rising again is not so clear: upon considering the erratic nature of the axis, its inferior extremity floating loose in a cavity, & the lower part of this cavity lying in a curved position,

it is clear that when the animal retracts itself, the axis must, from being forced into the terminal vermiform process, exert from its bent shape a considerable force; So that as soon as the animal ceases to contract itself, the stem would gradually rise:— But ~~(As~~ it appears improbable that the polypi in so large a part of the lower end of the stem should be buried in the sand.— ~~perhaps~~ hence when ~~in deep water~~ the tide rises I suppose that nearly the whole ~~body~~ stem is protruded. A very small force would be sufficient to enable the animal to ~~work its way back again~~ withdraw into the soft sand, for the spines under the folds would act as pauls[3].— but ~~in this case~~ I do not ~~exactly~~ see how the ~~animal works upwards~~ stem could protrude to so great a distance, as the elasticity of the axis would not in this case come into play ~~?)~~.—

In the vermiform process, at the very extremity I found several ova, in shape regular oval; they contained granular matter; was of an orange colour; & a length 1/1000 of inch.— I think they had only just been formed; this being the early part of Spring renders this the more probable; when I ~~having~~ examined some specimens a few weeks before I ~~anyhow~~ did not ~~then~~ observe ~~them~~ any.— Above the vermiform process, the 2 internal channels or cavities were for some length filled *[illeg.]* with yellow pulpy
Virgularia matter.— This examined under a simple microscope |**111**| presented an extraordinary appearance.— The Mass consisted of various ~~shape~~ sizes of irregular globular semi-transparent particles; the larger ones being merely an aggregation of the smaller ones. All these grains possessed a most distinct very rapid vibratory motion, generally round varying ax~~is~~'es, but sometimes progressive.— The motion continued for a long time, as long
(a) as I watched it. I first saw it with a simple lens of 1/3 of inch focal distance, but it would have been quite clear with a less power; ~~I accurately examined the particles with a strong light & 1/20 focal distance~~.— [two notes (a)] Does not the great size entirely separate this fact from the "Molecular movement" of Browne? The motion continued for some time in distinct parti<cles> (as long as I watched them) when kept in water.— [notes end] It seems probable that these particles ~~go towards~~ forming the ova & that ~~when ready for expulsion~~ when formed they ~~pass through~~ are pulled along the 2 longitudinal ~~cavities~~ channels to the upper part leading into the open sea.—

In this respect the animal (if my memory is correct) differs from the Virgularia mirabilis which I ~~saw~~ examined in ~~Edinburgh~~[4] *[word above illeg.]*; for in this species the ova were scattered in the fleshy part between the polypeferous folds.— ~~(Were they then passing through an internal vessel??)~~.— The above movement in the particles was more rapid & I think quite distinct from that of the particles in the elastic bag *[illeg.]* to lateral part of the *[illeg.]*.— ~~the latter would seem to bear some obscure analogy to a true circulation~~.—

I have called this animal Virgularia; but it *[illeg.]* seems to form a new genus: ~~it is most allied to Virg: juncea, but widely different in form of axis & in spines. According [to] Cuvier, the occurrence of spines being the leading character, it would be a Pennatula, from which genus it differs still more widely in habits & general appearance.—~~ |**112**|

1 The ova were those of the chaetognath arrow-worm *Sagitta*, discussed on pp. 68-71.
2 In *Journal of Researches* **1** pp. 117-18, CD has decided that this alcyonarian coral is the sea-pen *Virgularia Patagonica* of D'Orbigny. His first attempt at describing its anatomy and the mechanics of its growth has later been extensively revised with insertions in ink in smaller handwriting. Some further insertions have been made in pencil, and the letters B, C and D have been added, though with no indication as to their purpose, and finally the whole passage has been crossed through in pencil to indicate its publication. But the final account that appears in the *Journal of Researches* is substantially shortened and clarified.
3 The correct spelling of this word is 'pawl', but although CD was by now familiar with the naval use of such a device to prevent the slipping back of a capstan, he had possibly never seen the word in print.
4 As has been explained by Phillip Sloan in *Darwinian Heritage*, pp. 71-120, CD had joined the student Plinian Society at Edinburgh University in 1826, and had begun a close collaboration in research on marine invertebrates with its Secretary, Dr Robert Edmund Grant. He had gone with Grant on expeditions, and had accompanied trawlers in the Firth of Forth to collect live specimens of deep-water invertebrates. As described in his Edinburgh Notes (DAR 118:9) he returned with specimens of sea-pens for Grant on 15 April 1827. Preserved among CD's drawings (CUL MS DAR 29) is a note made at an unknown date that runs 'Phil Trans 1778 p. 178. An account of sea-pens at Sumatra — a wonderful account.— flesh over knitting needle'.

[**CD P. 112** commences]

Struthio rhea[1] This bird is very common in the sandy plains: in its stomach I have found roots of vegetables: at low water they come down 3 or 4 together to the
(a) sand bank, the Gauchos say for small fish; in their habits shy & wary, generally solitary; emit a very deep note: During September & Octob. we
(b) found an extraordinary number of eggs, in colour varying from pale yellow to white: the male eggs weight the most (am told so).— The eggs are
814 either found scattered about, when they are called Watchos[2], or collected in
(not spirits) circular shallow excavations or nest.— Out of the four which I saw, 3 contained 22 eggs each; & the other 27:— In one days hunting 64 were found; 44 of these were in two nests — the other 20 scattered about.— It seems strange that so many of the latter should be produced for no end, as
Cop Cuvier mentions the Gauchos state that many females lay in the same depository & that one male sits on them.— I can scarcely credit this; anyhow it is clear from the number of eggs that each female lays many eggs, & in the oviduct (it was told me by those who cut one up for the

ships company) that there was nearly 50, of a regular gradation in size.—

(a) 1833 Summer [note (a) added later] Mr King[3] tells me, that when in the Schooners on the coast of Patagonia, he & the others several times saw Ostriches swimming from one island to another.— This occurred at the Bay of San Blas, & at Port Valdes.— They took to the water when driven, & likewise of their own accord without being frightened.— The distance in both places about 200 yards.— When swimming very little of their bodies appears above water & their neck is stretched forewards, as a Goose or Duck in flying.— Their progress is slow.— Before hearing this account, everyone was surprised to hear of the plenty of ostriches & Guanacoes in the various small islands of San Blas.— The latter animals were often seen swimming.—

December [note (a) continues with a different pen] The male ostriches are easily distinguished by the Gauchos from the female, by the greater size of head & body & colour.— It is a most undoubted fact that the males sit on the eggs: the females never.— As the number of eggs in the belly & the nest seem to correspond, 20 to 40 or 50.— it would seem hard to be ascertained, but I was assured 4 or 5 females have been watched to lay their eggs one after the other in same nest, in middle of the day.— The reason seems obvious — if a female were to deposit 40 eggs successively one after the other, & then sit on them. The first would be so many days older than the last laid egg.— The same cause explains the male sitting, because it must ~~more~~ very often happen that the female has not ceased laying.— The male will not rise from the nest, without you pass very close.— They sometimes are dangerous; attacking, kicking, & trying to jump upon the horse.— My informer had seen an old man much terrified by one chacing him.— The eggs which I have called Watchos are supposed (Turn over) [note continues opposite **CD P. 113**] by the Gauchos to be laid first.— Perhaps before association *[illeg. words]* to the Male.

<u>Ostrich</u> (b) [note (b) continues] The ostrich, with a<ll> its swiftness, is easily balled, for they are simple animals & are easily turned & puzzled.— They generally run against the wind.— In fine weather, they will try to conceal themselves amongst the long rushes, & will thus lie till closely approached.— The noise of the ostrich (male I believe) is like a deep drawn breath.— it is neither easy to say where it comes from, or how far distant is the animal which makes it.— The first time I heard it, I thoug<ht> it was a Lion or other wild beast.

Wallis[4] saw Ostriches in Bachelor river in the Sts of Magellan, Lat between 53° and 54°.—

When at the R. Negro, I heard much concerning the "Avestruz petises", a species of ostrich ½ the size of the common one.— The following I believe to be a tolerably accurate description, colour mottled, shape of head, neck, body same as in ostrich.— legs rather shorter, feathered to the claws; feathers same structure as in ostrich; hairs about the head.— cannot fly, is taken more easily than other ostrich with the balls.— This bird is however more universally known by its eggs, which are little inferior in size to the Rhea, but of a blue green colour. It [is] generally frequent near the sea, frequently to the South of R. Negro, San Josè, & I believe near the Colorado, but not further Northward.— [pen changes] V **212** more particulars.— The Northern Gauchos know nothing about the Avestruz Petise, even at Bahia Blanca. [pen changes again] Albino varieties of the common Ostrich have been seen; it must be a most beautiful bird.— snow white, Gaucho at R. Negro told me. [notes end]

[1] Listed by John Gould in *Zoology* **3**:120-3 as *Rhea americana* Lath. The description of the habits of the species that follows is a slightly extended version of CD's notes here. See also *Ornithology Notes* pp. 268-71.
[2] In the published account, the correct Spanish spelling 'huachos' is used.
[3] Philip Gidley King was a midshipman on board the *Beagle.*
[4] See Samuel Wallis. *An account of a voyage round the world in the years 1766, 1767, and 1768.* Included in John Hawkesworth, *An account of the voyages . . . performed by Commodore Byron, Captain Wallis, Captain Carteret and Captain Cook* . . . 3 vols. London, 1826.

[**CD P. 112** continues]

Cavia patagonica[1]
Copy
817
(not spirits)

Frequent in the sandy plains, feeding by day; Azara[2] states that they only frequent the holes of the Viscaches[3], & that only when pressed by hunting (Griff: animal k:[4]). [note in margin] Found near Mendoza Traversia to the South [note ends] They certainly wander far from any holes; but they abound like rabbits in a warren, where there is a collection of holes.— I have watched them sitting on their haunches by the mouth of a burrow, which they will enter immediately they are frightened.— The dung is

Cavia patagonica
(z)

Cop

|**113**| of a remarkable shape, being an elongated regular oval.— now if the Viscaches were in sufficient numbers to dig the holes for the Agouti, some considerable quantity of dung would be lying about.— I did see some like (but smaller) an English rabbits, but I think it belong[s] to the Toco Toco, a small Rodentia which I know inhabits burrows in the same plain.— The manner in which the Agouti runs more resembles that of a Rabbit than a Hare. It consists in so many distinct springs.— The body weighs from 20 to 25 pounds.—

[notes added later for **CD P. 113**]

NB. For the future, the marginal letters will refer to notes on the back of page.—

(z) [note (z)] This animal is the most common characteristic animal of the dry plains of *[illeg.]* Patagonia: It does not occur to the North of the Sierra Tapalguen 37°.30′.— our officers have never seen it to the South of Port Desire 47°.— The Gauchos are of different opinions respecting its digging holes.— I have no doubt it uses ~~them~~ Biscatche holes where they occur, but I think certainly it must be its own workman in those parts where the Biscatche is not common, as S. part of Patagonia where I do not believe Biscatche is found [continued as marginal note] as the little owls do, which in B. Oriental are obliged to make for themselves: [marginal note ends] Two tolerably fast dogs often run them down.— Their flesh is very white & pretty good.— They bring forth two young ones in their holes.— Southern limits between Port Desire & St Julian (48°:30′). The Gauchos at B. Blanca say certainly that it digs its own holes.— [notes end]

[**CD P. 113** continues]

Lizard[5] Monte Hermoso.— In its depressed form & general appearance partakes of some of the characters of the Geckos.— Colours above singularly mottled,
454 the small scales are coloured brown, white, yellowish red, & blue, all dirty,
455 & the brown forming symetrical clouds.— Beneath white, with regular spots of brown on the belly.— Habits singular, lives on the beach, on the dry sand some way from the vegetation.— Colour of body much resembles that of the sand.— When frightened, it depresses its body & stretching out its legs & closing its eye tries to avoid being seen; if pursued will bury itself with great quickness in the sand.— legs rather short: it cannot run very fast.—

[1] Listed by George Waterhouse in *Zoology* **2**:89-91 as *Cavia patachonica*. See also *Journal of Researches* **1** pp. 81-2.
[2] See Felix Azara. *Essais sur l'Histoire Naturelle des Quadrupedes de la Province du Paraguay*. French translation Vol. 2, p. 41.
[3] This burrowing rodent related to the cinchilla is common in the pampas in the neighbourhood of Buenos Aires. It is listed as *Lagostomus trichodactylus* in *Zoology* **2**:88, and spelled as 'Bizcacha'.
[4] See Edward Griffith and others. *The animal kingdom arranged in conformity with its organization . . . with additional descriptions of all the species hitherto named, and of many not before noticed.* 16 vols. Edinburgh, 1827-35. Translated from Georges Cuvier, *Le règne animale*.
[5] Listed by Thomas Bell as *Proctotretus multimaculatus* in *Zoology* **5**:17-18.

[**CD P. 113** continues]

Hybernation of Animals Sept: 7th. Upon our first arriving here, Nature seemed not to have granted any living animals to this sandy country.— [note (a)] I must except

Sept: 7th (a) Trox[1].— This I observed also at M: Video. [note ends] By digging in the ground I found several Carabidous[2] & Heteromerous[3] insects, Mygalus[4] & some species of Lizards, all in |**114**| a half torpid state. On the 15th

Hybernation — different animals began to appear & by the 18th everything announced the commencement of Spring.— The plains were ornamented with flowers;

(a) birds were laying their eggs; [note (a)] Such as Parrots, Swallows, Hawks, Partridges: Ostriches were laying when we first arrived:— [note ends] numbers of Heteromerous & Scarabidous[1] insects were crawling about. The Saurian tribe, the usual inhabitant of a sandy district, were darting in every direction. For the first eleven day[s], from the 7th to 17th (both inclusive) the mean temperature from ~~bihoral~~ observations at 2 hours interval was 51°.3.— & I see that generally in the middle of day thermometer was from 52° to 55°.— In the 11 subsequent days, in which Nature became so animated, the mean was 58°.1.— Thus giving a difference of nearly 7°.— the general range of Temp: in middle was even in *[illeg.]* greater than this, varying from 60° to 70°.—

At M: Video, every animal was hybernating (Vide **P 120**) when the mean Temp: was 58°.4 & in the day Therm: often rising to 70°.— The difference of Latitude between the latter & this place is four degrees or 240 miles; Thus showing how much the general annual Temp: affects the degree at which animals reassume their living process.—

Entomology — By far the most abundant order is Coleopterous: In this Heterom & Lamellicorn[5] were in numbers of individuals by far the most prevalent. in species the first contained about 10, the latter 9.— |**115**| Amongst the

Entomology — Carabidous (or more properly Harpalidous) there were 7 distinct species: but all very rare.— in my collection I have every individual I have seen[6].— The other Coleopterous insects make no figure. I found one Staphylinus, Colymbetes, 2 Crysomela, Elater, 2 Coccinella[7].— Amongst the Diptera Musca was abundant & a Bombylius.— The orders Orthoptera, Hemiptera,

Vide Collection (a) — Neuroptera, produced scarcely anything.— In Hymenoptera, a large Pompilus[8] was common, as was its prey Mygalus[4]:— also a large humble bee feeding on the wild pea.— I saw three species of Lepidop: diurniæ: the Nocturnæ were more abundant.— [note (a)] Ants are very common: on Sept 22d Swarms were on the wing.— [note ends]

[**CD P. 115** continues]

Entomostraces[9] (Lophyropes) (b) — This animal does not come in any of Latreilles families.— In general appearance most closely resembles Ostracodes; but in structure very different.— [note (b)] It did not occur to me at first that by counting the rudimentary legs there will be 24, & that in its other characters (2 pediculated eyes, flat calcated legs &c), it must belong to the division Phyllopes.— Eggs in this one were irregular, numerous in the dorsal

457 [in spirits] (c)

posterior part of shell.— [note ends] Shell bivalve, gaping at each end from the approximation of the central parts of lower edges.— Back ~~round~~ curved, posterior extremity rather pointed, the other rounded: The animal could not completely close the anterior & posterior longitudinal orifices: Eyes 2 pedunculated, formed of a transparent substance enveloping dark central mass; their eyes were in constant motion. Between & beneath there were two antennæ, & in their structure most singular. Peduncle thick strong, nearly the length of the shell *[3 illeg. words erased]*.— terminated by a large circular transparent sucker; on the internal edge, there was a small branch with setæ, & on the posterior a bunch of Setæ.— the |**116**| whole organ instantly reminded me of the front leg of male Dyticus, only that in the latter the Tarsi (answering to the joint with setæ) incline outwards.—

Entomostraces (Lophyropes)

The cups or plates adhæred firmly to glass or any other object: it was most curious to then see the animal walk; this it managed very deliberately with ~~with~~ long strides, the swimming legs helping to keep the lower edge of shell vertical.— Thus it walked up the side of a watch glass; but from the inclination, the shell often fell over & by so crossing the leg-like antennæ interfered with its motions.— The mouth is obscure, & is seated at base of Antenna, within the central parts of anterior half of shell.— The body

(a) seems to be attached [at] anterior half of shell, & the stomach &c lies above & behind the head, the posterior half of body is free: so as to be more or less drawn up, it is terminated by short simple jointed tails, with double bunch of few setæ.— [note (a)] There is no separation between head & thorax or body. [note ends] It & legs are protruded by the posterior opening.— 6 pair of similar equal natatory legs; ~~base jointed, with flat row of setæ~~; (acting like caudal swimming in the "Macrourus") each one formed of row of strong setæ, on jointed base; At origin of leg is a small projecting point, or rudimentary leg, with few bristles:— there were no branchial plates.— Animal could swim laterally very rapidly, generally in circular direction; antennæ retracted: Shell hard elastic.— Animal coloured blue; in open ocean South of Corrientes.— |**117**|

[note (c) on **CD P. 115** added later]

Octob: Examined another specimen.— Each of the 12 legs is bisected at its summit, from whence proceeds a bunch of setæ.— this is more true than saying a small external leg: Eyes are formed of ~~number (not many)~~ some small transparent globules, seated on a dark coloured pedunculated mass. The body attached to the dorsal part of shell by many parallel tubes or vessels: these perhaps act as Branchiæ.— Mouth obscure, with two curved pointed jaws (mandibulæ or maxillæ) united at base & forming a horse shoe.— ((?) At base of Antennæ there are 2 rudimentary palpi (?):) Shell has not a true dorsal hinge, but merely a line: shell very tough elastic, with numerous fine parallel vessels running in it.— The jaws resembled the mandibuliform horns of Branchippus[10] figured in Desmarets:

The posterior part of body was to certain extent divided by lines into 6 segments, which corresponded with the 6 pr[s] of legs: does this now show that this number is normal & that the bisection of legs at summit ought not

Decem 2[d]
Lat 40°.20′ S.
Coast of Pat.

to make the number 24.— Animal not uncommon in open ocean: I find my description very accurate; perhaps the antennæ are obscurely jointed: tail very small:— [note ends]

[1] Scarabaeidae, dung beetle.
[2] Carabidae, ground beetles.
[3] Tenebrionidae, darkling beetles.
[4] Mygalomorphae, a tarantula-like spider.
[5] Scarabaeidae, another dung beetle.
[6] For an identification of the insects collected by CD at Bahia Blanca see *Insect Notes* pp. 61-7.
[7] Coccinellidae, lady birds.
[8] Pompilidae or Sphecidae. See *Insect Notes* p. 56.
[9] See account by P.A. Latreille of branchiopod crustacea in *Cuvier* Vol. 4, pp. 149-71. The specimen sounds like a shrimp of order Conchostraca, but they are never marine.
[10] See Planche 56 of Branchippe des Marais by A.-G. Desmarest in *Dic. Sciences Naturelles* Plates for Crustacea, Zoea, etc.

[**CD P. 117** commences]

Gossamer
<u>Spider</u>[1]
(a)

462
[in spirits]

Sailing between M Video & B Ayres on Octob. 31st the rigging was coated with the Gossamer web: it had been a fine clear day with a fresh breeze.— The next morning the ropes were equally fringed with these long streamers.— On examining these webs I found great numbers of a small spider.— [note (a)] October, answering to our Spring.— when they are abundant in England [note ends] On the second day (which was calmer) there must have been some thousands in the ship.— When first coming in contact with the ropes, they were seated on the fine lines & not on the cottony mass.— This latter appears to be only the separate lines collected by the wind.— From the direction of the wind [they] must have travelled at least 60 miles from the Northern shore.— They were some full grown & of both sexes & young ones; these latter, besides being of a smaller size, were more duskily coloured.— Spider eyes 8 equal in size, seated on anterior end of thorax, viz. [see sketch in margin] the lateral eyes, or those on sides of the quadrilateral figure, are very close & seated on a common <u>small</u> eminence.— Cheliceres cylindrical, tapering at extremity; claw folding transversely & received between spine (with this it cleans its legs): Maxillæ, when mouth is closed incline on the Labium: when open are shaped each thus [see sketch in margin] inner side straight, summit

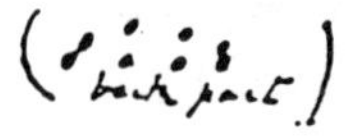

rounded truncate, outer inclined: Labium small, triangular, pointed: Legs four anterior ones longest, & 3d pair shortest; thin long:— ~~Thorax~~ Palpi, with organs in male much developed & coloured black:— Thorax heart shaped, truncate anteriorly, this part black, the rest red:— |**118**|

[further notes for **CD P. 117** added later]

November 25th — During the last week, every object both on the ship & on shore (Monte Video) has been occasionally covered with Gossamer: Invariably I have observed <u>great</u> numbers of the same small spider.— I frequently observed them sail away from any small eminence: I imagined that before protruding upward their abdomen & sending forth the web, they connected by delicate lines their feet together ?? I cannot actually say that the Spiders ever rose, but they laterally sailed from their position with unaccountable rapidity.— But even if they did ascend, I should almost imagine the ascending current on a calm & <u>hot</u> day would be sufficient to account for it.—

Decemb 4th — Lat 40°.20′ S. There were great numbers & spiders on the rigging, we being about 20 miles from the shore:

December 1833 — I saw at St Fe Bajada a brown coloured spider, I should think 3/10 of inch long (appeared very large) & from its general form a ~~Laterigrade~~ Citigrade.— standing on summit of Post it darted 4 or 5 lines from its anus, which glittering in the sun looking [like] rays.— they were a yard or two long & by a gentle current hardly perceptible were carried upwards & laterally.— The threads <u>curling</u> & diverging.— the Spider suddenly loosed its hold & sailed out of site.— The air is seldom so calm that a delicate vane like [a] spiders web is not affected.— on a hot day, would not the currents of air flow upwards; no ordinary vane would from its specific gravity would show a slight tendency to this motion.— [different pen] Yes, *[illeg. word]* mirage & tremulous shadows always occurs on any warm day.— [notes end]

[**CD P. 118** commences]

Gossamer <u>Spider</u> (b) — Abdomen pointed, oval, coloured dusky red: Filières projecting in a bunch at posterior extremity; each one cylindrical, short.— [note (b)] Body & legs covered with fine down [note ends] Length of body .1: When not moving, the legs are elevated: its motions rectigrade: I know not whether this spider
(a) belongs to the Tubitetes or inæquiletes: it does not agree with any of Lat: genera[2]:— [note (a)] From not clearly understanding the characters drawn from the Filières:— [note ends] These little spiders, after alighting on the ropes, were in their habits very active; They frequently let themselves fall from a small height & then reascend the attached line.— Occasionally when thus suspended, the slightest breath of air would carry them out of

sight on a rectangular course to the line of suspension.— I never saw them rise at all: They formed an irregular net work amongst the ropes: Could run easily on water: Lifted up their front legs in attitude of attention.— Seemed to have an inexhaustible stock of ~~line~~ web: With their Maxillæ protruded, drank eagerly water; this curiously agrees with an observation of Dr Strack[3]:—

[further note for **CD P. 118** added later] In the Spring of 1833 when about 60 miles off the mou<th> of the Plata, several came on board in their web.
(827) appear exactly same species: one is an old male [note ends]

[**CD P. 118** continues]

The above mentioned facts in the occurrence of numerous (sufficient I think to account for the Gossamer) spiders of same species but different sexes & ages, on their webs, & at a great distance from the land & therefore liable to no mistakes demonstratively proved that the habit of sailing in the air as much belongs to a division in Spiders as diving in the water does to Gossamer <u>spider</u> Argyroneta[4]: We may |**119**| then reject Latreilles[5] supposition that the Gossamer in the air owes its origin to the web of young ~~web of~~ Epeira & Thomisa.— Still more so that of Hermans (fils) that it belongs to an Acarus (Trombidium).— As far as the characters of the eyes goes, this Spider agrees with the sort described by Strack[3] as coating the ground.— I mention (**Page 49**) a spider under the name of Theridion (which shows the same position of eyes), as every where coating newly turned up ground.— I never however saw the aerial Gossamer ~~here~~ there: Kirby[6] thinks that Stracks spider & those Dr Lister[7] saw mount in the air are the same. Perhaps it may hereafter be shown, that if not identical the two sorts are closely united (viz aerial and terrestrial gossamer).— The celerity with which this spider voluntary can fall, shows, that light as its body seems, it must have considerable specific gravity: it is difficult therefore to understand in what manner the rapid rectangular motion was effected:—

[1] Probably a 'money' spider of family Linyphiidae, small black or brown spiders which disperse by air, attached to lines of silk (gossamer). CD's observations on the aeronautic or ballooning spiders encountered on board a ship at sea were of considerable importance.

[2] The tubitelae (correctly spelled 'tubitèles' in French) were spider families that build tube-like webs, while the inequitelae (inequitèles) build irregular webs. These terms are no longer in use.

[3] See Strack, C.F.L. 1810. Einige selbstgemachte Beobactungen über den Sommerflug und die Spinne, die ihm hervorbringt. *N. Schr. naturf. Ges. Halle*, Heft 5, Drei Abhandl, II, pp. 39-56.

[4] *Argyroneta aquatica* is the unique water spider of the Northern Hemisphere.

[5] See Pierre André Latreille in *Cuvier* Vol. 4 pp. 206-64.

[6] See account by William Kirby and William Spence in *An introduction to entomology*

(London, 1815-26), Vol. 2, pp. 334-46.
[7] See Martin Lister. *Historiæ animalium Angliæ tractatus de Araneis*. London, 1678.

[**CD P. 119** continues]

Chara[1]

476
[in spirits]

Common in running water: In the microscope could clearly perceive a slow circulation of round particles.— Branches finely striated, with distant spines, parallel to these the globules moved: In same manner as the Striæ, a colourless line encircled spirally the stem; but on one side of this the current ascended & on the other descended. So that in the equal spaces marked by the spine on the stem; the current alternately was |**120**| seen flowing upwards & downwards.— The axillæ of the branches are verticillate with pointed cylinders, in these the circulation was evident, but very obscure: Novemb: 20th M: Video.

Hybernation of animals

July 26th.-
Aug. 19th.

From finding Cassida, Crysomela, Curculionidous, Heteromerous, Lamellicorns, Carabidous beetles, & Epeira amongst spiders, under stones: from Vaginulus[2] & land shells with a membrane over the mouth being in same site; from finding Bufo[3] & Lacerta[4] half torpid; it is clear animals are now hybernating.— Considering the high temperature, this is curious.— From 276 observ: made at 2 hours intervals during 23 days from July 27 to August 19th (both inclusive), mean temp is 58°.4.—

Mean hottest day 65°.5
do. Coldest day 45°.8

The lowest point the Thermometer fell to was 41°.5; it occasionally in middle of the day rose to 69° or 70°.—
At (**P 113**) there are observations on the subject at Bahia Blanca & compared to those made at Monte Video.—

Gen: Observ:
Monte Video
July 26 to
Aug. 19th.

Birds are abundant in the plains & are brilliantly coloured.— Starlings, Thrushes, Shrikes, Larks & Partridges are the commonest.— Snipes here frequently rise & fly high up in great circles; in their flight, as they descend, they make that peculiar buzzing noise, which the few which breed in England are known to do.— On the sand-banks on the coast are large flocks of Rhynchops[5]; these birds are generally supposed to be the inhabitants of the Tropics |**121**| Every evening they fly out in flocks to the sea & return to the beach in the morning.— I have seen them at night, especially at Bahia Blanca, flying round a boat in a wild rapid irregular manner, something in same manner as Caprimulgus does.— I cannot imagine what animals they catch with their singular bills.—

Gen: Observ
M. Video
July 26th
to Aug 19th

The water of the Rio Plata at Monte Video is generally brackish, it is even sometimes fresh enough to drink.— It is not inhabited by many animals; a small Turbo[6] & a Mytilus[7] are nearly the only shells.— The occurrence of one of the Balanidæ Creusia[8] in quite fresh water is curious, for details see notes attached to (323 in Cat: for Spirits). On the shore, the genera Plagusia[9] & Grapsus[9] are exceedingly abundant.— indeed they are nearly the only Brachyures which I have seen between M. Video & Bahia Blanca.— On the beach are also great numbers of minute Crust. Amphipod:— which here assume the place which Ligia[10] takes in the Tropics.—

Amongst Arachnida by far the greatest proportion belong to Lycosa[11].— I found Mygalus & Dysdera under stones & Segestria abundant in crevices of rock.— Scorpio[12] & Gonoleptes are very abundant under stones.— In November an Epeira[13] with bright colour is abundant in every situation.— The Entomology is chiefly characterised by, as compared to Brazil, bẏ the
(a) great increase of Carabidous beetles: also by the comparative absence of the Orthopterous insects, which perform so essential [a] part in the latter.— |**122**| [note (a)] Amongst the Mammalia, the case is reversed; the carnivorous animals are as much more abundant in the intertropical regions than in the temperate, as the Carabidous amongst insects are in the latter compared to the former climate.— [note ends]

[1] Characeae, a green freshwater alga. See *Plant Notes* p. 221.
[2] Stylommatophora, slug.
[3] Procoela, toad.
[4] Sauria, lizard.
[5] As described in *Journal of Researches* **1**:161-2, and *Zoology* **3**:143-4, CD later saw that the Scissor-beak *Rhynchops nigra* scoops up small fishes from the surface of the water with its remarkable bill.
[6] Archaeogastropoda, turban snail.
[7] Anysomyaria, mussel.
[8] A barnacle allied to *Pyrgoma*, but probably misidentified here because as noted on p. 137, CD did not find the genus in the South Atlantic.
[9] Reptantia, crabs.
[10] Isopoda, woodlice.
[11] A still valid genus of hunting spider.
[12] Scorpiones, scorpion.
[13] The no longer valid name Epeira denotes an orb-weaving spider, probably family Araneidae.

[**CD P. 122** commences]

Crust: Decemb 1st.— South of Cape Corrientes, Patagonia.
Branchiopod Body composed of 7 pieces.— The anterior ~~one~~ case is rather narrower,
Cyclops[1] convex, rounded anteriorly & projecting over base of antennæ. in front it
478 terminates in a doubly pointed or forked rostrum, this projects downwards & gives the appearance of a sucking beak to the animal.— the posterior lateral part ends on each side in a point, projecting beyond line of body.

In shape this resembles some of the Cyclops: 2nd & 3rd segments are wider & longer & cover the body; 4th, 5th, 6th, 7th form the abdomen: the 7th is excised & ends abruptly.— Body cylindrical.— case horny elastic.— Beyond the last segment, there is (as in Cyclops) a narrow rounded tail, 4 jointed; extremity bifid; on each division are about 5 pair of setæ.— Within the tail, there was a pulsating ~~org~~ vessel.— Eyes, 2, seated on each side of the curved beak, rounded very distinct; Inferior to & between these, are approximate antennæ; which will be described presently. Mouth in its situation is pectoral & not produced; it is obscure.— the mandibles are flat plates with 6 teeth, the 2 inner ones largest:— They ~~precisely~~ partly resemble those figured by Desmarets of Apus[2]:— Tongue oblong, rounded at extremity; Maxillæ & Palpi doubtful:

Independent of the Antennæ, there are 10 pair of articulated organs: (1st) stem simple, bifid at extremity, with bunches of setæ on each; also a small external branch with setæ.— these are situated before the mouth; &
Crust: perhaps ~~compose~~ correspond to |**123**| a second pair of Antennæ: (2nd)
Branchio: stem short, bifid at extremity with setæ; at base a globular enlargement to
Cyclops which is attached the mandible already described: Are these Palpi? — (3rd) stem short, bifid with setæ: also about half ~~way~~ way up there is an external & internal tuft of bristles.— the internal are seated on a plate, which I should imagine acted as Maxillæ.— These 2^{d} and 3^{d} are seated close together.— The organs, hitherto described have simple setæ & when

collapsed point towards the tail; the two next pair differ in both respects; the setæ are feathered [sketch in margin] & the organs act towards the
(a) mouth so as to cross the others: (4th) stem very short, broard; with numerous long feathered setæ: (5th) Agreeing with the last, but much
(b) smaller: I should think these are the Branchiæ.—

[notes on back with accompanying sketch in pencil]
(a) The setæ arise at r^{t} L^{s} to the stem:
(b) I should think not from the ~~one~~ pair similar to 4th, but setæ not feathered in Cyclops (**P 134**): there would not be so much change in so essential an organ:
[notes end]

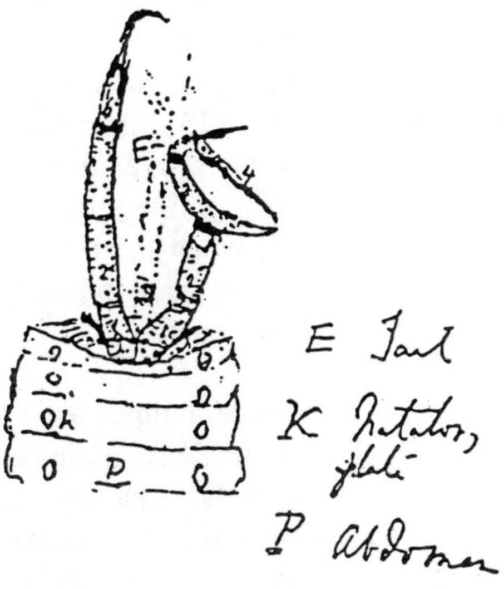

[**CD P. 123** continues]

All the organs, as yet, are seated on the three anterior grand divisions of body; the next 5 p^{r} are on the abdomen: (6th)(7th)(8th)(9th) are similar; they are natatory like in the Macrourus: Each one is jointed bifid.— the exterior branch longest, much flattened, toothed externally & ciliated internally; the other branch very much ciliated: The (10th) is most anomalous & extraordinary; I examined numbers of specimens: They are seated on the very extremity of body, beneath where the tail unites.— the two organs do not correspond in size or in function, although united at base:— |**124**| The left (I speak as with respect to man) organ is the most simple; it has 4 joints; (1st) basal one, short cylindrical, encased, with an

Crust: Branchio:

Cyclops

external tooth: (2^{d}) & 3^{d}, thick cylindrical, rather curved; 4th, finer, terminated by a very strong curved claw, lower than which is another straighter one:— The other & right organ, is of equal length & strength as the other, also the two first joints are the same; but the (3rd) differs remarkably; it is attached nearly transversely, & not by its extremity to the second.— the free end has a strong claw & smaller tooth.— to the other end the (4th) is articulated, it is curved, as is the last, & consists in a long tooth or spine; when drawn in, it reach[es> to the heel of the (3rd) so as to form an oval & hence prehensile: The animal frequently moves these organs & they retained irritability longer than any other part.— Generally the claw is retracted on the penultimate joint: (figured):— Length of these organs equals the tail or about 1/3 of the body.—

The antennæ of this animal are also extraordinary & agree in the curious circumstance of the two not corresponding: here also the left is simple; in length it equals the body; jointed, tapering, with fine setæ, colourless; the right one is thicker, crooked, coloured, strange looking; 5 jointed; (1st) basal one ¼ of whole length of antennæ, cylindrical: (2^{d}) short, much enlarged, flattened, with a long spine & tooth: 3^{d} very short, with group

Crust. Branch ~~Cyclops~~

of |**125**| short teeth & few long setæ: (4th) rather curved, with a group of short teeth & few long setæ: (5th) terminated by two strong claws & setæ: (6th), a fine joint, with setæ, behind the claws: This Antenna is projected [?] when the other is applied beneath the body.— Length of body .15, colour dark bluish green, occasionally with brown spots on the dorsal segments:—

These animals are truly pelagic: amongst them were some which agreed remarkably in almost all respects, even form of Mandibles & legs:— but differed in having 2 simple antennæ: in wanting the curious terminal organs: & in the tail, not having joints, terminated by two divisions with setæ.—

(a) What are we to infer from this?— are these most anomalous organs sexual? [note (a)] Cyclops is said in the males to have a singular antenna for clasping the females, & the generative organs lie where the curious claws in this animal are described:— [note ends] As far as regards this animals classification, in some respects it is allied to Nebalia[3] & Cyclops & in parts of mouth to Apus[2], but it is evidently distinct from every described genus:— In many respects it would come within the division of Lophyropes in which Nebalia stands; but then the flattened natatory plates seems entirely to be contrary to the general structure of those animals.

[1] Calanoida, Pontellidae, a copepod possibly of genus *Labidocera*, in which the male 5th legs are asymmetrical as shown in the drawing.
[2] See Planche 52 of *Apus cancriforme* by A.-G. Desmarest in *Dic. Sciences Naturelles*. Plates for Crustacés, Entomostracés.
[3] Nebaliacea, a malacostracan.

[**CD P. 125** continues]

Crust: Deca: <u>Notopod?</u>[1]

This crab would be a notopod; if it did not differ in the essential character of only having 5 joints, instead of seven:—

483

Crust. Dec: <u>Notopod</u>

Body, length 1/12 of inch; shape posteriorly heart shaped but anteriorly continued up in a straight line; much excised above the eyes; & between them produced forward & squarely truncate; the |**126**| anterior central part of thorax much elevated: case, thin transparent colourless:—

[date at head of page now changed to December 4th. **CD P. 126** continues]

(a)

<u>Tail</u> 2/3 length of body & 1/3 of its breadth; looks in proportion narrow.— Can be applied to the breast, ~~but does not lie close~~.— [note (a)] Having examined many specimens I have altered my opinion: the tail is applied close to the breast.— I did not see the animal alive.— I invariably found 5 pieces to the tail: I could not perceive sexual differences:— [note ends] It is composed of 5 joints; these are broarder than long & are terminated postero-laterally by a point: The extreme one is small, & has at extremity a rounded oblong simple plate.— Each joint carries a swimmer; these gradually decrease in size from the basal to the terminal ones.— The swimmer is formed of two ~~pieces~~ joints, the extreme one is a pointed oval plate, ciliated (with about 16 setæ) at extremity & internal edge.— at the ~~joint~~ articulation there is point, evidently the rudiment of a bifurcation.— The swimmers on the last piece of tail are small & but little developed:

Legs: 1st pair "en pince" ½ length of 3 following pair; 2nd, 3rd, 4th pairs equal, terminated by a strong claw, & in the Tarsus there is a single spine; 5th pair situated dorsally, when in inaction rests on the *[illeg. word]* of the other legs; slender, 2/3 of length of the others; penultimate joint (Tarsus Desmarets) ends in a point, from which arises 2 curved unequal fine bristles & near to these there is a third, which is rather shorter.— The longest equals the two foregoing joints in length.— These fine ~~spines~~ setæ are delicately (only visible with 1/10in focal d) serrated, the teeth pointing towards the base.— the curling extremity is flattened & on this part there are 5 most minute cups, which I should think acted like those in Octopus:

Crust. Deca: Notopod

|**127**| From this & the fine teeth on the three curved bristles, the leg must be able to adhere firmly to any object:—

Eyes, large, pedunculated, reaching width of body, ~~pupil~~ central part black:—

Antennæ. external ones seated behind peduncle of the eyes; straight, jointed, tapering to extremity, nearly half the length of the body: the peduncle formed of few large joints: extremity with ~~small~~ some irregular setæ.— Internal antennæ seated at base of a globular enlargement which separates them from the external: They are formed of 3 joints, extreme one large spherical, on this is a minute branch & several bunches of setæ.— the latter antennæ very short, approximate, curved:

Mouth, there was nothing particular; the external branch of the pied-machoires were very simple & they were all rather short:—

These Crabs were taken in considerable numbers (December 4th) at night, off the mouth of the bay of San Blas & several miles from any land: The structure of this animal is very curious; its pelagic habits require the high development of the caudal swimmers, & length of tail & the other points in which it agrees with the ~~Brachyures~~ Macrouri: but the formation of Dorsal legs is most remarkable: they are evidently fitted for performing their usual office of supporting the animal; but here instead of a Sponge, perhaps a Medusa; hence the change of structure:— This inclines me to think this is a new division amongst the Notopodes:— |**128**|

(Crust: Branchiopod? Latreille)

The description of this Zoea can be divided into two parts: the animal & its singularly shaped case or "carapace".—

Zoea[2] 486

Case oval, anteriorly ending in a very long pointed spear, which is serrated in a direction from the body: on the lower & posterior parts it is widely open, & from each side spears project.— The two are close together, & are in same straight line as the anterior one: they are shorter & are serrated from the case; so that the teeth on ~~both~~ anterior & posterior spears point from each other: The length of large specimen from extremity to extremity is .6;— of which the case is 1; & the posterior spears nearly .2; the anterior one being much the longest, rather more than .3.— Case, transparent, elastic, colourless: The head part of animal is intimately united with this case, but the tail & thorax (thorax known by supporting legs) is free; the tail can scarcely be retracted in case:—

Body: Eyes pedunculated.— 2 pair of antennæ seated beneath them & on same line; these are large but imperfectly formed, for size animal: the internal ones are divided, with setæ on the larger: external ones rather longer, simple, divided, with fine branch coming off low, cylindrical, pointed: these antennæ project straight forward: [note (a)] Reexamined the
(a) antennæ: the internal ones are anterior to the external, & the former are divided at the summit.— the outer branch thickest, ciliated on inside.— the inner is merely a point.— The external antennæ are bifid, the division being low down.— the outer branch shorter & much finer.— both quite simple, pointed.— [note ends] The mandible is attached close to the base & within the external antenna; it is of some thickness; toothed & one large one in the corner.— [note (b)] Also the mandibles.— they are very large for body.— the plate is curved & truncate obliquely.— the large tooth is
(b) at the upper end.— the base or *[illeg.]* gradually narrows in a point, with lateral smaller one:— [note ends] to the side of the mandible |**129**| the
Zoea palpi adhære.— these are very fine short, but with two long setæ at extremity:

The Labium is horse-shoe shaped, with each end rounded & ciliated, lamellar & coloured pink: On each side & close before it.— are 2 pair of organs, answering to its "Machoires": the first one is smaller & more simple, it is composed of three divisions, 2 square lamellar with bundle of setæ & one cylindrical; (this ~~second~~ would be more accurately said, if divided into two primarily, & one of them bifid): the second is also divided into two branches.— the larger one is divided at summit into 5 square, unequal spaces, each with bunch of setæ.— These organ[s] would close the mouth.— When the animal is at rest these are kept in a most rapid vibratory motion.— To these succeed 2 pair of large branched organs: answering 1st & 2nd pied machoires (or 3rd & 4th of Desmarets).— All four are similar; & nearly equal in length to the body.— on the basal cylindrical, so as to be bifid, joint are two equal branches, with setæ; external division has two joints (by joints I mean limbs or pieces & not its articulations), the internal 5 smaller ones:— At the base of ~~these~~ each there is ~~pair of~~ a very small organ, answering to external pied-machoire. they are bifid; the division being low down: the interior one is very fine, jointed, with setæ; the external simply pointed: Close to these come 5 pair of organs, very small & of a most rudimentary structure, |**130**| they are seated in a bunch
Zoea together: the first pair terminates "en pince", of an imperfect structure: the 2, 3, 4, 5, are equal, are cylindrical, curved, jointed & terminating in point.— These organs can be of no use in locomotion — There was no greater distance between pied-machoire & a leg, than between the bases of two of the latter.— Each of the next 4 joints of the abdomen.— has a pair of cylindrical points.— rudiments of swimmers: these caudal joints are

square (angles of course removed).— the next terminal joint has true spine & a large swimming plate at extremity ~~at extremity~~; in shape it is wedge-shaped, base highly convex.— [see sketch in margin] on the convex edge there are 13 long feathers bristles, but the central & 2 extreme ones are short.— The abdomen (as far as I could see) is composed of 7 pieces.— the one joining body & the 2nd support the pieds machoires & legs: the 3^{d}, 4, 5, 6th the swimmers & the 7th the tail.— On the inner surface of this is the anal orifice:—

These Crust: were found in great numbers at night at San Blas: There were specimens rather larger, & many much smaller.— in the latter the spears were flexible & case more globular & legs even more rudimentary: These animals could swim easily & looked most singular: For opinion about Zoea, V next animal: I have copied order of description from M: Edwards[3] — Dic. Class:— |**131**|

[1] Not identified.
[2] As CD was beginning to recognise, zoea is in fact a distinctive larval stage of large crustaceans such as crabs and shrimps; but this one is not identifiable from his description.
[3] See article on Zoé by H.-M. Edwards in *Dic. Class.* **16**:719-22.

[**CD P. 131** commences]

Zoea[1] & Erichthus

485

Found with the Zoea, just described; another differing in some respect; but in the important organs essentially the same.— Size nearly equal, but more globular.— & the spears not so long & not serrated: only one posterior one & not in same line as the anterior.— 2 short lateral ones.— Antennæ, mandible, machoires, nearly the same as last Zoea.— but 1st pair of pied machoire has only one joint in the external branch, in length equal.— 2^{d} pied machoire has internal jointed branch shorter than external.— the 3^{d} pied machoire & 5 legs closely agree with those of last Zoea.— There are 5 pair of short cylinders, or rudimentary swimmers.— Tail is spinose & its outline is concave instead of convex.— By reading over the description of the former Zoea & that given in Dic Class.— it will be seen how closely this one agrees with the one described by M. Edwards: The Swimmers here are rudiments instead of oval plates; & M. Edwards does not mention the ~~division~~ branch in the 3^{d} or external pied machoire.— Analogy would ~~point out~~ lead to the expectation of this, as the 1st & 2nd have the division so strongly marked.— and yet [it] is unlikely M. Edwards should have overlooked it.— I think it ~~probab~~ certain, whatever Zoea may be.— my two & the one in Dic Class must belong to the same ~~order~~ family of Crustaceæ: |**132**|

Zoea &

Amongst these Zoeas there was a single specimen of an Erichthus, which

Erichthus[2] appeared young & was imperfect in some respects.— The plate of the external antennæ was the only part developed, & the branch consisted solely
485 of a projecting point: also the third pied machoire & 1st leg (the 3 & 4th pied-machoire of Desmarets) terminated without a claw, but the last joint was rather enlarged.— In Erichthus (**P 88**) these limbs have a claw.— At the base of anterior pied-machoire were respiratory plates:— Before finding this specimen, I had thought these Zoeas perhaps belonged to the Stomapodes.— The close approximation of pieds-machoires & legs & these being placed on different segment of body from the head.— leads to this opinion.— Also, by considering Erichthus, the curious case of Zoea will require less change to resemble it, than any other crustaceous animal.— Upon seeing however the gradual change in case between the minute globular Zoeas & this Erichthus, I have no doubt but what this Zoea belongs to an Erichthus.— In confirmation of this it may be remembered that the two pair of most developed organs in Zoea become in Erichthus the
(a) 2 principal pied-machoire.— also that the two next, viz 3d pied-machoire & 1st leg, which have claws in Erichthus, are also more organized in Zoea; the 4 other pair in both animals are equally rudimentary.— [note (a)] If Zoea should be proved to be the Larva of a Stomapod, it would be curious to see the relation between this order & the Decapods, more clearly marked by the structure of the legs in the young than in the perfected animal [note
Zoea & Erichthus ends] |**133**| Again, shape of head, tail & especially terminal plate & spines are not very dissimilar & the resemblance of the 'carapace' has been shown.—

M. Edwards states Zoea has a double thoracic cavity something like the Decapod Brachyures: [insertion with different pen in margin] Branchiæ not existing in the two last thoracic segments [insertion ends] but from the greater similarity to the Macroures, he overlooks this.— In Erichthus the respiratory plates are seated at base of pieds-machoire, hence in anterior portion of thorax.— Is it not possible that these in Zoea were included in cavities?— From these considerations I imagine such Zoea, as mentioned
(b) in Dic Class & here, are young of that division of Stomapod in which Erichthus & Alima are: [note (b)] Of course I do not mean to say but what other animals which would come under the wide characters of Zoea, may be as Mr Thompson[3] states the young of Pagurus.— NB. it is odd, if so, that they should be pelagic:— [note ends] There is no reason to be surprised at the number of Zoea, as at **P. 88** the Erichthus was found in
(a) great numbers.— Not finding some of them more advanced is the most solid objection.— [note (a)] This particularly applies to the former Zoea of **P (128)** [note ends] perhaps like other Crustaceæ they retire during any changes of their cases. It has been remarked that Squilla[4] has never been found with eggs.— now if the young are pelagic Zoea, this would be accounted for.— M. Risso[5] supposes they go to deep water & sandy bottom.— These Zoeas were found in 7 fathom water & in sandy bottom — off the Bay of St. Blas.— |**134**|

[1,2] These are different stages in the larval development of stomatopods that are sometimes not readily distinguishable.
[3] According to J.V. Thompson *Zoological researches and illustrations*, Vol. 1, pp. 1-11, Cork, 1828, *Zoea taurus* is the young of *Cancer pagurus*.
[4] Stomatopoda, mantis shrimp.
[5] See M. Risso, possibly in *Histoire naturelle des Crustacés des environs de Nice*, published 1816.

[**CD P. 134** commences with an entry dated 7th December]

Cyclops[1] Body pointed, oval, colourless or faint red, integuments soft, length 1/20th of inch: composed of 6 segments, anterior one bearing organs analagous to
488 pied-machoire.— the four next the Natatory plates & 6 the tail.— Tail, very narrow cylindrical, 3 joints.— with a pair on the 3rd.— terminated by setæ: Anterior antennæ seated under extremity of body, much longer than the extremity of tail; tapering with numerous joints, extremity with scattered very long spines growing at rt angles to antennæ.— 2d (articulated organ).— seated before mouth, bifid, inner branch with fewer joints & setæ: 3d.— in line of mouth, close to base of mandibles, bifid with setæ.— short: 4th bifid, extremities with rounded ciliated plates.— & between them there arise[s] a trifid branch with setæ:— 5th base broard but short, with bunches of longer & more setæ than cross those of the foregoing organs:
(a) 6th: cylindrical, 3 joints, twice as long as the former ~~natatory~~ organs.— [note (a)] All these organs are in 2 straight lines on the thorax: [note ends] After these, on distinct abdominal segments, are 4 pair of swimmers; each one is bifid, flattened; outer plate broarder & longer:— These are the true legs; as for the other organs.— I suppose 2d pair are antennæ, otherwise there would be 5 pair corresponding to pied machoires.— The 2d, 3d, 4th, 5th pair generally correspond in structure & relative position to those described in an Entomostr. **P 122**: the 6th differs.—

Mandible not much curved, short; with large tooth at upper corner: the teeth gradually decrease in size from this to the other corner & base or fang.
Cyclops broarder.— on the inner side of both |**135**| the mandible, there was a most minute cylindrical organ, ½ the length of mandible, truncate at extremity, flexible.— I must certainly consider these contrary to Desmarets as Palpi:— Eye, very minute, dark red, within small transparent ball & seated between anterior antennæ:—

These minute Crustaceæ move by a jumping motion.— they were found Lat 40°S at a distance from land.— Depth 45 fathoms: truly pelagic:— Cyclops seems generally to be a fresh-water animal:—

[**CD P. 135** continues]

<u>Clytia</u>[2] or (a) <u>Campanularia</u> 2nd species 489

Coralline growing in short much branched tufts; branches irregular in shape, crooked, short; the articulations (or globular impressions V. Clytia **P 103**) very obscure.— [note (a)] Tufts scarcely an inch long: whereas in Clytia **P 103** the masses of coralline were <u>many</u> inches [note ends] Terminal cup bell shaped on a short peduncle, with the articulations obscure.— The whole Coralline is shorter in proportion, & characters not so much marked as in Clytia **P 103**:— Polype with body cylindrical on narrow base (as at **P 103**), mouth tubular, highly expansible, projecting.— Tentacula 28 in number, seated on outer rim of polype: every alternate one hangs down.— so that they appear in double row of 14 each.— (b) [note (b)] They are occasionally in a single row round the mouth:— The numbers of tentacula & alternate manner of arrangement best separates this species from that of **P 103**.— in the latter I put 16?: it is <u>quite</u> impossible that there could be a mistake between 16 & 28.— [note ends] Tentacula soft, formed of concentric layers of pulpy matter, hence semi-opake.— The central living mass is included in delicate case distinct from the outer horny one: itself consists in a central mass, distinguishable by its colour from the outer, & which communicates with each Polype.— I only saw unperfected ovarium, which resembled that of next species: for locality &c &c V next article:— |**136**|

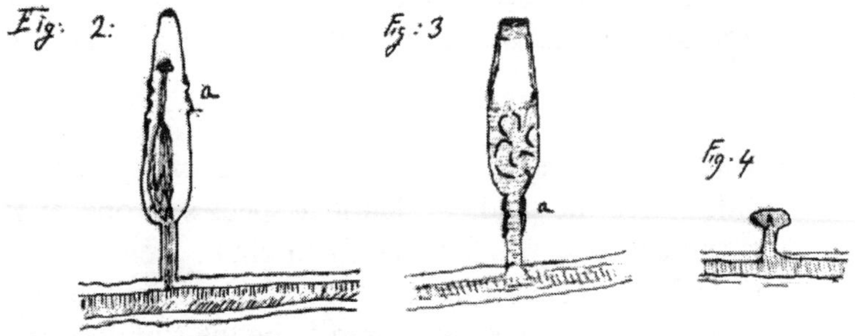

Plate 7, Figs. 2-4

[**CD P. 136** commences]

<u>Clytia</u>[3] the 3rd species 489

Stems creeping on a Fucus: 2 or 3 generally in parallel lines, for several inches.— central living matter the same as in last species.— From the stem, branches are sent off perpendicularly; length about .3: each of these is terminated by polype: Beneath the cup, there are obscure articulations as in former species: cup bell shaped, truncate obliquely, one side rather enlarged.— body of polype globular, not uniting with the living matter in centre; so that the greater part of body lies in the enlarged half of cup; from this cause also mouth & arms of polype are protruded in a slightly lateral direction.— Tentacula & mouth same as in last.— On the creeping stems there are also branches;— shorter & in shape a much elongated cone

& a peduncle.— These are of two sorts, viz Ovariums & Buds or young polype.— In the former case, they are sometimes truncate: Vide Pl 7. Fig 3: in the first state they are full of white pulp matter: this by degrees shrinks & is divided by reddish lines into rudimentary balls; the summit of ovarium being closed by an opake mass, which communicates by lateral vessels with the lower: in this state Fig 3 is drawn:— as maturity advances the upper mass is absorbed; the ovarium is seen to consist of a double case (as might be expected from nature of the integument of stem) open at
Clytia summit: & at bottom there lies |**137**| 5 globular eggs, enveloped in a viscous fluid; in colour white, diameter about 1/100.— Those which I found already expelled were smaller in size & of a darker colour.—

~~The buds or young~~ At the base of the ovarium were the globular impressions or articulations:— The buds or young polype were in the structures of their external cases, very similar to the Ovaria.— they are originally filled with pulpy matter, & I should think it was at this period fixed whether the young branch should ~~turn out~~ bear eggs or a polype.— The two sorts were generally ~~together~~ in distinct places & in groups.— The bud when half matured presented the appearance figured at Fig 2: Above the middle of the cone there were marks of the globular impression: at this place the included matter was contracted into a narrow stem & surmounted by a head: beneath this also the central matter was much shrunk.— I imagine this process continued till the regular branch is produced.— When first seen on the stem, these organs are irregular balls on a peduncle, Fig. 4:—

These 2 species were taken on Fucus picked up at sea, Lat 45° S. many miles from the land.— These 2 species & one of **P 103** evidently belong to same genus: which certainly might be included in Cuviers Campanularia[4] if such characters did not deserve a distinct genus.— When examining
(a) these corallines, they appeared to be the simplest of the Polype or Polypier, & most [continued on back at (a)] allied to the naked ones.— the central living mass is so much developed compared to the thin horny, transparent, & simply constructed envelope [entry ends] |**138**|

[1] Calanoida, pelagic copepod.
[2] Leptothecata, hydroid in family Campanulariidae.
[3] Another hydroid.
[4] See *Cuvier* Vol. 3, p. 300.

[**CD P. 138** commences]

Butterflies in grt flock December 4th.— About 10 miles off the Bay of San Blas, in the evening, the infinite numbers of Lepidoptera formed a most curious spectacle: They

were of various species, but chiefly a yellow sort.— with them were some
870...872 (not spirits) moths & Hymenoptera.— & even a Calosoma[1] [?] flew ~~with~~ on board.— The men all cried out "it is snowing butterflies"; at a distance it had this appearance.— the butterflies were in bands or flocks of countless myriads, & as far as the telescope reached, they might be seen fluttering over the water.— This took place in the evening.— the morning had been calm & the day before very light variable winds.— it is clear these insects had voluntarily come out to sea.— it was the last day for most of them, for a strong breeze sprung up from the North, which must have destroyed the greater number.— How are we to account for these flights, which others have also observed? Is it an instinct implanted in the animal to find new countries, its own one being overstocked by a particularly favourable year.—

Crust: amphipod Uropteres[2] 492

Abdomen composed of four cylindrical pieces, the last bears tail.— the other three the swimmers.— Tail formed of 6 pieces or 3 pair.— ~~central~~ in shape are flat, spear-shaped, pointed, sending off a small pointed external plate.— they have two articulations.— the central pair are seated more a posteriori than the others, but are of equal length in themselves.— the external pair are narrower than the others.— These organs when expanded form |**139**| a fan & are most essential to the animal in swimming.— 3 pair of swimming plates, these are bifid.— divisions equal with many joints.— Body with 7 segments, & lateral plates by the base of legs, coloured with stars of purple: Eyes exceedingly large; forming the whole anterior part of head.— transparent, containing an oblong opake part.— of fine purple colour.—

Crust. Amph. Uropteres

Antennæ, superior ones rather more approximate than the inferior, these latter are very fine, taper to a point & equal body in length.— the superior are much shorter, but the peduncle longer.—

Legs.— 1st pair (intermediate or 4th pied machoire) small simple with claw; 2nd at the base of the penultimate joint before the claw, there is a branch sent off with internal spines, hence closely resembles a true "pince".— (but evidently really is the palpus of pied machoire): 3rd & 4th equal & longer than than the last.— they have the penultimate joint very broard & flattened in order to receive en griffe the ultimate joint & claw: 5th pair strong, nearly twice as long as any other limb; the elongation takes place in the ultimate & penultimate joint, terminal claw small: 6th & 7th equal, strong, ending in claw:

Mouth: 3^{d} or external pied-machoire is composed of an open, hard, pointed fork.— with small internal plate: 2^{d} pair has a <u>small</u> & similar fork with setæ & larger concave plate: 3^{d} consists in two circular fringed plates: Mandible, with large distinct palpi; they are of singular shape, upper part a narrow plate with teeth (as usual), this plate is folded back, though not

Crust. Amph. Uropteres

parallel, |**140**| & forms as it were another interior & inferior mandible.— the edge is square & only ciliated: the palpi arise at the bend.— Labium bilobed.— pointed oval divisions.—

With these specimens ~~which did~~ were others which differed in the following respects.— Superior antennæ very short, curved, rudimentary: inferior also short straight pointed.— the external plates in tail were broarder: the legs varied in proportion.— parts of mouth the same.— in general appearance &c &c evidently identical.— These I have no doubt are the young ones.— If this Crust. belongs to the Uropteres, it is a new genus.— In its habits it is truly pelagic, occurring in deep, at great distance from land.— for several degrees.— North of St^{s}. of Magellan.— Swims fast by starts, rapid, in circles & back first.— uses its tail much.—

[**CD P. 140** continues with an entry on Planaria that has been crossed through vertically to indicate its publication elsewhere]

Planaria[3]

Omit (?)

Planaria

Lat:53°S

Body when crawling nearly ½ inch long; shape oval; very flat, edge thin: Beneath from the anterior extremity to beyond the half of length, within the body there is a white wedge-shaped mark.— Within this, one near the head & the other middle of body, are two minute circular apertures; which the animal occasionally opens & contracts.— Their situation can easily be seen from a white halo which extends round them.— Beyond the white space there is a third aperture.— this is very large (visible to the naked eye) & has folding lips: is highly dilatable.— from this within body |**141**| there runs two white lines.— Back coloured elegantly.— in centre is longitudinal band of "vermilion red".— this anteriorly sends off 2 pair of branches & terminates in three.— it is edged with white.— the rest of back is covered by dots of a purplish red.— At the point where the central band is trifurcate are two longitudinal groups of black spots.— also on anterior margin there are two groups of numerous eyes or black dots.— Animal in its habits, inactive.— found on Corallines, 30 Fathom water.— South of St^{s} of Magellan, Dec 15th.

[1] See *Insect Notes* p. 66.
[2] An amphipod of suborder Hyperiidea.
[3] Listed by CD as *Planaria (?) formosa* in *Collected papers* **1**:189, but later placed in sub-order Acotylea as *Leptoplana formosa*. If, however, CD's third aperture was in fact a sucker, it might better be placed among the cotyleans.

[**CD P. 141** continues]

Fistularia[1]
495

Habitat same as last animal.— Body: cylindrical, with thin cuticle of beautiful "Vermilion red": tapering towards both extremities but mostly

towards the Anus: length when crawling about .3 inches: body very soft:
(b) with obscure papillæ or <u>little</u> eminences chiefly on the tail: this latter part is also most strongly marked with transverse wrinkles: animal often irregularly dilates its body with water, but chiefly in posterior half.— On under side ~~there~~ is a linear space, more smooth than the rest, on this the animal generally rests.— Anus at very extremity circular.— Anterior extremity truncate, mouth in centre.— outer rim surrounded by 12 tentacula.— These & the mouth can be withdraw[n] in body:— Tentacula consists of cylindrical peduncle, bearing a disc or hand, from which about
(a) 14 fingers or papillæ diverge; central one longest.— at the base of these papillæ there is a connecting membrane: [note (a)] These little papillæ varied in number from 13 to 15.— [note ends] |**142**|

[note (b) added later] The animal may be called absolutely smooth, from examining a Holuthuria (January). I see what is meant by Papillæ.— The entire absence of true papillæ, would according to Cuvier rank this animal amongst "Echinodermes sans pieds".—
1834 June Port Famine.— Saw with strong power that on the surface, chiefly in posterior half of body, there were <u>many</u> little cups adhæring.— similar to those described at **P 261**.— With this difference, that each one was separate & not as these collected on a little eminence.— [note ends]

[**CD P. 142** commences with entry dated Decr 15th]

<u>Fistularia</u> The convex side of disc is turned inwards, hence the concave & connecting membranes of papillæ form a powerful sucking instrument, by which the animal can adhære firmly to glass.— In moving, a wave-like motion from the tail extends up the body & then the adhæsion of the tentacula allows the body to contract & ~~then~~ the process is repeated.— These tentacula perform another essential office, the alternate ones are widely extended & then drawn backwards so as to cover the mouth.— this goes on so steadily that it is difficult clearly to see the mouth: The animal voided great quantities of sand in their excrement & doubtless this is obtained by the action of the tentacula: they were found at the roots of Corallines, where the sand would probably contain nutritious matter.— At the base & within the Tentacula there is a fine bony collar: it is formed of 12 pieces, each of which is cylindrical, with a salient external angle.— hence the collar has a slight Zig Zag appearance: This species would appear to be closely allied to Fistu: digitata (Lamarck):—

Corall[2] <u>Celleporaria</u>? 889 (not spirits) Corall.— much branched, stony, fragile, colour "honey yellow".— 2 or 3 inches high; branches cylindrical, rather globular at extremities: surface covered with punctures & waved lines.— Transverse section composed of irregular tubes or cells, rather hollow in centre.— Cells not projecting placed irregularly ~~over~~ within branches.— aperture circular, lower lip rather prominent & at the |**143**| summit of branches ending even in a point: so

Celleporaria?

that these parts are rough with points: Polype with 16 delicate tentacula situated on a long tubular body, which is enclosed in transparent case.— This latter is protrudable & rather bell shaped, but contracted at orifice (as described in Corall **P 77**).— Found growing in 30 fathom water. Lat 53° S. Dec 15th: It is allied to that described **P 77**.— I am ignorant whether it exactly agrees with any described genus.— perhaps Celleporaria, Lamouroux.

Corall[3]
Celleporaria?

888
(not spirits)

Habitat & many characters agreeing with the last: Corall. with branches rather longer; centre more compact: colour pale "scarlet red", surface of branches granular.— & covered on every side by small projecting hoods; or they may be described as projecting slightly curved tubes, divided anteriorly, contracted at summit.— Scattered irregularly at the base of these are circular apertures for the Polype.— These hoods correspond to the truncate cones of the Corall **P 77**.— The branches ~~of~~ are essentially composed of these hoods.— so that looking down vertically on summit of young branch, a circle of these hoods are applied with their back towards the centre: [sketch in margin] : & there is no orifice for cell at the summit.— The cells seem to be in the central space when the branches have increased sufficiently in diameter.— These hoods are so numerous near tops of branches as to be imbricate:— |**144**|

Favosites[4]
(a)
890
(not spirits)

Habitat same as last: Corall, stony, hard, strong, white coloured: growing in very short vertical curved thick plates: ~~short~~, height about ½ an inch, breadth of plates varying from 1/10" to 1/20".— sides smooth, most finely punctured.— Extremities truncated, slightly convex.— entirely composed of the orifices of cells.— these are of different sizes; properly hexagons, becoming however circular.— a little way within each ~~cell~~ orifice is a plate with small aperture, which leads into cell of polype. This Corall appears to be a Favosites of Lamouroux:

2010
not spirits

[note (a) added later] May 19th 1834. Procured specimen, 48 fathom: the plate within orifice of cell is a mistake (probably the Polypus itself: [)] cells not being truly hexagons, there are spaces between tubes, sides of interwall tubes perforated with puncture, but more especially the external ones. Branches entirely composed of these hexagonal tubes: the plate-like masses of tubes spring & branch from a short stem.— [note ends]

[**CD P. 144** continues]

Corall[5]
? ?

892
(not spirits)

Habitat same as last: Corall, much branched, about 2 inches high, white: branches flattened, on one side they are rugose, with ribbed lines running lengthways.— on the other are the orifices of cells — these are placed irregularly & consist in short tubes truncate obliquely: these project also laterally from branches: the termination of branches [is] rather wider & consists of an aggregation of angular tubes, generally hexagons ~~& in~~ but not orifices of polypeferous cells: Corall elegant, very strong. For remarks about its classification see next Corall, which is of same nature:

Corall[6]
? ?
892
(not spirits)

Habitat same as last.— Corall. Much branched about 2 inches high, white, brittle: Branches with one side punctured & with longitudinal lines.— the other long curved punctured tubes, which are the orifices of cells.— These are placed most symetrically on the branches, in parallel oblique rows & tubes equidistant: [sketch in margin] : These tubes project |**145**| laterally so as to give a toothed appearance to the sides of branches.— Extremities rather wider & composed of numerous angular orifices of tubes, generally hexagons.— & out of these the regular projecting tubes are formed: the oblique line might be perceived amongst them: reminding one of the formation of vessels in the cellular system of Animals!— I do not think this corall agrees in its characters with any genus.— Catenepora is described as composed of parallel tubes, arising through plates anastomizing in net work: This would appear to bear an analogy to the formation of the present Corall:—

All these 5 specimens of Coralls were taken by swabbing the bottom; hence rather injured & Polype would not show themselves:—

Cryptogamic Plant[7]
503
&
980 (not spirits)

In general habit resembling a moss.— colour pale green.— peduncle of capsule transparent, colourless.— capsule oval, dark brown, tough.— containing an infinite number of globular, light brown sporule[s].— diameter 1/2000 of inch: with these were bits of fibres, resembling necklace (each bead being about ¼ of the sporule). I should think these acted as placenta to the sporules.— Capsule opens into four longitudinal pieces, which curl backwards.— When placed in Alcohol no action, but the specimen was not fresh.— The immature capsules, when first bursting from sheath, appear involved in gelatinous matter: Grow in tufts in wet places. Near a cascade, in mountainous woods. Hermit Isle Decr 25th.— |**146**|

[1] With its fifteen digitate tentacles and no regular podia, this animal might be a holothuroid of order Molpadiida, or belong to order Apodida, family Chiridotidae, possibly *Taeniogyrus contortus* Ludwig. Lamarck's *Fistularia* is the European apodid *Labidoplax digitata* in family Synaptidae.

[2] *Celleporaria* is a bryozoan of suborder Ascophora.

[3] Specimen 888 (not in spirits) was identified by George Busk as the ascophorans *Adeonella atlantica* and *A. fuegensis* now preserved in the Busk Collection at the Natural History Museum.

[4] See *Lamouroux* p. 66. Specimen 2010 (dry) is listed as *Fasciculipora ramosa* 1875.5.29.58 in the George Busk Collection.

[5] More bryozoans.

[6] do.

[7] This is an unidentified liverwort, held in the Cambridge University Herbarium. See *Plant*

Notes pp. 166-7.

[**CD P. 146** commences]

Fish[1] 515 (a) Myxinus — (Cyclostomes ?) Caught by hook amongst the Kelp, Goree Sound & other parts of Tierra del F.— Above coloured like an earth worm but more leaden; beneath yellowish & head purplish: very vivacious & retained its life for a long period: had great powers of twisting itself & could swim tail first: when irritated struck at any object with its teeth, & by ~~opening~~ protruding them, in its manner much resembled an adder striking with its fangs.— Head most curiously ornamented with tentacula: Vomited up a Sipunculus when caught:— [note (a)] This fish is abundant amongst the rocky islets, having found one on the beach nearly dead.— I observed a milky fluid transuding through the row of lateral pores or orifices:— It would appear to be Myxinus ~~with no lateral branchial orifices~~.— [note ends]

Coralline[2] 512 Clytia — Coralline, transparent, colourless, delicate & most elegant.— Stem short erect with simple alternate branches; stem jointed, each joint bearing a branch.— Branches with simple small terminal cups, ~~also~~ as likewise on the upper surface at regular distances. in these latter the cup is applied to the branch or rather the branch passes through it.— so as to resemble the cell of Sertularia: Internal semi opake vital matter not filling up the transparent case.— Polype with long body, not retractile within ~~cell~~ cup; mouth broard with no fine tentacula around it.— This very beautiful little coralline from its general habits & structure is allied to the Clytias **P 145** &c &c.— Growing on Fucus, 6 fathom water, Goree Sound:— |**147**|

[1] Identified by Leonard Jenyns in *Zoology* **4**:159 as *Myxine australis* Jen.
[2] Leptothecata, a hydroid in the family Campanulariidae.

[**CD P. 147** commences]

Edible Fungi ~~Excrescences esculent~~ 528 (a) — In the Beech forests, the trees are much diseased: on the rough excrescences vast numbers of yellow balls grow.— These are of the colour of yolk of an egg.— & vary in size from a bullet to a small apple.— in shape globular, but a little produced towards the footstalk or point of attachment. They grow both on the branches & stems in groups.— When young.— they contain much fluid & are tasteless, but in their older & altered state they form a very essential article of food for the Fuegians.— The boys collect them, & they are eaten ~~raw~~ uncooked with the fish.— When we were in Good Success Bay in December, they were then young.— in this state, externally they are quite smooth, turgid & of a bright color, & with no internal cavity.— ~~Upon keeping are~~ The external surface was marked with white spaces, as of a membrane covering a cell (in this state, but rather more advanced, the specimens 528 are).— Upon keeping one in a drawer

my attention was called after some interval by finding it become nearly dry.— the whole surface honeycombed by regular cells & possessed of the decided smell of a Fungus.— & with a slightly sweet mucous taste:— In this state I have found them during Jan: & Feb over the whole country (with the exception of specimens 528, which were found in Feb, high amongst the mountains).— Upon cutting one into two |**148**| halves.— the
Esculent excrescences centre part is found partly hollow, & filled with brown cellular fibrous matter.— this evidently merely acts as a support for the elastic semitransparent ligamentous substance which forms the base & sides of the external cells.— The development of these cells would appear to be [the] main end to which the growth tends: It is however especially to be noted I cut open great numbers & scarcely ever found the central cellular part without one or more larvæ of the same sort.— In the young state I unfortunately neglected to examine them.— Now I am in doubt whether it is an excrescence formed for the nourishment of some insect or a true cryptogamic plant[1].— The very general occurrence of the Larvæ may be explained by observing how universally Larvæ occur in the Boleti in England: Some of these balls remain on the trees nearly the whole year. Capt. FitzRoy has seen them in June.— but great numbers fall on the ground.—

1834 [note (a) and another added later] Feb. Port Famine. Color "ochre yellow & dutch orange" of the Wernarian nomenclature. when young, or central part soft & *[illeg.]*, strong fungus smell, & sweet taste.— no larvæ.— From the root a hollow vessel passes to the centre, from which white ligamentous rays pass through the semi-gelatinous mass to the bottoms of the cells.— I can have no doubt it is a Crypt: plant.—

1834 June

Found some more very turgid ones, highly elastic; a section of the central parts white: the whole under a high power looks like a Vermicelli pudding from the number of small thread like cylinders.— at about 1/20 of inch from
984 exterior surface, there were placed at regular intervals small cup shaped balls 1/12th in diameter, of a bright "dutch orange".— the cup was filled with adhæsive, elastic, colourless, quite transparent matter (hence at first appeared hollow).— the upper edge of cup was divided into conical points about 10 or 12 in number [see sketch in margin], & these terminated in an irregular bunch of the above threads; the cup was easily detached from surrounding white substance excepting at its fringed superior edge.— Right over the cup there was a slight pit in the exterior surface: ~~Which~~ This afterward became an external orifice to the cup (where the gelatinous matter perhaps has formed seeds(?))— Some of the balls were attacked by Larvæ, but their entirely irregular course showed that they had no connection with the structure.— [notes end]

[**CD P. 148** continues]

Fuegian Paints 974 (not spirits) (a)

The Fuegians paint their faces, bodies & hair with white, red & black in various figures & quantities. The red is the oxide of Iron & is prepared by being collected near the streams, dried & burnt. The White is of a more curious nature — in the state fit for use it is of very little specific gravity.— it is collected from under water, is made into balls (as J Button expressed it, 'all the same Ostrichs egg') & burnt: did not effervesce with acids.— & with bit of cobalt gave a permanent |**149**| blue.— I suppose therefore it [is] nearly pure alumina.— It occurs in the Slate Mountain, I imagine from the decomposition of the beds of Feldspathic rock.— The black I have not obtained: the black is I believe only charcoal & oil:— [note (a)] I found some of the feldspathic greenstones decomposed into a white substance to the depth of 3/10 of inch.— [note ends]

Heteromerous insects[2] 1021...24 (not spirits) copied

The habitat of these insects was the most singular I ever observed: it was in the fissures of slate rock & in which the genus Capulus [Limpet] was adhæring to the stone alive, & therefore of course beneath high water mark.— from the wet condition of the insects & their inactivity I do not believe they remove themselves.— There would appear to be two sorts, or in different states of maturity.— from the soft state of some specimens, the larva must have undergone its metamorphosis in this site:—

[1] The idea that an insect was responsible for the existence of the edible excrescences was quickly abandoned by CD. As explained in *Plant Notes* pp. 221-4, the fungus was duly classified and named *Cyttaria darwinii* by M.J. Berkeley in 1845.

[2] The insects in question were the coleoptera listed in *Insect Notes* p. 71, and the specimens are now in the British Museum.

[**CD P. 149** continues]

Gasteropod[1] 559 (b) (a)

March 7th.— Falkland Islands: As far as I was able to observe without dissecting the specimen, ~~this would appear to be a curious animal~~.— Mantle orbicular, much convex, bordering ~~over~~ the foot on all sides: it evidently contains within it a *[illeg.]* much developed shell.— [note (a)] The ~~right~~ left side of the mantle is largest: [note ends] On the anterior surface, near to margin, there is a projecting tubular orifice, formed by the division & overlapping of the mantle.— (perhaps would be better described as anterior part of mantle echancrè [échancré = hollowed out]; but in its action it is a perfect tube). this conducts to a large cavity, lying behind the head & extending down a short way the right side: it is open, as in Crepidula for instance |**150**| ~~for its whole length~~.— At the bottom of the cavity there is (I think) 3 rows of tapering, simple, white branchial fillets: on the right side within the cavity, there is (I think) anal orifice: Pænis very large, lying in the Branchial cavity, curved, flattened, & tapering.— seated behind the right antenna.— Foot oblong.— anterior margin truncate, scalloped or grooved with the corners recurved like horns: Head

flattened, long, extensible, in front square, with antennæ on each side & eyes at the exterior base.— Mouth seated between the foot & head, longitudinally folded: Antennæ short, simple, cylindrical: Mantle covered with pointed papillæ: diameter .6: colour pale yellowish, with marks of flesh colour: in centre an irregular oblong mark of dark brown from which are sent off a reticulated vein & *[illeg.]* of same color:— Has power of considerable adhæsion to a smooth surface: can roll itself into a ball: was
(a) found at low water mark under a stone:— This animal would appear to belong to Pectinibrands; although most probably to Tectibrands: [note (a)] ~~It would appear to have an intimate relation with the family of Capubridis: thinking the Branchiæ are arranged in several rows is a more wide difference than the Shell being internal instead of external.~~ [note ends]

March 25th
584
[note (b) added later] Colour uniform "orpiment orange", with "vermilion red" brighter in regular spots.— length 10½ inch: foot larger, anterior part with not so large lateral horns:— head forked in front, antennæ more approximate.— body very convex & smooth.— I think it is a distinct species.

These animals are closely allied to Sigaretus, perhaps differ in spite of shell not being so lateral.— Shell highly developed, spiral.— sexes distinct: Branchiæ obliquely transverse, basal row with long fillets; the two superior rows with minute fillets:— I could not clearly see anus.— [note ends]

[**CD P. 150** continues]

<u>Gasteropod</u>
570
583
March 9th. Caught two specimens of same genus: but I think different species: Habitat &c same: Body rather more oblong: length one inch: colour dirty pale yellow, thickly clouded & veined with purplish brown: surface smooth with <u>few small</u> papillæ.— This seems the most specific difference:— Also another smaller; dirty yellow with dark brown dots, surrounded by a halo of light brown: Are these Species or Varieties? |**151**|

[1] Mesogastropoda, Naticidae, probably *Sinum*, moon snails.

[**CD P. 151** commences]

<u>Doris</u>[1] March 7th. Under large stones, East Falkland Island.— Shape elongated oval; length 3¼ inches; breadth 1 & ½; flattened; mantle much projecting over foot & covering head & tail; colour <u>uniform white</u>, with a faint tinge of yellow; Surface, smooth to the touch, but thickly ~~were~~ studded with minute cylindrical papillæ.— This gives a fine fimbriated appearance to branchial orifice:— Branchiæ very large, frondescent, beautiful; primarily divided into eight divisions; each of these like a folded leaf.— divided into

tufts, which are again subdivided: surrounding anus: Generative orifice large: Superior or dorsal antennæ, short, thick pointed, horn-shaped, faint brown colour, surrounded by concentric oblique membranous ridges, which are divided anteriorly by a white space: summit uncovered white: frontal antennæ small. I could not see the eyes:— but under (by dissection) mantle & behind the dorsal ones are two black dots resembling eyes: Digestive tube, gullet muscular, surrounded by (vermiform salivary glands?), entering between the Generative organs into centre of liver, is slightly enlarged & ~~turns~~ removed backwards, & reaches the anus in an oblique direction on the dorsal surface of liver.— Liver very soft, white & red:—

Eggs deposited in a ribbon. this adhæres by its edge to the rock in a spiral oval of 4 or 5 turns. is evidently formed by the turning of the animal on its centre.— & the distance of axis is the length from generative aperture to centre of revolution in the foot: Eggs in diameter .003, are collected in number from 2 to 5, generally in a[n] oval transparent case or |**152**| ball, length .012: These balls are arranged, two deep, in transverse rows in the ribbon:— In a large collection, the ribbon must be 20 inches long, in breadth it is .5 of inch; from counting how many balls in a tenth of inch & how many rows in same length, at the smallest computation there could not have been less than the enormous number of six hundred thousand eggs.—
(a) This is a wonderful instance of fecundity: yet the animal is certainly not common: I only saw seven individuals:— [note (a)] Especially when it is recollected every individual is an Hermaphrodite & lays eggs:— [note ends]

[1] The only Magellanic nudibranch of this colour with minute tubercles having such an enormous egg ribbon is the cryptobranch doridacean, Discodorididae, *Anisodoris punctuolata* D'Orbigny.

[**CD P. 152** continues]

General Observations — Jan & Feb Tierra del Fuego (South of Lat 54°45′)
Before mentioning any of the effects of climate, I will state, what I know, of its nature.— Capt King[1] has observed during Autumnal & Brumal period.— & the thermometrical observations made in this ship include the hottest part of the year.

From the 18th of Decemb: to the 14th of Jan (a period of 18 days) the mean from 332 observations made meanly at every two hours interval gave

temp:	44.92	
Mean daily Max:	47.98	Range
________ Min:	41.28	6.7
Mean of extremes	44.63	

During this time half was on the outer SE coast & half at sea, sometimes one or two degrees to the South of Cape Horn.—

From 15th of Jan to 20th of Feb (a period of 37 days) the mean from 161 observations, mostly at 6 AM: 12: 6 PM & some at the two hour interval give as:

Mean	49.9	
Mean of Max:	55.54	Range
.... Min:	45.36	10.18
Mean of extremes	50.45	

During this time the Ship was in different harbors, in Nassau bay & in Goree Sound. |**153**|

The mean from these two sets of observations from the 18th of Decembr to 20th of Feb. (a period of 65 days) gives

temp:	47.41	
Mean Max:	51.76	Mean Range
Min:	43.82	8.44
Mean of extremes	47.54.—	

The accuracy of this mean is affected by several causes.— the first set was out at sea & sometime at a higher latitude; it may therefore be supposed to be too low.— the second set is calculated from observations made in the day time (6 AM. 12. 6 PM) & the weather was, from the report, of those who have known the climate for some years, most unusually hot & fine; this second set gives perhaps too high a mean; it is to be hoped the mean of both may be near the truth.— These 65 days, judging from the appearance of the Vegetation in first part, & from the weather in Falkland Islands at the latter, includes the whole summer.—

Mr. Daniell journal[2]

In the years 1820, 21, 22 in London, the mean of Extremes of June, July, August, which months correspond to the Fuegian summer, was 60.93: so that the English summer is 13.39 hotter than the Fuegian.—

Capt King from observat: at 6 AM. 9. 12. 3 PM. 6, makes the mean of May, June, July, the brumal period in Tierra del F. 34.49.— In London from same years as above & corresponding months, it is 41.34; making the winter of the latter 6.85 warmer than the Fuegian: From these facts, we may form some judgement of the climate.— |**154**|

General Observations

(**P 152** VB)

I was surprised to see in Lat 55° & near to West ocean, magnificent glaciers forming perpendicular cliffs into the Sounds: this was in the end of January: Mr Bynoe[3] has actually seen a glacier reaching to the sea & in the summer in the gulf of Penas in Lat 47°.— This is a most singular fact when we recollect that Von Buch[4] first found glaciers on West coast of Norway in Lat 67° at Kunnen.— This gives a difference of 20 degrees for the same phenomenon in the Northern & Southern hemispheres.— It may

be here noticed that Capt. King gives the line of perpetual snow to St of Magellan (a little North of parallel of 50°) to be between 3000 & 4000
(**P 227** feet.— Von Buch in Norway says in Lat 70° the line is ~~about~~ 3000 above
V Buch) the sea.— Again there is this difference of 20° degrees.—

At the end of December large patches of snow were lying on the East side of hills at about 1700 feet elevation: these had disappeared by the end of
(a) February (answering to our August).— The Westerly winds have the constancy of the trades. it is clear the snow lies longest on the ESE side from being most protected from WNW wind, which of the prevalent ones would be the warmest.— [note (a)] Jemmy Button[5] said 'when leaves yellow, snow all go'.— Capt Fitz Roy states that in April the leaves of the trees which grow on the lower parts of the hills turn colour; but not those high up.— I recollect having read a paper to show that in England warm
Not Copied Autumns hastened the falling of the leaves: that the process is regular part of the vegetation: This fact would seem to show the same law.— It was in
(B) January in these very hills, abo<ut> 1400 feet high, that a snow-storm destroyed two of Mr Banks party[6] & caused so much suffering to the whole of them.— [note ends]

[**CD P. 154** continues]

At the height of about 1400 feet I found dwarf Beech trees, (about a foot high), in sheltered corners.— the main line of separation between the trees & grass is perhaps 2 or 300 feet lower. Within the Beagle channel this line was so horizontal & wound round in the vallies in so straight |**155**| a direction as to resemble the high water mark on a beach.—

General
Observations
Vegetation
(a)
(b)

The extreme dampness of the climate favours the coarse luxuriance of the vegetation; the woods are an entangled mass where the dead & the living strive for mastery.— Cryptogamic plants here find a most congenial site.— Ferns however are not abundant.— The Fuegians inhabit the same spot for many years; in one place I found 10 inches of fine vegetable mould over the layer of muscle [sic] & limpet shells, in consequence of this these mounds may be told at a distance by the bright green of the vegetation.—
1076. ~~amongst~~ the concomitant plants are mostly the wild celery, scurvy grass,
984. 985[8] black-currant tree; these, although not used by the Fuegians, are the most
(not spirits) useful plants in the country & seem placed to attract attention.—

[notes added later] (a) The appearance of these ~~woods~~ forests brought to my mind the artificial woods at Mount Edgecombe[7]: the greeness of the bushes & the twisted forms of the trees, covered with Lichens, in both places are caused by strong prevalent winds & great dampness of climate.— (b) It would be difficult to find a spade-full of earth in Tierra del F, excepting in the spots where the Fuegians have long frequented, & on the remnants of ancient alluvial formations, described in Geological Notes; but even this latter ground is, in some places, covered with peat, as in Goree Sound.— [notes end]

[**CD P. 155** continues]

Peat *[illeg. note in pencil]* In every part of the country which I have seen, the land is covered by a thick bed of peat.— It is universal in the mountains, above the limits [of] the Beech; & everywhere, excepting in the very thickest parts of the woods it abounds.— The beech often grows out of it & hence great quantities of timber must annually be imbedded.— ~~It flourishes~~ It increases most on the sides of hills & is I think of great thickness: the only section I saw varied from 6 to 12 feet. In more level sites the surface is broken up by numberless pools, which have an artificial appearance, as if dug for the sake of peat.— These are often close to each other & yet of |**156**| different
Peat levels, showing how impervious the peat is when acted on by water.— At the bottom of these shallow pools there is a quantity of brown flocculent matter in which Confervæ flourish & very little moss.—

1075 & 976 (~~not spirits~~) The great agent which forms the peat is a small plant with thick leaves & of a bright green colour (No. 976)[9].— The plant grows on itself; the lower leaves die, but yet remain attached to the tap root.— this latter penetrates in a living state to the depth of a foot or two.— & from the surface to the bottom the succession of leaves can be traced from their perfect state to one almost entirely disorganized.— Subterranean streams are common, these & the ~~pools of~~ *[illeg. word in pencil]* water by breaking up the upper peat & ~~dissolving~~ macerating the rotten leaves helps to form the more compact
(1073) not spirits parts.— Specimen (1073) is cut out of the surface of a peat Bog: This above plant is eminently social, few others grow with it: some small creeping ligneous plants bearing berrys (978 &c); another in its form, habits & colour strikingly resembling the Europæan heaths (1077); & a third equally resembling our rush (1045); ~~It would appear to be necessary under similar circumstances, the landscape should possess the same form & tints.~~— These latter plants & some others doubtless add their efforts: But the plant (976) & not any sort of moss is the main agent; On the sides of hills where it mostly abounded the surface of the peat was often convex.— By these gradual changes of level, water rests on different parts & thus completes the disorganization of the plant & consolidates the whole.— |**157**|

General observations Upon considering these facts, which show how inhospitable the climate of Tierra del is, we are the more surprised to hear from Capt. King[10] that Humming birds have been seen in Sts of Magellan sipping the flowers of
(b) the Fuchsia & Parrots feeding on the seeds of the Winters bark.— I have seen the latter South of the parallel 55°.— [note (b)] The tropical resemblance given by these birds & Plants is continued in the sea by the stony

branching Corallines, the large Volutans, Balanidæ & Patelliform shells.— [note ends]

Zoology (a) 1002 not spirits — Amongst the Mammalia, excepting Cetaceæ & Phoceæ, I saw a Bat; 2 sorts of mice; one of which I have (1002); [note (a)] The other mouse was of much larger size: but I could not catch it: [note ends] a Fox & a sea otter; & in Navarin island there were plenty of Guanaco.— the presence of many of these animals in these islands is accounted for by the probability of there being at one time an extended formation of Alluvium which connected them. Vide Geology[11]:—

Amongst birds (I refer more to numbers of individuals than species) Certhia was abundant in the woods; also Fringilla, Sylvia & Merlus: On the sea:— Petrels & Albatrosses, especially the first, exceedingly numerous:— Gulls not nearly so numer. or *[illeg.]*. I never saw any reptiles; Jemmy Button[5] states there are none.—

526 — Besides small fresh water fish (526), I have good reason to believe the genus Salmo exists:— There a few land shells; Succinea in the damp climate is common; in the pools I did not find any molluscous animals: the only inhabitants were Colymbetes & some small Hirudo's.— In the sea.— Capulus, Crepidula & Fissurella are all abundant.— the latter of great size.— But of all the Gasteropods: Cyclobranche, Patella & Chiton, in numbers of individuals & species are |**158**| beyond everything numerous.— The Chitons reach up to a large size.

General <u>Observations</u>

Crustaceæ 497:541 — Amongst Crustaceæ: Cymothoades[12] (Leach) take the lead.— the numbers of the genus Sphæroma are wonderful.— under every stone amongst the rocks at low water they swarm like bees: I was immediately reminded of the numbers of Trilobites in the Transition limestones:—

Insecta (b) — In the Coleoptera the only genera which are abundant are some few Harpalidous & some few Heteromerous — they are chiefly found under stones high in the mountains (such as Katers peak, 1700 feet high) together with Lycosa (Arachnidæ): scarcely any other Coleoptera, excepting a few Curculios are found: The tribe of Cycliques (Lat:) so characteristic of the
(a) Tropics is here absent; [note (a)] I must except one alpine Haltica. [note ends] whilst Harpalidous insects, as I have noticed are common.— In the hottest part of the year, the mean maximum (during 37 days) was 55.34 & the thermometer often rose to about 60°.— Yet there were no Orthoptera, few diptera, still fewer butterflies & no bees, this together with absence of flower feeding beetles (Cycliques) throughily convinced me, how poor a climate, that of Tierra del F is.— [note (b)] It will be curious to ascertain whether the plants of Tierra del bespeak as high a Latitude as many of the above facts point out.— [note ends]

[**CD P. 158** continues]

The sea is very favourable to the growth of Hydrophites.— Here grows Fucus giganticus in 25 fathom water:— the little pools abound with small
(c) species, almost to the exclusion of Corallines.— Corallina was present: & some species of Clytia (or allied to it) grew on the F Giganticus — They were the same species which I found floating in Lat 45°: V **P 135**.— [note (c)] The immense number of encrusting Corallines form the strongest
1834 exception to this remark.— I think a comparison of the Corallines of this country & England (nearly similarly situated) would be interesting in showing a very wide difference in the leading forms.— [note ends] |**159**|

[1] Capt. Philip Parker King R.N. was in overall command of H.M.S. *Adventure* and H.M.S. *Beagle* during their voyage to South America in 1826-30. His measurements of the temperature when the *Adventure* was anchored at Port Famine on the Straits of Magellan during May-July 1828 were cited in *Narrative* **1**:582-5.
[2] See John Frederic Daniell, *Meteorological essays and observations*. London, 1823.
[3] Benjamin Bynoe, Surgeon on the *Beagle* during 1831-36, had been the ship's Assistant Surgeon for the voyage of 1826-30, when he had visited the western part of Tierra del Fuego.
[4] See Leopold von Buch, *Travels through Norway and Lapland . . . Translated . . . John Black. With Notes . . . by Robert Jameson.* London, 1813.
[5] Jemmy Button was the Fuegian boy who had been taken back to England by Capt. FitzRoy in 1830, and was now being returned to his tribe on Navarin Island.
[6] See *Beagle Diary* pp. 121-2.
[7] Mount Edgecombe is an estate in Cornwall overlooking Plymouth Sound.
[8] See *Plant Notes* pp. 167, 170, and *Beagle Diary* p. 129.
[9] For an identification of the species of plants involved in formation of the peat see *Plant Notes* pp. 164-70.
[10] See *Narrative* **1**:134 for Capt. King's account of humming birds seen at Port San Antonio on the Straits of Magellan in the middle of April 1828 shortly before the winter had set in.
[11] CD is here referring to his *Geological Diary* and *Geological Notes* (CUL DAR 32-38), which are at the Cambridge University Library.
[12] Isopoda. The Cymothoidae are ectoparasites of fish, and *Sphaeroma* is a small marine crustacean similar to a woodlouse.

[**CD P. 159** commences]

Creusia[1] Shell formed of four quadrilateral pieces overlapping each other, with no calcareous support.— Operculum bivalv<e> each may be considered as half
574 an oblong, one end of which is bent obliquely at right angles: the line shows the direction of the bend: at the exterior corner there is a tooth

(where a dot is) [see sketch in margin]: it is at this end that the tail or Cirrhi are protruded.— Animal, to begin with the tail, there are here seated

(b) 3 pair of the usual bifid articulated arms, approximal, the central ones longest from which they gradually decrease in length on each side; stipes formed of 2 joints: between the central pair is the trunk (there is no projecting joint as described by me in Pyrgoma). behind, or beneath as relates to position of the animal is a longitudinal anal orifice:— 4th & 5th pair of arm are short, thick, conical, articulated, seated together & between 3 first pair & the mouth; 5th pair rather shorter than the 4th, as also the 6th than the 5th. This 6th pair is seated in same line as other but at base of mouth; this pair is remarkable by the internal branch being finer & nearly twice as long as the external.— it reminded me of the external "pied machoire" of the Crustaceæ.—

Mouth seated on a projection.— Part answering to the Labium (lower lip) formed of ~~two~~ a pair of closely approximate simple, short arms, with ciliæ.

N.B. (a) Labium formed of a hard plate, bilobed, & bent into an angle, so as rather to form half of the gullet:— within these organs are the pair of mandibular organs [sketch in margin], which ~~work~~ lie vertically & parallel to the side of the Labium.— the one nearest Labium seen vertically is an elastic plate enlarged at extremity, truncate & fringed with spines. |**160**|

[notes for this page have been crossed through vertically]

(a) On the Labium *[illeg. insertion]* or plate are attached two very small flattened arms or palpi with ciliæ, which fringing the sides of Labium close the mouth above.

(b)
March 31st Upon reexamining the animal, I suppose it is a Conia, but the descriptions are very imperfect; the anterior palpi are closely approximate & seated on a footstalk, the front part of which is flattened & closing the mandibular organs acts as a Labium as lower lip.— the bilobed organ ~~has~~ (called upper lip Labium) has within it a projection, which more truly acts as a labium.— the posterior palpi axis at base of mandibles, & continued & united along the edges (with bristles) of bilobed pieces, terminates as described in a flattened organ.— The generative trunk lies behind on right side of body & passes backwards between 5th pair & external pied-machoire, & thus separates the two groups, the cirrhi & Trophi.— If we consider the bilobed organ, as palpi or pied-machoire altered, we shall *[word missing]* 5 pair of such organs & mandibles; also these 5 pair of bifid cirrhi or legs.— It is impossible not to be struck with the analogy with the Crustaceæ as Schizopedes.— (Generative trunk with bunch of few setæ at extremity & few scattered ones of sides.) [notes end]

[**CD P. 160** commences, entry again crossed through vertically]

Creusia They more resemble the Maxilla of Crustaceæ: the other pair is stronger

(a) & larger & true mandibles, hard with 5 strong teeth; superior one very large & decreasing in size to the 5th, which is rather a crenated surface.— So that of the Trophi, we have ~~5~~ 4 pair of articulated organs, & the part called Labium.— Of the arms (legs or Cirrhi) then *[illeg. deletion]* 6 pair; one of which rather acts as external palpi for the mouth & indeed the intermediate ones I should think conveyed food from the beautifully constructed extreme 3 pair to the mouth: The body is attached behind the mouth to the Operculum: The trunk arises from the very extremity of body, where it is much contracted, varies in length.— sometimes 3 times as long as the arms, elastic with fine rings, tapering with internal tube.— appears to lie on right side of body:— This animal would be a Creusia of Cuvier in Dic Class.— The habitat is remarkable, it was found but little below high water mark, about 15 inches at most, & in a stream of fresh water.—
(a) In a note attached to (No 323 in spirits) I mention a Creusia found at M: Video under similar circumstances.— & that the animal ~~then~~ would expand its arms in fresh water.— These facts would appear conclusive that this genus of Balanidæ is especially fitted for brackish water, & for a certain time even for fresh.— At M: Video I thought the habits of the ~~individual~~ species had been changed gradually by the less salt water of the Plata.— but here there was no gradual change:— the water emptied itself over rocks into the sea, & on these rocks the Creusia was attached.— |**161**|

[note for **CD P. 160** entered a few days later, but not crossed through]

(a)
March 31st
590
[illeg. pencil note]

There is a common sort in the sea, exceedingly like this one in general habit.— but differing in the operculum being quadratic.— Within there is the same external tooth, & in addition several processes.— the suture is simply serrated.— The sutures of the shell are plainer.— The animal is precisely the same in every most minute part even of the mouth.— It is clear if the bivalve & quadrivalve operculated shells are not the same species.— they are same genus.— The only difference is the sutures being united & processes connected with articulation removed; yet this character is used as essential amongst the Balanidæ! — Can the fresh water have any action in obliging the animal to keep its operculum more throughily closed?— There were however some quadrivalve ones near the Bivalve:—

1 CD rejected his initial identification of the shell as the barnacle *Creusia* of Cuvier, and plumped instead for *Conus*, which belongs to order Neogastropoda and superfamily Toxoglossa, and is a poisonous turret snail. But his doubts remained, and he finally reverted to *Creusia* once more in his entry of Specimen 574 in the Spirits of Wine List. When he concluded a few days later that specimens 574, 590 and 591 all belonged to the genus *Creusia*, he was still wrong, because writing much later in his monograph on barnacles, he stated on p. 469 of *Cirripedia* that the commonest species in the Falkland Islands was *Chthalamus scabrosus*, and on pp. 375-82 mentioned no species of *Creusia* from this area. However, in 1833 he had not yet made himself familiar with the niceties of distinguishing

between some of the most closely related members of the suborder Balanomorpha.

[**CD P. 161** commences: entry and accompanying notes crossed through vertically up to middle of **CD P. 163**]

Corallina[1] Linn: (inarticulata) 585 & 1153 (not spirits) (b) (z) (a)

Coralline, stony, brittle, inarticulate, encrusting rocks & sending forth lichen-shaped thin expansions.— Growth concentric, shown by lines & changes in the tint of colours; Colour darkish "crimson red" or that of Corallina officinalis: a section shows that the superior part is composed of horizontal layers of a stony & slightly coloured substance.— the other softer, white, & of a more granular nature:— the inferior surface is rougher (for attachment) & paler coloured than the upper: the border or extremity of the expansions is thickened; edges semipellucid, covered with a delicate transparent membrane, & containing a soft granular cellular tissue; in all these latter respects, the similarity of this with Corallina & its subgenera is very great.— On the superior surface, & in the more central parts, in some pieces there are numerous small cones or paps, with a minute circular orifice at the summit.— They precisely resemble those described at **P 56** in an Amphiroa.— [note (a)] The ovule-bearing cones are very uncommon; I only found one specimen with them, & out of many cones which I examined only three had the regularly formed ovules: The rarity of this generative process may perhaps explain the general ignorance of method of propagation in Corallina.— [note (b) added later] For similar particulars, in an Halimeda[2], V **211**. [notes end]

[**CD P. 161** continues with a paragraph in square brackets not crossed through]

[3][These cones are formed in any point by a separation in middles of the ~~superior~~ stony layers; & the upper part gradually assumes the conical shape.— At first they have no aperture, when it first appears it is small; but in time increases to a diameter of $1/500^{th}$ of inch; after this epock, the cone becomes white & brittle & its surface exfoliates.— the concavity on which the younger ones rest is partially filled up & it is clear the little cone has performed its office in the economy of Nature.—]

Corallina (z)

If the cone is removed in one of the early ones, the bottom is concave & on it there is a layer of the pulpy cellular |**162**| tissue or granular matter, such as occurs at the extremities of the branches.— this lies on the white softer substance of the Corall.— so that the stony layers are perforated.— At a later age, the granular matter is collected into [semi-opake spherical or oval bulbs, with a transparent case: these are slightly coloured & between 30 & 40 in number.— in diameter $1/500^{th}$ of inch.—] They are ovules & the cones ovaries.—

The simplicity of this generative process is shown by its ~~the~~ similarity to

ordinary growth.— the external border is thickened, composed of precisely a similar substance & enveloped in a transparent membrane; it may be considered as formed by a juxtaposition of cones, or rather the cone & ovules owe their origin to the creative power acting on a point where the
(a) growth or extension cannot take place, hence the granular matter is enveloped in a spherical case & seeks an exit through the stony layers instead of increasing laterally.— [note (a)] It is to be remembered that the cones do not occur near the margin, where the Corall is growing.— [note ends] In some specimens these cones were absent; in others there were white spots with the surface exfoliating, & there I imagine cones to have existed.—

The Corall abundantly coats the rock in the pools left at low water. According to Lamouroux[4] it would be in the III Sous Ordres Corallinées inarticulees; but from the description of genus Udotea it cannot belong to it.— Upon reading over description of Amphiroa **P 56**, it will [be] evident how very close a relationship in manner of growth & cones |**163**| there
Corallina (inarticulata) exists between that Corallina & this. The absence of articulations is the chief difference: I think we may hence expect that the propagation in the whole family Corallinaæ will be somewhat similar to the one described.— I have never been able to perceive any Polypus or true cell, & till I do I must rank these beings as belonging to the Vegetable rather than animal world.— the simplicity of the reproduction would seem rather to favor this idea.— I suspect the strongest argument against it is ~~the~~ a false analogy of form with respect to Corallines; in this case however there is a stronger one to Lichens.

[notes added later for **P. 161**]

(z) On tidal rocks at King George's Sound, found a Corallina[5] growing in
3557 nodules to a Granite rock: color such as is universal to the family in the Atlantic & Pacifick oceans, in T. del Fuego & Australia: consists of numerous, strong, cylindrical, inarticulate parallel small columns, partly adhæring one to the other. Many of them show an obscure globular necklace like structure, centre of each column white.— Some of the smaller & irregular arms were covered on all sides by the generative bladders. These in every segment resembled those already described: the older ones scale of ~~with~~ in form of an irregular particle of white crust.— Size of each pap or bladder rather more than the square of $1/100^{th}$ of inch & the circular aperture has a diameter a shade larger than 1/1000 of inch.— I was not fortunate enough to extract an ovule: This Corallina is evidently a connecting species, most closely allied to the division of Inarticulatas.—
N.B. I saw in a delicate transparent articulate Corallina that the branch appeared to be composed of several hollow transparent ligamentous vessels, which in
3557 the solid parts between the articulations were filled up with calcareous granular matter.— Species with flattened joints & symetrical lateral

branches. [note ends]

(z) [note (z) for **CD P. 162**] Decandoelle & Sprengel[6] Botany P. 92 Consider that propagation in Lichens & Confervæ is a kind of budding & not true generation. In Halime[d]a & in the Inarticulatæ such certainly I think is the process.— In the method described in Corallina of Hobart town of the extremities of branches being "laid" in branches of trees, & when from the foliaceous expansion buds appeared, perhaps in this method we see the only kind of propagation known to this genus, in which the bladder-formed cones have not been discovered.— [note ends]

[1] *Corallina inarticulata* (Specimens 585 and 1153) is the coralline alga *Amphiroa exilis* Harvey. See *Plant Notes* pp. 203-6.
[2] *Halimeda* is a green alga (Chlorophyta). See *Plant Notes* pp. 194-5.
[3] Square brackets inserted later by CD with a different pen.
[4] See *Lamouroux* p. 27.
[5] Another coralline alga, but precise species not identifiable. See *Plant Notes* pp. 197-8.
[6] See Augustin Pyramus de Candolle and Kurt Polycarp Joachim Sprengel *Elements of the philosophy of plants* (Edinburgh, 1821).

[**CD P. 163** continues]

Holuthuria (a) 586[1] Body cylindrical, transversely wrinkled, rather pointed at posterior extremity or anus:— length 1 & ½ inch: colour pale salmon: ~~covered~~ with 5 longitudinal irregular rows of (2 or 3 broard in each) long papillæ.— these rows extend whole length of body.— Mouth surrounded by 10 tentacula, these are unsymetrically branched & long.— much resembling a tree in (b) growth.— Not uncommon under stones.— same as (522) in Tierra del F.—

[notes added later on back of page]

(a) April 1st 594[2] Holuthuria closely allied to last: body more elongated, coloured "peach blossom red", coriaceous.— Tentacula long, irregularly branched, *[illeg]* like.— On side of body generally used as attachment there are two ~~clear~~ longitudinal spaces clear of papillæ; but on each side they are thicker, hence look like three rows of papillæ. On back papillæ scattered irregularly.— There is a short smooth neck free of papillæ.—

(b) Holuthuria (586) with short smooth neck with few papillæ; body coriaceous, transversely wrinkled. Bony collar round neck of œsophagus, simple, form of 5 double pieces, or 10, the alternate ones being slightly different.— the ~~parts drawn~~ rim only was white & calcareous, the intermediate parts cartilaginous.— Fig. collar cut open.—
One of the papillæ examined, shows its whole surface reticulated (rather

Port Famine June 1834 broard plates) with stony substance.— termination a saucer shaped depression.— I believe no sort of aperture: it is only a locomotive organ.— [notes end]

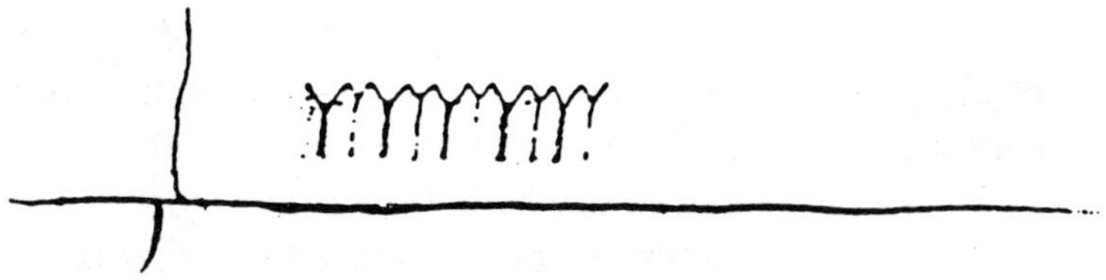

[**CD P. 163** continues]

Sipunculus[3] (allied to) 586 — Body cylindrical, smooth, finely wrinkled, colour "yellowish brown", posterior extremity suddenly & much pointed; trunk about ½ length of body, total length between 4 & 5 inches. Mouth surrounded by several rows of small, short, flattened lancet-shaped tentacula, closely approximate so as to form a tuft.— Anus minute white speck at base of trunk; internal anatomy precisely as |**164**| described by Cuvier for true Sipunculus.— Body was exceedingly distended by water, so as when dead to squirt it out with force.— Animal was under stones in sandy mud.—

[1] Dendrochirotida, Cucumariidae, sea cucumber, possibly *Pseudocnus dubiosus leoninus* Semper. The correct spelling of this animal, never used by CD, is Holothuria.
[2] Same family, possibly *Cladodactyla crocea* Lesson.
[3] Phylum Sipuncula, burrowing marine worm.

[**CD P. 164** continues: next entry crossed through vertically]

Corallina[1] (inarticulata) 1153 (not spirits) — This species somewhat resembles in appearance that of **P (161)**. Corall, exceedingly hard, stony, compact; a section shows no horizontal layers & no great difference of hardness in different parts: is coated by thin layer of the soft cellular tissue, of which the cells are very minute.— The covering is so thin that it requires a microscope & lancet to procure any.— Superior surface coloured blackish "crimson red"; smooth, very regular:— expansions thick (about $1/10^{th}$ or more), strong:— grows in large circular patches, when two interfere the junction rises in a crest; these were nearly the only ones which I could procure as specimens. Is not very common, chiefly distinguished from that of **P 161** by the much greater thickness of expansions.— Amongst organized beings, few could be found which would show fewer of the signs of structure & life.—

Corallina[2] (inarticulata) 1153 (not spirits) — This, as that of **P 161**, most abundantly coats the rocks, or growing on itself forms bosses: in its structure it is likewise closely related, although different in external form.— Corall mamillary, composed of numerous small oblong pieces, with globular heads; these often grow into each other & are always close together, so that the surface is very irregular. the summit of ~~each~~

nearly all its rounded heads is marked by an irregular line or suture, as if originally formed by the |**165**| junction of two pieces; colour pale with

Corallina (inarticulata)

faint tint of purple.— Structure same as others, central parts of nearly uniform hardness; external coat of cellular tissue (or granules, for I am not yet sure whether each hexagon is a cell or grain) is thin, (but thicker at summits), but composed of rather larger cells, than the other species: If that of **P 161** from its figure called to mind the Lichen which grows on rotten wood: this is equally like to a dry crumbling sort which grows on stone.—

Corallina[3] (true)

1143 (not spirits)

Trichotomous, joints nearly cylindrical; those which give off branches triangular. others round; articulations semi-pellucid; colour same as usual, grows in small, low, tufts:— A longitudinal section of extreme part of limb gives following appearance: beneath a thin transparent coat is a mass of cellular tissue (such as so often described) & within this, parallel longitudinal darker coloured fibres surrounded on all sides by the cellular tissue: the extremities of these follow the same arched line as the external surface, & it is probably by the successive hardenings of these that an occasional appearance of concentric lines is seen in a section of older ~~joint~~ limb.— At base of ultimate limb, the outside part first becomes stony:— A section of old limb gives first a very thin coat of cellular tissue, & I think the external transparent membrane.— then ~~then~~ a semi-pellucid hard stony case, which by the appearance in microscope appears to be part of cellular tissue of young extremity filled |**166**| up with stony matter; the lines are

Corallina (true)

rather transverse in it.— the central part is white, softer, yet calcareous & with longitudinal lines; this is clearly the horny fibres of extremities also hardened.— The distinction between the central & external stony parts is best seen in the penultimate limb.— as the external case becomes perfect before the former.— The connection between the whole Coralline must chiefly be carried on by the external soft cellular tissue:

(Vide infrà)

~~The articulations have not much motion, & that must only be from increased elasticity: within these is a largish cavity, with arched roof & filled with a soft substance, which I imagine to be the central mass, not lapidified:—~~

I am convinced that it is out of the question to suppose these beings have any connection with Polypi.— What claims have they to be considered as animals?—

At the articulations the stem is contracted & the external stony case bends in & is not continuous with that of the adjoining limb.— A Section gives the appearance of a cavity; but is really formed of a globular mass of tough semi-pellucid inelastic matter. This at its base unites with the central softer stony part, & above articulates into an arched cavity in the next limb.— hence motion is tolerably free.— |**167**|

[1,2] Coralline algae related to *Amphiroa*, but Specimen 1153 has not been specifically identified

in *Plant Notes* pp. 186-206.
[3] Specimen 1143 is preserved in the Herbarium of Trinity College Dublin as *Corallina officinalis* Linn., the coralline alga *Amphiroa caloclada* Dne. See *Plant Notes* pp. 191-4.

[**CD P. 167** commences]

Balanus[1] Shell depressed quadrivalve: base membranous with narrow calcareous rim:
Linnæ:  externally rough, irregular.— Operculum quadrivalve, suture doubly
591 serrated.— [Sketch in margin] within plain, & no external tooth. Animal, with 4 pair of the usual bifid articulated cirrhi (like the three pair of animal **P 159**): 5th pair short strong.— Then the generative trunk passes on right side backwards, it is rather short & rings plainly marked on it.— External pied machoire with equal arms.— Maxillæ with the truncate spinous edge irregular: Mandibles with superior tooth not larger than others, than in regular proportion.— In other respects mouth agrees well with animal **P 159**.— Mouth, Trophi & Cirrhi all coloured dark greenish blue.— This is the commonest sort, which at low water mark covers the rocks.—

Synoicum[2] April 2^{d}.— Aggregate body, oblate spheroid, seated on a footstalk, which
595 tapers at root to a fine point: gelatino-membranous, external parts yellowish transparent, internal reddish orange. Formed from the aggregate of numerous animals, the bodies of which point towards common centre or footstalk, hence the central ones are longest & others gradually decrease in length towards the sides.— They adhere side by side, & from each a narrow elastic ribbon ~~goes~~ proceeds to the footstalk & passes down to the root.— External surface slightly mamillated, with apertures each of which is common to the branchial cavity & other orifices of |**168**| each
Synoicum animal.— Orifice bean shaped, edge slightly fringed.— near to convex side, there is a white internal mark formed of collection of dots.— [note
(b) (b)] I do not know what to make of these white dots, which are universally present: they can easily be separated.— numbers also occur about the region of the stomach, but in this latter place they are not constant in numbers or site.— [note ends] this side is external to pole of sphere on inferior.— From these white marks & shape, consistence & colour, body resembles some fruit, such as a Strawberry: size of large specimen, breadth of sphere .8: height .6; length, including head & stalk, about 2 inches.— Grows on the leaves of the Fucus giganticus.— .separated

Body of animal ~~may be~~ is divided into two parts;— branchial cavity,— & abdominal viscera.— Branchial cavity bell-shaped, furnished with slightly
(c) tubular lip, on which are two rows of differently sized papillæ, ~~about~~ 16 in number (?): these project across the expanded aperture.— [note (c)] The papillæ resemble on a small scale those on the arm of an Asterias.— When the animal is undisturbed, the branchial cavity is widely open & a slight circulation of water may be perceived at the aperture.— [note ends] the

(a) sides on mantle is [are] divided into two halves [note (a)] not separated or cut [note ends] by vessels running up on each side; in both there is a most beautiful & symetrical trellis work of branchiæ. ~~They~~ It consists of 5 concentric rows (or combs) of parallel filaments, which are vertical; they are attached at each extremity to mantle; in middle rows they are attached to bands.— Perhaps they might be described as four concentric bands with filaments above & below, but where opposite united.— The filaments
V. Pl: 7 towards each end of the comb decrease in ~~size~~ length.— When the animal is undisturbed the two upper ~~& larger~~ rows can only be seen, the others
¼ & ⅓ focal distance were discovered by difficult dissection; On these filaments, with a high power, a rapid vibrating motion is visible, as if of ciliæ, clearly a function of respiration.— |**169**|

[next two pages have been crossed through diagonally in pencil. **CD P. 169** commences]

Synoicum The vessels which divide the mantle & the two sets of trellis work; are very clear near the aperture but by no effort could I trace them to a junction with others of the viscera.— On the external side, a clear space runs up, to which the concentric bands unite.— & in this is a vessel, containing
(a) another, which ~~runs up~~ seems to unite to the white space by branchial aperture.— [note (a)] Is it impossible that this vessel is connected with base of tentacula or papillæ & from thence leads to mouth of œsophagus at base of branchial cavity.— animal would then live solely by absorption!? — it is the simplest method of joining the vessels: [note ends] I could not see any orifice.— I could trace these vessels down the side of cavity, but not across it, which direction it must pursue if it unites to any of the viscera.— On the anterior & superior side there is a minute vessel, which seems also to terminate in a yellow dot by branchial aperture & right opposite to white space.— the interval between this vessel & intestine is so small, that I have no doubt that it is the anus.—

Near base of Branchial cavity the œsophagus enters, & ~~proceeding~~ descending a short distance, bends nearly at right angles & passes under & through the liver.— forming together large dark reddish orange unequally sided oval.— the intestine taking a sweep ascends close by the œsophagus to near aperture of ~~bran~~ Mantle.— between the stomach & bend of intestine the heart lies, appears elongated & very transparent; pulsating
X strongly; I could trace the oscillations to within the Branchiæ, I imagine therefore the circulation is simple:— Resting on & beneath the intestine & stomach: there is a large sack of white pulpy matter, which ~~generally~~ often is divided internally into a star like mass.— it is in this state when most undeveloped. |**170**| When a little more advanced, the white matter is
Synoicum collected into globular ova.— from the centre of this sack a vessel descends & bending suddenly ascends close by the intestine & therefore on the outside of animal.— I could trace it as far as the end of intestine, but from these vessels & œsophagus all lying close between the trellis work of

branchiæ, I could by no effort trace them to their orifices.—

This last vessel is clearly the oviduct: I will first describe the most extraordinary ovules & then the process of generation.— From the first rudimentary globular collection of white matter, they pass into (2nd state) defined reddish orange spheres: 3rd with a point on one sides: 4th.— surrounded by ~~clos~~ a transparent band in which are transverse opake partitions: 5th a rounded oblong, with central dark mass enveloped by ~~gelatinous~~ transparent matter, furnished with a long tapering tail.— Tail has numerous transverse partitions, & in 4th state was curled around ovum.— it terminates by a fin[e] hair & in different times is either filled with homogeneous matter or opake partitions.— Total length .11; breadth of head .015, so that the tail is about 5 times as long as head.— ~~it~~ Ovule is capable of rapid vibrating motion & hence progressive: it is evidently a young Synoicum in search of a Fucus on which the tail will be fixed & become a footstalk: From appearance of head it is a single |**171**| animal.— This gemmule resembled in its habits some Infusoria[3], as Circaria.—

Synoicum

In the described ovarium, only those ova in 1st & 2d state are found.— For independent of this organ, there are, when the aggregate body abounds with ovules, two intestine shaped sacks, longer than the body & attached near to extremity of intestine, or supposed anus.— I never saw these except when with eggs. At lower extremity the ovule appears to be much in same state as in the true ovarium, but at the upper end or mouth they are in state 4th: ~~& some even with~~ when fused their tails uncurl: I should suppose that ovules pass down the oviduct & enter the two additional ovaria & there remain till ready to become independent animals.— In same proportion as the two additional ovaria contain many ova, the central one contains few & the whole animal becomes exceedingly shrunk; so that the aggregate body is of a darker reddish orange & appears to be composed of intestine shaped sacks with ova.— The number of eggs in each animal vary according to its size, so that those near the footstalk only contain a few, whilst the large central ones very many.— The ovules in same aggregate body were nearly in same state.— some with central ovarium only containing white pulpy matters, others filled with large bright coloured ovules: |**172**| Aggregate bodies of different sizes (therefore ages?) contained ovules; otherwise I should have thought from shrunk state of bodies that after parturition animals had died.— [pen changes] The footstalk is enveloped in strong membrane & consists of the elastic ribbons & some granular balls, the nature of which I am ignorant of, enveloped in gelatinous matter:

Synoicum:

I have called this animal Synoicum, as in external characters being nearest, but it is evidently distinct.— In the anatomy the generation is very curious & one more instance of ovules having a motion of which the parent animal is not possessed.— the number of tentacula round edge of mantle, & the curious trellis work of Branchiæ are all remarkable facts.—

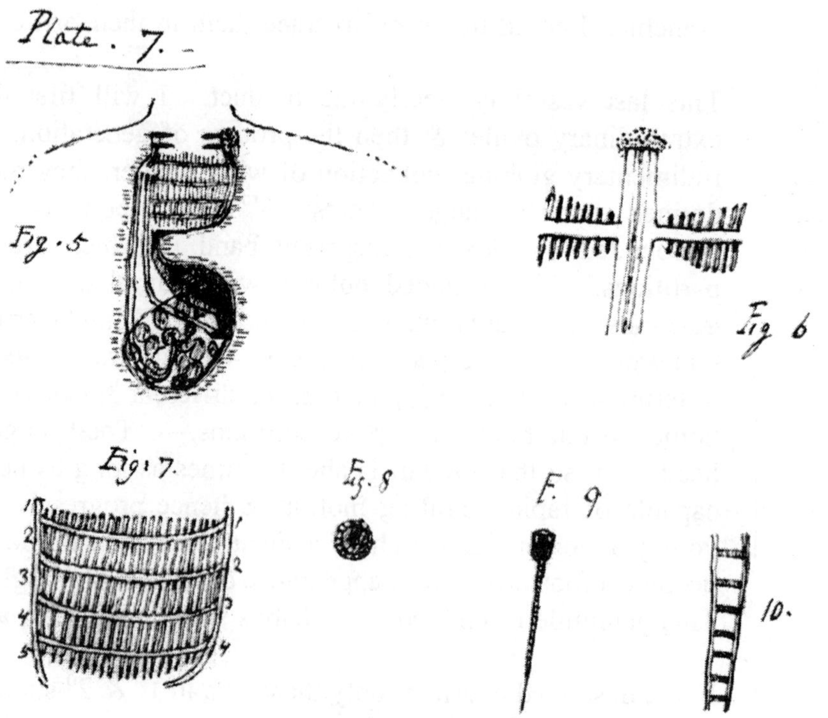

Plate 7, Figs. 5-10

Plate 7, Fig 5. represent, but stiffly drawn, an animal with branchial cavity expanded: tentacula about aperture: the vessels within branchiæ are disjoined from fault of observ: heart lies on under surface just by function of stomach & intestine, not (drawn): ova just formed: Fig 6 is the vessels which lead towards collection of white dots, with upper band of branchiæ of the two trellis works.— Fig 7. one whole set of trellis work expanded; miserably drawn, filaments longer & far more numerous & regular.— Fig: 8.— Ovule in 4th state: Fig: 9:— Ovule in 5th state.— Fig 10, piece of tail much magnified:— |**173**|

[1] In footnote[1] on p. 137 it has been seen that in his monograph CD said that the commonest barnacle in the Falkland Islands was *Chthamalus scabrosus*. It may also be noted that although on p. 136 he had commented on the strong affinities between barnacles and crustaceans, in his Specimen Lists he always classified barnacles as molluscs, being unaware until the end of the voyage of the discovery that had been made by J. Vaughan Thompson in 1830 of their metamorphoses.

[2] Ascidiacea, a tunicate or sea squirt. CD has well observed, independently of its discovery by Milne Edwards a few years earlier, the brooding larvae typical of the animals in cold waters, and the way in which the tails of the tadpoles curve around the trunk. But it is their heads rather than their tails that become attached to their supports.

[3] Infusoria are ciliated protozoa.

[**CD P. 173** commences]

Holuthuria[1]

596

April 4th. Body very smooth, soft, with three rows of papillæ on under side & on back two sorts of low crests formed of more rudimentary papillæ united by a membrane:— 10 tentacula round mouth.— simple short, irregular but not much branched, or rather tuberculated: colour "dutch orange", often with much darker specks: when ~~first~~ taken out of water, quite shapeless:— Bony collar round œsophagus exceedingly rudimentary: there is a mere vestige of calcareous matter on upper rim.— the rest being cartilaginous:— there are 5 double pieces [sketch in margin]: at base of grand division is point of attachment for the longitudinal ligamentous band. 5 of these extend whole length of body & thus differ from Sipunculus.— Also intestine not spirally convoluted.— Found in great numbers between the roots of Fucus giganteus.—

Obelia[2]

(a)

597 &

1161 (not spirits)

Corall, stony, brittle tender, growing in mass specks [?] like incrustations of Fucus gigantes; polypiferous tubes, curved cylindrical growing in united groups, from 2 to 4 on stony plate, nearly in a direction from one centre; & pointing upwards nearly vertically; tubes & plate thickly covered with punctures; colour very faint yellow:— Polypus I only saw by dissection: tentacula 10 in number, fine simple, seated on a neck, which joins cylindrical body with central vessel.— nearly at base it contracts & is bent; perhaps lies in curved position in tube; body terminated by mass of reddish matter & above this (which is a very curious but certain fact) there was a collection of reddish grains, enveloped in transparent matter, which possessed a |**174**| rapid revolutionary motion; each separate grain might be seen with ⅓ focal D: revolving: when cut out of body they mingled with the water: the exact position of the ball seemed to vary & in one I thought there were 2 or 3, although only one with motion.— What is this, a heart? or preparation of ova?—

Obelia

(a) On same Fucus there was what appeared to be a different species of Obelia, only differing from the last in the puncture being smaller, colour white.— tubes not so high & generally united in rows, which, like fibres from the mid-rib of a leaf, branch off on each side; several of these leaves sometimes form a star: Both these species belong to Obelia of Lamouroux[3]:— Are very abundant:

[notes for **CD P. 173** added later]

(a) March 1834.— on Fucus leaves, in Ponsonby Sound, were minute specks of Coralline.— which perhaps may be same species as this in young state,

890 when the punctures are not developed.— Arms 10(?), terminal red, viscus nearly sphærical, at one side small enlargement near junction of basal vessel of tentacula, evidently *[illeg.]* organ, as mentioned in note (a) to next

species.—

1834 May I examined a small species of Obelia: its body has the true structure of the Flustraceæ: as this was one of the first I examined I am not surprised at

971 overlooking the curved vessel with the (Liver?) attached at both extremities: it was probably ruptured in detaching the Polypus.—

[note for **CD P. 174**]

(a) 1834. AO. March 1st. East entrance of Beagle Channel; there is an abundance of these white stars on the Fucus. highly polypiferous Polypus, with 10 or 12 arms, very delicate, only the arms were protruded; body resides in the tube: body len[g]thened cylinder, which near base (as described in other species) contracts & slightly bends & in extremity contains a red viscus is of an oval shape.— there is a central vessel.— just before the bend, this vessel seems to pass by another & smaller viscus also of a red color. (liver?) This same vessel or another conducts to the main terminal reddish mass; in this were two spots where rapid revolution of the contained fluid was evident.— one & the upper centre of motion was most energetic.— the site of both is close above the terminal viscus where the arms are extended; whole body moves.— [note ends]

[**CD P. 174** continues]

Phocœna[4] Body, above & before dorsal fin, depressed, before tail compressed & arched:— belly tapering gradually to tail: Head forming about an

711 equilateral triangle, conical.— Upper part slightly "bombè": Outline of the junction of the upper jaw with head straight, but on each side ~~junction~~ there is a slight depression:— Eye with iris dark brown placed above & behind corner of mouth: Teeth slightly curved, placed regularly; in upper jaw 28, (on each side) in lower 27; the two most anterior teeth are in the ~~latter~~ lower jaw: lower lip projects beyond upper; Eye & breathing vent in same circle around head; concavity (or horns) of vent point towards anterior extremity of body: Dorsal fin posteriorly simply excised: Pectoral, placed rather below a line joining under lip & tail, posteriorly doubly excised. Tail, between extremities straight, with central deep division.

Phocœna The specimen appeared to be of the common |**175**| size:

Length (following curvature of back) from tip of nose to end of tail	5ft	: 4 inches.—
From do to Anus	3	: 10.9
From do to anterior base of dorsal fin	2	: 6.5
From do . . . . pectoral . .	1	: 4.5
From do to eye	0	: 9.9
From do to vent (following curve of head)	0	: 10.7

From do to corner of mouth	0 : 7.9
Girth of body	
Before dorsal fin	3ft : 0.6inch
. . . . pectoral . .	2 : 8.2
. . . . . . tail fin	0 : 7.8
Over the eyes	2 : 0
Length of dorsal fin following convex ~~each~~ edges	1ft : 0.5in
a perpendicular dropped from tip to the back	0 : 6.4inch
Length of pectoral following anterior or convex edge	1 : 2.8
Width of tail from tip to tip	1 : 4.5.—

(a) Colour: beneath resplendent white, above jet black, ~~most of junction~~ the two generally shading into each other by grey: extreme of snout, edge of under lip, ring round eyes, & tail fin, jet black: dorsal & pectoral fins dark grey.— this latter colour is continued from corner of mouth to the pectorals; but above them there is an oblique white band, which gradually shades into a pale grey above the eyes.— Again the dark grey is continued from back in an oblique line to anus.— but within this tail part, there are two white & grey bands which run parallel to that above the pectoral; thus forming the diagonal white & grey bands on the side: the two posterior ones Phoccœna I should think would |**176**| occasionally coalesce & be subject to variation: [note (a)] There were several small Crust. Læmodipodes[5] adhæring to the skin near to the dorsal fin: colour dark reddish brown with white spot near base of leg.— By mistake these were lapped [?] up & put into spirit without number being attached to them.— [note ends]

This specimen was a female & harpooned out of a large troop which were sporting round the ship in St Josephs Bay; Lat 42°.30′ S.— April 17th.— Vide drawing of animal by Capt.— Fitz Roy.—

[1] Apodida, Chiridotidae, a sea cucumber, possibly *Trochodota purpurea*.
[2] Not the thecate hydroid *Obelia*, but might be a calcareous hydrocoral of order Stylasterina. On the same *Fucus*, specimen 1161 (dry) in the George Busk Collection had in addition bryozoans listed as *Porella margaritifera* and *Tubulipora phalangea*. On specimen 1877 (dry) in the Busk Collection CD later found the same hydroid together with *P. margaritifera* and *Diastopora tubuliporide*.
[3] Probably *Obelia geniculata*. See *Lamouroux* p. 81.
[4] This porpoise was named by George Waterhouse *Delphinus FitzRoyi* in *Zoology* **2**:25-6, where a lithograph after FitzRoy's watercolour was included. Specimen 711 was the head only of the animal, and the dimensions cited by Waterhouse were supplied by CD's measurements.
[5] See A.-G. Desmarest *Considérations générales sur la classe des crustacés* (Paris, 1825) pp. 272-80.

[**CD P. 176** continues]

Snakes taken at Maldonado for May & June.

Bipes[1] 608 _____ Copied	Beneath white gradually shading into a light brown above, with four dark brown lines.— the 2 central ones being the broardest was caught near the water of a lake.— motions inactive.—
Coluber[2] 623 & 702 _____ Copied	Above of a uniform blackish lead colour, with an opaline bluish gloss; beneath pale, at the junction of the two sorts of scales the gloss is least seen; differs from the following one in shape of scales, & proportional length of tail &c
Coluber[3] 624 _____ Copied	The commonest species in this country; is it not same as taken at Bahia Blanca, reaches 3 or 4 feet long.— The first maxillary tooth is very large: by aid of microscope I saw a narrow deep groove running down on convex surface.— Is it for conveying poison?— Specimen of tooth is in pill-box (1320)
Coluber[4] 639 & 705 _____ Copied	Beneath cream-coloured with irregular rows of blackish dots as if of interrupted chains; above all the scales, "yellowish" ½ "wood brown", with lateral darker band on each side; chiefly on anterior part of body, the interstices between scales are coloured in symetrical small spaces of white, "tile red" & black, (the latter most strongly marked), this gives a singular mottled appearance to the animal.— Inhabits not uncommonly the sand dunes.— |**177**|

[following entries are dated May 14th]

Limas[5] Ferussae[?] (a) 614	Body narrow, of a uniform black-lead colour, beneath & sides paler.— Superior antennæ short, thick, blunt, with terminal eye, same colour as body; inferior as usual, much shorter, rugose: Branchial orifice seated on right side of shield & about 2/3 from the its anterior margin.— Shield covering about half the body; leaving a little of the neck exposed when the animal crawls — on the shield there are parallel furrows, following its curvature:— tail moderately pointed; body length .1 inch, but slightly wrinkled, found crawling in a field near head of the R. Tapes.— North of Maldonado.— [note (a)] May 29th.— Found some more specimens crawling on plants in a very wet place; their length, colour, & general appearance the same, so that I have no doubt they are full grown:— body very narrow linear, when crawling .9 long & sup: antennæ (protruded) 1/12th of inch long.— [note ends]

[1] Anguidae. *Ophiodes vertebralis* Bocourt.

[2] Colubridae. *Clelia occipitolutes* Duméril.
[3] Colubridae. *Philodryas patagoniensis* Girard.
[4] Colubridae. *Lystrophis dorbignyi* Duméril.
[5] Stylommatophora. Land slug.

[**CD P. 177** continues: the entries up to p. 164 which follow were copied by CD with a number of small changes in *Ornithological Notes* pp. 214-25, and the majority of the birds were identified by John Gould in *Zoology* **3** as noted in brackets. A list of modern synonyms of Gould's names is given by Pete Goldie in *Darwin 2nd edition*, Multimedia CD-ROM, Lightbinders Inc., San Francisco, 1997.]

Ornithology — The following are a few scattered observations on the habits of various birds in the vicinity of Maldonado during the months of May & June.

Furnarius rufus (b) — No. (1200) is commonly called the oven bird, from the form of its nest.— this is composed of mud & bits of straw, & in shape about ⅔ of a sphere: with~~in & much afterward is~~ a large semicircular opening; within & fronting this there is a sort of partition which reaches nearly up to the roof, so as to form a sort of passage to within the nest.— The bird is very common, often near houses & amongst bushes, is active in its habits, & utters loud reiterated peculiar & shrill notes.— The nest is placed in the most exposed situation on the top of a post, stem of cactus or bare rock. [note (b)] Is now (end of May) working at its nest: it walks on the ground like a dove; & thus feeds on Coleoptera:— Is called "Casera" [copied as "Casita"] or house maker. [note ends] [listed as *Furnarius rufus* Vieill in *Zoology* **3**:64, and see *Ornithological Notes* p. 214]

Copied

(1201) Icterus. Exceedingly abundant, in large flocks, generally making much noise, in habits resembling our starlings: Found also at R. Negro. [listed as *Leistes anticus* G.R.Gray in *Zoology* **3**:107]

(c) (1202) Anthus. resembling in most of its habits a lark, very common; not in flocks; alights on twigs:— [note (c)] Eggs, spotted & clouded with red. nest on ground, simple. No^r (1592) [note ends] [listed as *Anthus furcatus* D'Orb. & Lafr. in *Zoology* **3**:85; numbered 1202? at NHM] *

(1203) Scolopax. flight irregular as in Europe; makes |**178**| a singular drumming noise as it suddenly stoops downwards in its flight; this it frequently repeats whilst flying round & round in a lofty circle.— [listed as *Scolopax magellicanus* King in *Zoology* **3**:131]

Ornithology

(1204) Lanius. (I call all these birds thus, although I believe the greater number belong to Muscicapa, & this species is not very common.— Iris bright red coloured.— [listed as *Xolmis nengeta* G.R.Gray in *Zoology* **3**:54]

(b) (1205) Lanius.— very abundant, most beautiful; sits on a twig or thistle & habits like a true Lanius, but more quiet & not noisy.— [note (b)] the female has some grey on its back & shoulders. This & the foregoing bird seen to catch most of their insects in the air: they frequent the open camp & sit on thistle or twigs.— [note ends] [listed as *Fluvicola azaræ* Gould in *Zoology* **3**:53]

(1206) Muscicapa. common, sits on thistles & habits like English fly catcher, but does not so generally return to same twig; also feeds on the turf; in stomach coleoptera, chiefly Curculios.— beak, eye-lid, iris beautiful primrose yellow:— [listed as *Lichenops perspicillatus* G.R.Gray in *Zoology* **3**:51-2, numbered 1206D at NHM] *

(1207) Fringilla. common amongst the reeds in swamps, loud shrill cry: flight clumsy as if tail was disjointed: base of bill dusky orange.— [listed as *Emberizoides poliocephalus* G.R.Gray in *Zoology* **3**:98; now No. B19600 at the Victoria Museum in Melbourne it carries CD's original label] *

(1208) Arenaria. on sea beach [not located in *Zoology* **3**]

(1209) Fringilla. very abundant in large flocks, female specimen: male with head & ~~throat~~ gorge jet black, colours more brilliant.— [*Chrysomitris magellanica*, *Zoology* **3**:97]

(a) (1210) Alcedo. with long tail, frequents the borders of lakes; sits on a branch or stone & taking short flights dashes into the water to secure its prey.— as might be expected, it does not sit in that upright manner as the European Alcedo [Kingfisher], ~~& the~~ neither is the flight remarkably direct & rapid; but rather undulating as one of the soft billed birds.— [notes (a)] not uncommon; flight weak & short: note low like the clicking of two small stones: in stomach fish, internal membrane of stomach bright orange colour.— Stops in its flight & hovers over one place, as European, also when seated on twigs perpetually elevates & depresses its tail.— Exceedingly abundant on the R. Parana. said to build its nest in trees.— [notes end] [*Ceryle americana*, *Zoology* **3**:42]

(c) (1211) Icterus pecoris, common in flocks, often with Icterus (1201);
(d) frequently alights on the back of cattle: in the same flock there are frequently many brown specimens (1212).— are these one year birds as amongst European starlings?— Females? |**179**| [note (c)] A flocking, when basking in the sun, in a hedge. Many of them sing, but the noise is most curious; resembling bubble of air passing through water from small orifice & rapidly, so as to give an acute sound. I at first thought it was a frog.— [note (d)] egg snow white. Found at Bahia Blanca [notes end] [this parasitical bird was listed as *Molothrus niger* Gould in *Zoology* **3**:107-

8, and as explained in *Journal of Researches* pp. 60-2, is closely related to *Molothrus pecoris* of North America]

[**CD P. 179** commences]

Ornithology (1213) Lanius with a long tail; very active in its habits, in its motions expands its fan tail in same manner as English magpie: is exceedingly
(d) abundant:— often near houses, ~~from~~ at which it feeds on the meat hung up & chaces away other small birds.— harsh note: generally in thickets. [notes] (d) Besides the harsh note this bird has a short warbling song: & is the most musical of any I have heard in this country: yet it only deserves the name of song relatively to the other birds:— More generally frequents thickets & hedges; [added later] (d) August 10th.— Shot at R. Negro specimen (1461).— Inhabiting wild desert plains: manners apparently rather different, wilder, does not seem to use its tail so much.— Alights on summit of twig & enlivens by a very sweet song the dreary plain.— Song resembling the sedge-warbler, but more powerful.— some harsh notes & some very high ones intermingled with a pleasant warbling.— Called by the Spaniards Callandra.— Also found at St Fe Bajada. [notes end] [from *Zoology* **3**:60, specimen (1213) appears to have been the mocking bird identified by John Gould as *Mimus orpheus*, NHM 1855.12.19.227, while specimen (1461) was the closely related *Mimus patagonicus*] *

(1214) Limosa. legs yellow; shot near a lake [? listed as *Totanus flavipes* in *Zoology* **3**:129]

(1215) Sylvia. shot in a Garden.— [possibly the specimen of *Trichas velata* identified in *Zoology* **3**:87]

(c) (1216) Lanius. exceedingly abundant, in habits generally like a butcher bird; also I have often seen it hunting a field by fluttering in one place as a Hawk & then proceeding onwards: it does not, however, stoop so suddenly.— it often frequents the neighbourhead of water, & will in one place remain like a King fisher stationary; it thus catches small fish which come near the margin.— In the evening this bird seats itself on a branch & repeats continually a shrill rather agreeable note without any alteration; & which somewhat resembles some articulate words.— [note (c)] flight undulatory; head as if weighed down by the bill.— When hovering much resembles Hawk: [note ends] [listed as *Saurophagus sulphuratus* Swains in *Zoology* **3**:43]

(e) (1217, 1218) Xanthornus. common in large flocks.— [note (e) added later] Found at Bahia Blanca [note ends] [listed as *Xanthornus flavus* G.R.Gray in *Zoology* **3**:107; labelled 1217D at NHM] *

[**CD P. 179** continues]

(1219) Psittacus. common in small flocks; feeds on the open plain; there is also in this country a wood-pecker: one would not expect to find these two climbers common in a country where there are no trees.— 2500 said to be killed in one year on corn land near Colonia. build conjugating in trees. vast heap of sticks form joint nest; many in islands of R. Parana [listed as *Conurus murinus* Kuhl in *Zoology* **3**:112]

(b) (1220) Turdus? in small flocks; feeding on the plain, in its flight & habits resembling our field-fares.— [notes (b)] Hops, not walks: in stomach seeds & ants: iris rich brown: (b) I have seen this bird at Bahia Blanca; pursuing & catching on wing large Coleoptera.— [notes end] [listed as *Xolmis variegata* G.R.Gray in *Zoology* **3**:55]

(f) (1221) Himantopus. legs rose pink.— [note (f)] This bird is very numerous in the swamps & Fens between Sierra Ventana & B. Ayres: its appearance is by no means inelegant when walking about in shallow water, which is their proper position, wrongly accused of inelegance.— Cry curiously alike to a little dog barking while it hunts.— at night often paused to discriminate [note ends] [listed as *Himantopus nigricollis* Vieill in *Zoology* **3**:130]

(a) (1222) Furnarius(?) common amongst the sand dunes. a quiet little bird.— I do not believe this bird is found South of R. Negro.— [notes] (a) also frequent in the camp: walks, but not well: in stomach Coleoptera, chiefly Carabidous insects.— (a) When disturbed flies but a short distance; set down alights near bushes; is quiet & tame; is it a Furnarius? if so, habits very different from the active habits of "rufus".— (a) At certain times it utters a peculiar shrill reiterated cry (I especially noticed it at Bahia Blanca) in this respect its habits are similar in a small degree to the noisy Oven bird.— (a) Dusts itself; in action in the evening; always very tame:— [notes end] [listed as *Furnarius cunicularius* G.R.Gray in *Zoology* **3**:65-6, NHM 1855.12.19.57. See also extended entry in *Ornithological Notes* pp. 217-18. Labelled 1222D at NHM.] *

(1223) Perdrix. very abundant; does not live in covies: runs more & does
Ornithology not lie so close |**180**| as an English partridge; note a high shrill chirp; but not so much of a whistle as the other greater species.— Flesh most
(Egg 1378) delicately white when cooked; more than a Pheasant.
(1224) V **P 193**.— [listed as *Nothura major* Wagl. in *Zoology* **3**:119, NHM 1855.12.19.34; labelled 1223D at NHM.] *

(1226) Certhia; does not use its tail much, but alights vertically on the reeds & other aquatic plants, which grow round the borders of lakes & which are its resort:— iris rusty red:— [listed as *Limnornis rectirostris* Gould in *Zoology* **3**:80, NHM 1855.12.19.77] *

(1227) in same habitat as last; is in small flocks: in its stomach various Coleoptera [not located in *Zoology* **3**]

(a) (1228) ~~Sylvia~~ Certhia. tailless: same habitat: conceals itself:— [notes] (a) Certhia: turns out not to be tailless: vide specimen in spirits (630). The tail would appear very liable to fall out; even in this specimen it is imperfect: there would seem a great degree of similarity in the construction of this birds tail & that [of] the two Certhias (1226 & ??), as there is in the loosseness of their attachment (a) iris of eye yellow.— legs pale coloured. [notes end] [listed as *Synallaxis maluroides* in *Zoology* **3**:77-8]

(1229) Fringilla. feeding on the fruit of a cactus.
(1230) . . . . .
[listed as *Aglaia striata* D'Orb. & Lafr. in *Zoology* **3**:97-8]

(c) (1231) Muscicapa, not very common.— iris yellow; small eyelid, plain color [note (c)] Generally frequents the rushy ground near lakes: base of bill, especially lower mandible, bright yellow.— eyelid or cere blackish yellow: walks. [note ends] [listed as *Lichenops erythropterus* Gould in *Zoology* **3**:52-3]

[**CD P. 180** continues]

(1232) Emberiza, in very large flocks, feeding on the open plains on the ground: as they rise together, they utter a low shrill chirp.— [listed as *Crithagra ? brevirostris* Gould in *Zoology* **3**:88-9]

(1233) Turdus, ~~not very common~~.— Note of alarm, like English one: [listed as *Turdus rufiventer* Licht. in *Zoology* **3**:59, but NHM 1855.12.19.235 lists it as *T. Albiventer*] *

(1234) Fringilla: not common: in stomach seeds.— [listed as *Pipillo personata* Swains in *Zoology* **3**:98; labelled 1234D at NHM] *

(b) (1235) Rallus; easily rises on being disturbed.— [note (b)] Base of bill, especially lower mandible, fine gree<n> colour.— [note ends] [listed as *Crex lateralis* Licht. in *Zoology* **3**:132]

(1236) Tringa: on the Camp [not identified in *Zoology* **3**]

(e) (1238) Picus, not uncommon; frequents stony places & seems to feed exclusively on the ground.— the bill of this one was muddy to the base:

in the stomach nothing but ants:— cry loud, resembling the English one, but each note more disconnected: also flight undulating in the same manner: they are generally by threes & fours together.— tail does not seem to be used: the tongue is in spirits (620) [note (e)] When it alights on branch of a tree, not vertically but sits horizontally *[illeg.]* very like common birds:— I have since seen it alight vertically; in old specimens a little red in corner of mouth, & tail seems to be used.— Also scarlet tuft on to head. [note ends] [listed as *Chrysoptilus campestris* Swains in *Zoology* **3**:113-14.]

[Writing in *Proc. Zool. Soc. Lond.* 1870, pp. 705-6, CD was rather indignant when in *P.Z.S.* 1870 p. 158, W.H. Hudson disputed the accuracy of the statement that this woodpecker, now named *Colaptes campestris*, often lives in the open plains far from trees, as does the related *Colaptes pituis* of Chile. See p. 244 and *Collected papers* **2**:161-2]

(d) (1239) Lanius; not common; cry rather loud, plaintive, agreeable.— |**181**| [note (d)] Iris reddish orange, bill blue especially lower mandible; there are specimens in which the narrow black & white bands on breast are scarcely visible, & what is more remarkable the under feathers of the tail are only most obscurely barred.— as this absence varied in extent, I imagine it to be the effect of age not sex.— [note ends] [listed as *Thamnophilus doliatus* Vieill in *Zoology* **3**:58]

[**CD P. 181** commences]

<u>Ornithology</u> (1240) Muscicapa, in stomach chiefly Coleoptera [? listed as *Alecturus guirayetupa* Vieill in *Zoology* **3**:51, NHM 1855.12.19.245] *

(1241) Fringilla, not common

(1242) Icterus in small flocks, in marshy places, not so abundant as the other species.— [possibly *Molothrus pecoris*, as discussed in *Zoology* **3**:107-9]

(1243) Scolopax, differs from (1203) in being rather larger & different colours.— it is this bird which more especially makes the drumming noise, & is then very wild.— it is also more abundant.— [listed as *Scolopax magellicanus* King in *Zoology* **3**:131]

(c) (1244) Icterus, not common, marshy places, utters a loud shrill reiterated cry, with beak largely open;— tongue cleft at extremity.— [note (c)] the note of this bird is plaintive & agreeable & can be heard at long distance, is sometimes single, sometimes reiterated; flight heavy, is a much more solitary bird than most of its family.— I have since seen it in a flock, young birds with head & thighs merely mottled with scarlet:— [note ends]

[listed as *Amblyramphus ruber* G.R.Gray in *Zoology* **3**:109-10]

(1245) Fringilla, does not ~~appear~~ to go in flocks [? *Anthus correndera* Vieill in *Zoology* **3**:85]

(1246) Anthus, rare.— [? *Cyanotis omnicolor* in *Zoology* **3**:86]

(1247) Fringilla. in small flocks, amongst bushes, females with very little yellow.—

(1248) Certhia. legs blueish [listed as *Limnornis curvirostris* Gould in *Zoology* **3**:81, NHM 1855.12.19.56 and .74 type, labelled 1248D] *

(a) (1249) Certhia [note (a)] iris bright yellowish orange, legs with faint tint of blue.— [note ends] [listed as *Anumbius ruber* D'Orb. and Lafr. in *Zoology* **3**:80, NHM 1855.12.19.53] *

(1250) Certhia. legs blueish. These three birds together with (1226 &
(b) 1228) are very similar in their habits & general appearance; they all frequent & conceal themselves amongst the rushes & aquatic plants on borders of lake.— the tongue of all of them is bifid & with fibrous projecting points: legs all strong: iris of eyes all yellowish red.— tails have a somewhat similar structure; the note of those I have heard are somewhat similar, a rapid repetition of high chirp.— Yet how different their bills.— Are they not allied to the genus Furnarius?— [notes (b)] These numerous species & numerous individuals seem to play the same part in Nature in this country which Sylvia does in England, feeding on small insects which are concealed amongst the bushes & plants near the margin of water.— (b) When winged crawl with great activity amongst the thickets: tail curiously loose.— I have seen individuals of most of these species flying about without tails. [notes end] [listed as *Limnornis rectirostris* Gould in *Zoology* **3**:80, NHM 1855.12.19.77 type, labelled 1250?. See also *Ornithological Notes* pp. 218-21.]

(d) (1251) Certhia: have never seen more than this one; flight different from length of tail & it alighted on the summit of a thistle in an open & dryer site.— legs blueish, very pale: |**182**| [notes (d)] I have since seen others: they do not frequent the thickets on borders of lakes & especially differ in feeding on the ground.— Furnarius? S. Covington saw the nest of this bird (I recollect seeing one which I then believed to belong to the above). it was made of a vast number of sticks in a thick bush, in length between one & two feet (nearer 2), with the passage vertical, or up & down, making a slight bend both at the exit & entrance of nest itself, lined with feathers.— [listed as *Oxyurus ? dorso-maculatus* Gould in *Zoology* **3**:82, NHM 55.12.19.177] *

[**CD P. 182** commences with entries dated May (latter half), June]

Ornithology (1252) Certhia: legs pale colour, iris rusty red; exceedingly like to (1226), differs in ~~that~~ depth of lower mandible & curvature of upper; I scarcely

June believe it to be a different species, more especially as I found one specimen which was intermediate in character between them both.—

(1255) Certhia. only differs from (1248) in shape of bill. Upper mandible in the latter is longer, & the symphysis of the lower one is of a different shape in the two specimens: Are they varieties or species?

(1256) Certhia: iris yellow reddish; legs pale with touch of blue [? listed as *Synallaxis ruficapilla* Vieill in *Zoology* **3**:79]

(1257) Parus (?) in very small flocks, habits like Europæan genus [of tit]: there is specimen (650) in spirits, because the beak of this one is imperfect.— [listed as *Serpophaga albo-coronata* Gould in *Zoology* **3**:49-50]

(1258) Sylvia, not very common

(1259) Sylvia, uncommon, amongst reeds

(a) (1260) (Furnarius. same genus as (1222)?) This is a common bird: & is always easily distinguished by the double reddish bands it shows in its flight.— Note like (1222) is a succession of high notes quickly repeated; they are here higher: flight similar; but does not walk:— not very tame: chiefly abounds on margin of lakes, amongst the refuse; but also common in the camp: in stomach nothing but insects & almost all Coleoptera; some of them were Fungi-feeders: often picks the dung of cattle. tongue of a bright yellow colour:— I know nothing of the nidification of this bird or of (1222); but it [is] clear they do not make nests like Fur: rufus; for they could not escape notice in such open countries as that of Falkland Is^{ds}. — Bahia Blanca & the country:— |**183**|

[note (a) added later] This species & (1222) make their nest by boring a hole said to be nearly 6 feet long in a bank of earth. A thick strong mud wall, round a house at Bahia Blanca, was perforated in a score of places by these birds, thinking it to be a bank or cliff: curious want of reasoning powers, since they were constantly flying over it.— The species (1222) I hear is found at Cordova, as I have seen it at St Fe.— I know not how much higher it is found.— M. Lisson is curious about the nidification of these birds.— They are called Casarita, as the Oven bird is called Casar<a>. the Spaniards have observed their alliance, although their nidification, the original cause of name, is different.— [note ends] [listed as *Opetiorhynchus vulgaris* G.R.Gray in *Zoology* **3**:66-7. There was a copy of René-Primevère Lesson's *Manuel d'ornithologie* (2 vols. Paris, 1827) in the *Beagle*'s library]

[**CD P. 183** commences with date now altered to June (early part)]

<u>Ornithology</u> (1261) Lanius (?). Legs pale blueish; iris reddish: I have never seen but this one specimen: Coleoptera in stomach. [listed as *Cyclarhis guianensis* Swains in *Zoology* **3**:58]

(1262) Fringilla. uncommon.— [listed as *Ammodramus manimbè* G.R.Gray in *Zoology* **3**:90; No. B19633 at Victoria Museum, Melbourne]

(1263) (Charadrius) legs "crimson red"; toes leaden colour, under surface most remarkably soft & fleshy: in small flocks common in open plain; often with Turdus (1220); as they rise utter plaintive cry: iris dark brown:— [listed as *Oreophilus totanirostris* Jard. & Selb. in *Zoology* **3**:125-6]

(1264) Rhyncops: base of bill & legs "vermilion red". This curious bird
(a) was shot at a lake from which the water had lately been drained & abounded with small fish.— They were in flocks: I here saw what I have heard is seen at sea: these birds fly close to the water with their bills wide open, the lower mandible is half buried in the water. they thus skim the water & plough it as they proceed: the water was quite calm & it was a most curious spectacle to see a flock thus each leave ~~the~~ on the water its track: they often twist about & dexterously manage that the projecting lower mandible should ~~trip~~ plough up a small fish, which is secured by the upper.— This I saw as they flew close to me backwards & forwards as swallows: they occassionally left the water, then the flight was wild, rapid & irregular: they then also uttered a harsh loud cry: The length of the 1st remige must be very necessary to keep the wing dry: the tail is most used in steering their flight: It appears to me their whole structure, bill weak, short legs, long wings, appear to be more adapted for this method of catching its prey than for what |**184**| M. Lisson states, viz. that they ~~catch~~ open & eat Mactræ buried in the sand.— [see R.-P. Lesson *Manuel d'ornithologie* Vol. 2, p. 385]

[**CD P. 184** continues]

I have stated that at M: Video, when these birds are in large flocks on the sand banks, that they seem to go out to sea every night.— now if I were to conjecture, I should imagine that they fished at night, when their only method of catching prey would be by thus furrowing the water: it is probable that they eat other animals besides fishs; & many, for instance

Crustaceæ, come to the surface chiefly at night.— It would be curious to note whether the lower mandible is well furnished with nerves as an organ of touch.— I imagine these birds fishing by day in a fresh water lake an extraordinary circumstance, & depended solely upon the myriads of minute fish which were jumping about.—

October [note (a) added later] These birds are common far inland near the R. Parana. They rest on the grass plains, in same manner as in day time near the sea on mud banks: are said to stay whole year & breed in the marshes. One evening near Rozario, as it was growing dark, we were anchored in a narrow Riacho or arm; here there were many smaller fry, & I saw one of these birds rapidly flying up & down ploughing the water as described at Maldonado. Class. Dic. is aware of this habit.— I think these & other marine birds perhaps enter far inland the more nearly from its extreme flatness. [note ends] [the Scissor-beak *Rhynchops nigra* Linn. is discussed in *Zoology* **3**:143-4, and it is mentioned that Richard Owen had dissected the head of a specimen brought home by CD in spirits, but had not found any special innervation in the lower mandible. See also *Ornithological Notes* pp. 221-3]

[**CD P. 184** continues]

(1268) Larus. common in flocks near a lagoon

(1269) Ardea. not uncommon, also in Patagonia: hoarse cry: iris & cere, bright yellow — bill waxy colour.— [listed as *Egretta leuce* Bonap. in *Zoology* **3**:128]

(1270) Owl. uncommon: in long grass, flew in mid-day:— [listed as *Otus palustris* Gould in *Zoology* **3**:33. Labelled 1270D at NHM] *

(1271) Sylvia. (male of 1259?)

(1272) Palombus. uncommon.— [listed as *Columbina strepitans* Spix. in *Zoology* **3**:116. Carries CD's own label numbered 1272 at NHM.] *

(1273) Perdrix — Scolopax. male of (1224)

(1274) Turdus [listed as *Turdus rufiventer* Licht in *Zoology* **3**:59, NHM 1855.12.19.235, labelled 1274?] *

(1275) Alecturus; sits on a thistle, from which by short flights catches prey: in stomach Lycosa & Coleoptera. tail seems useless in its flight.— [listed as *Alecturus guirayetupa* Vieill. in *Zoology* **3**:51, NHM 1855.12.19.245] *

(1276) Alecturus. is this different species?

(1277) Parus (?). most beautiful. amongst reeds. very rare.— Soles of feet, fine orange: |**185**| [note (a)] This bird is also found at Bahia Blanca [note ends] [listed as *Cyanotis omnicolor* Swains in *Zoology* **3**:86]

[**CD P. 185** commences]

Ornithology (1293) Owl.— Excessively numerous. mentioned by all travellers as a striking part of the Zoology of the Pampas & live in burrows especially where the soil is sandy. in B. Ayres seem exclusively to use holes of the Biscatche: stand on the hillocks near their hole & gaze on you: are generally out in the day, but more especially in the evening.— flight remarkably undulatory: very frequently utter shrill harsh cries on the wing & occassionally hoot: in stomach of one, remains of mice. if I had not known by my traps, the extraordinary number of the smaller Rodentia, I should have been puzzled to have conjectured on what food such great numbers of owls could live on.— I saw one kill a snake; said often to do, cause of appearance by day. [listed as *Athene cunicularia* Bonap. in *Zoology* **3**:31-2, labelled 1293D in NHM]

(a) (1294) Vulture. very abundant around the Ranchos & towns: these small carrion feeders, in large flocks, finish what the host of large black ones (called Cuervos & Carranchas) have begun. Called Chimango. [listed as *Milvago chimango* in *Zoology* **3**:14-15]

[note (a) added later] This Vulture & the Carranchas (saw a Carrancha at Cape Negro in the St[s] of Magellan.—) frequent the dryest most sterile plains, & feed on the animals which dies; in such passages as between R. Negro & Colorado.— the (Gallinoras?) or black Cuervos always frequent damp places. I have seen them at the Colorado &c &c. they would seem to require animals in a more rapid state of putrefaction; & do not like picking dry bones.— it is natural for I believe they are are more abundant within the tropicks.— They are certainly pretty gregarious; on fine day ~~which~~ wheel at great height in graceful turns in bodies, uttering short cry.— clumsy near the ground, but run fast:— Carrancho utters very harsh cry like Spanish G & rr.— very crafty, steal eggs.— do not run fast, or soar, or gregarious.— ~~build in cliffs~~ if a person lies down in the plain, one of these birds will soon appear & patiently watch you with an evil eye.— (V **P 239** & **P. 260**).— more particulars.—

[note continues with different pen] Chimango very abundant archipelago of Chiloe (known by diff name), will eat bread: often injures potatoe fields by scratching them up & devouring them!.— Is a great enemy to the Carrancha: When the latter is seated on a branch, the Chimango flies in a semicircle ~~backwards & forwards~~ upwards & downwards, trying to strike at each turn the other. Will continue thus flying for a long time [note ends]

[**CD P. 185** continues]

(1295) Water hen. bill fine green: legs brown, toes with much membrane.— [listed as *Crex lateralis* in *Zoology* **3**:132]

(1296) Parus (?). common on the borders of lakes or ditches with water; frequently alights on the aquatic plants.— expands its tail like fan when seated on a twig.— [listed as *Serpophaga nigricans* Gould in *Zoology* **3**:50]

(1297) Rare & beautiful Fringilla.—

(1340) Palomba.— legs coloured dull "carmine red". frequent the Indian corn fields in large flocks.— [listed as *Columba loricata* Licht. in *Zoology* **3**:115]

(1349) Thalassidromus shot in the bay being driven in by gale of wind; walks on the water, very tame:— [listed as *Thalassidroma oceanica* Bonap. in *Zoology* **3**:141]

(1382) Perdrix. much rarer than the other species: they are generally
(b) found several together; flesh [when cooked] snow white; are unwilling to rise, uttering a whistle shriller than in species (1223) whilst on the ground.— Generally frequent marshy places on borders of lakes.— In the common [continued on **P. 185(bis)**, on back of **P. 185**] partridge, the habit of uttering a whistle before rising on the wing, is different from the English one.— [note (b)] Found also at B. Blanca. [note ends] [listed as *Rhynchotus rufescens* Wagl. in *Zoology* **3**:120]

(1383) Ostralogus — Guritti Island

(1384) Sterna do do

(1385) Palomba.— exceedingly abundant, living in small flocks in every sort of situation.— [listed as *Zenaida aurita* G.R.Gray in *Zoology* **3**:115]

(1390) Larus. soles of feet deep "reddish orange", legs & bill dull "arterial blood red". <u>Breeds</u> & frequents fens <u>far</u> inland. in B. Ayres: slaughtering houses. [listed as *Xema (chroicocephalus) cirrocephalum* G.R.Gray in *Zoology* **3**:142]

(1396) Falco. not very uncommon |**186**|

[**CD P. 185(bis)** having ended, **CD P. 186** commences]

<u>Ornithology</u> The following observations are necessary to complete the Ornithology in the neighbourhead of the town.— There are several sorts of Hawks which I have been unable to procure: of the carrion feeders there are three which I have not.— ??? ~~The brown sort, which is so plentiful at the Falklands is not very common here~~, but the large black ones (here called Cuervo) are excessively so.— I have never seen the Turkey Vulture:

(b) Amongst the Passerinæ my collection is very perfect: day after day & walking long distances impossible to procure any others.— Amongst birds which I have not, a sparrow (there was a specimen (683) & (1615)[1] at M: Video.) this bird is excessively common.— often near houses; but not in flocks: they have not that air of domestication which the English ones have:— *no more than the gorged Vultures, of a blackish color, resemble*
|| *Rooks*[1]. There also is a black bird with rusty back & long claw (903)
|| common on sand dunes:[2] Also Sturnus ruber, not very abundant: ~~I have never seen the Cardinal~~ *The Cardinal is found here*[1]: There is a larger species of Kingfisher, *same as in T del Fuego*[1]: ~~a large partridge~~: Ostrich:
(c) a Vanellus[3] (1602) with horn to wings is exceedingly abundant: is called "pteru-pteru" from their incessant & odious harsh cry: always seem to wish to attack you: give notice to all other birds of your approach.— [note (c)] The bird seems to hate mankind: shams death like the Peewit.— eggs pointed oval, brownish olive thickly spotted with dark brown. [note ends] There is a large sort of Water Hen: There are some duck, & black necked swan & others with black tips to wings:— [note (b)] Capt Fitz Roys collection has another ~~Cassicus~~ Icterus & another Parus(?): evidently both rare birds: Decemb. Icterus is (1418): *Also Certhia (1451) occurs at Maldonado*[1]. [note ends]

The birds generally are very numerous in the camp: especially Cassius & Lanius (or more properly Tyrannius).— It is impossible not to be struck with great beauty: the most general colour is yellow, & it is worth noting that from the prevalence of certain flowers this is the general tint of the
(a) pasture.— [note (a)] As Songsters they are miserably deficient: I have never heard one which could compare with one of our English performers, although of a low class.— [note ends] |**187**|

[1] The words shown in italics were added later with a different pen.
[2] Listed as *Muscisaxicola nigra* Gray in *Zoology* **3**:84, now *Lessonia rufa.*

[3] *Vanellus* is the English peewit. Specimen 1602, the bird called pteru-pteru, was listed in *Zoology* **3**:127 as *Philomachus cayanus* Gray.

[**CD P. 187** commences]

Hyla[1] 606 Copied	Hyla.— above emerald green, beneath white, on sides a black & silvery stripe, also a shorter one at corner of mouth.— under side of hinder legs & side of abdomen marked with black spots. tympanum brown, iris gold-colour. Hind feet semipalmated.— They frequent in great numbers the open grass camp, also marshes.— These can never ascend trees, for they are entirely wanting.—
Rana[2] 607 Cop.	Brown, with circular & asymetrical marks of black.— always in immediate neighbourhead of water.— Same as in Brazil?
Rana[3] 631 Copied	Eye very prominent; behind & by the side of them fine green markings; body brown with black markings; beneath silvery, with lateral band do:
Coluber[4] 644 (a) Copied 645	Above "clove brown", shading beneath into pale; on the sides & back, there are regular black spaces with yellow specks; likewise whole length of body two narrow dorsal ribbons of "saffron yellow": on under side of tail a broard central band of "tile & ½ scarlet red".— there is also on the back a faint trace (chiefly shown by interrupted chain of specks) of a similarly coloured band. [notes (a)] Upon taking this animal out of spirits I observed in its ~~worms~~ mouth several small worms; as there was a tight ligature (to kill it) round the neck, they could not have proceeded from the stomach. In the mouth of another Coluber (623) I noticed one alive (the animal being strangled as the former one), & if I remember right it crawled like a leach by the aid of its extremities. Common in the swampy plains between Sierra Ventana & B. Ayres. [notes end]

[1] Hylidae. Listed by Thomas Bell in *Zoology* **5**:46-7 as *Hyla agrestis* Bell. Currently *Hyla pulchella pulchella* Duméril.
[2] Leptodactylidae. *Leptodactylus mystachinus* Burmeister.
[3] Leptodactylidae. *Leptodactylus ocellatus* Linn.
[4] Colubridae. *Liophis anomalus* Günther.

[**CD P. 187** continues]

Cavia cobaya[1] 1266 (not spirits) Head 1318 (not spirits)	This animal called the Aperea is exceedingly abundant.— it inhabits the sand dunes, hedge rows of Cactus, & especially marshy places covered with aquatic plants. On gloomy days & in the evening they come out to feed, are not very timid & can easily be shot. In dry places they have burrows, but in swamps the mud is so soft that it is impossible. They are very injurious to young trees in the garden.— The hair is remarkably loose on

Copy (b) their bodies.— An old male weighed 1 lb. 3 oz. (Imperial weight) |**188**|

___
Cop

Jan: 1834

[note (b) added later] Killed in August at R. Negro another species[2] (1471); besides the difference in colour & fineness of hair: it is smaller; & in habits is tamer, more of a day feeder: frequent dry hedges, produces two young at a time (good authority). I have specimen of its head (1587): Generally called Conejos[3].— Old male, Port Desire, weighed 3530 gr[s].— [note ends]

[**CD P. 188** commences]

Rodentia
Talpiformes
Toco Toco
1267[4]
(not spirits)
Head (a)
1311
(not spirits)

659
Spirits

This curious animal is abundant, but difficult to be procured & still more difficult to be seen at liberty:— it lives almost entirely under ground; prefers sandy soil & gentle inclination, as for instance where the sand dunes join the camp, but they are often found in other situations.— it is not often that there is an open burrow; but the earth is thrown up as by a mole & generally at night.— the burrows are said not to be deep but of great length.— they seem gregarious.— the man who procured my specimen found six together; in many places the ground is so much undermined that the horses hoofs sink into it.— They are well known & take their name from their peculiar noise: the first time it is heard, one feels much astonished, as it is not easy to judge where it comes from & it would be impossible to guess what made it:— It consists in a short nasal noise repeated for about four times in succession: the first time the noise not being so loud & more separated from the others: the musical time is constant.— This noise is heard at all times of the day.— It is said that they come out at night to feed; that they come out is certain for I have seen their tracks, but I must think that their principal food is roots; it is the only way of accounting for their extensive burrows.— In the stomach of one there was a yellowish greenish mass, in which I could only distinguish fibres.— |**189**|

Toco Toco

When kept in a room.— They move slowly & clumsily, chiefly from the outward action of their hind legs: cannot jump: their teeth (of a bright wax yellow) cannot well cut wood: when frightened or angry make their peculiar noise; are stupid in making attempts to escape: When eating biscuit, rest on hind legs & hold it in fore paws; appeared to wish to drag the food away: Many of them are very tame, & will not attempt to bite or run away, others are a little more wild.— The man who brought them [asserted[5]] that very many are always blind: specimen (659 for dissection) would appear to be so; did not take any notice of my finger when placed within ½ an inch of its head.— it made its way about the room nearly as well as the others.— An old male weighed *[no weight given]*

[note (a) added later on back of **CD P. 188** is headed: Covington — Copy all this out at end of regular account]

(a) At R. Negro (in August) an animal frequents the same sites & makes the same burrows: but the noise is decidedly different: it is more distinct, louder, sonorous, peculiar, much resembles the sound of a small tree being cut down in the distance.— the noise is repeated twice & not 3 or 4 times as at Maldonado.— At Bahia Blanca the animal makes a noise repeated at single intervals, at equal times or in an accelerating order.— I was assured these animals were found of different colours.— Having caught one of the Gerbillos (1284) I was assured that this was the Toco Toco which made the noise.— Very many people said the same.— What is the truth? Monsieur Dessalines d'Orbigny[6] who collected many animals at R. Negro must have specimens of them both &c &c.— Immense tracks of country between R. Negro & Sierra Guitro-Leignè are curiously injured by these animals; the horses fetlock sinking in every 2 or 3 steps.—

Feb 3rd.— 1834 At Cape Negro, the last of Patagonia, where features of Tierra del F are present, the ground is a warren of holes: several heads were lying about, of which (1795) may perhaps be sufficient to recognise identity of species. [note ends]

[1] Described by George Waterhouse in *Zoology* **2**:89 as *Cavia cobaia* Auct.
[2] Described by George Waterhouse in *Zoology* **2**:88-9 as *Kerodon Kingii* Bennett.
[3] A modern Spanish dictionary gives the translation of 'conejo' as a 'rabbit'.
[4] Described as *Ctenomys Braziliensis* Blain. by George Waterhouse in *Zoology* **2**:79-82, where an extended account of the species is given, based on this entry and the slightly revised version copied out later, not by Syms Covington but by CD himself.
[5] The word 'asserted' was originally omitted by CD, but was inserted when he recopied the sentence.
[6] Alcide d'Orbigny was a palaeontologist sent out by the French government to South America, who as reported by CD to Henslow in a letter dated 24 November 1832 had just been working on the Rio Negro for six months. A report on his labours reached CD in 1835, and a full account was later published in Paris. See *Correspondence* **1**:280-2.

[**CD P. 189** continues]

Lycoperdium[1] or Phallus 647 Copied — Nearly all my specimens are in their young state.— They then look like the bulb ~~of~~ from which the Phallus springs, only with the difference that the outer coat is penetrated with apertures.— This outer coat seems to expand untill it becomes a bag of trellis work.— There is a fragment showing the structure.— They are of a salmon colour.— but through the aperture the internal parts are brownish green.— They grow on the sand dunes & near to a Phallus, but appear to be uncommon.— Did not possess any strong odour.— |**190**|

Lacerta- Sides of body light rich brown, with black marks, a longitudinal white line

Ameiva[2] 648,649	on each side; Within these & the inner brown for ⅔ of anterior part of body there is a fine emerald green colour.
Hyla[3] 652 Copied	Above coppery brown, mottled with black, which latter colour is most distinct on hinder thighs & sides of body extending over the tympanum a blackish brown band; iris coppery on edge of upper mandible white line.— Caught under stone
Rana 653 Copied	Body above light greenish yellow, with lateral brownish black band & distinct circular patch on sides before the thighs.— There are obscure longitudinal marks on ~~under~~ upper surface of thighs & the under is tinged with reddish orange.— Caught under a stone
Bufo 654 Copied	Above yellowish green, with central line on back more bright: *[illeg.]* brown; beneath yellowish.— Under stone. Same as at M Video?.—
Coluber 663 Copied	Above dark "Pistachio green", with central narrow dorsal line of brown: beneath "Aurora & ½ Vermilion red" but mostly on posterior half of body, altogether very beautiful. open camp.—
Lycoperdium[4] 664 Copied	This curious fungus consists of a dark brown bag containing powder, like a common Lycoperdium: but instead of growing on the ground, it is seated on a circular flat disk (of a lighter colour) the superior & inferior edges of which are cracked & curled.— They would seem like sphere burst through, especially the lower one: which latter is slightly attached to the soil.— Grow in damp & rather shady places:— |**191**|
Bufo[5] 613 Copied	This is the same extraordinarily coloured animal which I found at Bahia Blanca (**P 99**).— They were not very uncommon amongst the sand-dunes: the quantity of marks of "buff orange" varied, in some individuals ~~being~~ these being more, in some less than at B. Blanca.— Eye jet black.— When placed in water could scarcely swim at all.— & I think would shortly have been drowned.— They crawl about during the day & frequent the driest places.—

[1] Identified as *Clathrus crispus* var. *obovatus* Berkeley in *Plant Notes* pp. 224-5.

[2] Listed by Thomas Bell as young specimens of *Ameiva longicauda* Bell in *Zoology* **5**:29.

[3] Listed as *Hyla Vauterii* Bibr. in *Zoology* **5**:45-6.

[4] Identified by M.J. Berkeley in *Annals and Magazine of Natural History* **9**(1842):447 as *Geaster saccatus* Fries.

[5] Identified by Thomas Bell in *Zoology* **5**:49-50 as *Phryniscus nigricans* Weigm.

[**CD P. 191** continues]

Insecta
610
or
328[1]
June 1833

The following facts I have noticed at M. Video & frequently in this place:— After a heavy thunder storm in a little pool in a court-yard which had only existed at most seven hours.— I observed the surface strewed over ~~the~~ with black specks; these were collected in groups, & precisely resembled pinches of gunpowder dropped in different parts on the surface of the puddle. These specks are Insects of a dark leaden colour; the younger ones being red.— Viewed through a microscope, they were continually crawling over each other & the surface of the water; on the hand they possessed a slight jumping motion.— The numbers on each pool were immense:— & every puddle possessed some of the pinches.— What are they? & how produced in such countless myriads? We have seen their birth is effected in a short time, & their life, from the drying of the puddles, can not be of a much longer duration.—|**192**|

[1] Identified in *Insect Notes* pp. 40-3 as *Collembola*, or springtails.

[**CD P. 192** commences, crossed through vertically to foot of page as are previous entries on Planaria, and continued for the entry on capybara]

Planaria[1]
627

I found under stones, on rocky hills, great numbers of terrestrial Planariæ.— in same manner as mentioned **(P 71)** at M: Video.— There are two species they seem to be the same as there described.— I observed two of them in perfect close contact on the under surface.— Is it a generative process?— On opening the body at the situation of orifice, there was a hard white cup-shaped organ with a sinuated margin.— The animal not being quite dead.— This expanded & contracted itself.— I have not the slightest doubt, if this organ was protruded & perfectly expanded it would present the appearance described in Planariæ **(P 15 & 21)**.— When the contraction was most it might be described as being star shaped, from the sinuated margins being drawn in to central point.— On the surface I noticed the corpuscular motion.— [note written vertically in margin] (Ocelli numerous black round anterior extremity & foot) [note ends]

Cavia
Capybara[2]
(a)

Copied

These animals are abundant on the borders of the lakes in the vicinity of Maldonado, & occassionally frequent the islands even at sea: During the last voyage two were shot on Goriti.— At Maldonado Three or four generally live together; in the day time they are either lying amongst the aquatic plants or feeding openly on the turf plain.— When viewed at a distance, from their manner of walking & colour they resemble a pig; but when seated on their haunches & watching with one eye, they reassume the appearance of their congeners the Agoutis.— Their great depth of jaw gives to their profile & front view a quite ludicrous appearance.— They are very tame, by cautiously walking I approached within |**193**| three yards of four large ones: As I came nearer they frequently made their peculiar noise; it is a very abrupt one: there is not much actual sound, but

Cavia
capybara

rather the sudden expulsion of air.— The only noise I know at all like it is the first hoarse bark of a large dog. Having watched them (& they me) for several minutes, almost within arms reach.— They rushed into the water with the greatest impetuosity at full gallop; & emitting at the same time their bark.— When three or four thus dash in together the spray flies about in every direction.— After diving a short distance, they come to the surface, but only show just the upper part of the head.— In the stomach & duodenum of one there was a great mass of a yellowish liquid matter, in which nothing could be distinguished.—

[note (a) added later] These animals I believe do not occur South of the R. Plata.— I could not hear of any at the R. Negro.— Number in islands of Parana & Uruguay. chief food of the Jaguars.— where there are ~~not~~ many Capinchos there is ~~most~~ not much fear of these animals. In the water the two young of the Capincho often sit on its back.— N.B. There is a Laguna Carpincho East of B. Ayres, at the higher part of the Salado.— [note ends]

[**CD P. 193** continues]

Perdrix—
Scolopax(?!)
1224 & 1273[3]
not spirits
707 spirits
(a)

Copied
Perdrix-
Scolopax

At **P 99**, I have mentioned this bird.— They were more abundant here.— They generally frequent the same spot; & that always a dry one.— I have repeatedly noticed them in a particular part of a dry road.— They are either in pairs or in small flock; when in the latter they all rise together, when in former one waits (even when one is shot) for the former: As they rise they utter a cry like a Snipe & in same manner fly high & irregularly & generally a long distance.— they however occassionally soar for short distance like a partridge.— Their general habits so much resemble a snipe |**194**| that our sportsmen call them "short-billed snipes".— their real connection is marked by the length of the Scapulars.— When on the ground, they squat close to escape observation & are not easily seen; in this position, & when walking from the width which their legs are apart, they resemble a Partridge.—

In the stomach of several which I opened there was nothing but pieces of rushy grass, the summits of which were pointed, also small bits of some leaf & grains of quartz: the intestine & dung were bright green.— In another (killed at different time) there were seeds & a dead ant.— The specimens have either black markings round the neck or not.— They are specimens in Spirits of both.— Male & female?

[note (a) added later] This is perhaps the most common bird in the dry plains between the R: Negro & Sierra de la Ventana.— it runs in flocks from 3 or 4 to 30 or 40 in number.— ~~it~~ is said to builds on the borders of lakes & has 5 or 6 eggs in its nest, white spotted with red.— In its nidification & flocking resembles Snipes; is called by the Spaniards.— Avescasina.—

[**CD P. 194** continues]

Bufo[4] 665 Copied	Elegantly marked with black & pale green; colours most vivid on the lumbar glands; hinder thighs with little tinge of orange on softer parts.—
Coluber[5] 673 Cop.	Above "sage green", shading into beneath "siskin green": most beautiful:
Coluber 674 Copied	Scales generally dirty "oil green", the interstices on the sides & edge of ventral plates, dark brown. these brown interstical [sic] spaces likewise form numerous irregular transverse bars on the back; the ~~sides~~ scales themselves in these parts being brown; beneath with dirty "siskin green".—
Coluber 675 Copied	Ventral plates fine "Vermilion red["], becoming paler towards the gorge, with black specks on each side; sides "greenish grey", back reddish grey, with central "blackish grey" line: head & upper side of neck, "umber brown".— |**195**|

[**CD P. 195** is headed: Specimens collected by the Officers in Schooner, Coast of Patagonia. Note in margin says 'Copied all on this page']

Agama[6] 681	General colour blueish grey with tinge of rust colour on back. broard transverse bands with white undulation behind them.—
Agama 682	General colour not so blue, with pointed, bright yellow undulations in hinder part of brown band
Agama 683	General colour rather darker; back dark brown with central light reddish longitudinal band with small transverse ones branching off.—
Agama 684	Pale reddish grey, brown transverse bands, yellowish white posterior undulations
Agama 685	General colour especially tail much redder:
	All these Lizards were caught at Port Desire in beginning of January by the officers in the small Schooners.—
Lizard 686	On back transverse rows: each with 3 semilunar rich brown marks, edged with cream colour. Lateral line of same colour; about head traces of bright green.— Port Desire
Lizard 687	Mud colour with lighter lateral line.—

Bufo 689 — Head remarkably flat, dark grey, with much blacker & symetrical markings. Rio Chupat

Bufo 690 — Slate colour, with dark markings.— Rio Chupat.— B. Engaño Bay.—

All the above specimens were collected by the officers in the Schooners under the command of Mr Wickham[7], during the summer of the year: the colours of each were stated not to have altered, only to be less vivid.— |**196**|

[1] Identified by Darwin (1844) (*loc. cit.*) as *Planaria pulla*, currently *Pseudogeoplana pulla* Darwin. The animals might indeed have been copulating, though another possibility is that one was eating the other.
[2] Identified in *Zoology* **2**:91 as *Hydrochœrus capybara* Auct. See also *Journal of Researches* **1**:56-8.
[3] Specimen 1273 was the male of 1224. CD was mistaken in identifying the bird as *Scolopax*, and it was identified in *Zoology* **3**:117-18 as *Tinochorus rumicivorus* Eschsch. For an account of its anatomy see *Zoology* **3**:155-6.
[4] This frog was said by Thomas Bell in *Zoology* **5**:36-7 to be 'remarkably bufonine', and was listed as *Pleurodema Darwinii* Bell.
[5] See list of Specimens in Spirits of Wine for Thomas Bell's identifications of specimens 673-6 and 682-90.
[6] From the description of its colouring given by Thomas Bell in *Zoology* **5**:21-2 this lizard is confirmed to be *Diplolæmus Bibronii* Bell.
[7] Lieutenant John Clements Wickham was second in command of the *Beagle*.

[**CD P. 196**, dated June-July at its head, commences]

Mus[1] 1280 not Sp^s. — Was killed at the Island of Goriti where they are said to be common.— They are also said to ~~be~~ occur in numbers at East Point.— They inhabit burrows in the sand dunes.— It is a likely place for ships to leave this animal, if they are infected with such monsters.— But I think from habits it is an aboriginal.— The occurrence at Island Goriti is no difficulty as a reef now connects ~~the~~ it with mainland, probably was once continuous.— The ears were whitish & oddly contrasted with rest of body.— An old male weighed 15 & ¾ oz:

Cervus[2] 1292 (N^t Spirits) (815) (b) not spirits Horns (z) 1440:1441 (old in front) Cop — Are very abundant in the mamillated plain round Pan de Azucar.— Manners resembling those at B. Blanca.— This specimen was shot out of a herd of seven.— The Gauchos say he is nine years old:— teeth all decayed.— Smell intolerably strong & offensive, almost creating Nausea.— this seems to occur at seasons when the Horns are perfect: Out of same herd (without moving I shot three. from having crawled a long distance the deer did not know what I was & as usual advanced to reconnoitre me) I shot another & younger buck.— Horns (1337 & 1337)

[notes added later] (b) The officers of the Beagle have never seen this animal to the South of the R. Negro.— The smell is most offensive.— I have often perceived the whole air impregnated, when the distance of Buck could not have been less than ½ a mile to Windward.— Are said by Gauchos to change their horns annually. Seem to like mountains. excessively numerous near the Sierra Ventana.— But they are spread more or less over the whole country.—

(z) A pocket handkerchief, in which I carried on horseback the skin; has constantly been in use since & therefore repeatedly washed; not withstanding this, now 13 months have intervened I know this handkerchief from the others by its smell.—

Cervus Campestris. It will be seen in my journal[3] when I shot the deer at Maldonado: a pocket handkerchief, in which I carried the skin, has been in constant use & repeatedly washed, yet in December 1834 the odour was very perceptible.— [in different pen] do in Jan:— 1835 [notes end]

[**CD P. 196** continues]

Procellaria[4] I took a specimen at Maldonado which I suppose is "gigantea", appear however to differ in colour: colour "greyish black", or shade darker above
(c) & one lighter beneath.— The following measures may help to point out differences with any future specimen: Extreme points of Tarsus of legs 3.4^{inch}, measured on outside: Fibula from centre of articulations 10.8^{inch}:
— Lower mandible from feathers to extremity 3.15: nose on central part from
Cop a membrane at base to concavo—truncate extremity, 1.65^{in}: depth of bill, including nose, 1.2^{in}: 16 rectrices:.— |**197**|

[note (c) added later] Specimen (2080) procured at Port Famine.— Mr Low[5] says it [is] the young one of the common grey sort.— Their flight however appears rather more elegant, & the distinction of color strongly marked. I have long notice<d> this bird & thought it was a different species. They build at ~~Malaspina~~ Sea Lion Is^{d}, S. Cruz & other places on coast of Patagonia.— The officers have seen them at P. St. Antonia pursue & kill some sort of Coot.— The latter tried to escape by flying & diving: but was continually struck & beaten by its enemy. at last when rising from beneath the water the Nelly cut ~~his~~ its head off with its bill. At Port St Julian there was the bill of a very large Cuttle fish in the stomach: flight very like albatross; often settles & rests on the water: frequent inland ba<ys> & as well as open sea.— I think not generally very far from the coast.— Specimen (2080), bill wax white: legs black, upper surface greyish.—

June 15th I saw two this very day, 80 miles from West Coast of Patagonia.— [notes end]

[1] Identified in *Zoology* **2**:33-4 as *Mus (decumanus* var. ?*) maurus*.
[2] Identified in *Zoology* **2**:29-31 as *Cervus campestris* Cuvier.
[3] See entry for 20-28 June in *Beagle Diary* p. 160.
[4] This large petrel known to the English as a "Nelly" was listed in *Zoology* **3**:139-40 as *Procellaria gigantea* Gmel.
[5] William Low was a Scottish trader and sea captain for many years in the waters around Patagonia, who provided CD with much valuable information.

[**CD P. 197** commences]

Siliceous tubes from Lightning[1]

In the great sand dumes which separate Laguna del Potrero from the sea.— I found numerous fragments of those siliceous tubes, which are supposed to be formed by lightning entering the sand.— The dumes are not protected by vegetation & are in consequence perpetually moving their position.—
1375 From this cause I first observed the tubes projecting out, & fragments which
1376 clearly were parts of the same broken off, & strewed immediately around.— [note added at top of page] circumference of biggest smooth one 4:(.2 inches) [note ends] I found four of these entering the sand perpendicularly & going deeper than I could trace.— By clearing away the loose dry sand I traced one for two feet, & close to this there were fragments, which placed together, formed a tube 3^{ft}.3 inches long; So that here the tube must have been 5^{ft}.3^{inch} in length, & as the diameter was the same throughout, probably extended to a far greater depth.— At the level of about 12 feet below these were pools of water, left by rain: it is probable that these tubes penetrate to where the sand is of so damp a nature, as easily to conduct the electric fluid. Besides the four tubes which I found vertical & traced beneath the surface, there were several other groups of fragments, the original site of which was doubtless near.— The situation was upon a level piece of bare sand & amongst lofty sand-dumes; at about ½ of a miles distance there was a chain of hills of 400 or 500 feet in height. The internal surface of these tubes is vitrified; the external is very rugged with longitudinal furrows: the grains of sand which adhere to it are the same as the surrounding mass.— This sand is peculiar in possessing no |**198**| scales of mica.— The diameter of different sets varied; in shape more generally compressed, sometimes circular.— They entered the sand vertically, in some however there were slight bends.— In one case, which was much more irregular than the generality, the deviation at the bend from a right line amounted to 33°.— In this same one, there were two small branches which gradually tapered to a point; they were about a foot apart, & one pointed downward, the other upwards.— In this latter, the branch with the stem included an angle of 26°, this is remarkable as one would not expect the Electric fluid to make [an] effort to return at so acute an angle.—

Lightning tubes

V: Fig:

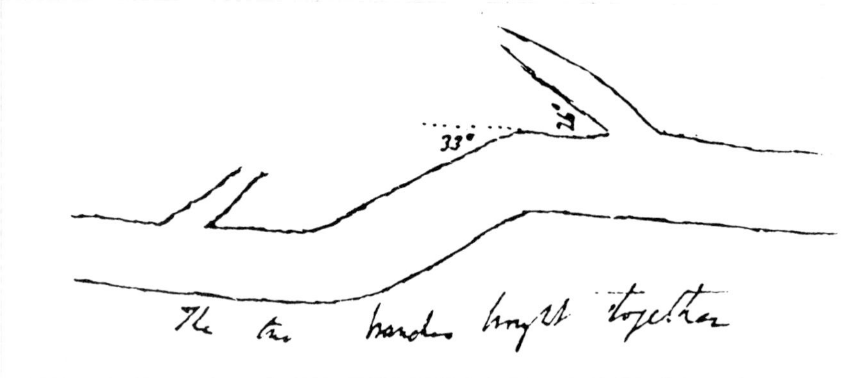

(a) I do not see any cause which could have produced these curious tubes excepting lightning; The sand hillocks were bare & evidently of short duration: Certainly the neighbourhead of the R. Plata is a likely place to meet with such phenomena; for the number of electrical storms is astonishing.— Twice on entering the river St Elmos light has shone on the Beagles mast head.— It is a curious circumstance the occurrence of so many groups of tubes, within a space of 60 yards ~~square~~ by 20.— Were these the result of one shock, & the electric fluid dividing itself shortly before entering the ground? or of distinct & successive ones? I should think the former the more probable case.— |**199**|

[note (a) added later] In 1793 A.D. Lightning struck B. Ayres in one storm in 37 places: & killed 19 people.—
Encyclo: Brittanica.—
Cause of furrows.— smooth internally.— *[two illeg. words]*.— sand siliceous black glass.— air bubbles.— fibrous appearance.— [note ends]

[**CD P. 199** commences]

General <u>Observations</u>

June 25th.— The Temperature of two Springs situated of South side of low rocky hills; & tolerably well protected from the effects of radiation; but not emitting much water: one gave 56½°. the other 57¼°. I should imagine the mean 56⅞° somewhere near the truth, & therefore as mean of year.

The nearly entire absence of trees in such a fine climate & in such deep rich a soil is a very surprising & inexplicable fact.— Some have explained it from the strong winds, but in the neighbourhead of Maldonado this is quite insufficient. the number of rocky & abrupt hills rising out of the plain render ample protection for the growth of the most tender.— This same paucity ~~extends~~ is common both to the modern beds of the Buenos Ayres country & to the granitic rocks of Banda oriental.— Can it originate in the covering of Alluvial soil being of very recent origin.— It is clear that the latter has been formed over a large extent at same time & beneath water: from not containing organic remains, probably suddenly.—

(a) I was told that near the Arroyo Tapes there was a wood of Palms. From the number of leaves (used in thatching) it is very probable.— One I saw which appeared about 20 feet high & thick in proportion.— They grow at Pan de Azucar: on the West bank of the Uruguay they are not found untill you arrive at the Arroyo del Palmas |**200**| in Latitude 32°. Here likewise a sandy Granitic soil commences.— This would appear to be adapted to them.— [note (a)] These Palms & some semi-aquatic trees, which follow the courses of the streams, are nearly the only exceptions to the general & entire absence: it is said that forest timber does not occur for a long distance N of Rio Plata.— In the mountainous country on the Northern half of the Laguna de los Petos, there is an abundance.— [note ends]

In Lat 35°!

General Observations

[**CD P. 200** continues]

After being accustomed to the great numbers of Coprophagous insects in England.— It was at first with surprise that I here found the ample repast afforded by the immense herds of horses & cattle almost untouched.— Aphodius[2] (?) (No 1181) is the only good exception: this insect amongst the sand dumes burrows holes beneath Horse dung:— Aphodius (1225) I have only observed once under very old dung.— Any other Aphodii which I have taken have been wandering.— It is curious to enquire what animal (No 1181) belonged to before the introduction of horses.— All the larger animals here, such as Guanaco, deer, Capincho, have dung in the form of pellets, which must be of a very different nature with respect to insects. M. Video was founded 1725, it is said the country abounded with Vicunnas. Cattle & horses have perhaps only abounded for about 80 years. This absence of Coprophagous beetles appears to me to be a very beautiful fact; as showing a connection in the creating between animals as widely apart as Mammalia & ~~Cole~~ Insects. Coleoptera, which when one of them is removed out of its original Zone, can scarcely be produced by a length of time & the most favourable circumstances.—

(c)

(b)

The same subject of investigation will recur in Australia: If proofs were wanting to show the Horse & Ox to be aboriginals of great Britain I think the very presence of so <u>many</u> species of insects feeding on their dung, would be a very strong one.— |**201**|

(a) [notes added later] <u>Turf or Peat</u> is not generally supposed to be formed within the Tropics; as the Latitude of this place is under 35° I thought it worth while to enquire respecting its occurrence.— In many marshy places the earth is very black, & contains much vegetable matter, in one place reposing on this there was another of much less specific gravity & so penetrated by roots & fibres as almost to be capable of burning.— (leaving however great quantities of ashes). This I was assured by a person well capable of judging was the nearest approach he had ever seen to the Turf of Ireland. As there are an abundance of situations favourable for the production [of] this substance, its existence only in the above imperfect state shows that this Latitude is too low for it.—

(b) At Bahia Blanca (September) there were very great numbers of Copris (1491).— Almost every heap of <u>horse</u> & <u>cow</u> dung was undermined by a deep circular hole, as is seen in England.— It is clear this beetle is partly omni-stercovorous.— & that before the introduction of latter animals *[7 illeg. words crossed out]* to S. America could not have been in any numbers.— At the *[illeg.]* Guitro-Linguè there were countless numbers of Aphodii (1492) flying about in the evening.— there was much horse dung, but I never saw one alight upon it.— The troop of horses had not been there more than 5 months & before that the plain was desert.— What dung or other substance could these ~~Animals~~ Insects feed on?

(c)
1833. March
1834. April

At the E. Falkland Islands there are no beetles under the dung.— Here perhaps the Climate so damp would be highly unfavourable to them (& all insect) life.—

Chiloe & Hobart town V. **264** (a) [notes end]

[1] Described by CD at greater length in *Journal of Researches* **1** pp. 69-72, while for a modern account of the phenomenon see W.B. Harland & J.L.F. Hacker (1966) 'Fossil' lightning strikes 250 Ma ago. *Advancement of Science* **22**:663-7.

[2] According to *Insect Notes* pp. 76-81 and 103, no specimens of the Scarabaeidae collected by CD would appear to have survived, so that an exact identification of the species cannot be provided. But the *Aphodius* that he later found in St Helena might have been *A. (Nialus) pseudolividus* or *A. granarius.*

[**CD P. 201** commences]

<u>Salmacis</u>[1]
1392
not spirits
(a)

Growing in abundance in pools of water. Guritti Island. Colour "sap green".— Diameter of filament .004 or rather more.— length of each from a little greater than this, to double: Very transparent containing but little internal matter.— The spires (with hyaline globules) close, each globule however not approximate to the others, the whole having net appearance.— There were 6 or 7 spiral lines: I could only count these by observing the apparent angle one made with a transverse line & thus guess its point of reappearance on upper surface, & then noting how many lines were included in this space.— In each cell about 9 lines encircled it.— [note in margin] there were about 12 hyaline dots in one complete spire [note ends] The gemmules were semi-opake, dark green & slightly oval.— The tube which connects the two ~~the~~ filaments was longer than that figured in Dic:

class: & not cylindrical, the central parts having a larger diameter; & evidently formed by two slightly ~~well~~ funnel-shaped tubes having joined.— The mark or lip where these would arise was visible in the cells with spiral lines of globules:— The necessity of the connection of two filaments to produce gemmules was clearly proved by the occurrence of occassional cells with spires surrounded by those with gemmules, & which had not, from the varying length, an opposite one to unite to.— One end of a filament would often contain gemmules whilst the other had not been joined & therefore remained in its original state.—

[note (a)] Having kept the plant for four days in a <u>dark</u> & warm place.— I noticed the following fact.— The gemmules are circular & much flattened. They lie in a plane in which the connecting tube is.— The stem or filament is cylindrical. In the interval of these days the gemmules had altered their position, They were now inclined in different planes, so that of course I immediately saw they were not spherical.— I found filaments (which appeared young ones) with the middle of each cell marked with cross lines, of a green colour & not extending whole length of cell. These cross lines were really each a part of a spire & from transparency of stem & their shortness appeared like cross bars.— They evidently ~~were~~ extend
∴ not 9 as before stated — till those of different cells nearly join. The number of these lines or vessels in each cell is 7; the hyaline points have not appeared, but even then the lip of where junction would take place was evident.— The appearance is of a set of spiral lines, alternately erased for an equal length. In some specimens these lines were quite rudimentary & short & others those of two adjoining cells were almost united. Then the filament or stem must exist previously to their perfect formation.— [note ends]

[**CD P. 201** continues]

<u>Arthrodièes</u>[2] 1393 not spirits — In same pool there was a genus belonging to this family: joints or cells cylindrical, about ½ ~~&~~ inch long & 1/18 in diameter; extremities rounded: it forms a trellis work, either pentagons, hexagons or square; three limbs articulating together being most common.— |**202**| Limbs are transparent,
<u>Arthrodièes</u> — turgid & elastic with water, appear to have no communication one with another: outer case colourless, no organization; is lined with thin layer of soft tender gelatino-granular matter, which is grouped into small numerous irregular dots.— Colour pale yellowish green.— Floats on surface with the above Salmacis, & in large net or trellis work pieces, several inches square.— I know not to what family this belongs

<u>Daphnia</u>[3] — Daphnia & an Ostracodes were in great plenty amongst the above described plants.— Daphnia of usual shape; with spine at posterior <u>extremity</u>: length
727 .8, breadth .5: colour light brown transparent: case very thin marked with

regular cross lines. like *[illeg.]* etching or fishing net:— eyes large black with irregular motions: Antennæ bifid, each division with 3 joints; & terminated with 3 setæ, the outer branch has likewise two lateral ones the inner only one.— Mouth with mandible formed of a narrow plate more

Plate 8, Fig. 1

Daphnia

bent at extremity & rounded, overlapping each other. & tips coloured brown; under surface with 3 raised rough lines or ridges:— In my imperfect examination did not see Maxillæ:— 1st pair of legs seated at base of last organs. peduncle very short, with few irregular long setæ.— 2nd pair is divided into two parts (perhaps 2 distinct legs) which do not act in ~~for~~ same plain, one semicircular with close even pecten of setæ; the other with few long scattered setæ & a jointed peduncle bearing setæ: 3rd, layer semicircular even pecten of setæ with few irregular ones at one corner: then three pair of *[two words lost]* act towards the tail: the 4th pair is very |**203**| similar to & approximate to the last, but seems to act or fan towards the head or in opposite direction: Heart in strong action: Intestine bright green; with Cæcum very plain in head.— tail terminated by a narrow sort of foot: which is terminated by two claws.— heel with two long bristles: sole on each side with short curved spines: in this spaces or sole is anus.— sole of foot is turned towards back of shell:— Eggs lie in dorsal space & imperfectly formed young ones: Antennæ large in proportion:— The legs do not seem used in progression:— At very point of head there are two most minute bundles of setæ.—

Fish[4] (spirits)

747 Caught in October in the R. Parana — as high as Rozario.— The four first fish are the common fry of the river.— Back blueish silvery, with silver band on side: blueish black spot behind the Branchiæ.— Fins pale orange, tail with central ~~part~~ band black.—

748 Back iridescent greenish brown, silver band on side.— Fins dirty orange, tail with central black band, above & below bright red & orange

749 Silvery; eyes fine black, peculiar form of belly; grows to twice size of specimen.

750 Fish called Salmon grows to one or two feet long.— Above blueish gradually shading down on sides; fins tipped with fine red, especially the tail, which latter organ has central black band

746 Fish. not very abundant. Upper part of body with its fins with tint of yellow, but stronger on the head, with dorsal clouds of black.— tip of tail black. Beneath silvery white.— pupil black. iris white; usual size sometimes larger.— |**204**|

[1] The specimens of these filamentous algæ have not survived, so that the species cannot be identified (see *Plant Notes* pp. 190-1).
[2] But from *Plant Notes* (*loc. cit.*) this alga appears to be a species of *Halodictyon*.
[3] Cladocera, water flea.
[4] In *Zoology* **4**:123-5, Specimens 747 and 748 were identified by Leonard Jenyns as new species of Salmonidæ, respectively *Tetragonopterus Abramis* Jen. and *T. rutilus* Jen.

[**CD P. 204** commences]

Armadilloes I have had opportunities of seeing something of four species of this genus.— & hearing respecting their habitats.— The Taturia Pichiz[1] (375 Spirits); the T. Apar.[2] (403 spirits) called Mataco.— The T. villosa[3], called Paluda.— are all found in some numbers on the sandy plains of Bahia Blanca, Lat. 39°.— The three species show no difference in choice of situations.— The first Pichiz, or sometimes called Kerikincha [later spelling Queriquincho]; is excessively numerous in all the dry country of B. Blanca, Sierra Ventana, R. Negro &c. It appears never to be found on this East side of America, to the Northward of the Sierra Tapalguen in Lat: 37°.30' They are said to occur plentifully in the Laguna desagualero at the foot of the (b) Andes.— Some of the officers of the Beagle have seen it at Port ~~St Elena~~ Desire, Lat 48° ~~30'~~ I have frequently opened the stomach of this animal; generally it contains Coleoptera & various Larvæ.— I have found roots & an Amphisbœna.— When surprised, it either buries itself very quickly, or lies close to the ground to escape observation.— in loose dry earth it is necessary to get off your horse quickly in order to secure your prize, which when fat & roasted is most excellent eating.— it often frequents the sand dumes & can drink no fresh water for years together.— They bring forth 2 or 3 young ones at a time.— They are constantly wandering about by day.— The Mataco & Paluda appear to have a wider range.— they are found at St Iago in Lat 28°.— The Paluda is a nocturnal animal & is taken by going out at night with dogs.— The fourth species, T. hybridus[4] (1413) does not occur to the South of S. Tapalguen, Lat 37°30′; to the North of this it is common & supplies Buenos Ayres; near to which latter place it is (a) not found.— It seems rather to prefer rocky ground, |**205**|

(a) [continued on back of page] rocky ground occurring commonly in Banda oriental.— It & the Paluda occur both there & in Entre Rios. as high as St Fe 32°, how much higher I know not.—

(b) [note (b)] Not having specimen of the Paluda, I give an imperfect description.— Front legs with 5 toes; 2 middle claws longest very broard flat; 2 outer ones shorter, 1 inner one very narrow long.— the 2nd toe has a remarkable ball on the under side at its base.— belly with rows of stiff hair; back with 8 moveable bands long hairs scattered on back.— Tail half length of body 9 teeth in upper jaw; 10 in lower on each side.— Nearly 3 times as big as a Pichiz

[continued with a different pen] & at S. Cruz (1697) is Specimen Q. whether it is the same species with Pichiz?.—
The whole four species are found near Mendoza.— [note ends]

1 T. Pichiz was identified by Waterhouse in *Zoology* **2**:93 as *Dasypus minutus* Auct., a no longer valid name. It is the Pichi, sole member of the current genus *Zaedyus* (formerly *Taturia*) under the name *Z. pichiy*.
2 Specimen 403 (Spirits) of *T. apar* was listed by CD as *D. tricinctus*, and in *Zoology* **2**:93 as *D. mataco*. It is the Southern Three-banded Armadillo (Apara), currently named *Tolypeutes matacus*.
3 The species of which CD did not have a specimen was the Paluda, listed in *Zoology* **2**:93 as *D. villosus*. It is the Larger Hairy Armadillo, *Chaetophractus villosus*, and is the type species of the current genus *Chaetophractus*.
4 Identified in *Zoology* **2**:92-3 as *D. hybridus* Auct. In the list of Animals copied out by CD in CUL MS DAR 29.1, he says on p. 12 'The fourth species, T. Hybrida, is called Mulita or Mulillo (little mule)'. It is the Southern Long-nosed Armadillo, and the name *Dasypus hybridus* is still valid.

[**CD P. 205** commences]

Biscatche[1]
(a)
1442
(not spirits)

The Viscatche is exceedingly numerous in the neighbourhead, to the South it appears less frequently although it is found at the R. Negro.— Late in the evening they come out to play; but do not seem to wander far from their holes.— they run very awkwardly; from their tail being elevated & shortness of the front legs they resemble rats. In the evening are very tame, you may ride quite close, without disturbing the gravity with which sitting in the mouth of their holes they watch you.— They are abundant even in the great thistle beds where there are no other vegetables: are said to live on roots, which from great size of teeth I think probable.— They inhabit very dry regions.— I have been informed on the best authority, that quasi canes post coitum adnexi sunt[2].— Their flesh is very white & good eating.—

They have one very singular habit; it is the constant dragging of all hard things to their holes.— around every ~~hole~~ group of holes you will see many bones, thistle stalks, hard pieces of earth, dry dung &c &c collected to the amount sometime of more than a wheel barrow could carry.— The holes enter the ground at a small angle; it is above the mouth, ~~on the~~ that the greater quantity of rubbish is placed.— I cannot even guess for what reason they take this trouble; it cannot be defence, for they are not in front of the mouth.— the trouble must be considerable for not a bone or stone is left uncollected for many yards from the burrow.— I was told (on good authority) that a Gentleman riding at night dropped his watch; the next morning he went & examined all the Biscatche holes in the line of road, &

Biscatche as he expected found the watch |**206**| near the entrance of one.— The Biscatche is abundant in all parts of the province of B: Ayres & Entre Rios; it is very curious they have not crossed the R. Uruguay. In the Banda Oriental there is not one of these animals; there are plains with thistles exactly like B: Ayres & others equally well suited to the habits of the animal.— It is a puzzle in the geographical distribution of the Biscatche, which I cannot solve, & is no small advantage to B: Oriental.—

(a) [notes added later] They are numerous near the Sierra Guitro-Leigniè.

(a) The habit of collecting hard things round its burrow is seen near Mendoza.— This animal is very different from the mountain species.— the tail in this appears more bushy & the breast reddish — stony inaccessible spots.— [notes end]

[1] Identified by Waterhouse in *Zoology* **2**:88 as *Lagostomus trichodactylus*. Both in his account and in that given by CD in *Journal of Researches* **1** pp. 143-5, the Spanish name of the animal is spelt Bizcacha.

[2] In CD's notes on Animals in MS CUL DAR 29.1, the words in Latin were first copied out, and then deleted. However, they were included in the description of the Bizcacha quoted by Waterhouse (*loc. cit.*).

[**CD P. 206** continues]

Guanaco[1] These animals occur in the Sierra de la Ventana Lat: 38°.12′ S. I should

V (b) **205** think on this side of America this was the Northern limit.— They are found in the islands of Tierra del Fuego & particularly abundant on north side of the straits of Magellan.— When at B. Blanca, I saw the track of a herd of 50 or 60; they appeared to have come on an exploring party from the interior.— their line of march ~~was~~ had been in a direct line till they arrived at a muddy salt creek. Here they seemed to have found out that that the sea was near, for ~~they~~ the track wheeled like a body of cavalry, & returned in as straight line as it had arrived.— Byron[2] says he has seen the Guanacoes drinking salt water:— our officers saw a herd drink out of the brine pits or Salinas at Cape Blanco.— they swim readily, & were seen crossing at Port Valdes from one island to another.— on the mountains of Tierra del I have seen the Guanaco, when disturbed, not only squeak or neigh, but jump & prance in the most ridiculous manner, apparently in

(a) defiance as a sort of challenge.—[3] It is commonly believed amongst the Gauchos, that where there are Guanaco, there is Gold. |**207**|

(b) [note added later on back of **P. 205**] Excessively abundant in central

May-1834 Patagonia; banks of rivers; the herds are much larger. I saw one with I should think 500, & many from 50 to 100.— The Southern part of

S. Cruz
Guanaco

Patagonia, here & at Str^{s} of Magellan, their more favourite haunts.— Measured the foot of one from the Lava country[4]: width of sole 26/8 inches: of one claw of fork 1.4 of hind leg: color of hair on upper parts dark "clove B with blueish grey": Saw a heap of dung 8 feet in diameter, it was suggested to me they sleep in same place, & in a circle with their head outwards, to keep watch for the Lions, & hence the heaps of dung.— The Guanaco seem to come to particular places to die; the ground in some low bushy places near the river is white with bones, in circumscribed spaces; the animals have crawled under bushes & bones are not torn by Pumas; I have seen 10 to 20 heads in one spot.— Mr Bynoe has noticed the same thing at R. Gallegos.— A wounded Guanaco immediately walks to the river: The Guanaco often dusts itself in saucer-shaped cavities in the dry plains.— [note ends]

[note (a) for **CD P. 206**] Guanaco seem particularly liable to have in their stomach Bezoar stones.— The Indians, who come to trade to R. Negro, bring great numbers to sell as remedies, quack medicine.— I saw one man with a box full, large & small.—

December 24th

[further notes added later] Shot at Port Desire a Guanaco; without blood, lower lungs or intestines weighed 170 pounds: From tip of tail to nose 7^{ft}.— circumference of chest, 4^{ft}.8^{inch}: Tail in length 9^{inch}: from extremity of nail to joint (hind leg) 6 & ½ inches: from this joint to extremity of Tarsus, 11 inches.— Most wide part of sole of foot 2.7/8th inch.—

The Guanaco at Port Desire & St Julians are excessively numerous. They are very wary when in a flock (generally from 10 to 30) & see very great distances. Mr Stokes[5] saw through a glass a herd of Guanaco evidently running away from us when they were not visible to our naked eyes.— When in pairs or single, not infrequently may be approach<ed> or suddenly met with.— If by chance you get within a few yards even of a herd, they will stop some time to graze, but if seen at a couple of hundred yards, the whole herd go off at a canter.— is this from mistaking at a distance a man for the Puma.— the footsteps of which animal are often-times to be seen.— The males seem to fight together. I shot one of two, who came squealing close to me, & another was marked with deep scars.— The Guanacoes have the habit of dunging on different days in the same place, & evidently more than one.— the heaps of dung from this cause are very large. Dung is oval pellets, rather larger than a Sheep. Frezier[6] remarks that Guanacoes & Llamas dung in heaps & that the habit is useful to the Indians as it saves them the trouble of collecting ~~fire~~ them for fuel.— [note ends]

[1] Identified by George Waterhouse in *Zoology* **2**:26-8 as *Auchenia llama* Desmarest.
[2] See J. Byron *The narrative of the Honourable John Byron Commodore in a late expedition*

around the world . . . Aberdeen, 1822. In seeking for water along the coast, Byron's officers had observed guanacoes drinking at the salt pans.

[3] The immediately preceding passage on **P. 29** of CD's Animal Notes (CUL MS DAR 29.1) runs as follows: 'Frequently the sportsman receives the first intimation of their [the guanacoes] presence by hearing from a long distance the peculiar shrill neighing note of alarm; if he then looks attentively, he will perhaps see the herd standing in a line on the side of a distant hill. On approaching, a few more squeels are given, & the herd set off, at an apparently slow, but really quick canter, along some narrow beaten track to a neighbouring hill.— If however by chance, he should abruptly meet a single Gaunaco, or a herd; they will generally stand motionless & intently look at him — then perhaps move on a few yards, turn round & graze again.— What is the cause of this difference in their shyness? Do they mistake a man in the distance for their chief enemy the Puma? Or does curiosity overcome their timidity? That they are curious is certain, for if a person lies on the ground, & plays strange antics, such as throwing up his feet in the air, they will almost always approach by degrees to reconnoitre him. It is an artifice, which has been repeatedly practised by our sportsmen: it has moreover the advantage of allowing several shots to be fired, which are all taken as parts of the performance.'

[4] According to a footnote in CD's Animal Notes (CUL MS. DAR 29.1), the guanacoes from Tierra del Fuego had been reported to have broader feet than others.

[5] John Lort Stokes was Mate and Assistant Surveyor on the *Beagle.*

[6] See A.F. Frézier *A voyage to the south-sea and along the coasts of Chili and Peru in . . . 1712-14.* London, 1717.

[**CD P. 207** commences]

Puma (b) — Very numerous in some parts of the province; I was told that near Tandeel [Tandil] 100 were killed in three months.— They are by no means a dangerous animal to man, excepting when a female has young, when I believe they will (very rarely) attack a man; of course when wounded they must be avoided. They are easily taken by being balled & then lassoed.— They live in the open plains, either amongst the reeds, or in a hole in a cliff.— It is a very silent animal, never roaring, even when lassoed.— They chiefly live on small quadrupeds, Deer, Biscatche, Ostriches &c.— The former they catch, sometimes in the middle of the day, when the deer is resting from the heat.— They but rarely kill colts or young oxen[1]. When they do it, it is by springing on their back & pulling the head back so as to break the neck. This latter is what all the Gauchos say.— For some particulars about their flesh see **P 376, 482** Chili [added above], Private J.[2]

[note (b) added later] Very abundant banks of S. Cruz: live solely on Guanaco, kill them by breaking their necks; live in the valleys amongst the bushes; do not retire from man, but look at him; the marks of their claws on the hardened clay are very frequent, as if scratching the ground like the Jaguars do the trees:— I have seen the footsteps of a Lion in the

Cordilleras of St Jago, not much below the line of Perpetual snow, the height must have been about 10,000 ft.— [note ends]

[**CD P. 207** continues]

Jaguar — This is a far more dangerous animal; kills many young oxen & horses by same method as the puma. If disturbed from their prey will not, unless

copied — much pressed, return to it.— The Jaguar seems to require damp places with trees, such as the streams & islands of the Parana.— I have heard of them living amongst the reeds on the borders of a lake.— It is said the foxes plague the Jaguars at night by continually barking: in same manner as

no danger — Jackall does the Tiger in India.— It is a very noisy animal, roaring much before bad weather.— Is decidedly very dangerous to mankind.— When hunting for one on the coast of the Uruguay, I was shown certain trees on which they are said to sharpen their claws.— In front the [continued at (a)

(a) — on back of **P. 207**, treated by CD as **P. 208**] trees are worn smooth & on each side deep scratches (or rather grooves) a yard long.— It is clearly done, in same manner as a cat with protruded claws, sometimes scrapes the legs of a chair.— The scars were of different ages.— it is common method of discovering the Jaguar by examining the trees.— In the course of the ride we passed 3 well known trees.— The object I should think was rather to blunt, than to sharpen claws so seldom used.—

The Jaguars are killed without much difficulty by dogs baying & driving him up a tree, where he is ~~easily~~ dispatched with bullets.— for anecdotes of these attacks V **387** private Journal[3].— I heard of Jaguars, though uncommon near the Sierra Guitro-gugo [?] (N of the Ventana) & believe they certainly (though very rarely *[closing bracket) omitted]* are found in the islands of the R. Negro, Lat 41.— Falkner[4] says, the Lake Nahuel-Naupi[5], from which this river rises, takes its name from the Indian name of Tiger. Its Latitude is 42°.— The same author talking of the many tigers at South entrance of the Plata says they chiefly live on fish.— I was told the same thing in the Parana & it well explains their great abundance in the islands of this river[6].— |**208**|

[1] In his Animal Notes (CUL MS DAR 29.1), CD states: 'In Chile however, probably from the scarcity of wild animals, it destroys very many young cattle & Colts; I have moreover heard of several instances where men & women have so met their fate.'
[2] For CD's comments on the palatability of the puma's flesh, see *Beagle Diary* p. 189. The second reference is to P. **483** (not **482**) of his journal, which describes the manner in which a puma hunts his prey (*Beagle Diary* p. 259).
[3] See *Beagle Diary* p. 195.
[4] See T. Falkner *A description of Patagonia, and the adjoining parts of South America . . .* Hereford, 1774. Copy in *Beagle* library.
[5] In a modern atlas the lake in the Andes from which the Rio Negro arises is spelt Nahuel

Huapi.
[6] **CD P. 208** is missing, but it concerned an attack of rust on wheat on the north bank of the Rio Plata, as explained in *Plant Notes* pp. 174-5.

[**CD P. 209** commences]

Lizard[1] 764	Back with double semilunar transverse marks of "gamboge yellow": ~~above~~ before which, irregular patches of black, intermediate spaces, blueish-greenish-grey, mottled with black & rust colour: belly "primrose & gamboge yellow.— Common genus.— Sluggish, often asleep:
Lizard[1] 765	Back with 13 snow white transverse lines; intermediate spaces most beautifully sparkling with green & orange: iridescent.— centre of each scale black: belly orangish "tile red", clouded & net work black.—
Lizard[1] 766	Numerous jet black transverse bands, intermediate spaces, grey, & very pale reddish brown, belly grey
Lizard[1] 767	Blackish grey, with medial line black; row of marks of same color on each side of this, & marks on the sides.—
Lizard[1] 768	Whole body & tail ringed with "french grey", before which salmon colour, with anterior edge indented with "primrose yellow.— before this dark brown. anterior edge jagged.— then as before french grey &c &c.— Beneath whitish except tail with rings.— under the chin spotted with white.
Gecko[2] 769...771 (a) 803	Centre of back "yellowish brown" sometimes with strong tinge of dark green, sides clouded with blackish brown.— in very great numbers under stones.— cannot climb up glass.— makes a grating noise when ~~dis~~ taken hold of.— After death looses its darker colours.— [note (a)] A specimen, being kept for some days in a tin box, changed colour into an uniform grey, without the black cloudings.— I thought I noticed some change after catching & bringing home these animals; but could observe no instantaneous change.— Under same stone found a very black variety & another one "Hair brown" with tinge of green; mottled on sides of back with "Oil & Pistachio green" centre of each patch *[illeg. word deleted]* brownish black.— Being kept for 3 or 4 days, not the slightish change of color.— [note ends]
Lizard[1] 772	Three whitish grey longitudinal bands, between which there are <u>square</u> black-brown marks in pairs, which together with lateral marks, form transverse bands: intervals grey & pale rust; belly grey & black, mottled & with a tinge of orange.— |**210**|
Lizard[1]	Seven or eight very irregular transverse rows of dirty white, intervals

773 — blackish brown, grey & rust.— sides more mottled with yellow.— Belly blackish grey, scales of belly orangish

Rana[3] 774 — Back blackish; flanks with three or four circular marks of black.— young individual.— is bred in & inhabits water far too salt to drink.—

Lizard 793 Copied — Belly rather silvery white, with very fine waving lines of black: back with very indented brown bands, between which spaces grey with stains of "lemon yellow". Head figured brown do do yellow.— upper surface of feet yellow, tail ringed brown, white & grey.— Port St Julians.—

Ornithology (b) — On the dry sterile plains of Port Desire & St Julians birds are infrequent: even the Carrion Vultures which are tolerably common at B. Blanca & the uninhabited wild plains of the North are here excessively scarce.— I saw two or three Carranchos & small vulture (1772).— [note (c) added later] (c) This small Vulture[4] is common on the banks of the S. Cruz [note ends] But the Guanaco left for a long time uncovered were never touched.— The commonest bird is a sparrow[5] (1704), & this is seen in every place: we have also the Sturnus ruber: the Solopax-perdrix (1224), the Lanius (1220), the Charadrius (1623), ~~Furnari~~ are all present in small numbers. Furnarius (1698) is not uncommon & Furnarius (1702) amongst bushes takes the place of (F. 1222).— In the bushy valleys, some Lanii, a Fringilla (1701) are present, though uncommon.— The Ostrich is not abundant.— Ibis (a) An Ibis[6] (1773) in pairs frequent the desart plain; builds its nest in cliffs on sea shore: egg dirty white freckled with pale reddish brown. ~~length~~ circumference 7 inches. |**211**| [note (a)] In its stomach Cicadæ, Lizard, (Scorpions !!) Cry very singular.— I have often mistaken it for the distant neigh of the Guanaco.— Legs "carmine & scarlet red", iris scarlet red.— [note ends]

May 19th — [note (b) added later] Some miles near the head of inlet, where there are lofty precipices of Porphry, there are many Condors[7].— It appears to me that, that a mural precipice determines the presence of these birds.— I have seen them in sandstone cliffs at the R. Negro Lat. 40° & not further to the North (400 miles from the Cordilleras, their supposed residence), at St Joseph cliffs, & here: where the Ship was anchored, there are no precipices, hence the Condor seldom comes so far to the coast, but as stated 15 & 20 miles up the creek they are numerous.— At S. Cruz, near anchorage there are cliffs & Condors; proceeding up the river, there were none, till we first again met Lava perpendicular cliffs, where Condors were again abundant.— Condors are generally seen in pairs, & a single or 2 young brown birds (in winter season) are seen with them.— They breed in the cliffs, & many together; in one place there could not be fewer than 20.— (100? miles distant) They must at S. Cruz in central Patagonia live entirely on dead Guanaco, those which die & are killed by the many Pumas.— When gorged they return to a pinnacle or ledge in their favourite cliffs:— A female, I shot: 8 ft tip to tip; length 3^{ft}:8^{inch}: Iris scarlet red: *[3 illeg. words]*

[1] Some difficulty was experienced by Thomas Bell in sorting out these closely related lizards, and there are discrepancies between his MS notes as transcribed on p. 344 and the listings of *Diplolæmus Darwinii* and *D. Bibronii* in *Zoology* **5**:19-22, and of *Proctotretus Fitzingerii*, *P. Kingii* and *P. Darwinii* in *Zoology* **5**:11-15.
[2] Identified in *Zoology* **5**:26-7 as *Gymnodactylus Gaudichaudii* Bibr.
[3] Described in *Zoology* **5**:39-40 as *Leiuperus salarius* Bell.
[4] From *Zoology* **3**:13-14, the small vulture taken at Port Desire was *Milvago pezoporos*.
[5] Described in *Zoology* **3**:91-2 as *Zonotrichia canicapilla* Gould.
[6] Identified in *Zoology* **3**:128-9 as *Theristicus melanops* Wagl.
[7] Identified in *Zoology* **3**:3-6 as *Sarcoramphus gryphus* Bonap. See also *Ornithological Notes* pp. 240-5.

[**CD P. 211** commences with an entry on *Halimeda* that is crossed through vertically in pencil]

Halimeda[1]
1770
(797 Spirits)

Considerable quantities of this Corallina was thrown up on the beach: on each side of the limb were little pustules; such as described **P 161 & 56**. They varied in number from one to four.— when old they became white & exfoliated.— Aperture beautifully round.— When the pustules were broken open ovules were found in three states; sphærical & opake; lengthened & pointed oval, where the internal matter was clearly seen separate from the transparent case.— & 3^{d} where this pulpy matter was divided into distinct articulations sometimes 2, 3, or 4.— the shape of ~~articulation~~ the limbs even were clearly visible, one basal one was largest. (b) the transparent case was in this case very delicate, the slightest touch rupturing it.— color dark "crimson red".— in short a small Halimeda ready to float forth was indisputably evident.— the longer limb probably becoming the point of attachment.— As all the pieces I picked up of this Corallina were furnished with these ovules it may be suspected that the parent plant is easily torn from its root & like Fungi perishes after reproduction.— I have now seen this process in a Halimeda, Amphiroa[2] & one of the inarticulata.—

[note (b)] This observation appears to me of considerable importance in settling the long disputed point, whether the genus Corallina belongs to the grand division of plants, or to that of animals being included in the Zoophites.— The gemmules containing several distinct articulations, I believe is entirely contrary to any analogy drawn from the propagation of Zoophites: I am ignorant what relation it bears to any of the articulated Cryptogamic plants such as the oscillariæ.— But, anyhow, we should certainly expect that one gemmule would produce only one young Polypus [in all Zoophites] & we might as certainly expect that each inarticulation one (or pair or some definite number) would contain & be formed by a Polypus, neither of these expectations are realized in the manner of propagation of the Corallina. Therefore, I do not believe Corallina to have any connection with the family of Zoophites. [note ends]

[1] CD's conclusion that Corallinas were not plants like *Halimeda*, which as explained in *Plant Notes* pp. 194-5 is a green alga (Chlorophyta), was a significant one. However, his arguments were confused because included with 1770 and 797 (spirits) were specimens of *Cellaria*, a true bryozoan later identified as *Menipea patagonica*, which survives in the Busk Collection of the Natural History Museum.
[2] The specimen of *Amphiroa orbigniana* Harvey ex Decaisne, collected at Port Desire and included in CD's jar no. 1770 with *Halimeda* and *Cellaria*, is preserved in the Cryptogamic Herbarium of the Natural History Museum, and is illustrated on *Plant Notes* p. 193.

[**CD P. 211** continues]

Sea weed[1] First narrows St^{s} of Magellan: Branches very fine bifurcate. colour "Hyacinth red with little Aurora". Extremities of branches finely pointed, with tranverse divisions; shortly then are divided by longitudinal plates making double set of cells, as long as broard.— in mains stems, 6 (or 10?)
(a) oblong cells, six times as long as broard; [note (a)] often enveloped by fine transparent epidermis seen at junction of cells.— [note ends] side by side, extremities of cells not united in a straight transverse line; at stem junction
Sea weed of ends |**212**| of these oblong cells, there are small globular bodies.— Many of the branches are changed into a short, bluntly pointed, very slightly oval cases.— this at first is full of red pulpy matter, which subsequently contracts & forms only ¼ of bulk at upper extremity.— in this state it is an aggregation of small sphæres, which in a more mature state, are quadrifid, that is they present the appearance of four short mushrooms growing from a common central root, (a flattened head on short footstalk) These are enveloped in a transparent case; which nearly fills up the small vacuity between the separate divisions.— diameter of whole .0025 or rather more than 1/500th of an inch.— color. dark red.— Are there four eggs or one singularly shaped one?—

[1] Identified on p. 226 of *Plant Notes* as probably a red alga, Rhodophyta.

[**CD P. 212** continues]

Avestruz Petises[1] **Page 112 (b)** there is some notice about a second species of Rhea.— which is very rarely found N of the R. Negro.— Mr Martens[2] shot one at Port
(a) Desire, which I looking slightly at it pronounced to be a young one of the common sort.— that is it appeared to be ⅔ in size of the common one.— I also [saw] some live ones of same size, but entirely forgot the Petises.—
1832...1836 I have since reclaimed the Head, Legs & several feathers. 1832. . . 1836.— The scales on legs are of a different shape, & is feathered below the knees,

this accounts for their being said to be shorter in the legs & perhaps for being feathered to the claws.— (it is a bird which the R. Negro Gauchos have only seen once or twice in their lives).— An egg was then found, which is more pointed & 2 or 3/8th of an inch less in circumference; it is an |213| old one, but yet retains a slight blueish-green tint, different from the yellowish one of the common one.— The feathers amply bear out the Gauchos expression of "overo" or speckled, & some added that it was darker.— With the Patagonians at Gregory Bay there was a semi-Indian, who had lived with them for four years.— He tells me there are no others, excepting the Petises in these Southern parts; that like the other ostrich many females lay in one nest, but that mean number of eggs in one nest is considerably less, namely not more than 15.— (The port Desire Egg was a Watcho[3]).— Whatever Naturalists may say, I shall be convinced from such testimony as Indians & Gauchos that there are two species of Rhea in S. America.

[margin: Avestruz petise]

[margin: Agrees with Gauchos stating them to be many in San Josè]

[there follow in a different pen two notes added later]

1837./1838.[4] I bought from the Chinas some feathers & a skin

April 1836 (a) In the plains of central Patagonia, I had several opportunities of seeing this Ostrich: it unquestionably is a much smaller & darker coloured bird than the Rhea.— it is excessively wary; I think they can see a person approaching, when he is so far off as not to distinguish the Ostrich; in ascending the river tracks &c &c were very abundant yet we saw scarcely any; but when rapidly & quietly descending, we saw many, both pairs & 4s or 5s together. It was observed, & justly, that this Ostrich does not expand its wings as the Northern one always does, when first starting at full speed: takes to the water readily; saw four crossing the river where 400 yards wide & very rapid; & another day. one very little of the body appears above water:—

[**CD P. 213** continues]

Puffinus[5] (1816) (a) (z) This bird is very abundant in the Sts of Magellan near P Famine.— It is particularly active late in the evenings & early in the mornings.— flies in long strings, up & down very rapidly, settles in large flocks on the water.— [note (a)] On the East coast of Tierra del Fuego single ones & Pairs may generally be seen flying about. [note ends] When slightly wounded could not dive.— The male & female are of the same plumage.— In the stomach of one, small fish & 7 or 8 Crust. Mac. same as (820 spirits). stomach much distended.— shot late in the evening in a boat.— very wary & shy, will not approach a ship.— Mr Bynoes has seen them in very great number in the quiet sea of straits & passages of the Western Coast.— [In foot] inner web "red lilac purple", edges of all & greater part of outer web blackish; legs & half of lower mandible pale "do purple".— |214|

[notes added later] (z) The Petrel[6] (1782) I saw between Falkland Islands & Patagonia. Mr Stokes says they build on the Landfall Islands, in holes about a yard deep, even ½ a mile, on the hills, from the sea.— somewhat like Puffins.— If a person stamps on the ground, many will come out of one hole: eggs elongated white, about size of pidgeon.— [correction] I find I am mistaken. this observation of Mr Stokes applies to the small blue petrel with waving dark line (like S) on the wings.—

1834 December — I never saw so many birds of any sort together as of this Petrel in the inland sea behind Chiloe. There were hundreds of thousands, flying in an irregular line, in one direction for ~~an infin many~~ several hours, & when on the water it was black with their numbers.— Said to be very irregular in their movements, appearing in certain places in number & on the next day not one to be seen.— The water here contained clouds of small Crustacæ.— The flock together made a cackling noise, somewhat like people talking at the distance.— [notes end]

[1] Named by John Gould *Rhea Darwinii* in *Zoology* **3**:123-5. Further material was added by CD to his account of this smaller species of *Rhea* when he copied it out later, and the question of the Avestruz Petise is discussed at length in *Ornithological Notes* pp. 271-7.
[2] Conrad Martens was the second official artist on board the *Beagle*. For the story of how he shot the ostrich, and it was partly eaten before CD realised that it was not a young *R. americana*, see *Beagle Diary* p. 212.
[3] A 'Watcho', correctly spelled 'huacho', was the term applied by the gauchos to an ostrich egg not laid in one of the communal clutches (see **CD P. 112**).
[4] This date suggests that the line must have been added to the text much later, but CD wrote again in 1836 that he had bought some feathers and skin from the Chinas (see *Ornithological Notes* p. 274).
[5] Identified as *Puffinus cinereus* Steph. in *Zoology* **3**:137-8.
[6] Identified as *Prion vittatus* Cuv. in *Zoology* **3**:141.

[**CD P. 214** commences with an entry dated Feb. 13th]

<u>Sigillina</u>[1] (832) — Brought up by the Anchor. 14 Fathoms. East entrance of the straits of Magellan.— Stem much flattened 3^{ft}:4 inches long: free extremity rounded, thickest & broardest from which it tapers to the root.— These two extremities are alone preserved in the spirit (832).— Orifices approximate, tubular, slightly flattened, simple, edges very thin, projecting 1/12th of an inch.— Color "Lemon. with little wax Yellow". section shows the individual animals to be of bright "sulphur yellow".— On cutting the specimen into two parts. I noticed <u>in</u> many of the animals (strong difference with Zoophites) a collection from 10 to 15 pale "auricular purple ovules".— They were enveloped in a mass in a gelatinous substance.— They were primarily sphæres, from which state they gradually altered (those

in same state being in same body) till they were the object figured (Plate 8. F 2.) This consists of an cup shaped capsule with tail about 3 & ½ times as long as body.— tail gradually tapering lower half & extremity are most fine, transparent natatory membrane or fin.— central vessel divided in lower part by transvers partitions.— upper end of cup (in the semi-developed specimens which I obtained) became ~~blended~~ blended in the gelatinous pulp, in which the ovules were irregularly placed. (tail not coiled). within the Capsule was opake body united with tail & having a neck in upper part which ended in 3 sorts of horns or processes. (these parts, although I do not quite understand how, would probably form the orifices): the chief part of opake internal body was formed by paralled

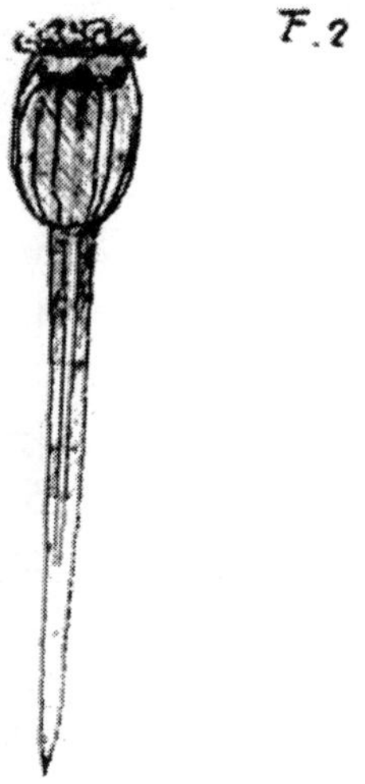

Plate 8, Fig. 2

(a) longitudinal vessels. (rudimentary Branchiæ?). Total length of tail & body, 1/10th of inch.— |**215**|

[note (a)] I have omitted to state the most curious part, that these young Sigillines by the aid of their tadpole-like tail & flat membrane could with a vibratory motion gain a tolerably rapid progress motion.— V. Synoicum Falkland Is^{ds}.— [note ends]

[**CD P. 215** commences]

Feby 25th Holuthuria[2]

(843)

Length of whole animal .7: head globular ⅓ greater diameter than body. length .2: body tapering to tail: 3 rows of papillæ (2 or 3 deep) on one side of body; few scattered round base of the anterior spherical enlargement or head.— This latter part is flattened on the top, round which are seated 10 ~~much but~~ irregularly branched tentacula; two approximate ones are very small & different from the rest.— In centre tubular, long projecting lip, with concentric lines. highly extensible & dilatable. Head obscurely lobate.— Surface covered with small oblong patches of fine punctures, which feel a little rough.— General color "peach blossom red". tentacular orange, with few brownish orange spots at their bases: central lip on mouth yellow.— Low-water mark.— Wollaston Island.—

Holuthuria[3] Doris-like

Body oval depressed, strikingly resembling a Nudibranch. Upper surface convex covered with scales, form truncated angular [sketch in margin] pointing from edges of body to central parts.— outer ones small (but not gradually) increasing towards the centre. Scales covered with punctures.— Lower surface soft concave.— The mouth is situated at ¼ length of body

854 (a) from anterior extremity; circular is completely closed by 5 pointed scales:
Feb. 26th.— [sketch in margin] Tentacula 10. long. ½ length of body: tapering, little branched, tree like (in contradistinction to bush-like).— Resemble that of  Holuthuria (**P 163**).— They surround the mouth.— The bony collar consists of 10 truncated gothic arches or rather 5 pair.— slightly stony.— When the Tentacula are retracted this collar is nearly in centre of body & lies in an inclined position with respect to the plain [*sic*] of body.— |**216**|

[note (a) added later] June 3d. 1834. Port Famine. Found abundantly in 5 & 6 Fathom water, adhæring to the large stones to which the Fucus gigant. grows.— They lie very flat & fill up any irregularities in the surface of the stone.— Removing one large one.— I found beneath 10 to 20 ovules beneath it; the animal being placed in water these were washed away: ovule — dark orange color. length 1/15th of inch; elongated oval soft sack, with several blunt, rough conical projections (doubtless in nature tentacula); by which it made attempts to crawl — Perhaps the lower membrane was ruptured or if not the mouth of the oviduct is on the inferior surface.— [note ends]

[**CD P. 216** commences]

<u>Holuthuria</u> The Anus is placed on the back, & in the same relation to the posterior extremity as the mouth is to the anterior:— it is closed by 5 unequal scales or valves & one small central one; nearly heart shaped.— The scales which lie between these orifices are larger & squarer than the others.— The viscera on lower surface do not fill up the whole concavity, but as far as where the small scales commence, which latter form a case over the body.— the inferior membrane is sprinkled over with hyaline ~~spots~~ points.— its outer edge has short striæ, pointing from the centre, of the same stony natures as the scales.— within these, there is a single row of papillæ, which extend round the ~~under~~ body.— the head of these is flat slightly coloured plate; the surface of which seen with a high power is covered with small suckers.— this plate is not contractile, but the long transparent footstalk is. Length of one specimen ¼ of inch; generally more than half this.— Edge of body from the scales, sinuous.— Color "Flesh & Aurora red".— under surface more orange".— They are found adhæring pretty firmly to leaves of sea-weed & in 16 Fathom water. NE end of Navarin Isd.— When the Tentacula are half protruded this animal most curiously resembles a Doris.— Can crawl, (but very slowly) by the aid of the Papillæ & the Tentacula which are adhæsive.— these when the animal moves are extended before it. ~~& can be seen~~. When detached from a leaf, the animal can curl in the edges of the shell to a small degree.— The animal is very pretty from elegant arrangement of scales & color.— |**217**|

(a) [note (a)] "Peach blossom with little Aurora red" is more accurate. Tree like Tentacula are coloured orange.— [note ends]

[1] Ascidiacea, Clavelinidae, another brooding tunicate, or sea squirt. See *Dic. Class.* **15**:421. CD notes that as in the Synoicum observed in the Falklands (see pp. 144-5) the larvae swim skilfully with the aid of their tadpole-like tails.
[2] Dendrochirotida, Cucumariidae, sea cucumber, probably *Pseudocnus dubiosus leoninus* Semper.
[3] Dendrochirotida, Psolidae, sea cucumber, probably *Psolus antarcticus* Philippi or *P. patagonicus* Ekman.

[**CD P. 217** commences]

Animal?[1]
853

Oval globules, with tough external skin: color dark olive brown — centre of sack filled with thick adhæsive brown substance, without vessels.— Adhæres to sea weed, by a flocculent substance at one extremity, through which a vessel might be seen.— No signs of irritability. when placed in fresh water, burst itself.— I believe several being detached & placed together in watch glass, reunited themselves one to the other.— 16 Fathom. NE of Navarin Is^{ld}, on sea weed &c &c.

Spongia (?)[2]
852
[in pencil] copied
Bowebank

Mass irregularly sphærical. general length of whole .3 of inch. sponge-color; thickly covered with numerous fine spines or hairs.— from centre of body a tube proceeds, length .1. formed of white approximate hairs; near base has delicate transverse partition, from this point the hairs slightly diverge, making the tube gradually widen at its mouth; tube conducts to central linear cavity, lined with hexagonal net work, which are the orifices of oblong spongy cells, which fill up the mass.— Could perceive no currents.— Adhæres to sea-weed.— Hab: same as above.—

[1] Not identifiable.
[2] Porifera, ? Demospongiae. *Bowerbankia* is a basic bryozoan.

[**CD P. 217** continues]

Crust. Mac:[1]
860

March 1st.— East end of Beagle Channel.— Roots of Fucus G. Back "Hyacinth & brownish red" with oblong marks & spots of gem-like "ultra-marine blue". one white transverse mark & longitudinal one on tail; 1st & great legs, same color as body, but penultimate limb centre part white edged with "do blue". anti-penultimate ringed with white, "do blue" & "do red". other ~~limbs~~ legs with basal limbs faintly ringed but ultimate limbs orange.— sides with oblique stripes "reddish brown".— Animal most beautiful.— |**218**|

~~Pleuro-~~

Length .6 crawling: breadth .3: color very <u>pale</u> dirty yellow: beneath

~~branchus~~ 861 Sigaretus[2] white — semi-transparent: very soft. (impossible to touch it after being killed in fresh water). mantle much depassing foot: superior feelers. approximate at foot, length .1, extremities square or truncate.— inferior feelers. extremities rounded. Seated wide apart, from tip to tip when extended .3: there is a connecting membrane which unites them half way up.— which has 3 sinuosities, central one ~~greatest~~ deepest; there are fine dots of black on it: Branchiæ. on right side, large, forming a pyramidal mass of tufts.— 10 Fathom: roots of Fucus Giganticus. East end of Beagle Cha[l].—

Octopus 862 General color. "Hyacinth red". which appeared when viewed through lens in fine dots: the animal being left in impure water & frightened, the arms & basal connecting membranes would become quite white, sometimes however leaving patches of the red on the arms or body: when irritated, or placed in fresh water, the red was driven to the surface in the space of 3 or 4 seconds: from which it might again be seen to retire, (as a blush from the face) but irregularly.— could swim backwards.— was very soon killed by fresh water; were found coiled up in roots of Fucus Giganticus. Hab as
863 above: ~~with~~ near them were small ones, spotted on upper-surface of body arms with a brighter red:

Crepidula[3] 864 The ovules or young shells were on a stone beneath the parent shell; were contained in 9 oblate sphæres or sacks which were connected by tubes in a circle to a common base.— There were about 12 to 15 in |**219**| each
Crepidula sack, sometimes more or less: the young shells were crawling about in the interior; every part seemed perfect.— the bars or lines of the Branchiæ were very much developed in superior part of shell.— Body large in proportion to shell: anterior part of foot much produced.— Eye black dots: general color, yellowish white. Hab: as above:

[1] Decapoda, Hippolytidae, identified as *Nauticaris magellicana* (M. Edwards) in *Oxford Collections* p. 212.
[2] Notaspidea, Pleurobranchidae, probably *Berthella platei* (Bergh, 1898).
[3] Mesogastropoda, slipper limpet.

[**CD P. 219** continues]

Flustra[1] (with moving beak) (874) (1874 not spirits) March 1st.— East entrance of Beagle channel: adhæring to roots of Fucus G:— I shall generally only mention those parts which are not preservable.— Cells spindle shaped. placed in straight rows — each cell adhæring laterally by 4 supports to others, forming a most elegant net work.— ~~the base~~ &c &c.— Polypus, with 26 arms which are very nearly length of whole cell.— These rest on an inverted cone (Pl. 8 Fig. 3). this cone acts as a mouth. a central vessel or opening may be seen closing, with
(a) a peristaltic motion; this again joins to a slight enlargement of the main red

viscus.— I believe just beneath in enlargement this (stomach? or œsophagus?) makes a bend; but this part is very difficult to [be] made out, for when the Polypus is protruded, this part is just in the aperture of cell; & when drawn back it is doubled at the very base.— These last parts are enclosed in a delicate tube: which & the arms are enclosed in a transparent case; which is protrudable:

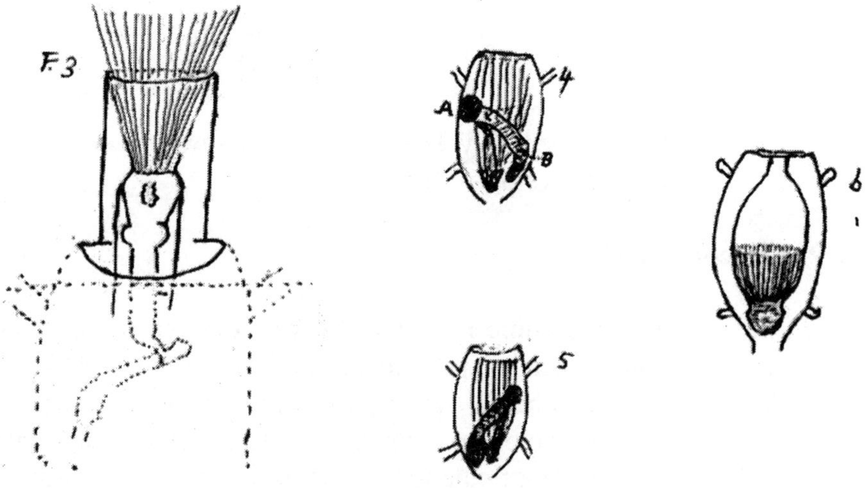

Plate 8, Figs. 3-6

Fig 3 is a drawing of the above parts: all which is beneath the dotted lines I have not actually seen in this position, but have no doubt it [is] the true one: For in (Fig. 4) we have a back view of polypus.— we here see a curved thick vessel, with more or less red granular matter, lying obliquely across the cell: (A) is rather globular, & the most solid viscus |**220**| in

Flustra

the body. Close to this at (K) there was a rapid revolution of small red grains, which apparently were contained in a sort of vessel or sack.— this sort of circulation sometimes extended ½ down the gut (B), the posterior end of which is full of red matter. Close to (K) a longitudinal red vessel (now seen over base of arms) contracts in diameter, bends & unites to the main one.— Fig 5. is same seen in front view, but is much more obscure.— The longitudinal vessel I think it is probable I have drawn too long — it is difficult to see distinctly.— When dying, the body is protruded as far even as in the circulating organ (K): but generally only beneath inverted cone.— Is the rapid motion of (K), that of the stomach (A) the liver & (B) the cæcum? The side of extreme part of (B) is attached to the middle of cell & all round liner. like those of muscles, are connected with the polypus. The position of all the above organs is not very constant, but subject to the will of the animal: the mouth of cell is composed of a moveable arched lip (like lower jaw of a bull-dog). Before the Polypus protrudes itself, this lip is lifted up & ~~backward~~ baseward.

[note (a) added April 1834] Also common in Falkland Is^{ds}. As every cell

grows before another in perfectly straight lines, & as the piece is irregularly circular; branch lines must frequently be sent off: one cell in this case producing two others, one in the usual place, another in the place of anterior connecting link or bracket: How completely does the Polypier produce cells
939 & Polypi.— Specimens (939) [in spirits] were attached to a Spider Crab.—

[note added July 1834 or later] I saw this species at Chiloe. on a stone, roots of Kelp.— [notes end]

[**CD P. 220** continues]

This Coralline is extraordinary from the presence of capsules resembling vulture heads[2], noticed in another coralline of the same family (**P 78**).—
|| Each cell has 2, seated at its anterior end, just above where the two upper
|| brackets go off to connect the lateral cells. These capsules have a peduncle,
|| with basal articulation: when at rest, they lie obliquely so as almost to meet at the very extremity of cell.— The peduncle is capable of being moved upwards & towards the base & nearly through 180 degrees; the |**221**|
Flustra with capsule lower mandible (keeping up simile with Vulture head but really superior) is kept wide open, so as to form a straight line with the upper one: it is occasionally closed, but not kept so; this motion is more frequent than that of the whole peduncle; both are rapid; chiefly take place when irritated by
(a) being touched, or fresh water: the mandibles firmly hold on to a needle: I never saw both capsules move at once, or any isochronism between different cells, excepting when affected together by fresh water or other cause.— There was an appearance of gullet at base of Mandibles, but I could trace no vessel or communication with cell.— (this can be investigated in the spirit specimens) I do not think these Capsules are exactly same shape with those of **P 78**.— [note (a)] The Capsule retained its irritability longer than the Polypus was dead & removed: this continued its rapid & starting motion.— This rapidity of motion is different from that of **P 78**.— [note ends]

[**CD P. 221** continues]

There is another curious organ[3]; In any row, the base of one cell is contracted & cylindrical & unites itself to the posterior one beneath the mouth.— Posterior to this point of junction, the greater number of cells have a thick, transparent, flexible, straight cylindrical vessel, projecting out.— it bends at right angles close to cell, & then continues parallel & beneath the row of cells; it is 3 or 4 times longer than cell, so as to project beyond the edges of Coralline. the extremity is rounded & impervious; it appears to me these stalks form a trellis work for the cells to lie, & perhaps also as means of attachment.— The connecting brackets appear hollow; where two rows of cells diverge, in the centre of an anterior bracket a globular enlargeme<nt> takes place, which afterward form a cell, so as to

Flustra with Capsule

|**222**| fill up the divergence between the rows.— In the young & extreme cells, the arms of Polypus do not reach half its length (Fig 6). they are enclosed in a bead, the neck of which is attached to anterior extremity of cell.— Here the four brackets are shown by knobs.— the capsules by a club-shaped mass with central little ball.— the posterior horn or vessel, & the site of anterior or other young cell is shown by short tube ending in a knob.— Before the arms of Polypus are complete & before any red viscera can be seen, the moving capsules are perfect.— The youngest form of cell, is globular mass with central spot or mark.— In some of the central & therefore old cells, I noticed (but did not examine sufficiently), a young Polypus — as at (F.6), ~~Above~~ anterior to which was a shrunk dark red viscus with central ball: it appeared as if the old Polypus had died (or produced an ovum) & a young one took its place in the cell. I could see no reproductive ovules.—

This coralline, when alive, from its extreme symetry, complicated Polypus, curious motion of capsules, was a most interesting spectacle: Coralline colored from Polypi dirty orange.— This Polypus is closely allied to that of Obelia, **P 174**. there the vessel which comes from the base of arms is elongated, possesses [?] a red organ, bends, contains a revolving mass & ends in a red-gut-shaped mass.— there is no difference, excepting that in this one, the longitudinal vessel joins an oblique one instead of passing by the Liver & then bending.— |**223**|

Flustra (encrusting) (a) 878

March 1st.— East entrance of Beagle channel.— (Pl. 9, F 1)[4] is drawing of Polypus from one of the cells, as I extracted it.— length from tip (if contracted), arm to end of cœcum or blind gut .015:— arms 16 in number.— they rest on footstalk in which an inverted conical space is contained.— there was here a small degree of the same corpuscular motion as will be described at (K).— It would seem to act as a mouth; just beneath this the stem contracts & bends.— & then proceeds in straight; it is generally full of reddish matter & is here (from A to B) much contracted. Above the centre of body is an irregular quadrangular body (K), more transparent than rest, formed of double edges, & revolving on its internal edges[5], especially & centre reddish granules.— From its external & lower edge, a line goes which seems to form the sack (cœcum?) (D), which contains reddish granular matter; a thick mass of which generally lies at the bottom, above the pointed extremity.— I do not know what the connection is between the red substance in (D) & in stem A & B.—

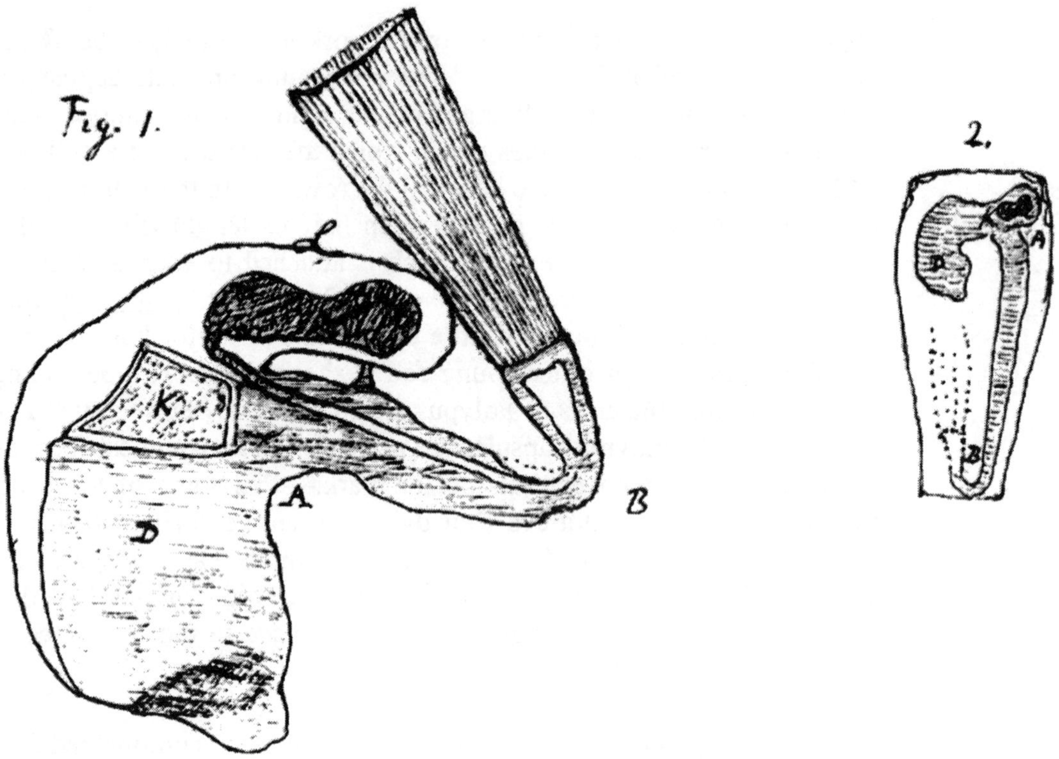

Plate 9, Figs. 1 and 2

[**CD P. 223** continues]

Above the revolving organ, the body takes a turn & forms an oblong case, which contains a dark red kidney mass.— this (I believe) is connected by its base with vessel running nearly to foot of arms.— the case is joined at its extremity to the case of the tentacula on arms & (I believe) at the base by a bracket with the stem (A B).— The position of these parts in the cell will be seen from the back view (F2). Here we have the stem (AB) much extended, & joining near the revolving organ (K) to the body, which lies underneath & close to mouth of cell: the other end of stem bends &
Flustra doubtless joins to base of tentacula, which are |**224**| represented by dotted
<u>encrusting</u> lines, for it is not possible to see them.— Cell is perforated by curiously shaped orifices [sketch in margin] .— Above the mouth of cells are long

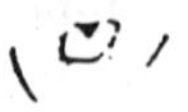

spines or hair & blunt points: which perhaps are young hairs.— There is also another very curious organ. it projects up like at hatchway on deck; is triangular; the door lies wide open on the surface, it appears to have a terminal tooth.— This door can be made exactly to close the hatchway, but
(a) immediately by elasticity or actual motion pulls open again.— This organ has some alliance with curious one of foregoing Flustra.—

[note (a) for **CD P. 223**] In Ponsonby Sound[6] procured more specimens:
887 did not here notice triangular hatchways; but before the mouth of cell there were circular hollows for orange ovules: [note ends]

(a) [note (a) for **CD P. 224**] Flustra with cells on one side of branch: Beagle
880 Channel: 15 Fathom: Polypus essentially the same as in the above animal; stem (AB) as might be supposed from form of cell is longer in proportion; near point of junction, revolving organ was visible, but the greatest difference was in the regular oval figure of the organ, which in the above animal is kidney-shaped, & in being much more distinctly divided from the cæcum: point of junction is merely a neck.— I could not count the arms or tentacula [note ends]

[1] The several species of *Flustra* described here are bryozoans of orders Cyclostomata, Cheilostomata and suborder Ascophora, and superfamilies Tubuliporoidea, Malacostegoidea and Cellularioidea, being sessile colonies formed of polymorphic zooids. 33 different species of polyzoa collected by CD during the voyage, of which 7 came from Tierra del Fuego, were listed by George Busk in his *Catalogue of marine Polyzoa in the collection of the British Museum*. 2 pts. London, 1852-4. Some 120 of CD's dry specimens are still held in the George Busk Collection at the Natural History Museum, and about 20 of those stored in spirits are still in the Zoology Museum of Cambridge University, where they were catalogued by S.F. Harmer in 1901. In Specimen 874 (in spirits) Harmer found *Tubulipora organisans* D'Orb., *Beania magellanica* Busk, and *Schizoporella hyalina* var. (= *Escharina brongniartiana* D'Orb.).

[2] CD's vulture heads in constant motion were specialized zooids now known as pedunculate avicularia, whose polypide are reduced but which have strong muscles operating a mandible-like operculum, also described on **P. 220** as the 'lower jaw of a bull-dog'. CD and Busk concluded that the function of avicularia might be defensive, but it has been pointed out by Judith Winston in an article entitled *Why Bryozoans have Avicularia - a Review of the Evidence* (American Museum Novitates, No. 2789. New York, 1984) that there is still little direct evidence in support of this or any other hypothesis.

[3] CD here describes the specialized kenozooids which form supporting and attachment structures.

[4] Plate 9 Fig. 1 shows the polypide of a feeding autozooid removed from its cell. A pencil note on the drawing states 'I believe L is not sufficiently circular & is attached too high to tube B', but the picture is not very informative.

[5] CD has observed correctly the rotating food-cord driven by the action of epithelial cilia in the pylorus of anascan bryozoans. The reddish food particles may have been phytoplankton.

[6] Ponsonby Sound opens out from the Murray Narrow running southwards from the middle of the Beagle Channel, and separating Hoste and Navarin Islands.

[**CD P. 224** continues]

<u>Polype?</u>[1] Stem creeping, throwing up upright footstalks, which bear at extremities,
881 each one animal. Whole substance membrano—gelatinous.— Animal cup shaped, one side being more convex than other, & considerably flattened: On the edge there are from 16 to 18 (17 common number) arms or tentacula; these are connected for about ¼ of their length, at their bases by a membrane. The summit of cup within arms is flat & oval; at one end,

there is a rather large transvers mouth, at the opposite small orifice of anus.—

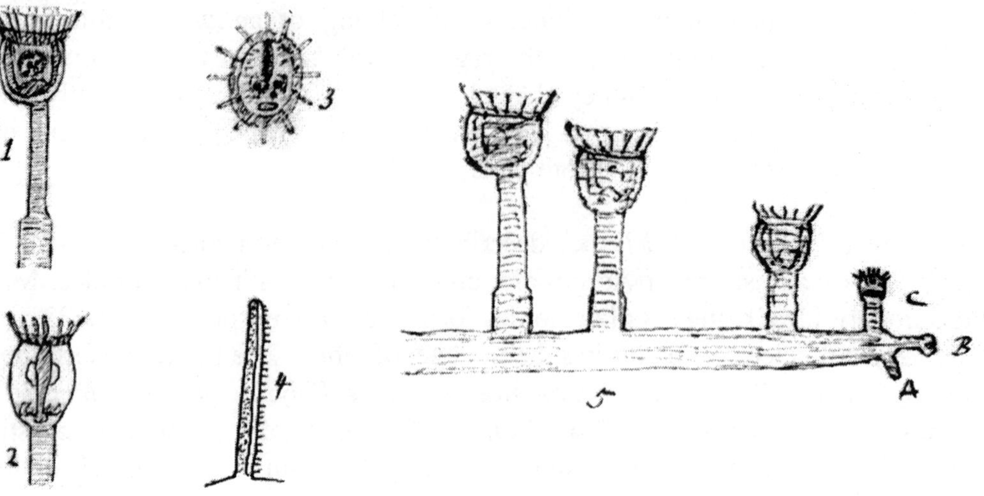

Plate 10, Figs. 1-5

[**CD P. 224** continues]

Pl. 10, Fig 1 is a ~~side~~ view of broard side of animal; The mouth conducts into straight, irregular vessel or sack; this possesses a peristaltic motion & another which resembles that produced by ciliæ. This sack contracts & enters in another & larger vessel, which varies in shape & dimension, fills up bottom of cup & generally contains some reddish granular matter; in this we see the rapid revolving motion, lately so frequently mentioned amongst the Flustræ.— This stomach leads into a |**225**| cylindrical vessel which lies in end of cup, opposed to (œsophagus?).— This generally contains pellets of dark red matter, which both by force & by animals will I have seen ejected.— they are fæces.— In centre of cup above the stomach, there is a transparent globular organ which contains (generally) from 4 to 5 small irregularly shaped bodies; these consist of central opake mass in transparent case, are irritable & highly contractile: there would appear to be two faintly coloured prominences, & between these there is a curved spaces covered by small vibrating fillets or ciliæ: These are sufficient to move the mass.— It is clear there are half-matured ovules.— Besides these the centre of cup, perhaps may contain some organs connected with the stomach.— All the above organs are enveloped in case independent of the outer one; which latter seems to form connecting membrane between base of arms.—

Polype?

(Fig. 2) is view of anus & intestine end of cup: F (3). vertically from above, mouth & anus.— (Fig 4) is one of the tentacula; they are lined on inner surface by numerous minute fillets, which are in incessant rapid vibration; & thus cause current in water.— the back part is filled with small globular

grains, between them & the fillets there is a clear space, which I think acts as a vessel & is connected with a circular one at base of membrane, which I believe emties itself near the mouth?— The connecting membrane is filled with grains, twice as large as those in the Tentacula: (Fig 5) shows manner of growth.— the first sign is then cylindrical projection: this soon
Polype? has a globular head, & even when very |**226**| small (C) little tentacula make their appearance. From this epock, they merely increase in size; in all the early stages the cup is very large in proportion to the stem.— The footstalk, in its lower part, has a shoulder, & increases suddenly in diameter.— this is .004, & is little less than that of the creeping stem.— They both contain semi-opake granular matter in a transparent case. Whole animal delicate, transparent; length of footstalk ~~.005~~ .05, of cup with collapsed arms .02. with those extended must be more than .03. Animal highly irritable; sensation evidently communicated from one to the other.— Beside the contraction & collapsement of the tentacula, the animal can move in all directions the footstalk: this it sometimes does in a circular manner & tolerably rapid.— When the cup was cut off & placed in water, it revolved steadily & slowly: power of motion must lie in that part.— Occurs plentifully, filling up the longitudinal furrows or wrinkles in the leaves of the Fucus Giganticus.— Ponsonby Sound.— March 5th.— What is this animal? Where does it come in the scale of Nature?—

[**CD P. 226** continues]

Cellaria[2] or Loricaria Ponsonby Sound.— Growing in small flesh-colored tufts on the leaves of the Fucus giganticus: Polypier. brittle very thin; each cell has its face excised by a shield shaped piece of thin membrane which extends ¾ length
885 of cell; the separation between the cells is of a soft nature &c &c &c.— Polypus 16 arms: in the young terminal, cells are seen pursed |**227**| up in
Loricaria a sack, as represented in Flustra (PL 8, F 6).— Body essentially same as in encrusting Flustra (**P 223** PL: 9).— The revolving mass was evident; when in cell, the cæcum (D) & organ (L) formed nearly a straight line, instead of lying obliquely.— Organ L. more circular & more detached from cæcum.— it was most evidently attached both to sides of cell & to the base of tentacula.— (I imagined I here saw external to the tentacula an
886.— orifice or anus!?).— Body altogether small.— On the same Fucus leaves there was another Cellaria, closely allied but I believe a different species.— (886).

Coralline stony (870. not spi^{ts}:) Small. white. branching stony Polypier; composed of central tubes encased in a stony net work, through which cells pass at right angles, these have a projecting tube, are placed in lines or irregularly on Coralline.— Extremity of branches flattened, rather dilated, composed of angular & circular net work of orifices, which would appear to be forming the cells.— Polypus I had very little opportunity (bad weather) of examining: possessed few tentacula, I believe 10 or 12; seated on long base, highly simple &

apparently not enveloped in case (?).— Growing in 54 Fathoms, some miles off Staten Land.— March 8th.—

[1] This animal belongs to the small phylum Entoprocta, similar to a bryozoan except that as CD's drawings clearly show, the anus opens within the ring of tentacles. In consequence, the feeding currents driven by the cilia are in the opposite direction to those in a bryozoan.
[2] In a modern classification, *Cellaria* is an anascan bryozoan in superfamily Pseudostegoidea.

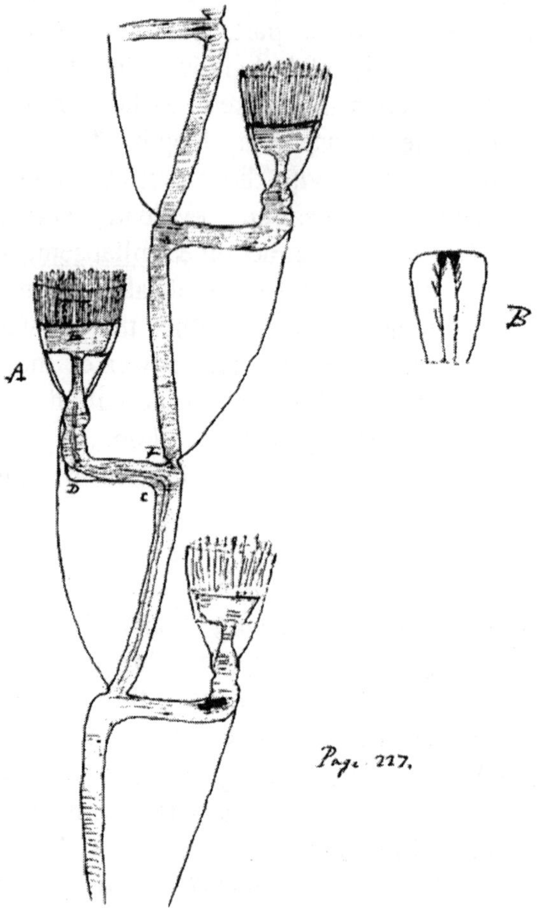

Plate 11

[**CD P. 227** continues]

Clytia[1]
5th species
894

Clytia

March. East Falkland Island: (PL. 11) will generally represent this Coralline.— the central living mass pursues a slightly zig-zag course, sending off rectangular branches, which bending upwards bear a cup & tentacula: it is remarkable by the |**228**| enlargements of the outer case, which cup-like contains the rt-angle of the living mass (C).— The living stem besides this outer case, which in the younger ~~stages~~ buds enclose even the mouth of cup, is enveloped in other & more close case.— this is best seen at the angle (D), & at base of cup, where it lies near to the outer

case.— it is traversed by central vessel, which being surrounded by granular matter forms the ~~cell~~ stem.— this granular matter can be forced to circulate in ~~the~~ its case.— the living stem having passed through the two semi-globular enlargements at base of cell, is much contracted, & chiefly consists of the central vessel; it is then suddenly enlarged into cylinder almost (E) filling the cup: which ~~contains~~ is filled by granular matter in which I twice perceived corpuscular motion.— In the middle of the tentacula there is the mouth; this when contracted is an inverted conical projecting tube, with round top & central vessel (B): when expanded it only forms a largely labiate mouth & this enters into the organ E.— the tentacula are 30 in number (am nearly sure) are short, thick, granular, with granulated surface are seated on superior & outer edge of (E).— We may imagine E & B to be enlargements of central vessel of stem & the tentacula, the coat of granular matter in a different form.— I consider E to act as a stomach.— The specimen was very poor; the tentacula, I think, could never be entirely retracted in cell.— [note in margin] Plate bad in this respect.
Clytia [note ends] |**229**| There is a retraction of the outer case above the angles (F).— At extremities of branches stem (DC) is not so horizontal, & after the cup is nearly perfect a fresh branch springs out at (F).— There were elongated oval ovaria attached by the enlargements at base of cell.— This Clytia grows on creeping stem in furrows of leaves of Fucus, throwing up short branches bearing alternate cups.—

Flustra[2] Hab: same as above Clytia: encrusting Fucus stalks: is very remarkable from
membranous being extremely soft & membranous (disproving classifications such as
895 Lamouroux[3]) Cells hexagons, with pretty regular cells; orifice tubular lipped.— Polypus in every main feature resembling that of (PL. 9, F 1, **Page 223**). ~~Arm~~ D & AB full of red granular matter; revolving organ K was evident, but I do not know of what figure.— the organ L was more sphærical & separated by very much longer & more narrow junction.— was united to case of tentacula, the point of union appearing near to upper edge: (the difference in this point in the various Flustra, is owing I believe to the transparency of the case & greater or less retraction of the arms).— the swallowing or peristaltic motion was present at base of tentacula; D & L were more in a straight line.— but every essential point is the same: Arms I am nearly sure 16 in number: delicate long (with central vessel?) the inner surface is lined with very fine, rapidly vibrating, fillets; which create rapid revolving motion in neighbouring fluid.— This is remarkable.— |**230**|

Cellaria[4] Growing in short branches. semi-stony Coralline, growing on Fucus G:
(a) branches, cylindrical composed of many cells placed in lines & each cell
915 placed between four others: allied to Cellaria cerealis[5].— Pl. 12, Fig. 1 is Polypus as seen far protruded out of cell: represented by dotted lines.— Arms 12. 14. 16. I know not which. I believe 14.— inside parts vibrating, especially at base.— are seated on inverted cone in which swallowing

action may be seen.— unites at ~~base~~ entrance of cell with vessel which is enlarged (Liver?) into oval organ containing dark oval mass & this is attached to side of case, not far from its mouth.— Whether this attachment is tubular I do not know.— (But this is certainly its arrangement, which probably holds good in all Flustraceæ; but is difficult to be seen by dissection) The dotted bag is supposed place of cæcum.— The transparent cylindrical case (is not drawn sufficiently cylindrical) is first protruded, bringing with it the Liver & then almost at same time the arms.— Coralline dirty "flesh red".—

[note (a)] I recognise it.— it is bifurcate, cells placed in oblique lines: about 8 in the circumference: extremities of branches formed of the cells.— I have often found it on the beach & very seldom on the leaves of the common Kelp: but yesterday I pulled up other sort with smooth edged leaves & thicker tree-like stem, & this abounded with this Coralline.— (∵ Point of attachment stronger?) [note ends]

[1] This animal is a hydroid in order Leptothecata, similar to those described by CD on **PP. 93/4, 118/20** and **126**. But Plate 11 looks more like a very stylized drawing of *Obelia geniculata* than of *Clytia*.
[2] Another anascan bryozoan.
[3] See *Lamouroux* pp. 3-4.
[4] Another anascan bryozoan.
[5] Not listed by Lamouroux.

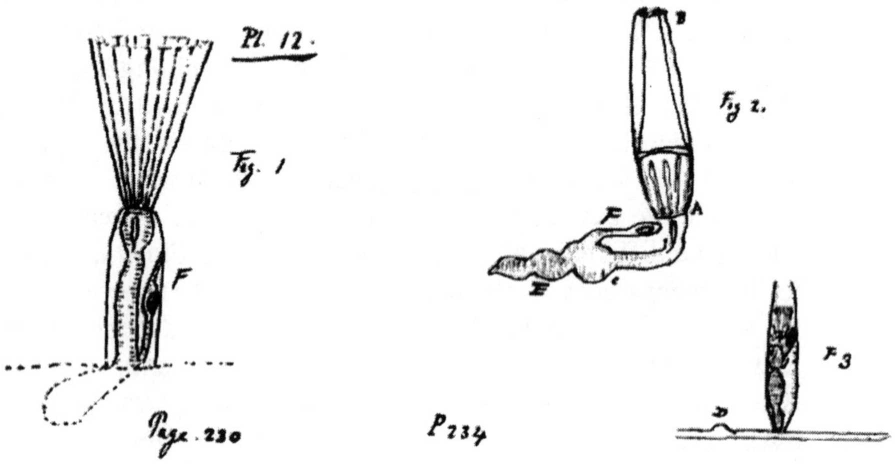

Plate 12, Figs. 1-3

[**CD P. 230** continues]

Flustra[1] with moving (or rather Cellaria). Is allied to that of **page 219**, but differs in many remarkable respects.— Polypus with 20 arms. body with essentially the

beak

920
or
1913
(n. sp.)
Flustra
with capsule

same structure: Each cell has one lateral capsule, these are squarer, or shorter footstalk. at hinge of lower jaw an excisement; it is very remarkable; although the Polypi were active I never could perceive the slightest motion in these beaks; again all the beaks were tightly closed: in these respects diametrically different from the two kinds: that the beaks are opened is certain, because I saw a fibre in one of them had been firmly caught & held. Mouth of cell |**231**| I believe labiate, protected by four spines: on each side of cell there is a straight line of short curved spines (like teeth of comb). there are 12 or 13 on each row, the points interfold & overlap in middle of front of cell.— These teeth are not very regular in their shape; often forked; extremities pointed: in young cell are blunt: give very curious ribbed appearance to cell: must form protection to it.— each one can with a needle be moved separately: ~~It appears~~ At back of cell there is a branch or vessel as at (**P 219**). but in this case is terminated by root for attachment; resembling the root in miniature of F. giganticus.— This proves that the simple organ in other kind was, as I supposed for this end.— There the attachment is much slighter, growing in irregular patches, edges free.— Here in circular patches much more firmly fixed to leaf of Fucus.— The cells are more adnate: ~~the basal brackets are shorter than the anterior~~ grow as before from the back part of inferior one. these brackets seem to be divided in middle. I could only trace a connexion of the capsule, root & brackets with the polypier & not body of polypus.— The young terminal cells grow as in other; arms of polypus in case, beak & brackets knobs &c &c.— In many of the basal (∴ old) cells. have a large dark oval organ in oval transparent case, close to which is a revolving organ: nothing else is clearly distinguishable: but I believe it to be a highly developed Liver for an immature Polypus. Young terminal cells have it not so much developed, or regular cells: What causes the absence of Polypus in these cells?— Is it connected with generation? This family of Flustræ is most truly remarkable.— |**232**|

[**CD P. 232** commences]

<u>Flustraceæ</u>[2]
927

Stony: in more or less globular masses: formed of cells united by their sides, not closed at base, but attached to Barnacle — colored dirty "flesh red".— Orifice of cell thus shaped [sketch in margin]; on each side of hinder part is small projecting orifice: & in front there is one or two others (this can easily be afterwards ascertained).— This gives the Coralline a most remarkable appearance.— Polypus dark orange color: arms (certainly) 14: body in every respect as in the family.— peristaltic motion, revolving organ, liver & cæcum: total length about 1/25th of inch: it is however peculiar: 1st in having a thin transparent plate thus shaped with edges orange [sketch in margin], & which forms a valve some way down in the orifice, the hinge being at parts A: it is attached to the case of arms, so as to be pulled close down.— 2^{d} at the base of arms, about ¼ diam: of arms, there is a nest of delicate white transparent vessels or threads, external

to the body & round the tops of inverted cone.— Have these & the lateral small orifices any connection? When the valve, as in defence, is closed do these supply communication with the water? Anyhow no other part of body passes through these orifices; but I could not trace these threads into them:— Mem: the Flustra with open capsule like Vultures head, had its cell closed by a lip.— Is there any parallelism in the cases?— Ova oval. with dark included, kidney shaped mass.— I believe generated between the cells or internally in the Coralline.— I was glad to find this change of structure in the Polypier followed or produced by one in the Polypus.— & the valve is an important one.— |**233**|

Flustra[3]
<u>with capsule</u>
932
4th species

Cells — pear-shaped encrusting; placed without order at base of cell there is a <u>fixed</u> "vultures head". (of less regular form than hitherto): the lower beak or jaw is generally open & but rarely closed itself, excepting when touched & then it firmly seizes the object.— I believe there is a membranous valve to the orifice.— Polypus with 14 or 16 arms). There is a human-ear-shaped indentation or orifice on each side of cell.— Upper rim of mouth transversely ribbed. Body of Polypus same general structure as in the family.— This Coralline in the simplicity of its structure: fewness of arms of Polypus. fixed "vultures head". evidently approaches to the common encrusting Flustræ.— Grows on a smooth leaved Fucus.—

<u>Flustra</u>[4]

 (a)
1424 (not spirits)

Encrusting: upper surface of cell with stony ribs, projecting like rays from the sides: orifice of cell thus shaped [sketch in margin].— Polypus. with 18 (certainly) arms; vibratory at base: Cæcum & Liver remarkably small & globular in proportion to length of arms.— I believe shape & size of viscera depend much on quantity of food &c &c.— It is interesting seeing this Coralline, so closely allied to many other species, with its Polypus with 18 arms showing of how little consequence is their number in distinguishing genera.—

[note (a) added later] I mention these particulars about <u>Polypier</u> from my own recollection [note ends]

<u>Flustra</u>[5]
931

(b)

Cell oval. Mouth square, with membranous valve, hinge superior: surface of cell with symetrical arrangement of hyaline stary points.— a centra[l] small orifice into cell, with its edges toothed.— Covering ovum, beautiful radiated structure: Polypus same general structure 14 or 16 arms:— Coralline orange color.— ~~Growing~~ Encrusting leaves of smooth leaved Fucus: |**234**|

[note (b)] The number of arms of Polypus is in all these cases difficult to be counted.— [note ends]

(c)

[note added later][6] Found a better specimen in which the greater number of cells had anterior to the orifice an enlargement containing dark orange ova.— (These ova appeared in early stages to be connected by a vessel to to the Coralline beneath them?). The cells with ova possessed equally active Capsules with those which did not: This proves they have no direct connexion with the ova: from the great similarity of this Coralline even in external character & much more in body of Polypus with common Flustræ, it is certain that the Capsule is connected with any important viscus.— In this species where there is only one kind (viz lower beak) of motion it entirely resides in the organ, for when separated it continued for some time to open & close itself — I believe there to be a direct communication between it & cavity of cell: Where there were Ova, the Polypi were not visible: Where there were none, I think they had lately burst forth, these cells contained a dark oval organ (as in **P 231**), which I believe to be commencement of growth of a fresh Polypus.— If this is the case how completely is the Polypus the flower of the Polypier.— (NB. The Capsule in this case more resembles a Crabs pincers than Vultures beak) [note ends]

[**CD P. 234** commences]

Flustraceæ[7]
937
(b)

Cells nearly cylindrical, ~~nearly~~ a little narrower at summit & enlarged in lower half; substance entirely soft membranous transparent: length nearly .04, breadth .01 invisible to the naked eye from transparency.— the cells are attached by a narrow junction to straight cylindrical creeping stem or vessel .004 in diameter. The cells are entirely separate, excepting by their springing from the same stalk: the structure of the body proves it to belong to the Flustraceæ.

[note (b) added later] Speaking merely from recollection I think (891) is perhaps of same genus with this, but a different species.— [note ends]

[**CD P. 234** continues]

(a)

PL 12. F 2. represents as forced out through the base of the cell: the only difference is the greater size of (c) where all parts unite, & the partial separation of cæcum (E) into two parts: [note (a)] The cæcum has a pointed termination & lies at very base of cell.— [note ends] The liver F is precisely the same as is ~~others~~ general.— the revolving ~~organ~~ motion was visible in two parts of (c) & I am not sure about its exact site: the swallowing motion was seen at base of arms: Arms 8 in number, rather short & thick: When as thus drawn, an inner case is seen much stretched.— When the arms were protruded (giving them a total length of .06), a transparent case was also protruded with its included vessel: Now from this (V Fig 1) I think in all these cases of the Flustraceæ the structure must be a cylinder at one end united to orifice of cell, at other to base of arms.— As represented in the Flustra Fig 1.— Now this turned inside out lying close on the arms would show the liver as in F 2.— Hence there always

Flustraceæ

is a resistance in forcing out the Polypus at base of cell & we explain the protrusion of case & arms where the latter, untill fully expanded is ½ enveloped in the case. |**235**| F. 3 is drawn too narrow as F 2 is too broard: the former represents the Polypus as seen quietly in its cell. (C & E) are in one line from which vessel of liver runs transversely.— the base of arms (perhaps from their shortness) are not, as is generally the case, drawn down to the bottom of cell, from which œsophagus (C to A) ascends, but this part would appear rather to be contracted & coiled in middle of the cell.— The very orifice of cell is slightly colored red, & when Polypus is withdrawn is contracted.—

[indistinct pencil note in margin]

This curious little Coralline generally is attached on Cellepora (933). the stems run in straight lines, sending off at intervals little groups of four or 6 cells: it appears that when the stem crosses the Fucus, on which the Cellepora adhæres it does not often bear cells.— I have seen stems crossing each other thus: they extend for several 1/10ths of an inch.— the structure is excessively delicate & tender.— The stem must I think contain granular matter for at the cut extremity there was an exuded mass.— there would appear to be an internal tunic: this best seen at the first enlargement, where a cell commences (D).— Perhaps the development of this forms the Polypus. The cell in a more perfect state than at D is oval, with rounded summit & broard junction with stem. (chief difference with old cell). Polypus can be indistinctly seen within.—

This Coralline by Lamx. arrangement would be one of the Sertulariæ[8]. Yet how truly different from the only one I have examined, the Clytias.— At first sight however it resembles in its appearance the creeping sorts.— |**236**|

[1] Cellularioidea, bryozoan identified by S.F. Harmer as *Beania costata* Busk.
[2] Anascan bryozoan.
[3] Coelostegoidea, bryozoan identified by S.F. Harmer as *Micropora uncifera*.
[4] CD's little marginal sketch nicely depicts the sinuate (sinus possessing) orifice found in many bryozoans in suborder Ascophora.
[5] Another anascan bryozoan.
[6] CD has omitted to mark the entry on his **P. 233** to which this note refers.
[7] This specimen may have been a stoloniferous bryozoan in order Ctenostomata such as *Bowerbankia*, rather than an anascan like most of the others.
[8] The 'Sertulariæ' of early 19th C. authors comprised many families of hydroids, one of which was Sertularidae in the modern sense. CD's description could apply to any of several families as defined today.

[**CD P. 236** commences]

Lepus <u>Magellanicus</u>[1] 1885 (not spirits) 1902

The black rabbit of these islands has been described by M. ~~Lesson~~ Rang[2] as a distinct species, the Lepus Magellanicus.— I cannot think so: my reasons are.— The Gauchos, who are most excellent practical naturalists, say they are not different: & that they ~~breed~~ & the grey breed together: that the black are <u>never found</u> in <u>distinct situations</u> from others: they have seen piebald ones: then other varieties such as white &c but not common (it would be curious to see how long varieties have remained, if the time of introduction was certain; the same idea applies to the cattle & horses which are of as varying color as a herd in England): there are no black rabbits on any of the small islands:— These rabbits do not travel far of their own accord, the Gauchos have transported black & others together to different places & hence know they do ~~not~~ breed.— I saw none to the South of the main chain of hills.— The spots on head of the specimens on board are not the same one with another nor with M. Rang description.— I have a head (1902) with broard white band, <u>the sides</u> of which do not correspond: this was a young animal, it had grey & brown hairs on its back, & a white patch on <u>one</u> thigh: Weight of my specimen 3 £b: of another 6 £b.— M. Rang[2] states that Magellan found this animal in his Straits.— Is it not the wild Guinea pig or Aperea, which is of a dark color & is to this day <u>very</u>
(a) frequently called a Conejos (rabbit): these are very abundant on N shore of St^s^ of Magellan: I have seen a small mantle made of their skins with the Indians. [note (a)] A Sealer has taken some of these rabbits to an island in Skyring Water in Patagonia.— [note ends] |**237**|

Vulpes <u>Antarcticus</u>[3]

Common in both islands. (M. Rang[2] states only in one) They are extraordinarily tame. The Gauchos have frequently taken them by holding a piece of meat in one hand & a knife in the other: they are inland as well as on the coast: dig holes in the ground: do not hunt in flocks: are generally very silent: but in the breeding season make a noise, like a Fox.— Gauchos & Indians from nearly all parts of Southern part of S. America have been here & all say it is not found on the Continent: an
(c) indisputable proof of its individuality as a species.— It is very curious, thus having a quadruped peculiar to so small a tract of country: [note (c)] Gauchos state there is no other quadruped whatever: With respect to the fish the Grebe (1918) was plentiful in a lake where there was no communication or <u>very</u> small streamlet with the sea.— [note ends]

The rat (1159) is also an aboriginal: it is evidently become partly domesticated & attached to the houses: There certainly are field mice, (I could not procure one), besides English ones now living far from the
553....555 houses: The fresh water fish (which are found in <u>inland</u> lakes) & the
(a) number of common earth worms probably belong to the same class.— [note (a)] Earth worms, from salt water being so deadly a poison (hence probably to the eggs?) is a difficult animal to account for accidental transportation? [note ends] The plants & insects might easily be transported
(b) from Tierra del in the SW furious gales!— [note (b)] I may mention

besides my collection plants as common to this island & Tierra del F. 1157: 1163: Bog plant: Rush-looking plant: tea plant: Celery: [note ends] Rats occur on the small islands.— The Sealers say this Fox is not found or any other land quadruped in the other Islands, as Georgia, Sandwich, Shetland &c &c.— ~~Very few of~~ No foxes are found in the NE peninsula of the East island (between St Salvador Bay & Berkeley Sound).— very soon these confident animals must all be killed: How little evidence will then remain of what appears to me to be a centre of creation. |**238**|

[further notes were made by CD on the back of **P. 237** with changes of pen, all but the last apparently while he was still in E. Falkland Island]

[1st note] The Gauchos state there are no reptiles now that this place is settled, in a few years this animal [the Falkland fox] will add one to the list of those perishing from the inhabitants of this globe.—
[2nd note] Out of the four specimens of the ~~eyes~~ Foxes on board, the three ~~larger ones~~ are darker & come from the East; there is a smaller & rusty coloured one which comes from the West Island: Lowe states that all from this island are smaller & of this shade of color.— There is a specimen of eyes from the East Island to show whether Fox or Wolf.—
[3rd note] I have seen the Culpen of Chili mentioned by Molina[4]. it is quite different from this Wolf-like animal.— [notes end]

[1] These remarks are quoted by George Waterhouse in *Zoology* **2**:92, where *Lepus Magellanicus* is listed as a black variety of the domesticated species.
[2] CD is evidently mistaken in referring here to the work on molluscs and their shells by *Rang*, which was in the *Beagle* Library, but did mean to refer to René-Primevère Lesson's *Manuel de mammalogie* (Paris, 1827), which was also on board. Lesson was co-author of the section on *Zoologie* in L.I. Duperrey *Voyage autour du monde . . . sur la corvette . . . La Coquille 1822-5* (Paris, 1826-30).
[3] Identified by George Waterhouse in *Zoology* **2**:7-10 as *Canis Antarcticus*. CD comments that the species is confined to East and West Falkland Islands, and that because of its tameness it is threatened with extermination by the settlers. This premonition proved to be correct, and the Falkland Fox, later renamed *Dusicyon australis*, is now extinct.
[4] See Juan Ignacio Molina *Compendio de la historia geografica natural y civil del Reyno de Chile*. Part 2 (Madrid, 1795) was acquired by CD when he arrived in Valparaiso.

[CD P. 238 commences]

Ornithology Caracara N. Zælandiæ[1] 1882 (a) (not spirits)	~~Is a young~~ Specimen (1882) is a young bird: but there are old birds precisely colored in the same method: the proportional length of wing feathers is different (specimen *[no number given]* of wing of old bird) & the skin about beak is quite white. There are others, but in considerably smaller proportion, where the legs & skin about beak is bright yellow, thighs rufous &c &c as described it is rather larger: now the Gauchos state this latter is the female & the grey legged one the male. The only one old

one ~~male~~ I dissected confirmed this.— It appears to me that all naturalists have ranked these latter as young birds.— They build in the cliffs on sea coast, but only in the islands: an odd precaution in such very tame birds.— They are excessively numerous in these islands; are said to be found on the Diego Ramirez & Il Defonsos. (hence live entirely on dead marine animals), but never on Tierra del Fuego: Are not found in Georgia or the Orkneys:— They are true Carrion feeders; following a party & rapidly congregating when an animal is killed; are extremely tame, especially when gorged with their craws projecting: in general habits much resemble the Carrancha; same inelegant flight & patient watching position: they however run much faster, like poultry or like the Cuervos (Cathartes atratus?). They have several harsh crys; one very like an English rook; when making this, they throw their heads quite backwards on their back.— are very quarrelsome, tearing
Ornithology |**239**| the grass with their passion: are commonly said to be very good to
(d) eat; flesh quite white.—

[note (a) for **P. 238**] Mr Mellersh[2] having wounded a cormorant, it went on shore & immediately these birds attacked & by blows tried to kill it.— Connection in habit as well as in structure with true Hawks.— I have now seen the bodies of three specimens which the Gauchos would call male birds, & which were so.— in some, as in (1882) the feathers appear young, but in others they were old.— Capt F. & Mr Bynoe have such.— Specimen (1932, unfortunately injured by fire) was a female with eggs as large as goose-shot; it generally agrees with the specific description of C. novæ-zelandæ:— legs & skin about beak bright "dutch orange", beak "ash-grey", in the male it is nearly black:— Specimen (1933) is remarkable, it is like the female, larger: ~~black~~ back blacker; thighs & under parts of wings partly rufous: tail <u>without</u> white band: feathers on neck same shape.— soles of feet slightly yellow, legs ash-grey, skin about beak with yellow margin. Beak lower mandible grey, upper black & grey.— By dissection could not see an<y> granulated surface in generative organs, so must be male or more probably young female — (bones rather soft, but feathers completely developed). Perhaps this bird, among the females does not acquire full plumage for 2 years, which together with males will account for larger proportion of grey legs over orange.— [note ends]

(d) [note (d) for **CD P. 239** added later] From the accounts brought by the Adventure[3], these birds in winter are very bold & ravenous: they come on board to steal from the vessel; & will pick up anything laid on the ground:
<u>Cop</u> a hat was carried a mile; a pair of balls: & a Katers compass.— they picked the very hide from the ropes on board.— It is said these birds wait, several together, at mouth of rabbit hole & seize the animal as it comes out.— They frequently attack wounded geese, & seized hold of a dog which was asleep.—

[**CD P. 239** continues]

Vultur Aura[4] (a) The Vultur aura? (1915) is tolerably common; is rather shy. may be known at a great distance from the Caracara by its lofty soaring elegant flight: I may notice, that for many days I saw scarcely one near the settlement, when suddenly one day I observed considerable numbers, as if they moved in bodies.— Is found near Port Famine. [note (a)] This bird if at all found in La Plata must be very rare.— for I have never seen one.— [note ends]

Caracaras (c) Copy (b) The Carrancha[5] does not come from Patagonia to these islands.— It is found but very sparingly on that coast: it there builds in low bushes: generally however in cliffs or banks: I have seen this bird tormenting horses with sore backs, trying to pull off the healing skin: the horse stands with back curled & ears down & the hawk hovers over his back.— Mr Bynoe once saw this Carrancha seize a live partridge, which escaped from his hold & was again pursued but on the ground.— This is very rare: the Caracaras, although placed amongst the Eagles, are in their habits inactive flight, cowardly disposition, protruding craw ~~are~~ true carrion feeders.— The Carrancha must be the Caracara vulgaire or Braziliensis of Dic Class:—

Cop [notes added later] (b) North of B. Blanca, I saw (& believe one or two others) a Caracara in figure & shape like the Carrancha, but differing entirely in color; legs & skin about bill blue: whole body light brown, excepting crown of head & round eyes which are dark brown.— I believe this to be Caracara shot at R. St Cruz (2028).
(c) All these particulars refer to the Carrancha of M: Video [in pencil above] Tharu of Molina [notes end]

[**CD P. 239** continues]

I do not believe the Chimango (1294) is found South of the R. Negro, without the one Caracara seen & shot at Port Desire (1772) is the same: anyhow it is very rare.— For more particulars V **185(bis)**.—

[1] Identified by John Gould in *Zoology* **3**:15-18 as *Milvago leucurus*.
[2] Arthur Mellersh was a Mate on board the *Beagle*.
[3] The *Adventure* was a schooner purchased by Capt. FitzRoy at his own expense from Mr Lowe for assistance in the surveys from April 1833 to October 1834.
[4] Identified in *Zoology* **3**:8-9 as *Cathartes aura* Illi.
[5] Identified in *Zoology* **3**:9-12 as *Polyborus Brasiliensis* Swains.

[**CD P. 239** continues]

Ornithology M. Lesson states that three sorts of Penguins are found about these islands: Capt. FitzRoy has a fourth |**240**| which kind I have seen in the St^{s} of

Penguin

Magellan. I saw much amused by watching a Demersa[1], having got between the water & it.— it continually rolls its head from side to side (as if it could only see with anterior portion of eye), stands quite upright: can run very fast with its head stretched out, & crawls amongst the tussocks by aid of its little wings so as extraordinarily to resemble a quadruped: throws its head back & makes a noise very like a Jackass, hence its name: but when at sea & undisturbed its note is very deep & solemn, often heard at night.— When diving (can do so in very shoal water) uses its wings very rapidly & looks like a small seal: from its low figure in water & easy motion *[illeg. word]* crafty like a smuggler.— is very brave, regularly fought & drove me back till it reached the sea.— nothing less than heavy blows would have stopped: every inch he gained he kept, standing close before me erect & determined.—

Steamers[2]

A logger-headed duck called by former navigators ~~& now~~ race-horses & now steamers has often been described from its extraordinary manner of splashing & paddling along: they here abound; in large flocks: in the evening when pruning themselves make the very same ~~noise~~ mixture of noises which bull-frogs do in the Tropics: their head is remarkably strong (my big geological hammer can hardly break it) & their beak likewise; this must well fit them **|241|** for their mode of subsistence: which judging from their dung must chiefly be shell-fish obtained at low water & from the Kelp.— They can dive but little; are very tenacious of life, so as to be (as all our sportsmen have experienced) very difficult to kill: they build amongst the bushes & grass near the sea.— [note (b)] The egg is pale blueish white.— [note ends] Mr Stokes once shot one which weighed 22 £b

Steamers

Ornithology

(b)

Lark[3]

(1898) is tolerably common over the island. Mr Sorrell[4] states it is found in Georgia & South Orkneys; & that it is the only Land-bird: this may truly be called "antarctica"; reaching to Lat: *[not filled in]* beyond which in this pole perpetual snow must reach to waters edge.—

Goose[5]

[in pencil] also Anas Hybrida

The Upland goose is common in small flocks, 3 to 7 & pairs, all over the island; does not migrate, but builds in the small outlying islands, it is supposed from fear of the foxes: from which same reason it is perhaps wild in the dusk but very tame by day.— it lives entirely on grass & vegetables. is good to eat.—

The black-necked swan is an occasional visitor in winter.—

(a) [in pencil] It is proverbial

The extreme tameness of the Furnarius[6] has been remarked on by M. Lesson: it is common to ~~many~~ every bird: Geese, Hawks. snipe; the emeberiza, & the thrush in flocks will in the stony valleys surround a person, within two or three feet of him. This tameness is remarkably seen

tameness of certain birds (c) — in the water fowl, as contrasted with same species [in] Tierra del Fuego; where for generations they have been persecuted by the inhabitants.— many individuals there must have seen as little or less of man, than here |**242**| so that the wildness seems hereditary.—

[notes added later for **CD P. 241**] (a) I suspect this Furnarius is of different & much darker color than that of Tierra del F, (1823).— & sometimes frequents inland parts.— Did I send a specimen last year?— (I have now a Specimen (1931)[6], in its stomach there was a small Cancer Brachyurus & a Buccinum .25 of inch long.— I think my collection of land birds with the Troglodytes of last year is nearly perfect.—
(c) The goose or Duck which is so tame here, up the river of S. Cruz, where they are entirely unmolested by man, are very wild.— What can the cause be?— The Puma? or migrations to Tierra del Fuego.— [notes end]

[1] A slightly extended account of the jackass penguin *Aptenodytes demersa* appears in *Journal of Researches* **1** pp. 256-7.
[2] Identified in *Zoology* **3**:136 as *Micropterus brachypterus* Eyton.
[3] Identified in *Zoology* **3**:85 as *Anthus correndera* Vieill.
[4] Thomas Sorrell was Acting Boatswain on the *Beagle*.
[5] Identified in *Zoology* **3**:134 as *Chloephaga Magellanica* Eyton. The trachea of CD's specimen was dissected and described by Eyton.
[6] Identified in *Zoology* **3**:67-8 as *Opetiorhynchus antarcticus* Gray.

[**CD P. 242** continues]

Zoology (marine) — The Zoology of the sea is I believe generally the same here as in Tierra del Fuego: Its main striking feature is the immense quantity & number of kinds of organic beings which are intimately connected with the Kelp.— This plant ~~I believe~~ (the Fucus giganticus of Solander) is universally attached on rocks. from those which are awash at low water & those being in fathom water: it even frequently is attached to round stones lying in mud. From the degree to which these Southern lands are intersected by water, & the depth in which Kelp grows, the quantity may well be imagined, but not to a greater degree than it exists.— I can only compare these great forests to terrestrial ones in the most teeming part of the Tropics; yet if the latter in (a) any country were to be destroyed I do not believe nearly the same number of animals would perish in them as would happen in the case of Kelp: [note (a)] I refer to numbers of individuals as well as kinds [note ends] All the fishing quadrupeds & birds (& man) haunt the beds, attracted by the infinite number of small fish which live amongst the leaves: (the kinds are not so very numerous, my specimens I believe show nearly all).—

Amongst the invertebrates I will mention them in order of their importance. Crustaceæ of every order swarm, my collection gives no idea of them, especially the minute sorts.— Encrusting Corallines & Clytia's are excessively numerous. Every leaf (excepting those on the surface) is white with such Corallines or Corallinas & Spirobæ[1] & compound Ascidiæ[2]. Examining these with strong microscope, infinite |**243**| numbers of

Zoology marine (b)

minute Crustaceæ will be seen.— The number of compound & simple Ascidiæ is a very observable fact.— as in a lesser degree are the Holuthuriæ & Asterias.— [note (b)] The number of Corallinas inarticulæ, encrusting & coating rocks & shells both in & out of Tidal influence is very observable.— [note ends] On shaking the great entangled roots it is curious to see the heap of fish, shells, crabs, sea-eggs, Cuttle fish, star fish, Planariæ, Nereidæ[3], which fall out.— This latter tribe I have much neglected.— Amongst the Gasteropoda, Pleurobranchus[4] is common: but Trochus[5] & patelliform shells abound on all the leaves.— One single plant form is an immense & most interesting menagerie.— If this Fucus was to cease living, with it would go many: the Seals, the Cormorants & certainly the small fish & then sooner or later the Fuegian man must follow.— the greater number of the invertebrates would likewise perish, but how many it is hard to conjecture.

[notes for **CD P. 243** added later] (c) Mr Stokes states that the furthest point North he has seen the Kelp on the East coast is about St Elena in Lat 43°.— It not uncommonly grows in 10 & 15 Fathom water.—
It may be remembered, as rather curious, that the Kelp Fish so abundant in T. del F. here scarcely seem to be found.—
Near the I^s of Chiloe Lat 42°, Kelp grows with no great vigor — but it is very curious to see that here neither the numerous shells & Clytias & Isopod Crust are quite absent; some few encrusting Flustræ, but they are much rarer; & some different compound Ascidiæ.— [notes end]

1 Spirorbidae are fan worms, sedentary tube-dwelling polychaetes.
2 Tunicates.
3 Nereididae are freely crawling polychaete worms.
4 *Pleurobranchus* is an opisthobranch gastropod of order Notaspidea.
5 Trochacea, top snails.

[**CD P. 243** continues]

Time of generation

I may mention that last Autumn as well as this, I noticed that most of the marine animals had their ova nearly mature; for instance, very many encrusting Flustraceæ, Doris, Synoicum, Asterias, Shell fish, Crustaceæ & Corallina.— The motion of the sea seems necessary to the life of its productions: this island is much intersected by water (Capt FitzRoy has compared it to the arms of the Cuttle fish). these far inland seas are nearly motionless, they seem to produce scarcely any organic beings. Creusia occasionally encrust the rocks. even where streams enter: The grebe (1917) proves that some few small fish are present; the water instead of cherishing (a) the elegant forms of sea-weeds & Corallines throws up [continued at (a) on back of page] a putrid mass of rubbish.— The powers however of Geology are quickly covering up these unproductive specks on this our globe.— V **157** & **158** for more particulars. |**244**|

Polyclinum[1] 940

[in pencil] Synoicum Blainville

Didemnum Savigny

Very abundant: coating Fucus G in large irregular masses: when undisturbed in water, the superior surface is studded over with very numerous, circular more transparent spaces, rather less than .01 in diameter: within this is an hexagonal orifice, the sides of which are rather convex, giving it a star-like appearance.— the edges are composed of white dots like rest of ~~body~~ mass, hence when closed very difficult to be seen: The mass is colored pale "buff orange", is composed of transparent substance containing infinite minute globular granules: upper surface transparent membranous: the orifices lie in this: they are seated in valleys on the irregular outline of surface & without any fixed position: There was not the slightest sign of orifices being placed in pairs: three would be close together & no other near them. The thickness of the substance is from 1/10 to 1/20th of an inch. ~~it presents~~ a section presents this appearance.—

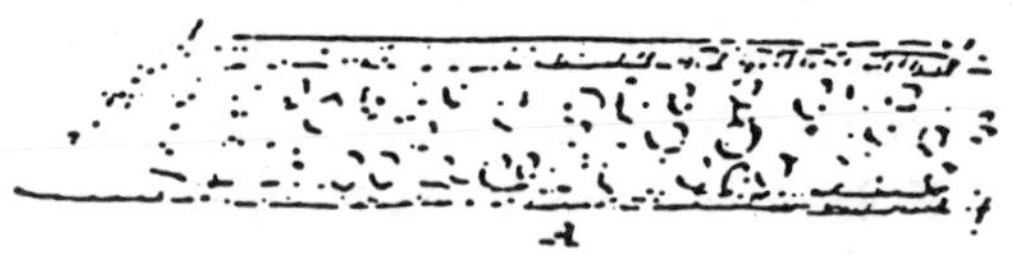

where (1) is transparent case in which the orifices lie: (2) granular matter: (3) cells with animals: (4) same as (2) & as partitions between the cells, these are generally very thin.— With respect to the cells I am much puzzled, the lower ones are in the most regular line & rather the largest being more than .01: their connection with the superior ones is obscure & must be by a very narrow junction: in not more than one or two case could I see appearance like (A). the upper cavities communicate directly through (2) with the orifices: each cavity having only one.— In the upper sack I believe I could perceive a most delicate pale orange sack, with 2 orifices, & a transparent globular organ with dark viscus, in the lower one an intestine shaped mass.— If each orifice has two cavities (Branchial & Polyclinum Abdominal) |**245**| from their very numbers these cavities would be packed irregularly: yet I do not understand not being able more clearly to trace the junctions. The membrane which forms the external orifice is highly contractile; & ~~animal~~ whole mass very sensitive: if one orifice is most lightly touched all close for some 1/10 of an inch round.— this appears to the naked eye like a white cloud passing over the substance; (from dark apertures closing).— In parts of the lower granular matter there were globular masses, dark red, few in number. I do not believe they were ova.

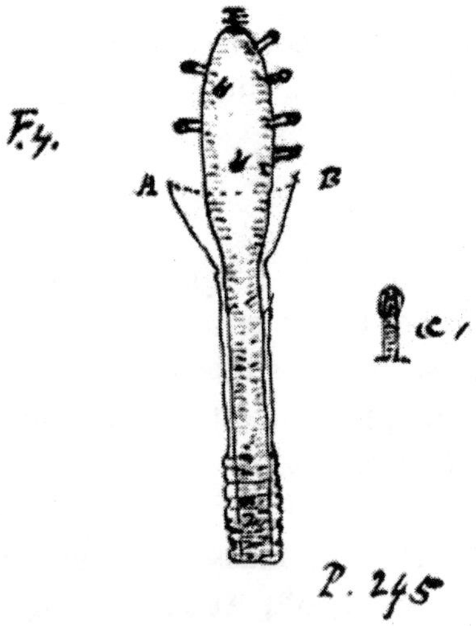

Plate 12, Fig. 4

[**CD P. 245** continues]

Tubularia-Clytia[2] 943 (a) Growing abundantly at the bottom of the Beagle; therefore an inhabitant of these latitudes: in general appearance resembles a Tubularia, but in the apparent articulations a Clytia.— From a very short examination I believe the structure of the Polypus to be very curious. PL 12, Fig 4.— The living stem is enclosed in transparent case which (I believe (AB) specimen not fresh) terminates in a small cup not large enough to confine polypus.— The polypus, or rather the enlargement of the central matter, is a very elongated oval: summit rounded with an orifice; contains red matter: surface studded with ~~numerous~~ cylindrical thick papillæ, transparent colourless, which have a granular slightly enlarged head.— These I believe to act as tentacula in the common polypus, & the whole mass to be a production of the mouth:— Amongst these papillæ, others may be seen, enveloped in transparent case (C), larger & containing central red matter: the superior extremity of which appear divided into papillæ. & there were young Polypi?— Santa Cruz.— April 16th.— |**246**|

[note (a) added later] After being a month in spirits I reexamined this most curious Coralline.— There is a cup but I could not see a perfect one: I think from its shape it never was intended to receive more than ¼" of Polypus: indeed from its oval shape the Polypus could not protrude & retract itself into any cell which at all fitted it; the arms or tentacula or papillæ are slightly enlarged, probably from contraction, caused by death: in each longitudinal row there are from 5 to 6 & from 10 to 16 of such rows, seen when held vertically.— length of oval or Polypus, $4/100^{th}$: breadth $3/200^{th}$.— length of arm rather more than $1/200^{th}$, probably when alive ~~nearly~~ longer.—
2009 Dry specimen.— [note ends]

[1] Aplousobranchia, Didemnidae, a colonial tunicate.
[2] Anthoathecata, Corynidae, but not necessarily from the southern latitudes because it might have survived being transported on the hull of the *Beagle* from warmer waters.

[**CD P. 246** commences]

Tubularia[1]

I procured off C. Virgins one single cell or stem of this coralline: the tube contracted towards its base, was horny, sides ~~covered with~~ contained numerous linear, slightly serpentine cavities, which were concentric & gave a ringed appearance to it. (contained a little red matter) The living stem, arms retracted, white & soft: by dissection I imagined I saw some arms or tentacula: stem itself is a circular aggregation of transparent sphæres, with a central opake mass [sketches in margin]: the coat is granular, interior matter pulpy; soft; At base, or near root these are easily detached from the viscous matter, in which they all are enveloped; at anterior parts they adhære much more firmly.— The sphæres with the highest parts showed no orifices.— I record this for any future dissection.—

Sertularia[2]
(Flustra)
PL 12, F 6[3]
2006
(not spirits)
(a)

972

Sertularia

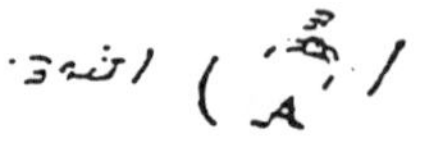

Off S. Cruz. I procured a bad specimen of this Coralline, which is miserably drawn.— The central living stem (which I believe is pulpy matter contained in a vessel) is slightly zig-zag & comes in contact with the base of each cell.— When first watching this Coralline, I was astonished at seeing, as I then thought, 2 different sorts of polypi protruding themselves, not only from different cells, but from the same: I presently saw two distinct Polypi, each furnished with eight arms, protrude themselves from a cell; the tubular case, which always in the Flustraceæ comes out with arms, was here dilated into a funnel about ¼th of length of arms; the membrane of which |**247**| this is formed is so delicate as scarcely to be visible, but it contains & is supported by at least 30 rays, Hence the exact appearance of a small Polypus with numerous arms.— AC[3] shows a Polypus partly protruded through the case with the funnel termination: There is an <u>appearance</u> of separation between the two Polypi when in the cell, but the cell itself is not divided by any solid substance such as the outer integument: the Polypi seem closely attached at their bases.— I in vain tried to separate a Polypus, case & viscera entire from its cell.— I detached organ A [sketch in margin] by pressure through the funnel & I could see a globular organ with an intestine shaped appendage [sketch in margin] filled with dark red pulpy matter, possessing peristaltic motion:—

The arms of the Polypus were vibratory on their internal surface.— By reading over my descriptions of the structure of the Flustraceæ, it cannot be doubted that this Sertularia belongs in its body to same divisions.— Shape of case & double polypus [are] strongest ~~are~~ difference; the connection of (A) with Case must be different from what I imagined (V **P 234**) in the

Flustraceæ (or that conjecture is wrong); because the brush-like termination of the case is the last part which is withdrawn.— (I forgot) — the vessels or organs (A) lie (I ~~imagine~~ believe I saw) for the two polypi on opposite sides of the cell; convex side outwards:— |**248**|

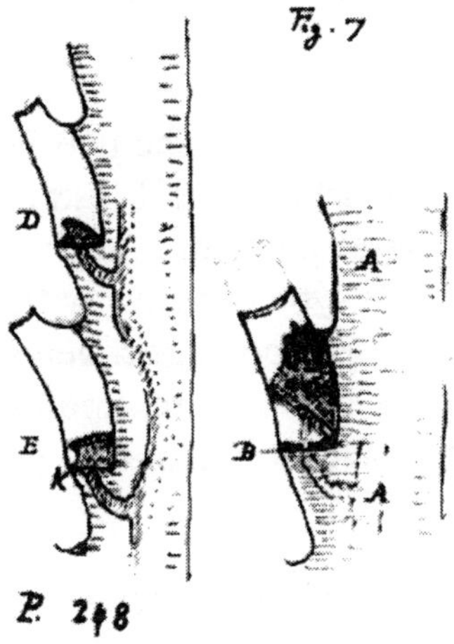

Plate 12, Fig. 7

May 27th [note (a) added slightly later] Off C. Virgins I had the luck to obtain some specimens, but could not examine them till they had been for some days in the spirit.— A slight pressure would force out the two polypus-like funnel shaped cups. the rays I believe are either 24 to 28.— Some other cells only contained one Polypus, in which case generally there was a dark red oval ball enveloped in a transparent case.— ~~in other~~ occupying the place of the other Polypus.— in other cells, there was the appearance represented

P. 248 in another Sertularia (Plate 12, F 7, D & E). I saw one where the footstalk was nearly length of cell, in the same cell with a fully developed Polypus: it occurred to me that very likely the red ball united itself to the base of cell & the living axis & thus grew into another Polypus.— this occurrence of a granular red ball in place of a Polypus has been noticed in some of the Flustræ (with moving capsule).— I could by no means (softness from spirits) detach an entire Polypus. I could see that the "case" was united to a long "œsophagus". I could see 2 dark red small globular viscera. I could see a transverse connection of two main vessels.— but I am not certain that the structure is the same with the Flustraceæ.— (although far most probable).— Each Polypus seems to be enveloped in transparent sack; closely connected at base.— (I should not be surprised if the viscera were united).— the central living axis is enveloped in case, is brown, & from central transparency must be hollow.— doubtless its structure is same as in Dynamena (PL 13, F 3)[4]; in some cases the axis seems to fill whole vacuity, in others sends off branches as in (PL 12, F 7).— The "case" is much enlarged & se<ems> [continued at (z)] globular beneath the funnel, more than represented: The coralline has numerous false articulations or

V (z) next Page

globular impressions ~~beneath~~ above root & at branches: Each ~~cell~~ branch has on each side a delicate ligamentous vessel, which forms a square at base of each cell & sends up a branch on external edge.— This band or vessel is easily detached from horny envelope of Coralline: In one place in specimen there ~~was~~ is a large bivalve-shaped capsule; it springs from between two cells.— is I suppose the ovarium; was empty & ruptured.— [note ends]

[note (a) facing **P. 248**] The two Polypi are withdrawn in parallel lines & apparently similarly to the Flustraceæ by the flexure & ~~turning~~ bending upwards of basal parts of œsophagus.— the funnel case is but smally irritable; the Polypus having withdrawn itself from a touch, the funnel remains protruded: upon again being touched it retires & from the ~~contract~~ approaching of the rays resembles a brush. The funnel often is seen projecting without its Polypus.— [note ends]

[1] Hydroid in order Anthoathecata.
[2] *Flustra* is a bryozoan, but at that time species of both bryozoans and hydrozoans were referred to *Sertularia*.
[3] This drawing has unfortunately not been preserved.
[4] See p. 225.

[**CD P. 248** commences]

Sertularia[1]
PL 12 F 7
(= Clytia)
959
2005
(not spirits)

Found a small fragment off C. Virgins, & from its great general similarity, thought it same as one last described, first found my mistake by wonderful difference in Polypi. These I only saw by dissection; (& this sufficiently imperfect) Stem filled with granular matter, which ~~at b~~ through base of cell ~~sends off~~ forms a narrow stem connected with Polypus (in last Sertularia I could not see this actual connection & branching off) Polypus (as drawn) lies obliquely in ~~the~~ a straight line across the cell, base reddish. over this is the mass of the Polypus, & the arms coiled up on it.— It is all represented as seen by strong light in cell.— Arms short 24 in number (I certainly believe this very number perhaps 22) seated on a wide extensible collar or ring: Polypus (I believe firmly) not contained in the case. I tore open & dissected many Polypi but could see no trace of the organs characteristic of the Flustraceæ: but all agrees with the Clytias.—

If my observations are nearly exact (& I have no reason for doubting it, for I was myself at first quite incredulous) it establishes a wide difference in material structure in the genus (~~even distinct from Dynamena~~) Sertularia of Lamarck[2].— Yet some of the familys Sertulariæ, Flustra & Celleporareæ
(a) have the same structure!!! Perhaps the ~~connection~~ junction of central living mass with Polypus may be an important character?— Coralline coloured yellow.— V next Page.— |**249**|

[note (a)] I omitted; Coralline much but irregularly branched; on basal parts of stem there are many false articulations & some on medial parts.— This ~~Coralline~~ Polypier most singularly agrees with ~~that~~ Sertularia = Clytia (V **250**) in external characters; excepting by comparative ones, it would be difficult to describe them specifically; only differences are stem of this is more thickly branched: broarder, cells more projecting, curvature of upper part of Cells rather different.— What trivial characters when we consider the wide difference of the inhabiting Polypus.— In the same manner, I think the Polypier of Sertularia (**P**. **234**) would be with difficulty distinguished as a genus from the creeping Clytias.—

[**CD P. 249** commences]

Sertularia
= Clytia
(1st species)

May 18th. Perused some more specimens (I do not know the reason, but all these & following species had the greater number of their cells empty, as if Coralline was dying). Coralline springs from a creeping stem: generally very little branched: basal part of stem with those false articulations (globular enlargements & contractions) [sketch in margin] which are so common in Clytia: stem with few true articulations, especially where branches occur; tufts about an inch high.— There is a[n] obscure central vessel in the pulpy axis.— The arms of the Polypus are certainly not contained in a case.— Many of the cells contained (as drawn at D Fig 7) instead of a Polypus a red mass of shape as drawn: in others this became more developed (E) into a broard stem or base, with a crown where rudimentary arms might be seen; the base continues to develop till the Polypus is inclined in an opposite direction as at (B) & is then perfect.— Young Polypi.— In one specimen many of the cells contained dirty orange egg-shaped ova (?), rather more than 1/100in in length. there were others about ½ this size, almost colourless.— a cell only contained one; they were easily liberated.— there was no Polypus in these cells.— I conjecture them to be ova.— When branches occur, they are formed at base of cell as at (K) by prolongation of the pulpy matter, & its central vessel, which ~~has~~ belongs to that cell.—

Sertularia
= Clytia
2d species
961
Sertularia
Clytia

Coralline, delicate white, not much branched, which spring from long creeping stem, adhæring to Terebratulæ[2]: stem zag-zig, many false articulations at basal parts; the medial ones obscure.— |**250**| Cells more detached from stem, spindle shaped, with obscure concentric lines.— The Polypi were in bad condition; but I could make out, that they at least possessed 20 thick short arms seated on large mouth or ring & not enclosed in a case: I could also see junction with central living axis.— 16 Fathom

Sertu: Clytia
3d species
900

Coralline pale yellow; much branched, generally in regular alternations; tufts inch & ½ high: cell different shape from both foregoing species: none or very few basal false articulations: many on branches.— I could make little for certain respecting the Polypus; but have not the least doubt from what I ~~saw~~ did see that the structure is same as in the 2 foregoing species.—

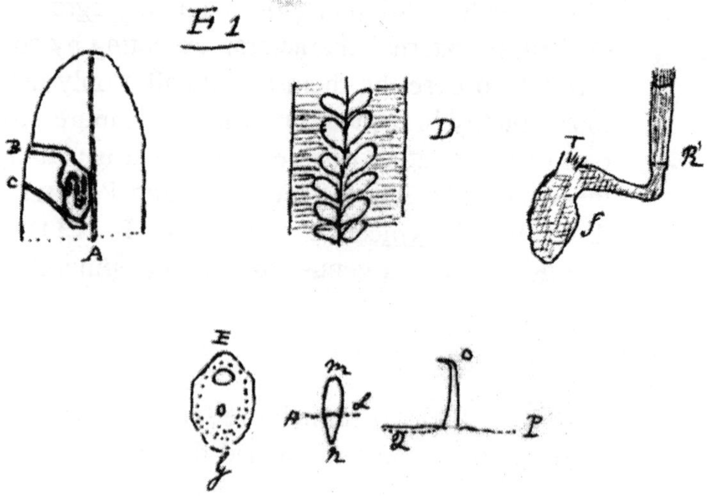

Plate 13, Fig. 1

[**CD P. 250** continues]

<u>Eschara</u>[3]
PL 13, F 1
963

Brought up from 48 fathom off S. Cruz: being kept a week in spirits, the description of Polypus is necessarily very imperfect.— Coralline, coloured "Brownish purple red", stem quite inflexible, very hard, branched (like a Flustra) 4 or 5 inches high.— Expansions or branches formed of a double set of small cells, placed back to back & in regular "seriales".— Cells as seen at surface hexagonal, about 1/50th in length, near anterior or superior extremity (E) is mouth, nearly circular; beneath this is a small oblong orifice which is furnished with a membranous, red, pointed lid, which works on a hinge, & opens downwards.— (m. n. represents it open; H. L being hinge): this orifice & its lid is generally rather indistinct from smallness in the middle cells: On the edges of the expansions, these lids may be seen projecting upwards; they are here are 3 or 4 times as large as the central

<u>Eschara</u> ones; they are membranous |**251**| with the extremity pointed & bent at right angles: (O. 2. is a side view; it can be moved backwards & downward to P, or upward to 2 so as to cover the orifice.— This organ being larger for the external cells than for the internal, is as it happened in the first of the moving-capsule Flustra's.— A transverse section appears as at (D) where pairs of cavities rest on a double plate: when a cavity is accurately divided, we see it as at A.— this cavity contains ~~its~~ the Polypus, with its arms is coiled up in the same manner as in Flustra; the pipe (B) opens into the mouth on external surface: I several times pretty clearly saw a very small stony vessel, running from the orifice with lid to base of Polypus cavity, where it becomes slightly enlarged, I could not trace any certain junction with it or with central plate.— The intervals between C & B is filled with stony plates & the pipes from the cells above & beneath them.— The Polypus has about 14 or 16 delicate arms, contained in a case,

which is contracted at bottom & joins on to œsophagus & intestine (R), both filled with dark red granular matter.— at T there was a general appearance of other vessel being torn off.— I never succeded in seeing the Liver! (organ so called) in its usual position.— But the Polypus was so soft & tender, it was impossible to detach it from cavity without tearing the body.— It is pretty certain that the Escharæ are allied closely, in the structure of their bodies, to those curious Flustræ which possess moving capsules or beaks.— |**252**|

[1] *Clytia* is a hydroid of order Leptothecata.
[2] Brachiopod, lamp shell.
[3] Bryozoan in suborder Ascophora, a bilaminar erect colony with zooids characterized by possession of a calcified frontal wall and an underlying sac, the ascus, opening to the exterior at one end. The structure in Plate 13, Fig. 1 is a frontal adventitious avicularium, CD's red lid being the mandible.

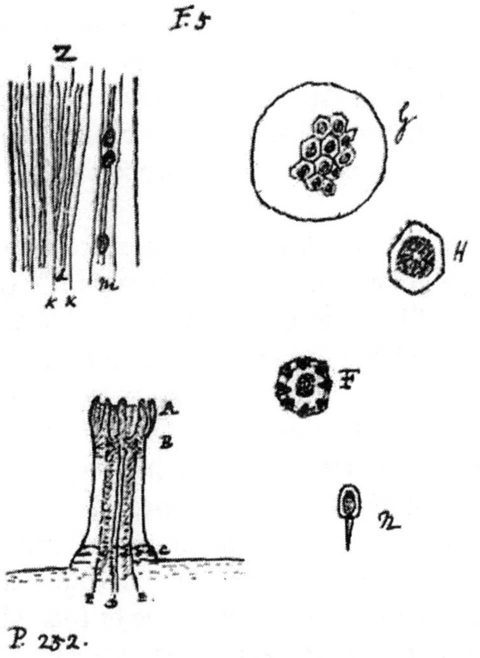

Plate 12, Fig. 5

[**CD P. 252** commences]

Alcyonium[1] Abundant in 9 Fathoms off C. Virgins, on shingle.— Colour white, with pale salmon color.— branched appearance very elegant.— extremities so
PL 12, F 5 thickly studded with Polypi, that the circles from which they protrude touch
962 each other.— surface coriaceous from white stony striæ.— Polypus consists of cylindrical slightly tapering transparent tube, surmounted by (A B) a ~~trans~~ crown, (F) viewed from above, formed of eight bluntly conical pieces, these rise from a collar formed of concentric stony striæ; the pieces

on the external surface are strengthened by longitudinal striæ, the curvatures of the striæ between these & those on the collar very graceful; the extremities of the internal surface of their arms, have a row on each side of short papillæ; these are seldom to be seen; but a contracted head being torn open a bunch of about 10 papillæ will be found at the base of each arm & attached to its membranous lining.— in the centre there is a large tubular mouth, which conducts to a passage, within a vessel containing pulpy matter, & all within transparent tube (AC).— At the base of tube, there is a collar formed of strong striæ; hence the transparent tube appears to rise through an orifice; When the Polypus is touched, the collars (B & C) are brought into close contact by contraction of tube, but there is no absolute withdrawal of Polypus.— A transverse section of any main branch shows a number of hexagons (G) packed closely together, formed of a ligamentous substance they contain (H) a circular opake mass, which |**253**| is roughly
Alcyonium divided by about 8 rays, & a central passage.— A Longitudinal section (Z) shows the sides of (KK) the hexagonal tubes each contains a delicate vessel (L) which contains & is enveloped in pulpy matter.— These tubes pass to the very base of the Alcyonium & are separated from the stone by no fleshy base: Each Polypus however, ~~doe~~ cannot send its tube to the very base, from their great number. They hence thin out, as represented at (Z):

In all parts of the Branches, there are ova, oval length about 7/200th, containing a yoke colored fine "Carmine R" (it is curious how general the assumption of brilliant colors for the ova is amongst all low sea animals):
(a) When procured by a transverse section & pressure, they are forced out through the central vessel in the circular dark mass (H).— There are immature ones, colourless, furnished with a pointed tail or placenta, which grows from a truncated extremity (n); the central yoke points to this; length with tail about ⅔ of full-sized one; without tail ½.— In longit: section, these immature ova may (m) be seen firmly attached to delicate vessel (L).— I believe they are enveloped within it; for the perfect ones when free, are ejected by this vessel; three or four adhære in a line:— The pulpy matter within (L) most likely forms the ova:— When examining the ova, I saw much granular-pulpy matter of different sizes & shapes, with a rapid revolutionary morion.— I only saw this |**254**| once.— V. a similar
Alcyonium appearance with a Virgularia at B. Blanca: there also connected with ova.— There is little communication of sensation between one Polypus & another; if one is cut off, his neighbour does not shrink; if however the whole mass is torn from the stone, the Polypi & whole stems contract & do not again expand; they then appear like cheese, & shaped like a Brain-stone.— When the Polypus is closely retracted — crown contracted, it can hardly be seen.— The Alcyonium is very tenacious of life, lives in same impure water far longer than most animals can do.

[note (a)] If the stony striæ had been so numerous as to form solid envelope, if the outer tunic of each body (which externally is striated) had been internally stony; this polypier would have formed an aggregation of stony tubes in a common stony envelope; allied to what Zoophite would it then have been? [note ends]

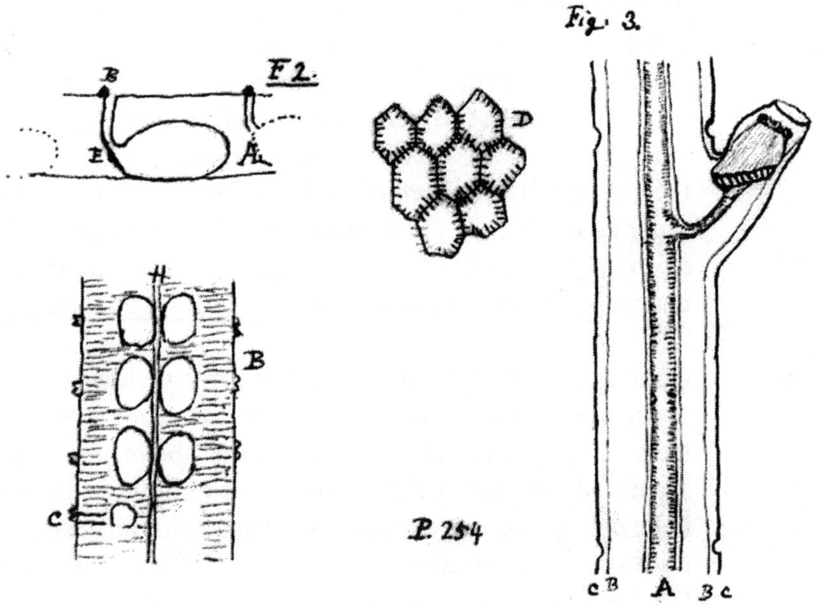

Plate 13, Figs. 2 and 3

[**CD P. 254** continues]

Escara[2] Coralline in foliaceous expansions, quite hard inflexible: Cells in regular
PL 13, F 2 series, on both sides.— A transvers section of branch shows (B) merely
2007 cavities, separated from each other by double plate & from external surface
(not spirits) by stony matter; here & there a smaller cavity (C) will be seen.— I was puzzled to understand this, but a longitudinal section showed (A) a longer, oval, cavity connected with mouth by bent tube: these cavities are placed length-ways in branch, so that small cavity (C) was section at (E).— All I could see of the Polypus (it had been for some time in spirits) were intestine shaped masses of granular red matter & in stony tube (B) a transparent cylindrical case as in the Flustraceæ: Mouth of cell with a pair of oblong points inclined to each other ∵ cells as elongated hexagons.— edges of cells, superior surface joined by crenulated suture: the double plate (H) being separated, base of cells is seen as at [D]: the short lines are vessels with red matter, they form ridges at internal base of cavity, &
(a) extend a short way [continued at (a) on back of page] up sides.— These perhaps strengthen (& produce?) the cells: I <u>believe</u> the Escara has Polypus of same general structure with the Flustraceæ.— Color "Brown purplish with some Cochineal R": 48 Fathom growing in.— |**255**|

[**CD P. 255** commences]

Sertularia[3]
Clytia
968
PL.13, F.3
Hab: V infrà

Coralline. white, branches proceeding from long straight creeping stems: Cells only on one side: False articulations at base of stem or tuft, at branches & in stem.— Polypus had 26 short arms on wide ring &c &c (if not 26, there were 24): As the stem was very transparent.— I have drawn what was apparent. The central line (A) is of a brown color, it must I think be a hollow vessel, because it is more transparent in the centre, as shaded: it is enveloped in a case, & both branch off to join Polypus.— We have then a tunic (B B), which is not affected by the false articulations in the outer coat (CC).— (BB) lines the cell of Polypus.— Polypus lies obliquely in cell, the basal parts reddish.— I saw in some young tufts Polypi appearing in cells in the manner described **P. 249**.—

Dynamena[4]
968
(b)

Coralline. long delicate branches; dirty yellow colored.— very few false articulations, excepting one or two just above roots.— The structure of stem & Polypus is the same as in the above: the tunics are not so clearly separated: the base of Polypus is not oblique in its cell.— Polypus has (I believe but am not certain) 16 arms without case, on larger ring. short. thick.— 8 Fathoms.— Off St^{s} of Magellan.— [note (b)] This Dynamena has sometimes (pear-shaped?) faintly purple Ovaria, attached between two of the cells.— [note ends]

[1] Alcyonacea, soft coral, dead men's fingers.
[2] A bryozoan listed as *Eschara gigantea* 1854.11.15.163 in the George Busk Collection at the Natural History Museum.
[3,4] Thecate hydroids in order Leptothecata, noted by S.F. Harmer as 'Sertularians'.

[**CD P. 255** continues]

Crisia (?)[1]
Lam.x[2]
970
PL. 13, Fig 4
(a)
Crisia

10 Fathoms.— off C. Virgins.— Colored "tile R with little Vermilion R". The structure of the Polypier is complicated; I have but very roughly examined it: Cells alternate, opening on one side: (A) represents this; (B) the ovary's: between which are the orifices of cells, irregularly semilunar protected by an inclined plate; within this a short truncate spine (F): the divisions of cells are but little shown, at the point there arises, a long moveable |**256**| bristle which will be particularly described.— [note (a)] There are punctures on this surface? this side of branches very complicated.— [note ends] (L) the back view shows these bristles & the cells divided by a double line, (which I believe is tubular). The Polypus lies at the bottom: at the back there are ligamentous bands, which I believe are connected with the roots.— The young terminal cells have on external angle two obtuse spines, internal angle one, & between them (2).— These spines (R) are hollow (proved by air bubbles).— they are lined internally by Membrane, which is suddenly contracted near base.— I imagine by the

growth of these spines the edges of the cell are formed.— The external ones spring from just above & upon the plate which protects the mouth.— The ovary lies directly over the basal parts of the anterior cell (represented by dotted line E in A).— I am doubtful whether the ovary & bristle belong to the anterior or (posterior & inferior) cells.— I believe to the latter; so that young cell in (L) could not have them.— The ovary opens towards the inferior.—

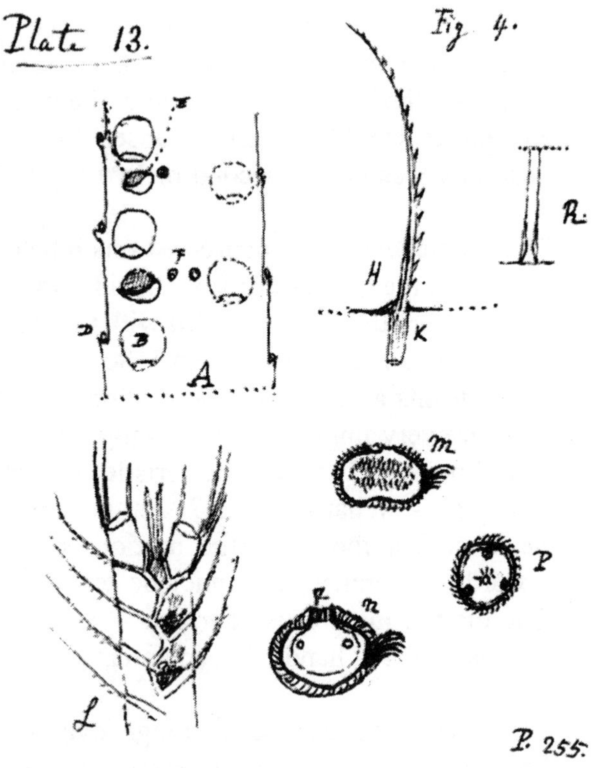

Plate 13, Fig. 4

[**CD P. 256** continues]

Common Character of structure of what I call Flustraceæ excepting no^{r} of arms

The polypus has 12 arms enclosed in case & has same structure as the Flustraceæ, that is the arms rise from a cylindrical long base, which joins to another transverse vessel (place of junction rather enlarged; one ~~end~~ half of this transverse vessel is an intestine-shaped mass with red granular matter: the other a long vessel ~~containing~~ having central enlargement containing red oval organ & united to (or about) the case of the arms.— The extraordinary organ the bristle is drawn up at (H). it is about $1/20^{th}$ long; arched, serrated on outer margin, supported on basal concave side by ridge: connected to its cell by a hinge, & has a membranous |**257**|

Crisia

appendage or vessel (K) leading into cell or polypier: These bristles stand out at right angles, on the outer edge of the ~~alternate~~ cells: I was perfectly astonished, when I first saw every bristle in one branch, suddenly with great rapidity, collapse together on the branch & one after the other (apparently by their elasticity) regain their places. Directly other branches commenced, till the whole Coralline, driven by these long oars, started from side to side on the object glass.— the motion of the Coralline & the setæ was visible to the naked eye: a bit of Coralline being dried on blotting paper, yet for a short time in the air moved its bristles.— Irritation would almost always cause the movement in a branch, & when one branch began, generally the others followed.— They likewise moved (even after being kept a day) spontaneously.— Any one bristle being forcibly moved, re-took its position & would move by itself.— The Coralline placed on its face ~~entangled~~ must entangle the bristles, they often made violent efforts to free ~~its~~ themselves: Generally the bristles on each side of a branch moved together, but one side sometimes would remain collapsed for a longer time than another: this generally was only a second or two.—

The bristle was never depressed much below the rt angle: when collapsed on branch the concave & smooth side was on the branch. & in the extreme cells, the bristles were mingled with the spines.— A bristle, when detached, never moved, the power must lie in the hinge.— Polypus sometimes

May 22[d] Crisia

protrudes its arms during the motion of |258| the bristles.— The above facts are very important as showing a co-sensation & a co-will over whole Coralline.— I think the bristle is not directly connected with the Polypus.— What is its use? As the serrated edge is external it ~~an~~ cannot be to collect food: as the motion is most vigorous & necessarily first towards the branch, it cannot be to drive away enemies or impurities.— The motion must cause currents.— Does it give warning to the Polypus that danger is at hand? When collapsed it does not protect mouth of cell.—

The ovaries contain dark orange ova; some of these I liberated, others liberated themselves; when immature they are simply oval, with included opaker mass.— When more mature the form varies.— (m) is a common form; when in this state, the ovum can move by starts in a zig-zig line & revolving very rapidly (so as not to be followed with ¼th focal distance): its length is about 2/300th: on upper surface there is a collar or projection: on one side four or 5 ~~long~~ curved setæ, (which sometimes seem to rise in a depression). ovum singularly resembles some of the Ostracodes.— the whole mass is surrounded by what appears to be a rapidly revolving transparent ring.— This is best seen, when the ovum is at rest, by the long setæ (which are prehensile) ~~being~~ adhæring to some fibre.— This appearance of a revolving ring (which is most faithful & exceedingly curious) is caused by numerous, curved, minute fillets, moving very rapidly

Crisia (a)

in one direction one after the other; |259| the motion of the longer setæ is totally distinct from this & often at right angles to it.—

(n) is a more developed form; the collar (z) is here much larger; the base is slightly pointed; an internal sack is visible, which contains three small dark red organs.— the apparent revolving ~~or~~ ring is present.— When in this state, the motion was (I daresay always) slower; the collar always was first; there was revolutionary motion, round an axis joining collar & base.— Viewed from right above the collar, we see it, as at (P), where the three small organs are placed, triangularly: here also, & from <u>every</u> point of view, the apparent revolving ring was to be seen.— Hence doubtless the whole surface is covered by fillets such as described: when at rest, it was curious to see the rapid, oddly curved, & extensive currents produced by these setæ.—

from impurity of water

(n) having died, the setæ appeared like a faint halo, (the appearance of ring having vanished). The collar, hence, appeared to project further; on dissection, ~~short~~ the collar was certainly formed of the short arms of a Polypus enveloped in membrane: I fancied I could trace some resemblance (it probably is the case) between the three little organs, with the three in the old Polypus.— The pointed base is probably point of attachment for young cell, which perhaps is formed by outer tunic of the ovum; when the setæ have dropped off.— The motion of the minute fillets is continued with equal rapidity, when the ovum is at rest & when moving.— (Respiratory?).— |**260**|

Jany. 1835

[note (a) added later] At Lowes Harbor — Chonos Archipelago. Lat: 43°.49′ Found some of this Crisia & again clearly saw the motion of the toothed setæ caused by irritation. [note ends]

[1] *Crisia* is a bryozoan in class Stenolaemata, order Cyclostomata, which do not have avicularia. CD's specimen, possessing a novel type of heterozooid nowadays known as a vibraculum, whose movements controlled by modified opercular musculature he describes brilliantly, was later identified by S.F. Harmer as the species *Caberea minima*, which belongs to class Gymnolaemata, order Cheilostomata, and superfamily Cellularioidea along with *Beania* and *Bugula*. The 'ovaries' were in fact ovicells, the cheilostome brood chambers.
[2] See *Lamouroux* p. 7.

[**CD P. 260** commences]

Zoology <u>S. Cruz</u>

During the expedition up the river[1] I ~~noticed~~ found the same animals, birds, insects & plants, which I have collected near to the coast: this extreme similarity in the productions of the sterile plains of shingle is a very striking feature in the whole of S. Patagonia. The geology likewise being similar, one view can hardly be told from another.— Amongst animals, the smaller rodentia, in importance, far takes the lead of all other animals; besides <u>several</u> sorts of mice: we have the Aperea[2].— Tocco Toco[3] & Gerbillus? (2032)[4] in great numbers; On such animals the Foxes, which are in considerable numbers, perhaps prey.— [note (b)] Then Taturia (1697)[5] exists thus far South.— [note ends] The skunk or Zorilla is found.— The Guanaco ~~are~~ abound in large flocks & thus support the Pumas: The number

(b)

(a) of Guanaco is the reason, why out of the few birds, four very striking ones
(z) should be Carrion feeders.— The Condor, & three Caracaras.

[note (a)] I am quite at a loss to know whether Caracara (2028) is the same with the Carrancha of the R. Plate[6]: I opened several, many were females, others had the ovaries (as for instance this specimen) quite smooth.— Yet I saw some more white ~~beneath their~~ on throats: Habits same as Carrancha — tolerably numerous all up country.— I saw one soaring at a great height — this very unusual in the whole tribe.— The Caracara (1772) is also pretty common.— & lastly a beautiful Caracara (Rancanea?) (2029)[7]: is abundant some distance up the river, but rare at coast.— I never saw it any-where else.— Skin about beak, yellow.— bill blue, black lines: Legs pale yellow.— [crossed through in pencil] Caracara (or Chimango) 1772, not uncommon at P. Famine [note ends]

[note (z) added later] Shot in Port Famine <u>decided</u> old female: Bill, Cere & legs as in descript. of "Vulgaris" in Dic. Class: Head "Lines & Blackish B^{r}" all the dark brown, are this color: pale are "yellowish B^{r}", Gorge rusty yellow — — breast & under tail coverts banded with (1/10th of inch) pale brown & do. yellow.— Back banded darker: wing coverts pale do.— .6 first ~~tectrices~~ remiges, central parts whitish. Rectrices broarder bands, outer margin of outer feather darker.— Under parts of secondaries broard bands: Length (full stretch) 18½ inch; breadth tip to tip 4^{Ft} 7^{inch}.—— [note ends]

[**CD P. 260** continues]

The Sparrow (1704)[8] is the commonest bird: & the Fringilla (2017)[9] is also abundant: there are the huge flocks of Sturnus ruber[10]: the Callandra often sings (2011)[11] amongst the spiny bushes.— Short billed snipe (1224)[12] inhabits the driest parts of the plains.— Three Furnarii are found: one 2025[13]: 1822[14]: & 1823: this latter[15] I was much surprised to see a hundred miles up the river; never before had I seen it distant from its favourite Kelp.— Alauda (1898)[16], Avestruz Petises: several hawk & insect birds. V. Collection:

Heteromerous insects are always numerous; a comparison with those of the N. Traversias will be interesting.— (I suspect Patagonia has but few productions of its own.— is the Botany sufficiently known to tell.— The
(c) extreme infertility, even close to running water, has [continued at (c) on back of page] has *[word repeated]* often much surprised me.— At different times I have attributed this general sterility to the salt contained in the sandy clay.— the extreme dryness of the climate, (which is an undoubted fact).— the poorness of the soil of the gravel beds.— and to no creation having taken place, since this country was elevated (I yet think this applies to the Northern parts): I am now most inclined to attribute it all to the poorness

of the soil.— Yet in the Lower country, where there was water, it was but little better!.— |**261**|

[1] After following the course of the Rio Santa Cruz for 245 miles, and approaching close to Lago Argentino, FitzRoy turned back on 5 May and travelling fast with the current reached the mouth of the river on 8 May. For CD's account of the journey see *Beagle Diary* pp. 231-9, and for pictures painted by Conrad Martens see *Beagle Record* pp. 199-213.
[2] Listed in *Zoology* **2**:89 as *Cavia patachonia.*
[3] Listed in *Zoology* **2**:79-82 as *Ctenomys Braziliensis.*
[4] Listed in *Zoology* **2**:69-71 as *Reithrodon cuniculoïdes.*
[5] Listed in *Zoology* **2**:93 as *Dasypus minutus.*
[6] Listed in *Zoology* **3**:9-12 as *Polyborus Brasiliensis* Swains.
[7] Listed in *Zoology* **3**:18-21 as *Milvago albogularis.*
[8] Listed in *Zoology* **3**:93 as *Fringilla gayi*, NHM 1855.12.19.42, though might perhaps be the smaller *Fringilla formosa*, NHM 1855.12.19.24 and 1856.3.15.12, now *Phrygilus* sp.
[9] Listed in *Zoology* **3**:95-6 as *Chlorospiza? melanodera* Gray, which also is Fringilla (1879), NHM 1855.12.19.50.
[10] *Sturnus ruber* is *Sturnella loyca of Molina*, listed as *S. militaris* in *Zoology* **3**:110.
[11] Listed in *Zoology* **3**:60-1 as *Mimus patagonicus* Gray, NHM 1855.12.19.221 and .311.
[12] Listed in *Zoology* **3**:117-18 as *Tinochorus rumicivorus* Eschsch.
[13] Listed in *Zoology* **3**:69-70 as *Eremobius phœnicurus*, NHM 1855.12.19.73.
[14] Listed in *Zoology* **3**:66-7 as *Opetiorhynchus vulgaris* Gray.
[15] Listed in *Zoology* **3**:67 as *Opetiorhynchus patagonicus* Gray.
[16] ? Listed in *Zoology* **3**:84 as *Muscisaxicola nigra* Gray.

[**CD P. 261** commences]

Holuthuria[1]
975
PL.13, F.5

Is pretty closely allied to Fist: (P 141) : found crawling amongst roots of Kelp in mud.— length in this state 5 to 7, narrow.— Color dirty pale flesh color.— semi-transparent, smooth; with transverse fibres in bundles.— 5 internal longitudinal bands.— Anus blunt with very extremity rather pointed; capable of much distention with water.— Mouth surrounded by 10 or 12 arms: from "alternate motion" difficult to be counted, white: each arm with 12 papillæ, central longest, gradually decreasing on sides; bases connected with membrane; The papillæ adhære on a longer base than in **P 141**: so do not have ~~not~~ so much the appearance of a hand: they adhære & crawl by their aid: When considerably contracted its length 3 inches, breadth .4:— in this state there may be seen in one of the intervals between the longit: ligamentous bands, & chiefly (not quite solely on the posterior half) 2 or 3 dozen small white very slightly projecting eminences or papillæ: about $1/20^{th}$ of inch: their shapes was [sketch in margin], probably owing to contraction of body: they were sometimes in single, double or more rows.— they consisted of numerous (20 or 30?) little cups overlapping each other (owing to of do contraction) placed on slightly convex surface & adhæring together on a gelatinous base.— Each little cup

(B)
miserably
drawn

(A) (a) (A) was most symetrical, more than 1/500th of inch in diameter, quite transparent, very shallow.— edges folding in, most finely serrated (to be seen with 1/10th or 1/20th" focal distance). bottom ~~base~~ of cup consists of 6 tapering spokes, perhaps connected by membrane, & uniting close to a puncture.— Are evidently used for adhæring.— [note (a)] (A) is the cup, seen from directly above; the central ring with the puncture in the bottom: the serrated edge is seen folding inwards.— it is a good likeness of all which is visible.— [note ends] |**262**|

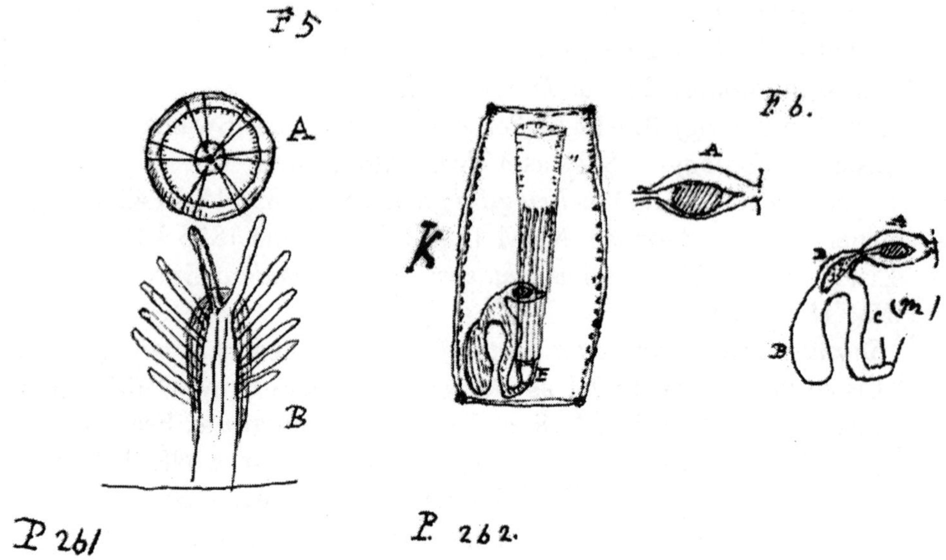

Plate 13, Figs. 5 and 6

[**CD P. 262** commences]

Flustra[2] encrusting 983 — I examined the Polypus of this very simple Flustra, so that I might errect at some future day, my imperfect notions concerning the organization of the whole family of Dr Grants Paper[3].—

Pl. 13 F. 6 — (K) represent[s] what is seen in a front view of cell, the transvers organ (A) lies attached to the case of the tentacula, & as here seen, beneath it.— (A) consists of a pointed oval case, one end simply attached to ~~trans~~ above case, the other joining on the main viscus.— it contains a small rounded oval red ball where granules are connected by transparent gelatinous substance.— this also lies in a very obscure vessel; by stretching the whole organ, I could pretty clearly trace it into the enlargement (B), where itself becomes enlarged.— This ball is very easily detached & will then keep its proper shape.— When first I examined the specimen, no motion was perceptible in any of the viscera, upon being kept for some minutes, in every polypus there was a rapid revolutionary & circulatory motion in the contained granules in (D). this motion seemed to be confined to the enlargement of the delicate vessel already mentioned.— the appearance was very strong of vibratory setæ which caused this motion.— Where this vessel led to or how it ended I cannot say.— In one, (which was rather injured & there the motion is generally most evident), I could clearly trace the same particles. revolve in (D), be circulated into (B); then into (C) & return to revolve in (D).— from this I am led to infer, that this vessel freely communicates with (B & C).— in this case viscera likewise had a peristaltic motion.

Flustra

there is an appearance of central vessel.—

Polypus has 16 tentacula, each one is provided on each |**263**| side with setæ, half its diameter long; these rapidly vibrate, in a progressive manner up one side & down the other.— Hence when all moving, complicated currents; I have seen particles at a little distance from extremity, revolving at one point from the different currents.— Where the tentacula join on the base, there is an inverted cone, in which a rapid vibration, as of setæ, is very clear; this unites directly with the intestine, or stomach.—

The tentacula are included in a case, one end of which is attached to the labiate orifice, the other I believe to base of tentacula, on the outside of this in central part. when the Polypus is quietly in the cell, the Liver (A) is attached.— The cells are separated by stony plates, each side of which is crenated with little points; in the centre I believe there is a vessel.— each cell has six projecting conical points [sketch in margin] (where plates from other cells meet), these have a corresponding (to one) hollow, filled with reddish matter & I believe connected with vessel in plates.— These vessels must be central axis which developes Polypi, from eggs builds cells; the Polypi are connected at base to side of plate.— the membranous covering of cell is easily detached from the sides.— At base of tentacula there was an appearance of fibres.— Flustra exceedingly abundant coating the Fucus Giganteus.— |**264**|

[1] Apodida, Chiridotidae, probably *Taeniogyrus contortus* Ludwig.

[2] Identified by S.F. Harmer as *Membranipora membranacea*, belonging like *Flustra* to superfamily Malacostegoidea.

[3] When CD was a student at Edinburgh in 1826, his friend Robert Edmond Grant had published a translation of a paper by August Friedrich Schweigger which supported the view that corallines should be regarded as plants rather than animals. See CD's letter to Henslow dated 24 July 1834 in *Correspondence* **1**:397-403.

[**CD P. 264** commences]

Actinia[1]

987

The superior & exterior margin of animal is formed into eight blunt foliaceous expansions. These are thickly covered with "Hair brown" papillæ, which are shorter but much [more] numerous at the extremities. The flat surface

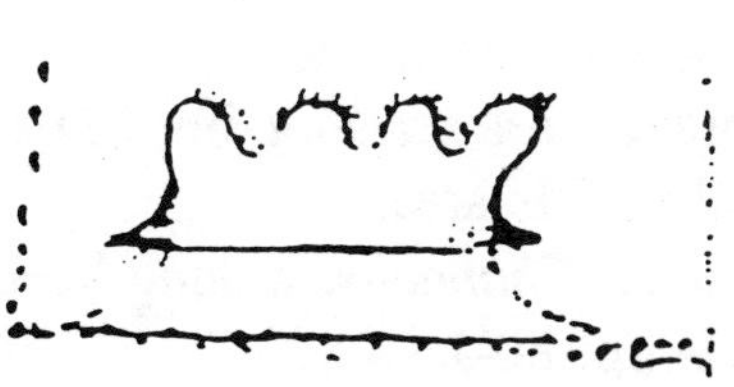

which surrounds mouth nearly free from papillæ.— Mouth small, with internal longitudinal folds.— From the structure of animal, it is with difficulty that the papillæ can all be concealed, the depth of the animal being small in proportion to width.— Color of surface "Orpiment orange" with short irregular concentric lines of "deep reddish O" which on the superior edge are so frequent as to form the prevailing color.— Round mouth narrow rim of do colour.— A specimen being removed from the rocks, during the night moved its position & firmly attached itself to the glass.— Tidal rocks.—

Gasterop=terus[2] 996 — Length of mantle 1.5 inch, breadth 1.1: far surpassing its foot, edge sinuous: inferior antennæ connected for ¾ of length by a membrane forming a larger angle with each other than the superior ones, which are slightly winged.— General color straw-yellow, branchiæ & viscera of rather a darker tint.— Roots of Kelp.—

Geotrupes[3] — This insect is excessively abundant, boring deep holes beneath every heap of horse dung (& once I saw sheep's).— Curious instance of increase in
(a) number & change in habit. no large quadruped in Chiloe.— At the depth of 2 & 3 feet, balls of earth, lined with a darker kind (dung?), containing larvæ, are very commonly found in gardens [continued at (a) on back of page] (when dung is not directly present) from what I can hear I have little doubt that no other beetle than the Geotrupes exists in numbers proportional
copied — to the balls.— I saw a man *[word missing]* 10 or 12 in a few minutes.— When first found they are not quite so hard as at present.— The Larvæ of many had eaten their way out & had escaped.— Vide **Page 200** for a discussion on this subject:— |**265**|

Febr^{y} 1836 — [further note on back of page added later] Hobart town, Van Diemens land. I carefully examined dung of the Herds: damper climate favourable to their
3446 increase as compared to New S Wales.— The Horses [&] Cattle have now
3504...... been introduced since 1803 (33 years) & I find Onthophagus (2 species)
......3512[4] very abundant under Cow's dung; there is a third species.— Al<so> 3 species of Aphodii; one of which was beneath hor<se> dung: Thus we have 6 species found even dur<ing> my short stay:— The dung of Kangaroos
copied — is in Pellets & there were no other large animals as in the case of Chiloe.— The subject is a curious one[5].— [note ends]

[1] Actiniaria, a sea anemone.
[2] Opisthobranchia, a sea slug, but CD's description does not fit with any species known today to be found in this area.
[3] Scarabaeidae, *Phanaeus*, a dung beetle. No specimen was found in CD's collection. See *Insect Notes* pp. 80-1.
[4] These specimens of Scarabaeidae from Tasmania have not survived. See *Insect Notes* p. 97.
[5] See Introduction p. xxi and long footnote in *Journal of Researches* **1** pp. 583-4.

[**CD P. 265** commences]

Ornithology Specimen (2127)[1] is a curious bird; is called by some of the officers the
2127 Robin, to which when hopping about the woods it bears some resemblance.— This bird frequents the most gloomy & retired spots in the
(a) damp intricate forests; it utters a loud, singular, repeated whistle; Can be seen only with some difficulty: & then perhaps by standing still, it will come quite close, busily hopping close to the ground amidst the impervious mass of Canes & dead branches of trees.— It is called by the inhabitants Cheucau[1].— The gizzard is muscular, it contained fragments of stones, hard seeds, buds of plants & vegetable fibres, but I could see no part of any
(b) insect.— Mr Stokes[2] has seen this bird near C. Tres Montes.—

[note (b)] This bird utters three very distinct & strange crys, one of which is called Chiduco & is a good sign; another Huitreu the bad sign; so called from a resemblance to the sounds — This bird is regarded with much superstition, & its noises serves for omens.— It is excessively tame & intruding to a person standing quite still: cocks its tail vertically like the Tapacolo[3] of Valparaiso.— is a most comical bird.— Is said to build its
C. Tres Montes nest in low bushes:— Very abundant at C. tres Montes: opened a Male specimen, found seeds & parts of insects & vegetable fibres in stomach: The Barking bird[4] is also very abundant here: habits exactly similar to Cheucau; but rather more shy.— Noise exactly like little dog yelping, & flies badly like the Turco[5] of Valparaiso.— [note ends]

[note (a)] These forest[s] wear from the climate a gloomy look: yet in many respects they have a more Tropical appearance than the latitude would
Copy ornithology lead one to expect.— The woods contain various sorts of trees: they are very thickly placed together: they are much covered with parasitical plants, many of them monocotylidenous.— An Arborescent grass jointed like bamboo, which intertwines the trees to the height of 30 feet is very abundant: the Ferns are singularly large.— I no where saw the Beech tree which forms the whole forests of T. del Fuego.— the Winters bark in common to both countries.— [note ends]

[**CD P. 265** continues]

2132 The Emberiza[6] (?) is very common in small flocks; the commonest bird in
2133 the island; by the manner in which it frequents the cleared land round the houses resembles the Sparrow: in the stomach much seed & sand.—

2134 The Trochilus[7] (2134) is very numerous; perhaps the next most abundant bird to the foregoing Emberiza: This little bird looks very much out of
(a) character amongst the gloomy dripping foliage & the endless storms of rain.— The commonest site, where these birds may be found, is on marshy open ground, where a Bromelia (?) (a plant bearing pine-apple sort of edible

fruit with long toothed leaves) forms thickets.— It frequently hovers at the sides of these plants & then dashes into them near the ground, but whether it alights on the ground I never could |**266**| see.— There are at this time of year scarcely any flowers, & none whatever where the above plants grow.— I was well assured that these birds did not feed on honey.— on opening the stomach (or rather duodenum) by the help of a strong lens, in a yellow fluid I discovered small numerous ~~fragments~~ bits of the wings & legs of most minute Diptera, probably Tripulidæ.— It is evident the humming bird[s] search these insects out, in their winter-quarters, amongst the thick foliage of the Bromelias.— [note (a)] I opened stomach of this species killed near Valparaiso, there were as much debris of insects as in a Certhia. besides Diptera I pretty clearly recognized remains & not so very small of ants.— [note ends]

[**CD P. 266** continues]

Besides the birds I have collected I know of the following birds:— the Condor, which seems uncommon: the Vultur aura: the Carrancha.— the Chimango, which follows in great numbers the plough, I suppose to pick up Larvæ.— Hawk (2014)[8]: Furnarius same as (1822):— Wren same as (1831): Icterus (1784), common: Scarlet-headed & black Woodpecker: all these same as at Port Famine: the birds generally being very similar in the two places.— We have also the noisy Pteree-Pteree: & the Barking bird.— The Scissor beak[9].

[1] Listed in *Zoology* **3**:73 as *Pteroptochos rubecula* Kittl. See also *Journal of Researches* **2**:288-9, where CD admitted to having forgotten what the third strange cry actually was.
[2] John Lort Stokes was Mate and Assistant Surveyor of the *Beagle*.
[3] Listed in *Zoology* **3**:72 as *Pteroptochos albicollis* Kittl. See also *Journal of Researches* **1**:329-30.
[4] Listed in *Zoology* **3**:70-1 as *Pteroptochos Tarnii* Gray. See also *Journal of Researches* **1**:352-3.
[5] Listed in *Zoology* **3**:71-2 as *Pteroptochos megapodius* Kittl. See also *Journal of Researches* **1**:329.
[6] Listed in *Zoology* **3**:93 as *Fringilla Diuca* Mol. NHM 1855.12.19.187.
[7] Listed in *Zoology* **3**:110-11 as *Trochilus forficatus* Lath., but in *Journal of Researches* **1**:330-2 as *Mellisuga Kingii*. See also *Ornithological Notes* pp. 251-3 for a fuller account of the humming birds.
[8] Listed in *Zoology* **3**:29 as *Tinnunculus Sparverius* Vieill.
[9] For an account of the Scissor-beak *Rhynchops nigra* see *Journal of Researches* **1**:161-2.

[**CD P. 266** continues]

Apple Tree — In Chiloe the inhabitants have a mode of propagating trees so that in three years it is possible to have an orchard of large fruit-bearing trees.— At the lower part of every branch, there are small (2 or 3 1/10th of inch), conical, brown, wrinkled projecting points; these are roots, as may be seen where any mud has fallen on the tree.— A branch, as thick as a man's thigh is chosen, & is cut off just beneath a group of points; this |**267**| is done in very early spring: the extremities of all the sub-branches ~~being~~ are lopped off, it is placed about 2 feet deep in the ground with a support.— the ensuing summer it throws out very long shoots, & sometimes bears a few apples (I saw one which had most unusually produced as many as 23): the 2^{d} summer, the former shoot throw[s] out others: in the third summer it bears a good deal of fruit & is (as I have seen) a small wooded tree.— Are the incipient roots present as trees in any part of England? or is this whole process owing to the extremely damp nature of the climate? it is a most valuable method, where applicable.— I have noticed that in the Apples, not above one in a hundred will have any seeds in its core[1].—

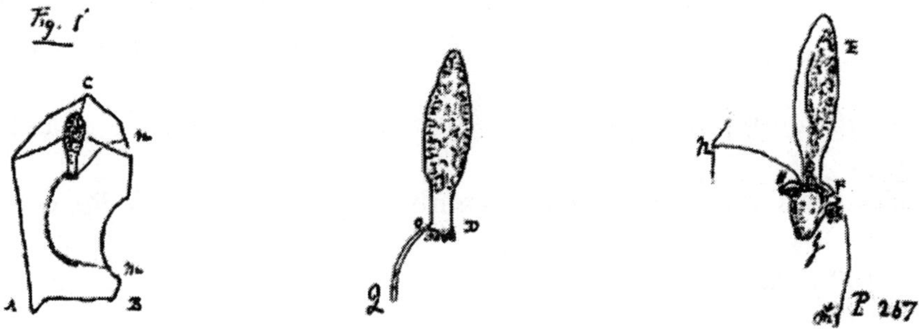

Plate 14, Fig. 1

Medusa[2]
PL. 14
Fig. 1.—

[July 22nd] About 10 miles off Valparaiso: the sea contained many angular Medusæ.— Body perfectly transparent, colourless.— rather hard, Length .4: Shape like a wedge, where the four corners of the head are "replaced" by four planes, (which form a vertex or point at the top).— Line (AB) is the edge of the wedge; two of the replacing planes are seen (drawn out of perspective), one side (the narrow side) of the wedge is scooped out ~~& form a~~: when the animal is turned on its narrow side, a slit or opening is seen, extending from m to n, & more than half way deep in its body. At the very back part of this cavity, but separated from it, there is a semilunar, thin

Medusa

vessel, which |**268**| is united just above mouth of another organ: This organ (D) lies in the very centre of superior part of body: (the mouth open[s] into the narrow slit or cavity).— it consists of an elongated oval

(a)

sack filled with semiopake matter, which seems divided into irregular sphæres; in the interstical spaces I saw a slow circulation: the mouth of sack is something like vermiform processes not well defined.— [note (a)] I could see no trace of communication with external surface from top of sack.— I have not attempted to keep any specimens. [note ends] the specimen being kept some time I saw several small bodies shaped thus

[sketch in margin], proceed from the mouth into the open cavity.— I

presume this sack to be the Ovarium.— The angular edge[s] are composed of numerous fine fibres.—

[**CD P. 268** continues]

I found other specimens which I believe to be the same species: essentially differing in structure: vessel (oq) is here coiled up (F), & lies within the cavity so as to be touched with needle & is then irritable: the sack is only partially filled with granular matter: ~~Near~~ Beneath its mouth (& within cavity) there is (when contracted) a heart shaped organ: this is highly irritable, coloured most faint red.— when expanded forms an Elephant-like long proboscis, open at extremity but very broard at base.— capable of much motion.— no apparent connection with sack: on the opposite to where (F) joins there is small sphærical organ, with small projecting point (or mouth? & central vessel?) Is this the mouth of sack? — Specimen about half size of last; same form.— Other (injured) specimen nearly same structure as this latter.— Caught on surface of sea.— |**269**|

[1] See *Plant Notes* pp. 177-8, *Journal of Researches* **1** pp. 363-4, and *Darwinian Heritage* pp. 101-3.
[2] Siphonophora, a eudoxid bract of family Abylidae, probably *Bassia bassensis*.

[**CD P. 269** commences]

July 25. Fish bought in market[1] [at Valparaiso].—

1008 Above blackish grey, indistinct bands of do on sides; beneath white.— are found 3 or 4 times as large.

1009 A uniform pale greenish tinge, most thickly mottled with "greenish black".

1010 Uniform pale flesh color (especially beneath), mottled with "deep reddish B" & transverse dorsal bands of do: Branchial covering yellowish.— inferior edge of Pectoral pink.—

1011[2] Above leaden colour, beneath paler; grow considerably larger.

1012[3] do do .— fins dark.

1013 do, slightly irridescent, do.— grow to 2 or 3 times this size.

1014[4] Under surface, sides, Branchial covering, part of fins "tile & Carmine R", dorsal scales pale yellowish dirty brown.—

1015 Uniform tinge pale dirty yellow with numerous angular spots of black.— Above clouded with pale brown.— Ventral & tips of pectoral & anal "reddish orange".— Common size.

1016 Sides "Cochineal red mixed with grey", an indescribable tint, belly strongly tinged with yellow, fins pale "blackish green", posterior half of body with numerous small scarlet dots.—

1017[5] Beneath brilliant white; head & back clouded with "purplish & Carmine R", longitudinal & transverse irregular bands of do.—

1018 Whole body silvery, back & fins with few clouds of leaden color. grows to 3 & 4 feet long.—

1019 *[illeg. pencil note]* Crust. Macrouri: sold in market; whole body & legs with "Arteri & Hyacinth R"; intermediate spaces paler; yellow & pale blue dots.— |**270**|

[1] Only four of the specimens in this batch were in good enough condition to be identified by Leonard Jenyns in *Zoology* **4** and CUL DAR 29(i).
[2] Listed as *Heliases Crusma* Val. in *Zoology* **4**:54-6.
[3] Listed as *Pinguipes Chilensis* Val. in *Zoology* **4**:22.
[4] Listed as *Sebastes oculata* Val.? in *Zoology* **4**:37-8.
[5] Listed as *Latilus jugularis* Val. in *Zoology* **4**:51-2.

[**CD P. 270** commences, entry as usual crossed through for a planarian]

Planaria[1].

PL.14, F.2 is central vessel (A)

2301 (not spirits)

Quite white, excepting the central vessel which is flesh coloured. Body very flat & thin, length when crawling 3 inches, breadth .2: Body seems composed of pulpy matter enveloped in transparent soft envelope; on the margins there are small black dots, placed at pretty regular intervals; they are not particularly numerous at anterior extremity. (Both extremities finely pointed). Through the whole length of ⅔ds of body, from ~~the anterior~~ extremity, there is a straight, gradually thickening, central vessel, this is coloured pinkish, from it (in some specimens) on each side, regular numerous vessels branch out & these being sub-divided at their extremities blend with the pulpy mass. For the remainder of the body, this vessel is divided into two; it here immediately encloses an oblong space (in which some organ may be seen), after this they run parallel to each other, again enclose a similar space, & then run paralle[l] (or rather approaching each other) & gradually become finer: ~~these two vess~~ till they reach to the very posterior extremity of the body.— These two vessels invariably throw off branches smaller than the previous ones, but of the same construction.— All the above facts are best seen when looking at the dorsal surface; on the under surface & corresponding to the two oblong spaces & therefore in the ⅓ posterior part of body, there are two small transverse mouths or slits.—

When the animal, being partly contracted, was only 1 & ¾ inch long, these apertures were .2 apart.—

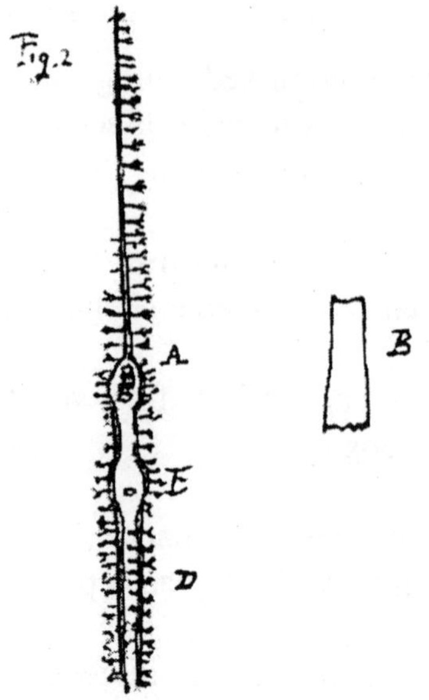

Plate 14, Fig. 2

[**CD P. 270** continues]

Planaria

By dissection, I found the anterior |**271**| enclosed space contained a closely contracted cup shaped organ, edges very sinuous: another animal having however been killed by spirits, the organ was as represented at (B), the lower end forming apparently a mouth with sinuous margin; as lying in the animal this mouth pointed towards the posterior extremity.— Can it not be protruded through external orifice? In the other space I could not trace any organ.— The animal always crawls on one surface that with the two orifices, & always ~~with~~ in one direction with respect to the extremities. Both these facts hold good even with small pieces of the living animal.— cut extremities in a few minutes become rounded.— Live under stones upon the hills, & amongst the pasture.— they have ~~burrows, like~~ sinuous chambers beneath the stones, like earth worms but not so regular, because (a) parts of the body lies in places coiled up in a knot.—

N.B. Colour evidently altered in spirits in *[illeg. word]*

[note (a) in different pen] Reexamined Specimens: Black points with strong power are not quite circular; they are within external transparent envelope & where best seen seem seated on extremities of the lateral ramifications of central vessel. they abound chiefly in anterior part & in groups of 2^{s} & 3^{s}, which shows some probable connection with above ramifications.— on very anterior extremity // together.— The term central vessel does not imply that I have any proof it is a vessel with darker contained pulpy matter, only its appearance is such.— Ramifications proceed externally whole length of vessel (but not between the posterior double parts.—) oblong spaces & double part (D).— When crawling, anterior oblong space much elongated is .2 long. The general structure of these spaces, apertures & vessels universal.— Specimen being in salt water, a white globular organ was <u>protruded</u> by anterior orifice, being touched retracted: was easily dissected out, consists of a white, most delicate bell shaped bag, the mouth of which when protruded points towards tail of animal: its length .15: breadth of broarder part .05: the upper end is attached to very point of bifurcation of vessels. Even when body was dead & motionless, this organ remained <u>highly</u> irritable: & assumed a thousand shapes, the mouth dilating & rapidly contracting: when within body, lies in oblong space, must act as sucker or mouth: (globular when quite contracted & harder).— I could find no organ in posterior oblong space, only the little transverse slit: These animals are certainly often found in pairs.— [note ends]

[**CD P. 271** continues]

<u>Bufo</u>
1023 Pale dirty yellow. Back blackish brown, over eye as far as behind front legs jet black band.

1024[2] "Yellowish & Broccoli B" with darker brown marks; broard medial dorsal band of pale "Gallstone Y". Lumbar glands "Saffron Y" with jet black marks.

1025 dirty "Wax Y" clouded above with "blackish grey": when taken or handled utters a gentle plaintive repeated note like some bird — Are found beneath stones. |**272**|

[1] Listed by Darwin (1844) (see *Collected papers* **1**:186) as *Planaria pallida*, currently *Pseudogeoplana pallida* Darwin. The pinkish colouration of the gut showing through the body wall is often observed in flatworms that have recently fed on an earthworm.
[2] Described by Thomas Bell in *Zoology* **5**:37-8 as *Pleurodema elegans* Bell.

[**CD P. 272** commences, using unlined paper]

<u>Vaginulus</u>[1]
1027 Colored above "Umber B. & oil green" <u>finely</u> punctured, many of the punctures black.— beneath pale, foot yellowish.— When closely contracted, breadth .7, length 1.3.— when crawling 2.1 & .6 ~~broard~~ broad: foot .2 ~~broard~~ broad.— ~~tail~~ posterior end of ~~tail~~ mantle abruptly rounded, mantle slightly narrowing towards the head where it is truncate: Mantle depressed, edges slightly recurved, forming a transverse section of this shape [sketch in margin].— Antennæ lead-colored, superior ones where extended more than .2 long, inferior ¼ of this.— Found under stones, in habits & form closely is allied to the M. Video species, differs as far as my memory goes, entirely in its color.— in its less regular & depressed form;

(a)
Vide Back
of Page

in the anterior extremity not being so broad & being truncate, & in being proportionally narrower.—

[note (a) added 6 months later on back of page] Specimen (1180) very
1180 nearly agrees with this, taken in the forest of Valdivia.— Colour "Oil green & Wood Brown", black minute punctures, chiefly at extremities, dimensions
Feb^{ry} 1835 rather greater.— colored do beneath.— Antennæ almost black, rather long, shape simply convex [see sketch in margin], tail not particularly abruptly rounded.— ? Same species? Locality & climate very different.—

1184 Specimens (1184) were also caught in forest under log of wood. color above uniform dark brownish black, beneath & foot do, but rather paler.— body narrow, mantle can hardly be said to be truncate anteriorly: in centre of back a ridge, <u>sharply</u> convex: antennæ very closely approximate.— These specimens appear to be grown, length where crawling 1.2:— I do not at all know how far all these are varieties or distinct species:— (1096) seems <u>very invariably</u> characterized.— [notes end]

[**CD P. 272** continues]

<u>Vaginulus</u>[2] no 1096 & 1160

Under piece of rotten timber in dark forest in Chiloe (SE extremity), December:— Whole mantle, upper & under surface, mouth, inferior & superior antennæ "Ink black". ~~Mantle~~ Foot dull "Saffron yellow". Mantle with angular edge; not very convex; nearly of same breadth ~~during~~ throughout its whole length, but slightly tapering to each extremity.— anterior end truncate; posterior bluntly rounded. depressed. Surface most finely punctured.— Dimensions when contracted, 1:.7 long; .9 broa~~r~~d; when fully crawling, 3.6^{in} long; .55 ~~broard~~ broad: superior antennæ .25 long.— foot .25 wide.— Same slow habits & torpid state as rest of genus.— I have no doubt a distinct species.—

The specimens (1160) certainly I believe are the same: on road to Castro from S. Carlos saw two specimens exactly similarly colored &c &c as this one: They were crawling about late in the evening.— |**273**|

[1] Stylommatophora, land slug. Specimens 330 and 471 had been collected in Monte Video in August and November 1832.
[2] More slugs. But how was it that specimen 1027 collected in Valparaiso in July 1834 came to be described on the same page with apparently the same pen as specimens 1096 and 1160 collected in Chiloe five months later? An examination of the paper on which **P. 272** was written provides the answer, since it was unlined, unlike that normally used by CD for these notes, which was faintly lined with a red margin. Hence the original **P. 272** was probably exchanged for a new one early in 1835.

[**CD P. 273** commences]

Asterias[1]

1031

Arms 44 in number, diameter from tip to tip 7 & ½ inches: arm one inch (the arms may be considered to be united at their bases): upper surface "chocolate R" thickly covered with globular, slightly tuberculated, pale-flesh-colored stony projections: these towards the arms are grouped together & form 3 lines on each arm; on the sides of arms are longer, more pointed, smooth, white spines, these are continued, on the under surface, towards the centre of body, & thus show the junction of the arms.— Under surface white; papillæ excessively numerous round mouth; adheres to tidal rocks, in stomach great quantities of small Balanidæ.—

Actinia[2]

1035

Whole body (tentacula & mouth rather paler) most intense, brilliant, beautiful "China blue", when contracted, rounded globular.— Is remarkable from outer surface being densely coated with little short, soft crests; these crests are either slightly convoluted or simple, they are placed so closely together that the real skin, colored coppery purple, can only with difficulty be seen by ~~forcing~~ pulling the little projection apart.— at first sight the outer surface appears smooth, from the closeness with which these lie to each other; a narrow line bordering the tentacula is greenish blue.—

Sertularia[3]

1034

Thickly coating the tidal rocks, body of polypus singularly elongated, as much protrudes beyond orifice of cell as is contained within it, cylindrical, suddenly tapering at base, tentacula about 20, round large mouth: the central living axis sends off little branch to Polypus, structure like Clytia [at] the base of cell apparently intercepts the communication. |**274**|

[1] Asteroidea, Forcipulata, starfish.
[2] Zoantharia, Actiniaria, sea anemone.
[3] Leptothecata, a hydroid.

[**CD P. 274** commences]

Ornithology
Egg (B)
(2427)
like that of
M. Video?

Partridge[1] (2159) in its general habits & appearance resembles strongly (its manner of running openly & not ~~easily~~ readily squatting) the partridge of the Plata: but I am sure it is different from the much shriller & different note it makes when rising.— Flesh white.— Tolerably numerous: not so easily caught as the Plata one.

[note (B) added later] Mr Dring[2] having specimens from each place, I carefully compared them. The color of both is palish "Chocolate", ~~mixed with little deep reddish brown~~: the La Plata one is a shade paler.— is smaller & more pointed at both extremities.— there is a close general resemblance.—

Dimensions	La Plata	Valparaiso	Diff
Longer axis	1.815 of inch	2.070	.255
short do	1.300	1.495	.195
	0.515	0.515	.060

[note ends]

[**CD P. 274** continues]

Woodpecker[3] (2161) is the "Pitui" of Molina[4]; I think the name must come from the curious noise, which somewhat resembles this word.— frequent the hills with bushes & low trees.

Lanius[5] (2169:70) called in La Plata "Callandra". here, as Molina says, "Thenca" very numerous. habits have formerly been described.— it only sings in the Spring (hence my surprise at R. Negro at the different note so as to think it different bird), beauty of song very much exaggerated by Molina.— I (good authority) am assured the nest is simply circular, but large & built externally of bits of prickly Mimosa.— Therefore Molina is wrong about its nest.

Caprimulgus[6] (2171) utters at night a simple gentle plaintive cry: which is regarded with much superstitious dread by the natives.— frequents the hills.—

Myothera[7] (2172) called by inhabitants "Turco", not uncommon, lives
(a) sheltered amongst the numerous thickets, may occasionally be seen hopping with its long legs & erect tail very quickly from bush to bush, as if ashamed of itself, & aware how very ridiculous figure it presents.— can hardly be made to fly.— its noises are as strange as its appearance.— is said to build its nest in a long hole in the ground. Gizzard very muscular, containing pebbles, beetles & vegetable fibres.— [note (a) When I first saw this bird, from length & strength of legs, membranaceous covering to nostrils, muscular gizzard, I thought it had some connection with the family of Gallinaceous birds.— [note ends] |**275**|

Ornithology Myothera[8] (2173:74) called by the inhabitants "Tapacolo" or "cover your posteriors" well deserves its name, as it generally carrys its ~~long~~ short tail more than erect, that is inclined towards its head.— it is very common, especially in the dry hills, over which bushes are scattered & where scarcely
(b) any other bird is present, hence this bird is very conspicuous in the ornithology of Chili. [note (b)] This hopping genus is admirably well adapted for the nature of the country, dotted over with low thick bushes.— [note ends] In habits of hopping, concealment, unwillingness to fly, nidification, resembles the "Turco", its appearance is not however quite so ridiculous, & the bird shows itself rather more: is very shy, will remain motionless in a bush & with much address will try to crawl, after some time, away on the side opposite to you.— It is very active, generally making a noise & these noise[s] are very various & strangely odd, some like cooing of doves, others like bubbling of water, & many defy similes.— The country people say it changes its note 5 times in the year, so that I

suppose they vary them according to the season.— Frequents the thickest hedges & thickets.— It is very odd that Molina takes no notice of this genus.—

[**CD P. 275** continues]

Emberiza[9] (2175)> with serrated bill (2175.2176) called "Rara" from its scarceness.— which however does not appear to be the case.— it is a quiet solitary bird: is very injurious to buds of trees.— iris of eye bright scarlet. "Phytotoma vera" of Molina.

(c) (a) Blue sparrow[10] (2177) called "Diuca", also of Molina.— habits have been described at Chiloe as very like the Sparrow: builds in trees: very abundant.— |**276**| [note (a)] Molina supposes it to be the bird mentioned by Capt. Cook at New Zealand. if so its Geographical distribution will be curious as not having crossed the Pampas.— [note ends]

[note (c) added later] Specimen (2320) will show nest & eggs taken first week in November: in a frequented path: male bird utters two or three notes like a Song: Molina talks as if it could sing prettily.— [note ends]

[1] Listed in *Zoology* **3**:119-20 as *Nothura perdicaria* Gray. The partridge of La Plata is *N. major*.
[2] John Edward Dring was Acting Purser on the *Beagle* at that time.
[3] Listed in *Zoology* **3**:114 as *Colaptes Chilensis* Vigors.
[4] See Juan Ignacio Molina. *Compendio de la historia geografica natural y civil del Reyno de Chile*. Part 1. Madrid, 1794. The copy in the *Beagle* Library was acquired by CD in Valparaiso.
[5] Listed in *Zoology* **3**:61 as *Mimus Thenca* Gray.
[6] Listed in *Zoology* **3**:36-7 as *Caprimulgus bifasciatus* Gould. NHM 1855.12.19.241.
[7] Listed in *Zoology* **3**:71-2 as *Pteroptochos megapodius* Kittl.
[8] Listed in *Zoology* **3**:72 as *Pteroptochus albicollis* Kittl.
[9] Listed in *Zoology* **3**:106 as *Phytotoma rara* Mol.
[10] Listed in *Zoology* **3**:93 as *Fringilla Diuca* Mol.

[**CD P. 276** commences with entries for August and September, although written in October]

Ornithology Large Humming Bird[1] (2179:80) This bird was here in middle of August, when it is stated it had just arrived, certainly afterwards in Sept. it became (b) much more abundant. [note (b)] Specimen (2319) will show the nest. [note ends] Its appearance is very singular when on the wing: its flight, like the rest of tribe from flower to flower is like lightning, but when hovering by a flower, the motion of its wings is slow, (not vibratory as the smaller sort) but each stroke very powerful, something like a butterfly; I never saw a bird where the force of its wings appeared so powerful in proportion to its weight.— The flight & hovering noiseless.— When by a flower, the body is nearly vertical, & the bird constantly expands & shuts like a fan its tail.— note very shrill.— It flies from flower to flower, yet in its stomach were abundant remains of insects.— This Trochilus seems to take the place

(2134) of the small one of Chiloe: this latter was most abundant in August, from

(a) which time they have gradually disappeared, so that now (October 12th) one only was seen during the whole day; on the other hand the larger sort is as much increased in numbers.— I suppose the smaller one goes Southward to avoid the heat of summer.— I do not believe it builds in Chili.

X(a) [notes added later] X(a) This is stated to be a regular occurrence.

XX(a) XX(a) Specimen (2425) Nest & egg of small Trochilus from the island of S. Pedro, Southern extreme of Chiloe; hence they migrate South to breed.— Eggs with partly formed young. Decemb: 8th — Even in the middle of

Chonos Archipel } summer they haunt the interior forests, where there are no flowers & where minute diptera must be the attraction.—

XXX(a) XXX(a) This species was plentiful at Valparaiso April 20th.— Saw the first on the 14th.— Perhaps they might have arrived a little earlier.— Humming Birds migrate to United States & Canada to avoid heats of Summer of the South.— Humboldt[2] Vol: V, P 1, P 352. Beechey[3] says that humming birds remain in N California all winter.— The two facts together make a case in the Northern hemisphere exactly parallel to that observed here.— [notes end]

[**CD P. 276** continues]

Woodpecker[4] (2185) is called "Carpintero" & by Molina.

Icterus[5] (2186) is the "Thili or Chili" of Molina. builds amongst reeds in marshy ground.— Common.

Long-tailed tit[6] (2193) not uncommon. cry shrill & loud.— builds (is said)

(c) its nest in prickly trees, is composed externally of prickly bits of stick, is

V 1469[?] very large & long (2 feet), with the opening |**277**| at the top, conducting

Ornithology into the vertical passage.— I recollect having seen such a nest at the R. Negro.— I believe Molina has described this nest as belonging to the Thenca.— [note (c)] In habits resembles our Tom Tits.— [note ends]

Wren[7] (2194) builds in holes in walls.— in month of October

Fringilla[8] (2195) Siu of Molina. often kept in cages

V. further Cordillera Muscicapa[9] (2197) in small flocks amongst the hills & not near the coast

Add Note Maldonado — Muscicapa[10] (2208) This bird which we have seen in all parts generally near coast, I have noticed inland but near the stony bed of rivers.—

1784 [n.s.] — I have seen the following birds. the black Icterus[11] of T. del F. is abundant in flocks: called by Molina "Cureu" — builds in bushes: can be taught to talk. frequents moist pasture.—

Sturnus ruber[12] abundant. called "Loyca" by Molina. builds on the ground.

(2125) — The Thrush of T. del F. abundant: can be taught to talk: nest lined with smooth mud (I suppose like English thrush)[13].

1602 Copy — The Pterec-Pterec of La Plata, called ~~here~~ by Molina "Theghel". it is quite false that the bird is silent by day[14].—

~~The Furnarius of La Plata (2297) is common~~

(d) 1823 — The black Furnarius[15] Is common in T. del Fuego on coast, is here found inland, generally near beds of rivers. I saw a nest built of coa[r]se grass on a ledge in a hole in a deep cavern, but generally in holes in banks.
[note (d) added later] Chonos Archipelago, (45° 18′), Midship bay; coarse nest in hole under tree (but not excavated like the Furnarius of La Plata) egg (2426): Nearly a degree South of Valparaiso, this bird had young ones Septemb 20th.— here it had eggs Decemb 15th.— What difference in climate! These Furnarii appear to me more to correspond in habits or place in nature to the water-Wagtails than any other bird.— [note ends]

[**CD P. 277** continues]

(c) The long-billed Furnarius[16] of R. Negro 1467. rare. [note (c)] Is said to build in low bushes. [note ends]

(b) The Scolopax-Perdrix[17] of Patagonia, not common. [note (b)] Renous[18] noticed to me that a pair of these birds might almost always be found in the same spot.— [note ends]

The little Certhia of T. del Fuego (2084)[19]

(a) Certhia of Chiloe (2129) |**278**|

[further notes added later] (a) All my information about the nests was obtained from a Gausso who had long paid attention to the subject.—
It appears to me surprising how many of the birds of T. del Fuego & Patagonia are common to Chili.—
NB There is at Valparaiso, Copiapo & Patagonia a brown Lanius[20] like the

Thenca, but with larger beak, which in habits appears a true Lanius, for it is said to kill young birds.— [notes end]

[1] The large humming bird is listed in *Zoology* **3**:110-12 as *Trochilus Gigas* Vieill., while the smaller species is *Trochilus forficatus* Lath.
[2] See Alexander von Humboldt. *Personal narrative to travels to the equinoctial regions of the new continent . . . 1799-1804 . . . translated into English by Henrietta Maria Williams.* 7 vols. London, 1814-29.
[3] See Frederick William Beechey. *Narrative of a voyage to the Pacific and Beering's Strait . . . 1825, 26, 27, 28.* London, 1831.
[4] Listed as *Picus kingii* Gray in *Zoology* **3**:113-14. NHM 1855.12.19.88 and .101.
[5] Listed in *Zoology* **3**:106 as *Xanthornus chrysopterus* Gray.
[6] Listed in *Zoology* **3**:49 as *Serpophaga Parulus* Gould. NHM 1855.12.19.98 and .161.
[7] Listed in *Zoology* **3**:74 as *Troglodytes Magellanicus* Gould.
[8] Listed in *Zoology* **3**:94 as *Fringilla alaudina* Kittl. NHM 1855.12.19.41.
[9] Listed in *Zoology* **3**:61 as *Mimus thenca.* NHM 1855.12.19.230
[10] Listed in *Zoology* **3**:48 as *Myiobius parvirostris.* NHM 1856.3.16.15.
[11] Listed in *Zoology* **3**:107 as *Agelaius chopi* Vieill.
[12] *Sturnella loyca* Mol., not listed in *Zoology* **3**.
[13] Listed in *Zoology* **3**:59 as *Turdus Falklandicus* Quoy.
[14] Listed in *Zoology* **3**:127 as *Philomachus Cayanus* Gray.
[15] Listed in *Zoology* **3**:68-9 as *Opetiorhynchus nigrofumosus* Gray. NHM 1855.12.19.244.
[16] Listed in *Zoology* **3**:67 as *Opetiorhynchus patagonicus* Gray.
[17] Listed in *Zoology* **3**:131 as *Scolopax (Telmatias) Paraguaiæ* Vieill.
[18] Herr Renous was a German collector working in Chile whom CD met in September 1834. See *Beagle Diary* p. 261.
[19] ? variety of *Opetiorhynchus nigrofumosus.*
[20] Listed in *Zoology* **3**:56 as *Agriornis gutturalis*, NHM 1855.12.19.344.

[**CD P. 278** commences]

Ornithology
(a) The Vultur Aura, the Carrancha & Chimango (they have different names here) are tolerably common but infinitely less so than in La Plata.— [note (a)] The Gallinozo does not seem to come so far South: it must be owing to dryness of country. We have seen them to the South of the Plata.— [note ends] The Carrancha, when uttering its harsh cry, throws its head far
(b) backward, like the Caracara of the Falklands.— [note (b)] This fact is stated by Molina.— [note ends] I see several of the pale varieties such as shot at S. Cruz[1].—

(2299) ~~Also (see 1615) the common Sparrow of La Plata abundant~~

Lanius[2] of T. del Fuego & Chiloe (2124) common: builds coarse nest in bushes. egg (2375)

Kingfisher[3] of T. del Fuego. (2122)

Fringilla of S. Cruz (2015) rare

Fringilla[4] (blue & orange) of T. del F. & S. Cruz (2017). Not uncommon.

2298 ~~Tufted Tit found in Patagonia & T. del Fuego is here tolerably common: found small soft simple nest at latter end of August.—~~

2198 ~~Muscicapa, called Silgaro~~

In my passage of the Andes, I noticed at heights which could not be less than 8000 ft; the following birds — the common Sparrow: Fringilla[5] (2015): The black Furnarius of T. del Fuego shores: Muscicapa (2197) common in all parts even in the utterly ~~dry~~ sterile Cordilleras of Copiapò: I also saw at an elevation of 10,000 ft a Humming Bird, am not sure of species.—

The ornithology of the valleys on the Eastern slopes differs to a certain extent from the Pacific sides; the resemblance is very strong in aspect & in zoology with the plains of Patagonia.— Of Birds we have the Furnarius (2025): Certhia (2020) — white tailed Callandra & Thenca.— Black & white Muscicapa of the Pampas & Gallinazo; it is singular this latter bird
(e) not being found in Chili. Diuca (2172)[6], although so very common all over Chili, does not appear to have crossed the Andes.—

[**CD P. 278** ends here, but CD's catalogue of birds is continued later at (e) on the back of the page]

Also the large tufted Partridge of the R. Negro.— The Ostrich is found on plain of Uspullata 6-7000 ft: it is odd it has not crossed to the other side.—

At Copiapò — Lat 27°20′

Common Sparrow: Diuca, 2177[6]: common Thenca 2169[7]: & white tail[ed] do of Patagonia:— Long tail-tit 2193: Wren 2194: Muscicapa 2197, expands tail like a fan:— Muscicapa 2208: Sternus ruber: Dove 2163: Lanius 2124: Icterus 2186 with little yellow patch on shoulder: Scolopax Perdrix: Fringilla 2017: Myotherus 2825: Turdus 2125: Furnarius 2297: & black one of T. del Fuego: Swallow 2200[8]: Hawk 2014: The Caracara of Patagonia, 2029: which I have seen no where else is found in country between Coquimbo & Copiapò.— Partridges although so abundant S. of Guasco are not found here:— Myothera 2172:

Lima Lat 12°

Sternus ruber: Furnarius (2297): common Sparrow: Thenca 2169: Carrancha or Caracara Braziliensis: Gallinazo & Vulture Aura (great limits Falkland I^{ds} & Lima) & Hawk 2014:— *in the deserts*

At Concepcion we see commencement of dry country: we have the

Furnarius (2297), Thenca: Sternus Ruber & Scolopax Perdrix: But we also have the Barking bird & Cheucau of the damp forests of the South.— [extension ends] |**279**|

[1] For a summary of all CD's observations on the carrion feeders of South America, see *Ornithological Notes* pp. 233-45.
[2] Listed in *Zoology* **3**:55 as *Xolmis pyrope* Gray.
[3] Listed in *Zoology* **3**:42 as *Ceryle torquata* Bonap.
[4] Portrayed vividly by John Gould as *Tanagra Darwinii* in *Zoology* **3**:97 (Plate 34), but listed as *Tanagra striata* Gmel.
[5] Listed in *Zoology* **3**:94 as *Fringilla fruticeti*, but this is not one of the specimens from Rio Santa Cruz and Coquimbo still held at the NHM.
[6] Listed in *Zoology* **3**:93 as *Fringilla diuca*, NHM 1855.12.19.187.
[7] Listed in *Zoology* **3**:61 as *Mimus Thenca*, NHM 1855.12.19.230.
[8] Listed in *Zoology* **3**:41 as *Hirundo cyanoleuca* Vieill.

[**CD P. 279** commences]

Mus[1] (2202) Excessively numerous in all parts of the country; frequent by hundreds the hedges, are very injurious to the young corn.— feed during the whole day — are very tame — when they run, they turn up extremity of tail, which gives them a very different appearance from true rats.—
1040 seem very subject to be pie-bald & Albinoes.— It is stated they are found on the Volcanic island of Juan Fernandez. if this is true, it is curious.— Called by Molina "Degu".

Corallina Examined carefully extremities of branches, they were covered by delicate
2151 membrane, beneath which is a cellular substance, irregularly hexagonal.
(a) each cell had a diameter from 1/3000 to 1/4000th of an inch. These cells appear gradually to become inspissated with calcareous matter till the above structure is no longer visible. Is plentiful on tidal rocks.— [note (a)] Encrusting Corallinas are present here. [note ends]

Corallina[2] [further note added later, and like the entry above was crossed through
3503 vertically in pencil] Feb^{y} — 1836. Hobart town, Van Diemen's land. On lifting up a fragment of Sandstone, which had lately fallen into a tidal pool, I found some ~~fragment~~ branches of this species of Corallina[2] attached to its lower edge.— These branches had been broken off by some violence from their present tuft; & the terminal joints being pressed against the stone had adhæred & expanded.— This foliaceous expansion had precisely the structure of the first growth of what I call "Corallinas inarticulata", but from it there were springing fresh buds.— Hence this joint would become the root

or point of adhesion to a new tuft: Thinking this manner of propagation was solely the effect of violence, I examined some flourishing tufts; but I there also found a few of the lateral stems, with their heads drooping & so attached to the stone.— Hence we have this novel method of extending the limits of any tuft in the family of Corallinas.

[final section crossed through again vertically] It calls to mind the propagation of trees by laying; & can hardly be supposed to take place in a true Corall, where each cell is inhabited by its Polypus.— The fact is of interest in showing the close identity in nature of the Corallina articulata & inarticulata: & is itself in as much as the observation is made in that part of the family, Where true propagation by ovules has not been observed. Is it possible that the terminal buds are periodically shedded? [note ends]

[1] Listed by George Waterhouse in *Zoology* **2**:82-3 as *Octodon Cumingii.*
[2] In CD's notes on plants copied out towards the end of the voyage, the words 'species of corallina' were altered to 'Nullipora', and 'what I call "Corallinas inarticulata"' to 'the encrusting Nulliporae'. Specimen 3503 was included in a shipment of *Beagle* corallines sent to William Henry Harvey for examination in April 1847, again described by CD as a Nullipora. As explained in *Plant Notes* pp. 186-206 *Nullipora* is in fact a symbiotic coralline alga *Bossea oribigniana* (Decaisne ex Harvey) Manza. See also *Correspondence* **4**:29 and Phillip R. Sloan in *Darwinian Heritage* pp. 104-5.

[**CD P. 279** continues]

Hot Baths of Cauquenes

Water & Gaz

The hot baths of Cauquenes[1] have long been celebrated. They are visited by numerous people affected with all sorts of complaints, but chiefly those of the muscles & skin. The patient is placed for several minutes in one of the baths, & then buried beneath blankets so as to induce a violent perspiration. The water is likewise taken internally.— These springs are situated at the foot of the Cordilleras in the valley or ravine of Cauquenes, about 22 leagues to the South of St Jago.— The surrounding district is composed of Porphyries, Breccias & greenstone, all of which have clearly undergone the action of violent heat, but have not flowed in a stream. The Strata are ~~inclined~~ aligned at ~~about~~ a varying angle from the Cordilleras: they are |**280**| traversed by dykes of greenstone. This however is the usual character of the geology of the low hills which immediately flank the Western slope of the Cordilleras.— [note (b)] There are no active Volcanoes in this part of the Cordilleras, but there is a group of peaks which perhaps forms part of extinct one.— A line however of singularly ~~uneven~~ irregularity in force, line of upheaval crosses the very springs.— I understand the Earthquake of Concepcion of 1835 stopped the water. [note ends]

Hot Baths of Cauquenes (b)

[**CD P. 280** continues]

The springs however burst through a mass of boulders & pebbles, cemented together by a crystallized calcareous base, which skirts each side of the ravine. There are several springs, but only a few yards apart. their temperature differs, this appears to be owing to a greater or less admixture of cold water: the water of the coolest springs has scarcely any taste: in all the springs there is an escape of gas, which ~~escapes~~ bubbles up by intervals. After the great earthquake of 1822 the Springs ceased & the (a) water did not return for nearly a year. It is stated that it never has regained its former volume or temperature. [note (a)] All assertions about temperature are to be taken with great caution: the proof they give is the comparative length of time which it requires to loosen by immersion the feathers of a fowl, in winter & summer & before 1822. (putting the fowl into boiling water is the universal method in S. America of removing the feathers, as we do the bristles of a pig) [note ends] The man who lives at the baths also assures me that in summer the water is hotter & more plentiful: the former I should expect by the partial drying up of the cold spring.— But the latter statement seems very strange, as I suppose the increased quantity must be owing to the melting of the snow in the higher mountains & these are distant at least 4 or 5 leagues.— The temperature of the hottest bath is such as to allow some people very slowly to immerse their bodies for a few seconds.— M. Gay[2] states that the water contains Mur. of Lime & Carb. of Magnesia!.— I hear of a Hot Spring higher up the valley simply acid.— Is the gaz, which escapes, Carbonic acid from the Mur. acting on the Carb. of Lime, which forms matrix of Conglomerate.— |**281**|

[1] See *Journal of Researches* **1**:320-2.
[2] See Claude Gay. *Aperçu sur les recherches d'histoire naturelle faites dans l'Amérique du Sud, et principalement dans le Chili, pendant les années 1830 et 1831. Annales des Sciences Naturelles* **28**:369-93, 1833.

[**CD P. 281** commences]

Fungus on Roble[1] | 1065 number on specimen | ~~Copied~~

On the hills near Nancagua & S. Fernando there are large woods of Roble or the Chilian oak; I was surprised to find a yellow fungus, very closely resembling the "edible ones" on the Beech of T. del Fuego. Speaking from memory the differences consist in this being rather paler colored, but the inside of the little cups a darker orange. the greatest difference is however in the more irregular shape, in place of ~~the~~ being sphærical ~~one~~ as of T. del Fuego. They are also much larger: many are 3 times as large as the largest of my specimens.— The footstalk appears longer, this is necessary from the roughness of the bark.— In the young state, there is an internal cavity.— The difference of tree & great difference in climate renders it certain that the Fungi must be distinct.— They are occassionally eaten by the poor people.— I observe these Fungi are not infested with Larvæ (so as to render their origin doubtful) as those of T. del Fuego.—

Condor (a)

Having an opportunity of seeing very many of these birds in a Garden.— I observe that all the females have bright red eyes; but the male yellowish brown: I however found that a young female (known by dissection, as this was in the Spring the bird must at least be one year old), whose back was brownish & ruff scarcely as yet at all white, has her eyes dark brown.— The young male has also its back & ruff brown.— & the comb simple.— These were fed only once a week.— The Guassos state they can well live 6 weeks without food.— They are caught on Corallitos or when roosting 5 or 6 together in a tree. They |**282**| are very heavy sleepers (as I have seen) & hence a person easily climbs up the tree & lazoes them: They are only taken in winter & Spring; in the summer are said to retire into Andes.— There are so many brought in that a live Condor has been sold for 6[d]. Common price 2 or 3 dollars.— They are wonderfully ravenous.— One brought in lashed with rope & much injured, & surrounded by people, the instant the line was loosed which secured the beak, began to tear ~~the~~ a piece of carrion.—

Condor

air calm a dog could not fail to have perceived it

The condors appear suddenly in numbers, where an animal dies, in the same unaccountable manner in which all Carrion Vultures are well known to do.— Tying a piece of meat in a paper, I passed by a whole row of them within 3 yards & they took no notice. I threw it on the ground within one yard, an old male Condor looked at it & took no further notice: placing it still closer, the Condor touched it with his beak & then tore the paper off with fury.— in an instant the whole row of Condors were jumping & flapping their wings.— I think it is certain a Condor does not smell at a greater distance than a few inches.— Mem: M. Audubon[2] in Wern: transactions, similar observ:— For more particulars about Condor, V. **P. 210(b)**.— Smelling powers of Hawks discussed, Waterton[3]. No[r] 32, Magazine of Nat. Hist:

V. **P.210(b)**

[**CD P. 282** continues]

I believe this from seeing no nests at Port Desire

The country people inform me the Condor lays two large white eggs in November or December; they make no nest but place the eggs on any small ledge.— I am assured the young Condors cannot fly for a whole year. At Concepcion on March 5th I saw a young Condor, it was nearly [continued on back of **P. 282**] full grown, but covered with a blackish down, precisely like a Gosling.— I am sure this bird would not have been able to fly for many months.— After the young birds can fly apparently as well as the old ones, from what I saw on coast of Patagonia, they appear to remain for some time with their parents. They hunt separately, before the ~~white~~ ring round the neck is changed white.— When at the S. Cruz river in months

(a)
(a)

of April & May, two old birds were generally perched on the ledges or sailing about with a full fledged young bird [added in margin] not white collar. Now I think it certain that this could not have been hatched during the same summer: if so the Condor probably lays only once in two years.— It is rather singular that the name Condor is only applied to the young ones, before the white feathers appear; the old birds being called "El Buitre" the Spanish of "the Vulture".—

The Condors attack young goats & sheep, I have seen dogs trained to chace them away.— It is beautiful to watch several Condors wheeling over any spot. Although you may never take your eyes off any one bird, for a quarter or half hour you can never see the slightest motion of their wings. I believe a Condor will go on flying in curves ascending & descending for any length of time without flapping its wings.— When the bird wishes to descend rapidly, the wings are collapsed for a second.— When soaring close above the beholder no tremulous motion or indistinct appearance can be observed in the separate feathers which terminate the wing.— The head & neck are moved frequently & apparently with force, as a rudder of a ship, but perpendicularly as well as laterally. by the former motion, the whole body seems to alter its inclination with the horizon & by action of contrary current of wind to rise.[4] The bird critically views the ground.— |**283**|

object of wheeling watching signs for animals & Lions

[miscellaneous notes on back of **P. 281** added later] Shortly before any one of the Condors dies, all the lice which infest it crawl to the outside of the feathers. Ricinus (2153)[5].—
~~Dogs taught to hunt Condor kill young sheep~~
Called Indian tongue Manque. Molina [notes end]

[1] Identified as *Cyttaria berteroii* Berkeley in *Plant Notes* pp. 228-9.
[2] On 16 December 1826, CD had attended a meeting of the Wernerian Natural History Society of Edinburgh (see Vol. 6 p. 562 of the Society's *Memoirs*) at which Audubon exploded the opinion generally entertained of the extraordinary power of smelling of vultures.
[3] Fierce controversies on this subject were reported at the meetings of the London Society of Natural History, as for example in *Mag. Nat. Hist.* **7**:164-75 (1834). But Audubon's views on the habits of the rattlesnake having been dismissed one day as 'a tissue of the grossest falsehoods ever attempted to be palmed upon the credulity of mankind', he was perhaps trying to get his own back on the occasion when CD recalled in his *Autobiography* p. 51 that Audubon had sneered somewhat unjustly at Waterton, author of *Wanderings in S. America.*
[4] CD's account of the aerodynamics of the flight of the condor was considerably extended on copying it into the *Ornithological Notes* (see pp. 240-5) and in *Journal of Researches* **1** pp. 219-24, but was shortened again in *Zoology* **3**:3-6.
[5] See *Insect Notes* pp. 81-2.

[**CD P. 283** commences]

Holuthuria=
Doris[1]
1097

Adhæring to a stone in 16 Fathom water; near island of Huafo.— an animal ~~closel~~ allied to the Holuthuria (Doris-like) described **page 215**.— Length .4: elongated oval: snail shaped, upper surface slightly convex covered with minute stony points, sides protected by four or five rows of scales, the rounded extremities of which point from each side upwards to wards centre of back: at anterior & posterior end of animal & above the margin, there is a projecting pap or cone capable of extension & retraction; round the base of these, small scales are visible & the stony points.— the whole surface of back is scattered over with cylindrical papillæ; extremities bluntly rounded. These are susceptible of motion, irritation & contraction but to no very great degree.— They seem to arise from between the scales.— The paps at extremities are equally covered with these, as rest of body; hence where drawn in, they appear as two bushes or groups of papillæ.— One of the paps (anterior?) is very much pointed, its terminal orifice is closed by 5 or 6 triangular pieces or scales.— I could not exactly see form of posterior orifice.— The paps highly irritable. It is probable tree-like tentacula are hidden within the anterior pap, but they were not protruded.— (NB. The animal had been kept a week in water, hence perhaps little irritability of papillæ) When the animal slowly crawls, it is a pointed oval, but when at rest it is nearly circular & the paps ~~are~~ projected as in Ascidia.— I do not know the use of the Papillæ, perhaps partly prehension.— The flat under surface is surrounded under margin or base

Holuthuria= (a)
Doris

of |**284**| scales [next 6 lines crossed through] by a row of adhæring papillæ, also down the centre there is an irregular double row.— These papillæ are organized precisely as described in animal **P 215** [see sketch in margin]. When in action they ~~extend~~ adhere without the edge of body.— Habits same as animal **P 215**.— Color "Flesh red".

[note (a)] The papillæ seem slightly protected by the Hyaline stony points.— the scales on centre of back are very obscure, as indeed all the scales are when compared to animal **P 215** [note ends]

[**CD P. 284** continues]

Peronia[2]
Blainville[3]
(1092)
(2421 dry)

Body when partly crawling ~~blunt~~ oval, posterior extremity truncate & retracted above Branchial orifice. This latter large, circular, widely open.— convex, when firmly adhæring conical.— Above ~~blueish~~ blackish blue, with pale projecting points & pale halo round each; edge with narrow alternate square spaces of white & blue; the latter color appearing Vascular: beneath white, excepting mouth.— Tentacula short with terminal black eye; beneath which a bifurcate ~~membrane~~ hood over mouth.— Inhabits in great numbers the tidal rocks where confervæ grow, amidst Balanidæ & the shells

Is^{d} of Tanqui>

(2364).— Surface of animal almost dry, from length of period during which it is uncovered.—

Doris[4]

Is^{d} of Caylen.— common under large stones; color pale yoke of

(1091) egg=yellow: foot & mouth darkest. Mantle far surpassing foot on all sides, surface with rounded ~~papilli~~ points of two sizes. Form of superior tentacula & Branchiæ exactly same as described in Doris **P151**.— (to which species this is <u>closely</u> allied). Branchiæ same color with mantle. Dimension, when partly crawling, 2 inches, breadth 1.1.—

<u>Cavolina</u>[5]
1091 Under stones.— General color "Crimson & Brownish purple R". Mouth & under side finer rose color. Branchiæ composed of conical fillets (basal parts leaden colored) arranged in numerous transverse rows on each side of back. Animal broard, truncate anteriorly tapering to tail hence triangular.
(a) Length, when crawling, 1.5, broardest .6: Anterior & inferior [continued
(a) on back of page] tentacula placed far apart, (at each corner of truncate extremity), very long, tapering, pointed, tipped with white; posterior & superior tentacula, blunt & much shorter, placed ~~between~~ behind some of the first rows of Branchiæ.— |**285**|

[1] Dendrochirotidae, Psolidae, probably *Psolus antarcticus*. Noted by S.F. Harmer as '1 ? *Psolus*'.
[2] Systellommatophora, the slug-like pulmonate *Onchidella marginata*.
[3] See Planche 63 showing Péronie de l'Isle de France portrayed by Henri Marie Ducrotay de Blainville in *Dic. Sciences Naturelles.* Planches. 2e partie, *Zoologie*, *Conchyliologie et malacologie*.
[4] Cryptobranch doridacean, probably *Anisodoris fontaini* D'Orbigny.
[5] Aeolidacean nudibranch, *Phidiana lottini* Lesson.

[**CD P. 285** commences, crossed through vertically up to the end of **P. 287**]

<u>Frog</u>[1]
1086
(a)
<u>Copied</u>

Under side: throat, breast & cheeks rich chesnut brown, with snow white marks; thighs ~~blackish~~ of hinder legs blackish with do marks. legs yellowish also with do marks.— Upper side, pale iron-rust color, with posterior parts of body, thighs & anterior marks (one triangular & other transverse) beautiful bright green.— iris rust color. pupil black.— eyes small.— appearance very pretty & curious.— Nose finely pointed.— Jumps like a frog. inhabits thick & gloomy forest. Is^{d} of Lemuy .—

<u>1835 Feb:</u> [note (a) added later] This species is excessively common in the forest of Valdivia. Seems subject in its colors to remarkable variation.— Specimen
1178 (1178) under surface posteriorly jet black & snow white marks, anteriorly rich chesnut brown: above cream color, with triangular slightly darker shades & small marks of green.— (There is a point in all at joint of hind legs.— iris of all is rusty red).—
1179 Above cream-colored, without shade of green: hinder <u>legs</u> yellow; beneath all black with different shaped marks of white.—
Another, beneath anteriorly the brown is replaced by bright yellow.— upper surface instead of cream color, rusty red — with darker triangular

shading.—
All die soon in confinement.— [note ends]

[1] Listed in *Zoology* **5**:48 as the only species in a new genus which was appropriately named by M. Bibron *Rhinoderma Darwinii.*

[**CD P. 285** continues, crossed through to end of **P. 287**]

Planaria[1]
PL.14, Fig:3

represents
lower
surface

2422

Found under round stone in a numerous group, in brackish water. Chonos Archipelago.— Length .2, breadth about .06.— The vascular system brownish purple, hence animal has this tint:— ~~seen from above~~ upper surface; there is a fine narrow inverted wedge shaped mark of dark color on anterior extremity; in centre of back a circular patch clear of color; color on back seems laid on in fine striæ.— Lower surface, white: Shape pointed oval, broard & rounded posteriorly; anterior extremity square, truncate & shouldered; this part (A) is prehensile by suction caused by folding edges towards each other; ~~thin~~ body much depressed, edges very thin: crawls something like a leach by adhæring with anterior extremes & dragging up body afterwards (in both last respects very different from the terrestrial Planariæ) can swim, back downwards, or rather it is crawling on upper surface of water.— As in the land Planariæ, there is in first half of body a central vessel, which sends off short, (moss-like) branching vessels full of granular matter.— The wedge shaped organ in head appears to be its altered extremity: on each side & attached to it (nearest to on dorsal surface) there is a black eye.— The central vessel, about middle of animal is joined to the foot-stalk of the protrudable ~~ball~~ |**286**| organ.—

Planaria

Planaria of
Ehrenberg[2]

This footstalk is much longer than in the land Planariæ, as ~~the~~ is the organ which is here nearly cylindrical.— indeed when fully extended, from base of footstalk to mouth it is one slightly tapering line.— hence the distinction is almost superfluous, being only seen when the organ is within body: it is here remarkably long, so that the animal can twist it over his back ~~& hang over on the~~ on one side & it will then project on the other. Weak Spirits of wine always caused the protrusion; shortly before death generally is retracted; is present & highly developed (so are the middle, lateral, branching vessels) in very young specimens.— I saw once a quantity of granular matter ejected from this organ, the animal being placed in *[illeg.]* Spirits of Wine.— The vascular system is continued in a perfect ring some distance from the margin of animal, round the posterior half of body.— On each side of the central anterior vessel & between the branches, there [are] about 7 or 8 circular cavities, which contain an opake spot.— (I never before saw this in the genus). Not all the Individuals ~~nor~~ or very young ones possessed them.— the anterior ones were best developed.— I have

X

some slight suspicion, they are ovules; it is rendered probable by the *[illeg. word deleted]* number of exceedingly young specimens found with the old ones.— On the under surface I most carefully tried to see <u>the 2^{d} or posterior</u> orifice; but quite failed; there is however a transparent spot just at (B).— One specimen being placed in strong Spirits of Wine, the "organ"

<u>Planaria</u>

burst forth through the circular dorsal clear space. |**287**| This must have been accidental, as by careful examination no orifice can be seen there.— This specimen is preserved & well shows the "organ".— A system of vessels sometimes lines the inside of ring by (B).— The drawing represents the organ, partly protruded.— The connection of organ & anterior central vessel (& hence whole vascular system) is very evident.— The organ & footstalk lie coiled up as represented in the clear space.— The motions of the "organ" are as described in Land Planariæ.—

Lichen= <u>Conferva</u>[3]
2377

Plate 14: Fig. 4 much magnified

copied

Consists of bunches of slightly branched hairs, colored "Reddish orange"; grows commonly on the dead twigs of trees, here at Chonos & in T. del Fuego.— The hairs when examined in cabin (from hygrometrical properties?) moved & started.— The hairs have their extremities rounded — truncate.— when examined in water, seem to consist of an outer vessel, containing an inner with red fluid; this fluid is divided transversely, apparently in very same manner as the green matter in conferva.—

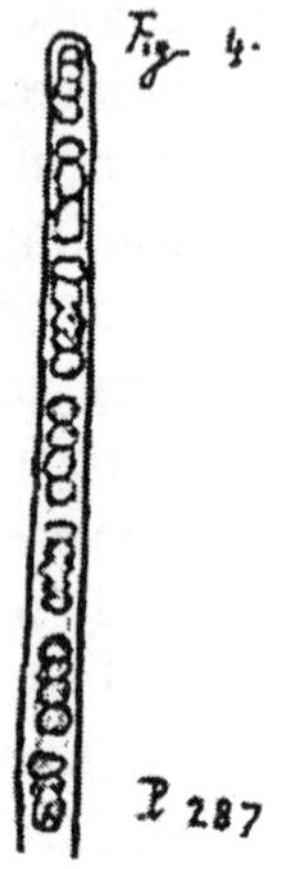

Plate 14, Fig. 4

each compartment ~~contains~~ is composed of 3 or 4 little sphæres of the red matter, which either only touch or run into each other more or less.— On the hairs there are irregular lumps which contain a particle of the red matter, separate from the column.— These are buds & thin young branches may be seen rising from them.— |**288**|

<u>Holuthuria</u>[4]
1099

In brackish water, adhæring to small stones, in the figure of an Ascidia.— Length 11 inches; apparent diameter 1¾ inch: when fully inflated, circumference 5 & ½ inches.— Body cylindrical, slightly tapering to each extremity.— Whole surface <u>most</u> thickly strewed with short, cylindrical, truncate papillæ. These closely examined have no orifice; may consist of a reddish colored, slightly convex cushion on transparent footstalk, which contains *[illeg.]* other ligamentous tunic, by which perhaps the cushion is made to act as a sucker.— Upper surface "Hyacinth & Art-blood R" shading at each extremity into white, lower surface & posterior extremity sooty.—

[**CD P. 288** continues]

Anterior extremity, tentacula, & mouth, dark greenish black.— broard.— mouth projecting, surrounded by 10, placed in pairs, large tentacula. These in shape are thick conical (not at all flattened) rest on cylindrical base, throw off short cylindrical branches with nobs or buds on them.— Within these are 5 pair of much smaller ones, bluntly conical, bush-like, fine, tops colored white.— Absorbs much air, sensitive of light & much irritated by warm breath: often sends a ring of contraction down whole length of body.— The mouth is withdrawn by the interior part of animal being turned outside inwards.— I omitted, that the surface or patch by which the body adhæred, was void of papillæ, it appears as if they had been removed by the friction.— |**289**|

[1] Described by CD in *Planaria* p. 189 as *Planaria (?) macrostoma.* It is now placed in order Tricladida as *Procerodella macrostoma.* Fig. 3 was excised from CD's Plate 14, and redrawn as shown here on p. 20, labelled 'Plate V. fig. 2. Under-side magnified'.
[2] It is mentioned by CD on p. 182 of *Planaria* that some of the terrestrial species are restricted to the genus of *Polycelis* of Ehrenberg.
[3] See *Plant Notes* p. 178. No such specimen was found.
[4] Dendrochirotida, Cucumariidae, probably *Athyonidium chilensis* Semper.

[**CD P. 289** commences]

Planaria[1]

(2440)

Found within ~~the~~ quite soft rotten wood, on a high hill within the Forest in St Andrews Harbor, C. Tres Montes; Lat. *[not filled in]* This is the furthest South I have seen this curious genus, & it is singular that this should be the largest sized species I have ever met with.— When closely contracted 1.4^{inch} long & .4 broard, posterior extremity very obtusely rounded.— When fully crawling length 5 inches, breadth .13.— I could distinguish no eyes.— Orifices on under surface obscure.— Almost killed by being placed to crawl on paper for a few minutes.— Colors; above "umber brown" with darker narrow medial line: narrow edges pale brown, bordered with the umber brown.— Beneath pale brown

Dyphyes[2]

PL. 15
Fig. 1

Caught in day time in harbor, C. Tres Montes. quite colourless, transparent.— length 1/5th, breadth 1/12th (therefore drawing rather too broard). Body flattened; outer envelope sharply conical; on right half of body, we see within a sack, of which the sides appear very thick; the mouth can be closed at will by horizantal membrane; the bottom of the sack appears to be double, as if ~~termin~~ dividing into two. The outer envelope projects on right hand corner; on the left is occupied by a ~~solid~~ square promontory, the sides of which & terminal edges are concave; this square; ~~this~~ is occupied by a sack, containing an ovary? to the base of this Ovary there is attached a cylindrical vessel, containing another interior one, which slightly enlarges, near extremity suddenly contracts, forming a little bag.— In all the specimens which I saw bubble of air was contained in the inner tube & a strong circulation might be seen, sometimes performing whole length, at others revolving in shorter distances; the particles had as well as progressive, a revolving motion; this vessel enlarges into what I have called the ovary . This when contracted appears like a bunch of opake little paps.

Dyphyes can |**290**| be extended, even to mouth of sack (A).— When in this state is seen to consist (z) of a tortuous vessel, bearing alternate little blindguts filled with granular matter, their footstalk is surrounded by ~~small~~ globular-mass of points.— These blind-guts appear less developed, at basal parts of vessel.— the air from the long vessel circulates through this ovarium vessel.— In quite young specimen, this part was but little developed, but otherwise similar.— The quadrangular projection blends its figure, with the flattened cone gradually, the lines of angles may however be traced to the very apex: in a like manner, from (B) where the keel CD joins on, ~~to~~ a line of projections on each side (or ridge) is continued to apex (D).— The animal moves by starts, quickly by the contraction of the ~~left~~ rt hand part of Sack; water is expelled & animal proceeds apex foremost.— Could also revolve itself.— Quickly perceived & avoided the approach of any body in the water: swam high or low in a glass.—

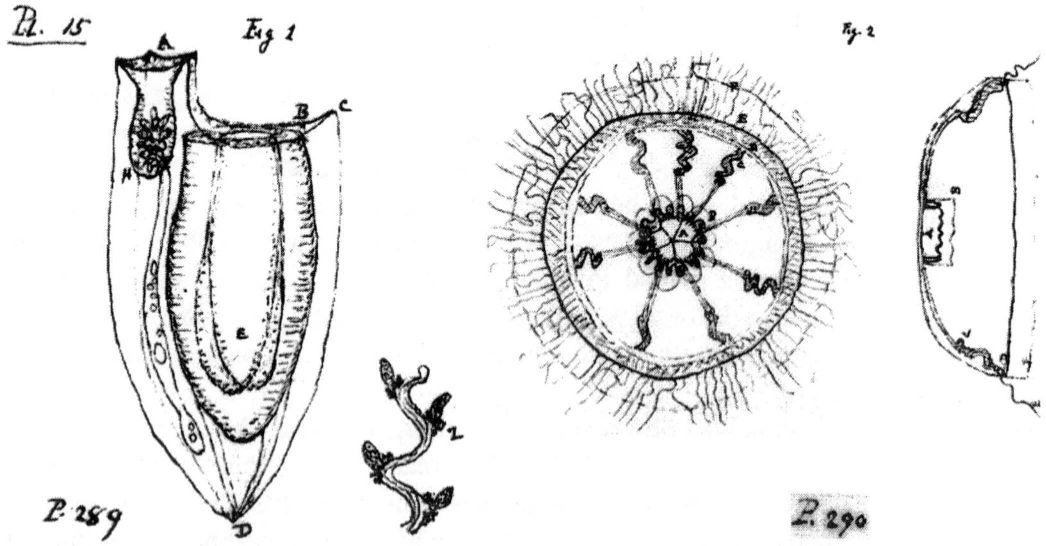

Plate 15, Figs. 1 and 2

[**CD P. 290** continues]

Equorea[3]
PL. 15, F 2

Diameter .6 to inch & some smaller; colourless exceptend [sic] the red tentacula on edge of ombrella.— Locality &c same as above.— Drawing represents the animal laid on its back & fully expanded. The tentacula are about 1/3 diameter, are placed so as to touch at their bases in the whole circumference; arise on the dorsal edge so that the veil (F) is within them (seen in the supposed section). The tentacula at base can be seen to contain inner vessel, these open into a space or vessel which surrounds the whole ombrella.— the tentacula taper to point, are adhæsive, red-colored.— In centre of body in dorsal parts, five very delicate vessels unite, without forming any sack; these bifurcate & become enveloped in |**291**| a thicker covering & so pass (10 in number) into the space on vessel (D) which surrounds ombrelli.— Above (as animal now lying) the exterior half of these spokes, there is a sinuous thick fold, which appears an enlargement of the vessel & contains granular matter, is probably an Ovarium.— Close to where the bifurcation takes place, there is a mouth formed of a sinuous fold which is intimately connected with the envelope of the spoke-like vessels; there are about 20 zig-zags; the bend of each fold being attached alternately above the vessel & in the interval between two.— Exterior to this there is a delicate scalloped veil.— No doubt, when the animal is well & swimming the body assumes the form represented in section, where one tectaculum on each side is seen.— I have said from very centre of body 5 delicate vessels branch off & bifurcate; this would appear to be normal; but sometime six meet, & one will trifurcate, in other instances bifurcate.— There was a rapid vibratory circulating motion within the base of of tentacula, circumferential space & spoke like vessels & it extended even in the zig zig folds which form the mouth.— (It may be doubted whether this is a mouth.—) Both this & the previous animal caught several yards beneath the surface, the weather being bad.—

Equorea

[1] Described by CD in *Planaria* p. 187 as *Planaria elongata*. Currently *Pseudogeoplana elongata* Darwin.
[2] Siphonophore of family Diphyidae, *Muggiaea atlantica*. The pulsating nectophore used as a swimming bell has been drawn upside-down by CD in Fig. 1 of Plate 15, and is attached to clusters of feeding polyps and gonozooids which are housed in the hydroecium shown as sack (A).
[3] Leptothecata, not *Aequorea*, but a hydrozoan of family Laodiceidae.

[**CD P. 291** continues]

Nudibranch[1]
(1106)
PL 16 Fig 1

~~Locality same as last animal~~ C. Tres Montes, Chonos Archipelago, 13 fathoms.— Length when extended one inch.— body very narrow, mantle not surpassing the foot.— tail very much & abruptly pointed; body slightly tapering towards the head; back convex. Mouth protected in a longitudinal fold of the mantle, circular, can scarcely be said to be seated on a proboscis.— No labial tentacula; anterior extremity truncate:— Dorsal tentacula two (there are no more), seated |**292**| near extremity of body: rather small: pointed club-shaped pectinated (no hole at extremity), with transverse waving plates or folds (as in Doris). These antennæ at their bases are enclosed in a case; which on the ~~two~~ ⅔ of its exterior margin expands out into a saucer, the edges of which are indented with about 8 points.— (B) shows this saucer & case with the tentaculum: The inferior part or footstalk of the case is alone retractable: the tentacula are withdrawn through the case.— But neither these or the Branchiæ are withdrawn, from irritation, for more than ½ a second. Half way between these Tentacula & Branchiæ & half way between foot & back on rt side there is a closed orifice: in death a double tubular organ was protruded to the length of .1: this tubular organ (c) consists of two tubes with orifices, united at base.— i.e. organs of generation.

Nudibranch

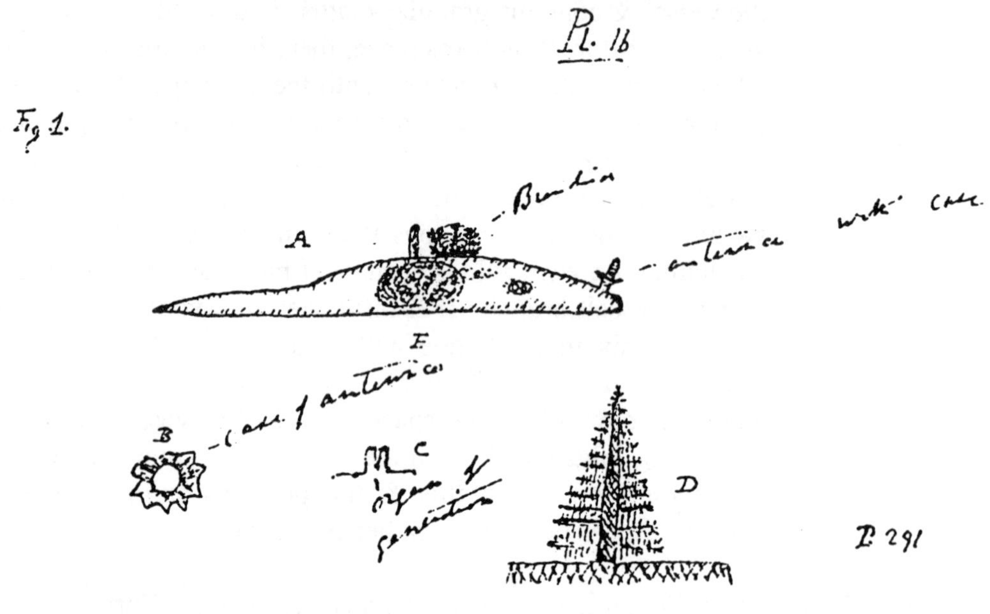

Plate 16, Fig. 1

[**CD P. 292** continues]

The Branchiæ are seated in very middle (in each sense) of back; consist of five trees; each tree (D) is subpectinated & like a fir: (D) represents back view; <u>every</u> line is curved towards the anterior surface, & these lines are the back edges of curved plates; so that the view in front would show a number of plates ~~passing~~ directed towards each other as represented. The trees are placed in a circle; towards the after part of the ring there is a tubular projecting anus.— fæces oblong, refuse of minute Coralls.— A little posteriorly to the Branchiæ, & a little exteriorly to them, there are two cylindrical, obtusely ended, tentacula-like organs rather longer than Branchiæ.— [note (a)] Are rather longer than Branchiæ: little more than

(a) 1/10th.— [note ends] These as well as Branchiæ are only slightly retractile. They do not appear tubular, & have no orifice; their use I am quite ignorant of.

[**CD P. 292** continues]

Within the body & a little posterior to Branchiæ there is a large white opake reticulated organ. anterior to this & close to dorsal surface the heart could be seen pulsating. Both these are drawn in. Body white, singularly transparent: scattered over whole surface there are circular & oblong regular

Nudibranch marks 1/30th to 1/40th |**293**| of inch in diameter, color dark brown, not placed quite symetrically 5 to 6 times their own diameter apart. Spirits of wine partially removed these spots.— The saucer-like case of Tentacula: the Branchiæ, excepting the very tips: & central broard band in the two posterior cylindrical organs (top & base white): & narrow margin at very extremity of tail, bright orange. Hence very pretty animal.— Foot narrow, can not adhære to flat-surface, excepting by the tail & anterior extremity, both of which are rather dilated: immediately adhæres to sea-weeds: often floats, back downwards: with edges of foot applied to each other for its whole length, so that body is a perfect cylinder: ~~frequently adheres even to sea-weed (or needle) only adhæres by foot.~~— Evidently adapted to crawl

(a) on Fuci & its habits beautifully agree with its living on small ~~microscopic~~ parasitic Flustraceæ, or (I know the species) rather Cellaræa (which in itself is curious fact as mostly these animals are herbivorous).— [note (a)] Its tail gives it the power of crawling like a Caterpillar from twig to twig: can probably swim well:— [note ends] The stomach should have some bony structure or teeth.—

This animal is allied, especially in habits, to the Molluscous one of Rio (**P.46**).— Comes nearest to Scyllæa.—

[**CD P. 293** continues]

Doris Same Hab & Locality: 2 Species, the larger one[2], pale yellow, with

1108 irregular brown spots, perhaps same species with that of **Page (284).**—

(b) [note (b)] ~~The Branchiæ however only have 7 principal divisions~~ [note ends] The smaller specimen[3] is different, but agrees in many respects: color bright yellow: smoother to the touch: mantle does not much surpass the foot: body more oblong & convex.— The chief difference lies in the Branchiæ, which here consist of ten, small, delicate, brush-like tufts; these are closely & finely subdivided.— Length .6 to .7.— [added in different pen] The Branchiæ have only seven principal divisions |**294**|

Tubularia[4] Tubes generally grow separate & distinct. same Hab & Local (tidal rocks

1107 as before): Each tube about 1 to 2 inches not branched high: terminal

PL. 16 polypus dark rose-color, not in the least retractile. Consists of a base, from

Fig 2 & 3.— which spring 16 long (.15 in length) tapering tentacula, are tubular within granular substance but only to be seen (tubular structure) in some specimens: These tentacula enclose, a pointed oval opake mass, which terminates in Mouth & Anus (for I saw small globular red fæces ejected). At about the middle of the oval there is a another ring of short cylindrical tentacula; 24 in number, but only ⅓^{d} length of the others. When the animal is at rest, the outer tentacula are curled backwards like petals of a flower; whilst the short ones enclose the cone or mouth.— A Section is shown at Fig 3.— The stem consists of an outer coriaceous case, enclosing another vessel; this near to the Polypus contracts very much; is narrow where ~~they~~ it joins the transverse base, leaving however a joint like appearance.— This neck is strengthened by a sinuous mass of gelatinous elastic substance, which appears a ~~continuation~~ transmutation of the coriaceous covering.— Within the inner vessel there was an appearance of 4 or 6 ligamentous bands.— The inner vessel contained red granular matter & passed on to the cone.—

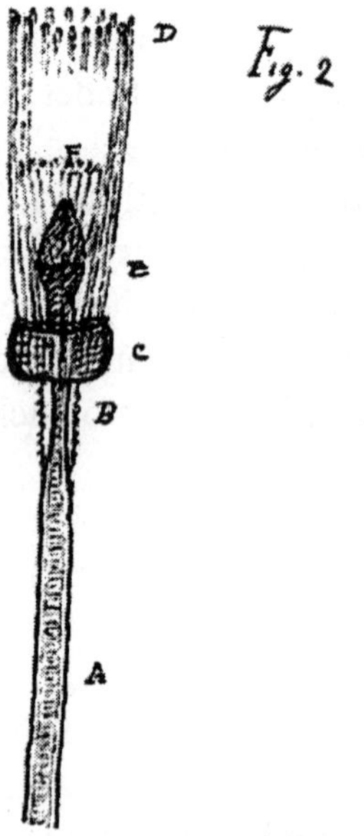

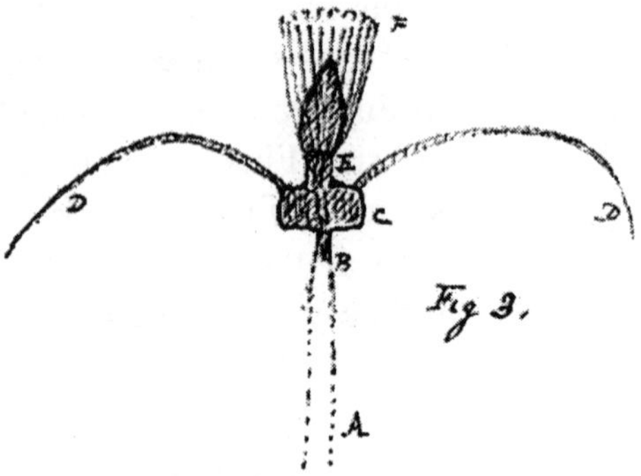

Plate 16, Figs. 2 and 3

The animal when touched covers itself with the outer tentacula, but does not otherwise move; if the head is moved recovers its position.— When taken out of water, Polypus droops, the neck of stem bending.— the neck contracts if vessels are emptied of their fluid.— Probably owing to elastic action of the ligamentous bands or surrounding jelly.— Polypus cannot be
(a) said to have any case or receptacle. Is allied to the Tubularia Clytia **P 245** of S. Cruz; through which a connection is [continued at (a) on back of **P. 294**] traced with Clytia; as shown by the numerous granular tentacula placed in one or more rows; the more or less projecting mouth, the non- or

imperfectly retractile polypus its simple structure; the immediate connection with living axis of the stem.— The S. Cruz species quite unites the two others, possessing the pseudo-jointed structure of stem, & small cup of the Clytia; the enlarged oval, non-retractile body of this Tubularia; & lastly differs from both in the many rows of tentacula. |**295**|

[1] Phanerobranch doridacean, Polyceridae, *Thecacera darwini* (Pruvot-Fol 1950), recently named after CD, a sea slug that feeds on bryozoans living on seaweed. In Fig. 1B of Plate 16, CD has correctly drawn the unusual saucer-like rhinophoral sheaths, and in Fig. 1C he has noticed the everted distal genital portions. See Plate 33 in Michael Schrödl 'Nudibranchia and sacoglossa of Chile: external morphology and distribution' *Gayana Zool.* **60**:17-62 (1996).
[2] Doridacea, *Cryptobranchia* sp.
[3] Cryptobranch doridacean, ? Platydorididae, *Gargamella immaculata* Bergh.
[4] Anthoathecata, a hydroid.

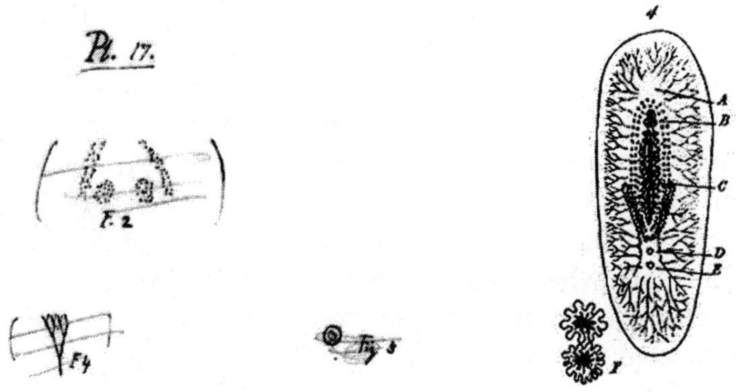

Plate 17, Figs. 2, 4, 5 and Fig. 3 as redrawn in 1844

[**CD P. 295** commences, the whole entry on Planaria being crossed through]

Planaria[1] [Decemb. 29th] Found under stones in tidal pool, Ynche[2] Is^{d}.— Body excessively depressed, edges very thin: broard in proportion to length,
PL.17:F.1=5 which is .55, broardest part .3 which is almost at the anterior extremity; tail close to tip .1 broard; hence three times broarder in front than behind.—
(2457) (b) [note (b)] The specimen is only in Fragments V the number [note ends] Color pale tile red, darkest on the ridge, with white narrow space over the posterior orifices & transversely oblong space where eyes are seated. These consist of black circular points, which are seated in two groups & also form two bands which are inclined to each other (Fig 2): it is to be remarked these latter are more deeply seated in body, near to inferior surface, than the circular groups. (Fig 1)[3] shows the animal with inferior surface turned upwards; there is a circular clear spot [labelled A], beneath where the eyes are; into this all the vessels from anterior part of body join.—

I must describe these vessels.— If we imagine, a thin cake of granular matter, which becomes coarser towards the centre, traversed by transparent lines or lines of division, which ~~continue~~ generally ~~tri=~~ bifurcate 3 times (Fig 4) before they reach the margin, we shall understand the appearance.— The black lines in my drawing Fig 1 represent these clear lines; so that it is hard to say which are the vessels, the granular matter or the lines, I suppose the (a) latter, because they join into a clear spaces such as that under eyes & round the central organ. [note (a)] In the dying animal, I noticed something like a rapid circulation of particles over the clear space, as if there was an aperture & currents of water flowing through it.— [note ends] The embourchure of the lateral vessels can hardly be distinguished.— Down the centre of body a white opake body ~~vessel~~ lies, which throws off on each side regular buds, which narrow off (as drawn) to point, on each extremity. When dying two orifices were widely opened over this organ, one anterior [labelled B] & the other rather posterior [labelled C]. These orifices closed Planaria so completely, that excepting when in the act of |**296**| opening I could never perceive them. Through these two orifices two cups were protruded; they differed from the general sort in being very shallow, more like saucers & margins narrow.— By dissection I procured the two organs separate from body (as in Fig 3).— (like lace round a cap) the margin was very sinuous & thin; this fringe is narrow; it seems to contain a sort of vascular system, somewhat similar to that of the body: is transparent, retains vitality & motion long after rest of body is dead; when pursed up, is very complicated from number of folds & like the section of a bud of flower.— in act of pursing up *[added in margin with different pen]* These two Saucer like organs touch each other & the *[3 illeg. words]*: When folded up in body, they produce the elliptic tree-like appearance described.—

This central line is surrounded in form of ellipse by an enormous number of small spherical bodies, arranged in packets of 2 to 4 each (in drawing I have only represented 2). By dissection these dropt loose; are sphærical with central opake mass (Fig 5): has diameter $3/500^{th}$ of inch: are manifestly eggs.— In dissecting the granular vessel-like masses in tail of body, it seemed full of partly-formed ova.— hence we must suppose they are matured all over the body, pass into the elliptical space, & from thence, probably, by the anterior orifice to open water.— In the drawing a double fork will be seen almost covering in posterior half the line of ova. This consists of chain of minute white opake bodies, partly or not united; the lower extremities of outer fork blend with the external vessels; the upper bends to form inner fork, which terminates abruptly: Has this arrangement any connection with the maturing [of] the ova: the forks do not appear Planaria connected with the lateral vessels: |**297**| Just at the termination of fork, there is a small orifice & again close beneath this another.— I am certain, from having seen them with high & low power, reflected & transmitted light.— that there are four orifices; two which emit the membranous saucers & two simple corresponding ones, the use of which I have never known[4].— Much as this species differs from others; yet the arrangement of vessels is almost similar in collecting in anterior ring in place of straight lines & from thence dividing into two lines, between which the cup-organ is placed; the ova are placed at the base of the lateral vessels. The forked arrangement of white opake matter is the most novel part.— & the doubling the common orifices.—

The animal crawls very quickly & adhæres firmly to stones: can swim well by action of thin edges of body; dissolved in fresh water from death, like butter in the suns rays.

[1] Classified by CD in *Planaria* pp. 191-3 as *Diplanaria notabilis*, belonging to a new genus.
[2] In Appendix p. 36 of *Narrative* **2** the *Beagle*'s position on this day is given as 'Off Ynchemo Island'.
[3] Fig. 1 of Plate 17 is missing from the drawings preserved in CUL MS DAR 29.3, presumably because it was removed and redrawn as Plate V, fig. 4 in *Planaria* p. 193, reproduced here in place of the original.
[4] On p. 192 of *Planaria* CD said that 'there are two minute, but quite distinct, orifices (D and E), which I do not doubt are the reproductive pores.'

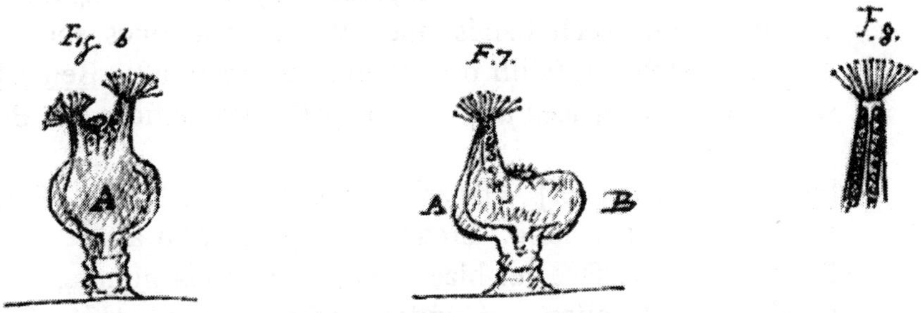

Plate 17, Figs. 6, 7 and 8

[**CD P. 297** continues]

Animal[1] (1112) (a) PL 17, Fig 6-7-8.— allied to Lucernaria?

Consists of an irregular globular hollow semi=gelatinous nearly transparent body, on a footstalk, which like an Actinia firmly adhæres to sea-weed.— [note (a)] The consistence of body is much that of some of the small transparent Actiniæ.— [note ends] (Fig 6) Seen with the back (A) in front. two short cylindrical branches, of unequal length, these have no orifice & are crowned by very numerous un-retractable papillæ, which terminate in a nob.— Seen laterally (as at F 7) these are seated on one side, in front of them & at their bases, there is a large orifice, which open into the inner tunic of sack: this great orifice is surrounded by numerous white, long delicate tentacula; & at the base of the branches is partially closed by a thin membranous moveable tongue.— The inner tunic of sack ~~sends~~ passes up

Animal

|**298**| into both branches.— These branches viewed from in front (B), instead of being cylindrical in whole circumference consist of two projecting longitudinal folds (Fig 8) which contain sphærical bodies (perhaps Ova?).— These are best seen in the larger branch, where the Ovary passes down into the sack.— Between the two branches & externally there is small white body on footstalk (Fig 6), with mark of orifice. Is not this a gemmule? Or young branch? — I should have said the branches (they ought not to be called so) are connected, up to some height.— The animal when touched near great aperture of sack doubles both ~~ar~~ branches over so as to protect it: the papillæ, ~~an~~ summits of branches are ~~moveable~~ irritable; the tentacula round aperture adhæsive.— Footstalk contractile (?).— (Fig 6) stands .4 high.— Colored dirty "Art Blood R".— What tribe does this animal belong to?.—

[1] Lucernarian, a sessile polypoid scyphozoan anchored by a contractile footstalk on the tentacles of which the animal is capable of creeping about.

[**CD P. 298** continues: entry and note on Planaria crossed through]

Planaria[1]
(2458)
(a)

Under a stone (on land) in Is[d] of Ynche, N. of C. Tres Montes: Above "Greenish Black" with minute white punctures; down centre of back two bands of "Gallstone yellow" separated by a narrow space.— On anterior extremity four such bands; the two external ones soon dying away.— Anterior extremity (with row of eyes on margin?) Beneath leaden color, with two white spaces in posterior part with orifices.— Body convex.—

Feb: 1835
2554

[note (a) added later] Planaria[2]. taken in the Forest of Baldivia: When crawling 1.7 or .8 long. breadth pretty uniform about .2: edges of body thin: Upper surface jet black, with numerous minute, oblong, variously sized spots of yellow.— under surface mottled white & black: Cup is protruded in the specimen.— [note ends]

Bufo[3]
(1118:19)
Copied
Bufo
Copied

Back pale "Chesnut B" with three longit bands of "Gamboge yellow" edged with black; marks of do on legs & on greenish sides. Under jaw "primrose Y", belly do with rings of black, or may be considered as black removed with very numerous circular yellow patches: Feet & very base of belly orange.— The stripes of Yellow often are irregular & become five in number. |**299**| Are exceedingly abundant all over the clear (from trees) damp mountains of Granite, "Anna Pinks Harbor[4]" or Pastel Harbor: crawl about actively during day time, & make noise like Englishman does to encourage horse.— When first touched, many close their eyes, arch their back & draw up their legs (as if spinal marrow was separated) I presume as an artifice.— They are chiefly remarkable from the curious manner of running, like the Natter Jack in England & scarcely even jump: neither do they crawl like a toad, but run very quickly.— Their bright colors give them a very strange appearance.— Abound at an elevation from ~~2000 to~~

~~3000~~ 500 to 2500 ft.—

Rana[5] (1120) From same great height & Locality, beneath a stone: on centre of back, strong tinge of grass green which shades on sides into a light yellowish brown.— Eyes very large. Iris coppery.—

Rana[6] (1117) Same Locality, but base of mountain.— above pale rust color, with obscure dark angular shadings.— Band of fine Chesnut B, reaching from nose, cross eyes & over the Tympanum.

[1,2] Classified by CD on *Planaria* p. 187 as *Planaria semilineata* and *P. maculata*, but now both in genus *Pseudogeoplana* pending further anatomical studies.
[3] Listed in *Zoology* **5**:49 as *Bufo Chilensis* Bibr. Specimen 1025 on p. 275 was the same species.
[4] See *Beagle Diary* p. 277.
[5] Listed in *Zoology* **5**:41-2 as *Alsodes monticola* Mihi.
[6] Not listed in *Zoology* **5**.

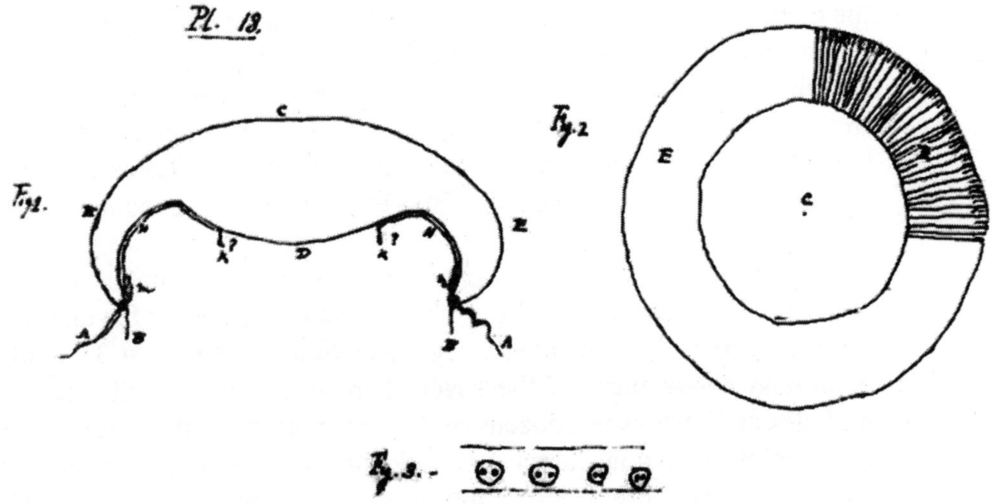

Plate 18, Figs. 1-3

[**CD P. 299** continues]

Cassiopœa?[1] PL 18 Fig 1-3 Common on all this coast; caught several feet beneath the surface. Fig 1. shows a section as in the water: diameter from edges (laid flat) 3.7 of which the margin on each side is one inch; the centre is an very oblate sphære, short diameter 1.1.— The margin thins to an edge & is recurved; from it a very narrow veil (BB) depends, outside of this are numerous tentacula (the structure of which is double) length .4; these however does not appear constant in all specimens: the convex surface is slightly depressed on margins with numerous lines; which probably are connected with the contractions of the Margins EE, by which the |**300**|

[now writing on New Year's Day for 1835, CD changes the page heading appropriately, and substitutes P. TRES MONTES for C. TRES MONTES. **CD P. 300** commences]

Cassiopœa animal possesses a limited motion: the outline of centre of ~~crown~~ back (C) is marked with slight irregular depressions something like a crown.— On the concave surface (HH) of the Margin there are very numerous semi-opake superficial vessels.— The animal seen ~~on its~~ from above shows appearance like Fig 2. The central oblate sphære being quite transparent.— These vessels are generally simple, sometimes they bifurcate & even trifurcate.— They terminate in the circumferential space from which veil depends & tentacula open into.— they arise in a delicate membrane which partially coats under surface (D).— I do not know how much is coated; for in all the specimens it was ruptured: I could make out that it must have been bordered by the sinuous veil & is highly irritable with motion, which probably depended at about (KK). These vessels are in short nothing but the membrane divided into numerous rays, & these have their edges united to the concave surface: Those vessels near to edge expand into small folds containing granular matter which ~~hang down~~ stand out (mm).— There are from 16 to 20 of these vessels (or ovaries) in every inch of circumference: yet I believe the tentacula & the external ribs of depression are more numerous.— The external veil (BB) is very narrow, is composed of very numerous parallel concentric fibres, towards base it has a reticulated structure: in the space, where tentacula & ovaries terminate, between each tentacula there are 3 or 4 little oblong capsules, each of which contains what appears to be two minute regular bubbles of air. Fig 3. And dorsal of ~~around~~ ~~above~~ them in the space there is a complicated circulation of particles as if there were dozens of distinct ~~on~~ centres of motion.— Animal quite colourless, transparent. I could not with high power perceive the
Cassiopœa slightest |**301**| organization in the oblate sphære.— The margins when rubbed by fingers, phos-phorescent with bright green light.— Could not perceive Sulp: of Magnes. had any effect on this appearance.— It is manifest this animal has same structure with Equoræa P 290.— I so called it there because I thought the internal veils formed a mouth.— In this case they cannot be said to do so; so that I am in doubt about the truth.—

[1] This jellyfish was certainly not *Cassiopea*, but appears to be a hydromedusa of suborder Leptothecatae, family Aequoridae, *Aequorea* sp.

[**CD P. 301** continues: first entry has been crossed through vertically]

Echinoderm
<u>sans pieds</u>[1]
(1122)
PL19,F1..3[2]

I found two individuals adhæring to the under surface of the foregoing Medusa.— Fig 1. Animal consists of an ~~inverted~~ funnel-shaped body ~~which rests on~~ *[illeg. passage of several words inserted above]* a circular fleshy disc.— This disc, close to its edge is surrounded by 12 equal, short cylindrical, obtusely ~~rounded~~ terminated ~~dorms[3] or paps~~ *[correction illeg.]* shaped as Fig 3.— The under surface of disc is mammillated, & has the power of adhæsion (of no very great power) in which manner the animal is fastened to the Medusa.— in the centre is a large irregular mouth.— Fig 2 is a section.— The stem or tube of funnel is terminated by small orifice, hence animal is open at each extremity: The funnel is lined by ~~a double~~ another tunic; the little dorms have no aperture, are lined with same tunic as funnel into which they directly open.— 10 or 12 lines proceed from margin to apex; perhaps these are of a ligamentous nature.— The inside of Body is filled with narrow intestines & a sort of fold or blind guts imbedded in reddish granular matter.— could find no sort of teeth.— The

(a)

animal distends itself with water, ~~when~~ then the tube is erect & the 12 little dorms become nearly transparent & stand distended & separate (like ornaments round a crown). [note (a)] I presume the distention with water is owing to a ~~process~~ method similar to that in Holuthuria.— The animal is not very irritable.— Its usual position must be upside down with respect to the drawing, because it adhæres by its disc to <u>under</u> side of Medusa.— [note ends] This is when animal is at rest; when molested, funnel shrinks, become striated longitudinally & some if not all of the dorms are

<u>Echinoderm</u>
sans pieds

contracted, when their inner tunic shows |**302**| itself more clearly & contain red granular matter. When at rest stood ½ inch high. color "Aurora Red". Body quite smooth, soft flexible (molluscous).—

The animal is evidently parasitic & belongs I apprehend to the family which the margin shows.— During the dissection I noticed that all the granular matter (at least I am sure of that in the dorms & in the blind guts) possessed a rapid revolutionary motion; it is similar to what has been described in Virgularia of B. Blanca[4]; the instant a mass of granular matter was broken, each little detached piece of whatever ~~figure~~ shape began to revolve.— the largest which I noticed was 1/100th in diameter & quite irregular in outline.— There could be no mistake: from the motion of the ship (in harbor) all loose particles vibrated from side to side, then quickly revolved on various axises & even progressed.— the more minute particles revolved the quickest: This power lay chiefly if not entirely in the reddish granular matter.— The field of view in Microscope appeared enchanted.— I cannot imagine what causes this motion; or what temporary organ[s] are thus employed.—

<u>Equorea</u>[5]
PL. 19,
Fig 4-6.—

Diameter of disc .6 to .1 in diff specimen; convex above, concave beneath: section like that of young moon. On edge or Umbrella a <u>narrow</u> depending veil, outside of this, there are 16 tentacula long tapering, which arise from a semi~~circular~~sphærical nob, & this opens into a circumferential vessel.

This nob contains an inner tunic with red granular matter.— hence, the rest of animal being colourless, appearance in water is that of a moving ring of red dots.— Between each pair of these tentaculiferous paps, there is a Equorea lesser, but similarly organized one; hence in |**303**| all 32 in number (Fig 6.) — In the concave disc there are four delicate vessels (Fig 5), which at exterior extremity open in circumferential vessels; & before this expand for short distance into a sinuous depending fold, which contains an inner tunic with opake granular matter (c): in this I could see a motion, also in circumferential vessel:— Near to where these vessels cross, they become thicker, are slit open on under surface; hence form a cruciform slit in centre of concave disc. This cross is surrounded by a membrane, with thin sinuous edges; the base of which appears like an expansion of the outer tunic of the four vessels. This membrane depends; & can contract itself or expand (as shown in F. 5). When contracted it is clewed in on four points & something resembles F 4.— the Cruciform slit being still preserved.— In Fig 5, the outline B only is supposed to show the thin edge of the membrane when expanded.— I am quite at a loss to know whether to consider this a mouth or not. The slit is superficial.— Animal abundant, caught by night beneath surface of water; evidently ~~allied~~ same structure with the last Equorea & Cassiopœa.— Only then It did not ~~notice~~ appear that the ovaries in centre were slit open.—

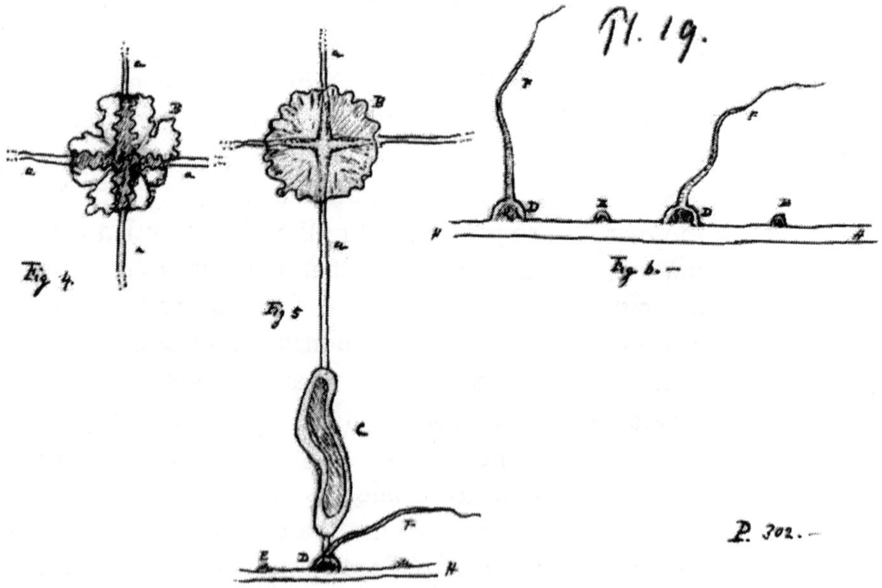

Plate 19, Figs. 4-6

Holuthuria[6] Hab: as above, "Patch Cove" (13 Fathoms): when contracted egg-shaped; length 1.7, breadth 1.1.— When at rest more than 3 inches.— Surface (1125) quite smooth to the touch, colored "Ochre & Gamboge Y". When at rest anus pointed; anterior extremity flat truncate, surrounded by 10 tentacula.—

tree-like, excessively long & tapering (1.4 length). Consist of tapering cylindrical stem, which throws off (at $\angle^{\circ}$ 45°) on all sides at regular but distant intervals, branches. [note (a)] Branches as in a tree, decrease in size towards the top: very base of tentacula almost void of branches. [note ends] These send off |**304**| in similar method other branches, which on each side are studded with little points.— Besides these regular branches, each tentaculum near base sends off a great branch.— Tentacula colored orange: truncate space in which mouth lies, purplish; ~~edge~~ margin of mouth deep reddish orange.— Body with 5 bands of papilli; each of these properly has 2 rows, but sometimes they are irregular & contain more.— Papilli, very long, when fully extended .4 in length.— Slightly tapering, terminated by a concave (adhæring) depression & no orifice; but when one is squeezed there appears to be a faculty of transudation, therefore perhaps of absorption; ~~the interior under~~ Beneath the surface of the Saucers, there is an irregular strong transparent fine net work; apparently for the purpose of strength.— This ~~reach~~ also encircles the upper part of footstalk; the stony vessel-like lines being in this part chiefly concentric, so as perhaps to allow contraction & protrusion.— Two of the bands of papilli are imperfect: the papillæ, being few, very small, & pointed, when retracted scarcely to be seen; They have however the stony net work, but not the terminal saucer.— Close to anterior extremity, on these bands the papilli are numerous & long, but yet pointed.— It is singular, that this should be (the side with imperfect bands) the adhæring surface: the animal however can well adhære by the 3 perfect bands.— It appears as if the two had been removed by attrition.— Habits, ring of contraction passing down body &c &c like others of the genus.— |**305**|

(a) Holuthuria

[1] Although CD identified the animals as 'echinoderms sans pieds', he recognised that they differed considerably from the Fistularia now assigned to order Apodida that he had found previously (see p. 125) in Tierra del Fuego. Neither do these animals attached to the undersides of jellyfish appear to have been narcomedusan parasites or platyhelminth flatworms. From their size and large mouths it seems most probable that they may in fact have been marine leeches.

[2] These three drawings have unfortunately not survived. Their disappearence, together with the crossing through of the accompanying text, suggests that CD might somewhere have published an account of the 'echinoderms sans pieds', but it is not known where.

[3] This word is definitely written several times as 'dorms', and could possibly be CD's aberrant spelling of 'domes'.

[4] See p. 199. Rotating food cords driven by cilia were observed by CD on a number of occasions.

[5] Leptothecata, Aequoridae, a thecate hydroid.

[6] Dendrochirotida, Cucumariidae, a sea cucumber. Species probably *Pseudocnus dubiosus leoninus* Semper.

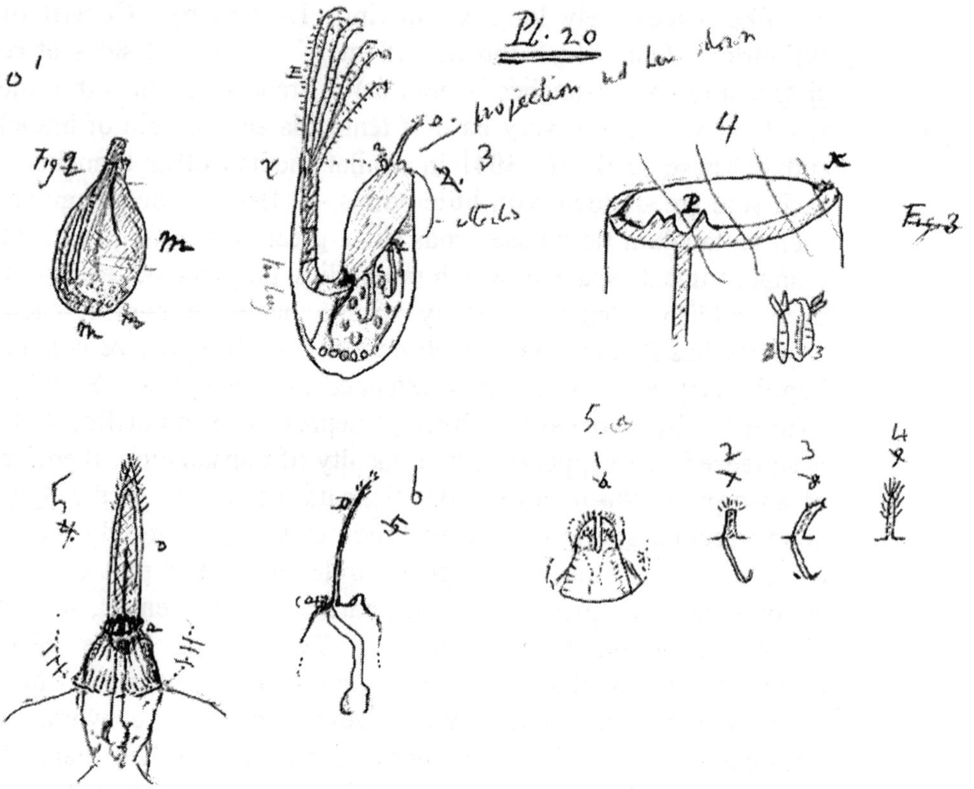

Plate 20, Figs. 1-9

[**CD P. 305** commences: the entry on Balanidæ is dated Jan. 8th, from Chonos Archipelago, and has been crossed through vertically to indicate its eventual publication in the monograph on *Cirripedia*]

Balanidæ[1]
(1131:32)

PL. 20
Fig 1...13.—

2495
shell of the
Concholepas

[illeg. pencil note]

The thick shell of some of the individuals of the Concholepas Peruviana is completely drilled by the cavities formed by this animal.— The cavity is oval, & lined by thin coating of Calcareous matter: small orifice in external~~ly~~ surface of shell: the base almost penetrates ~~to the~~ through the entire thickness.— The animal is orange colored & from 1/12th to 1/10th in length, is flask-shaped (Fig 1)[2]. The mouth is elliptic; with hinge at both ends; is edged with stony rim, which has outline as in (F3)[2]. (x) is the posterior half or that ~~by which~~ where the ~~chirri~~[2] cirrhi *[spelling corrected later with a different pen]* are protruded (I use posterior & anterior in reference to anus & mouth):— from the indentation (P) a line or rim of stony matter is continued down side of sack apparently for the purpose of strength:— The body lies in sack in something like the position (Fig 2)[2]: the great length of body is striking (or interval between anus & mouth), is perhaps owing to the ~~necessity~~ depth at which animal is buried in the Shell.— The body is terminated by only 3 pair of the usual double cirrhi;

these are but little curled, small, the centre ones slightly the longest.— on the hinder surface is a longitudinal slit, or anus.— it is very remarkable there is no true ringed trunk.— Following the body just beyond the bend, there is a tapering hook shaped organ (c) (containing inner tunic), & ~~behind~~ (a) before this, other corresponding one (b). [note (a)] Surface most finely hirsute. [note ends] Both of them depend amongst the ovules in bottom of sack.— Above these there lies the viscera:— & the body is terminated anteriorly by the mouth: The mouth is situated within & at the base of a narrow lancet-shaped horny, red closed ~~organ~~ projecting thin plate (Fig 4)^{2}: ((a) mouth in F2 & 4).— Length, 3/100th. At the tip there are few fine Balanidæ setæ.— |**306**| It is slightly curved: behind it there is a little rostrum:

Fig 5^{2} is a side view of anterior extremity of body: The mouth is composed of 4 pair of organs; The front (or exterior) pair (Fig 6)^{2} acts as a lower lip, is seated on a rounded cone, from which the lancet-shaped organ also arises; may not this latter be considered as the upper lip? Lower lip (F 6)^{2} is fringed with setæ on its margins & in front. Next to this is an oblong plate (F7)^{2} with setæ (Maxillæ) then we have (Fig 8)^{2} a longer curved plate, obliquely (Mandibulæ?) acuminated, the three points forming teeth; these two lie in rather a diverging line from lower lip; they are supported by a stony bar, which is imbedded in the fleshy cone from mouth. The last pair (Fig 9)^{2} is tapering upright with two terminal setæ & some on sides, is seated rather external to the others & has appearance of palpi: The 2nd & 3rd pair collapse together on the under lip (or 6)^{2}; the base of the lancet-shaped organ forms as I have said the upper lip.— All these organ[s] are very minute: We thus have 4 pair (including the lower lip) & the upper lip & 3 pair of posterior cirrhi.— All of which shows uncommon simplicity.— I omitted the Sack chiefly adhæres in its shells cavity surface (my Fig 1)^{2}; ~~at~~ its base is thickly lined by longitudinal vessels or bands, & others transverse, on side (m).— In the very base there are numerous ovules.—

[**CD P. 306** continues at a point in the text marked by a large double bracket]

These I saw, within the body in four very different states & their intermediate degrees.— 1st pointed oval, with included granular mass; 2^{d} (Fig 10)^{3} at one end, there |**307**| are two short club-shaped, transparent Balanidæ ~~organs~~ projections & on other a shorter one; length ~~about~~ of oval part about 1/100th: in 3rd state (Fig 11)^{3}, the two club-shaped organ[s] are very much longer, & contain a sort of limb within; the other is become a sort of pointed tail, within is inner tunic with sphærical-granular matter.— In the 4th state there is a great alteration, the length is now 1/50th; it is pointed coffin-shaped (Fig 12)^{3}; near extremity (G) there ~~are~~ is *[illeg. deletion]* a[n] open space through which two thick clumsy legs can be protruded; these are very big in proportion to whole size, three joints are quite manifest; the upper & last one is close beneath, a little plate, with circumferential curved spine (Fig 13)[4]. (in all directions) this joint has extreme play (like the wrist), the little plate hence can adhære & by retracting the legs can move the ovule. The legs ~~are~~ can be moved separately, have considerable powers of motion, are alternately retracted & protruded; singularly resemble the (a) misformed legs of some Crustacæ Entomostracæs; in basal part of ovule blend with the sphærico-granular in which no shape can be traced:— [note (a)] On the margins, chiefly anterior extremity few setæ:— It must be understood the legs do not come out at the very extremity, but on the one side of the flattened elliptic capsule:— Are the two terminal bunches of bristles young cirrhi? [note ends]

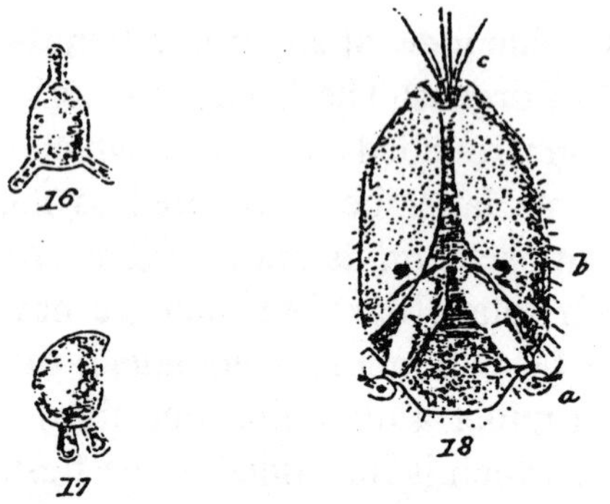

Figs. 16-18 from Plate XXIV of CD's monograph on Cirripedia

[**CD P. 307** continues]

At lower extremity ovule is slightly bifid, from this two short diverging bundles of bristles are protruded, capable of motion.— I feel little doubt this latter part is a developement of apex in Fig 11[3], & that the two legs were contained & hatched in the two cases (Fig 11)[3], which have subsequently dropped off, & the legs been retracted within body.— Although I have described the principal states of changes, |**308**| there were Balanidæ others intermediate: The last is the most developed; but I do not know it thus leaves the body of parent, it must clearly yet undergo a great metamorphis.— Who would recognize a young Balanus in this illformed little monster? Fig 12[3].— ? are the two strong legs, (with spiny plate capable of rotatory & other motion) for boring holes in the shell?—

It is manifest this curious little animal forms new genus.—

[1] This, the smallest known cirripede, was later named by CD *Cryptophialus minutus* (see *Cirripedia* pp. 566-86). Plate 20 served as a partial basis for the engravings in the monograph. In a footnote on p. 566 CD says 'I am greatly indebted to Dr. Hooker, for having several years ago, when I examined this my first cirripede, aided me in many ways, and shown me how to dissect the more difficult parts, and for having made for me several

very correct drawings, which, with some subsequent alterations, are now engraved'. However, he ignores the fact that his love affair with the barnacles must actually have begun in the Chonos Archipelago, some years before he met Hooker, when it appears that he managed the dissections fairly well without any guidance.

[2] These numbers are those originally entered by CD, and later crossed out.

[3] Figs. 10-12 have not survived, perhaps because they were redrawn to illustrate the states of development of the ova, providing Figs. 16-18 reproduced here from Plate XXIV of the monograph.

[4] Fig. 13 has also not survived.

[**CD P. 308** continues]

Holuthuria[1]
(1135)

Lowes Harbor[2]: seems to differ from the one described at **P 163** chiefly in its color, which is white, with tinge of flesh red: whereas the former is here seen of its usual salmon color.— Tentacula 10, small tree-like: 5 rows of papillæ, ~~placed~~ each in double order: papilla terminated by saucer like depression, which together with the sides are strengthened by a very close net-work of stone: it might be described as stony plate with circular holes:— Length 10½ inch, ~~breadth 1/15th~~ when crawling.— Some of the individuals were closely adhæring to stones & bodies contracted; beneath there was a nest of loose ovules, 30 or 40 in number: colored orange, quite sphærical, diameter 1/20th of inch: contain coarse spherico-granular matter (possessed no motion).—

Planaria[3]
(2510)

[paragraph crossed through vertically] Hab: do amongst stones in shallow water: body depressed convex above of nearly equal breadth: length 1 inch, breadth 1/15th: tail obtusely pointed: head distinct from body, in front slightly bifid, posterior corner angular: on which & anterior margin are imbedded numerous eyes.— head placed on a neck of rather less diameter than body [see drawing in margin]: internal structure consists of slightly tortuous central clear space or vessel, on each side, with very |**309**| short & dense ramifications of the granular matter as so often described: ~~could~~ this clear space in posterior part of body become opaque & narrower so that lateral vessels arise nearer to each other: could see no orifice on under surface; on upper surface near tail through a square irregular orifice 2 white projections were protruded.— I do not know whether naturally.— Colored above dull "Art Blood R" beneath white. could crawl in either direction.—

Hirudo[4]
1137 (a)
(& 1094
from Chiloe)

Jan 10th. Lowes Harbor.— These leaches bite peoples legs even when wearing boots, & are excessively abundant in some long grass near the beach.— When placed in a bottle with water, carefully adhæred to the dry parts: crawl very quickly, like geometric caterpillar.— Length varys much, generally about an inch, & narrow in proportion:— posterior end rather broarder.— inferior surface flattened, upper convex.— Anterior sucker formed on a superior pointed oval entire lip & a semicircular lower one [sketch in margin]; in middle is the orifice, within which are the three jaws,

the true mouth is in middle of these & rather triangular in shape. Each jaw is an elongated ridge, truncate at one extremity (a) & rounded on the other: there are on the edge of ridge about 50 most minute teeth (I believe each tooth is transversely double [drawings in margin]) they can only be made out with 1/10" focal distance lens. these are largest at truncate end, which is close over the mouth, & decrease (but very little) gradually to the opposite & rounded end.—

Hirudo

On the superior lip there are 10 black spots [see drawing in margin]; these are not homogenious, but have small transparent aperture through them.— 6 are placed equidistant & follow curvature of very extremity; four other, two on each side are in |**310**| a straight line & rather further apart. The posterior cup is placed rather obliquely on the extremity, but nearly parallel to inferior surface. is pale colored diameter 1/12th.— Skin strong leathery, with strong marked regular equidistant concentric segments: each of these segments is mamillated with small points.— There are also less marked longitudinal lines; hence the tegument is divided into squares (like back of armadillo): general appearance rough more like that of Pondbella muricata[5].— Under surface blackish brown, on each side a pale yellowish brown longitudinal band.— above figured with three shades of brown.— When crawling amongst herbage, surface rather dry.— Are said only to be present in the summer (V note to 1094), are a great pest from their eagerness for blood & great numbers.— Is a true Hirudo, differs chiefly in having upper lip entire?— The cause of its strange terrestrial habitat must be the dampness of the climate & forests.—

[note (a) added later] These animals are exceedingly numerous in the forests of Baldivia, & far from any water: they were sucking the blood in numbers from the bare legs of my guide: I found one young one adhæring to the body of a frog.— Perhaps these reptiles, which are very common in the woods, is the proper object of prey to the leach.— [with different pen] Bishop Heber[6] alludes to Wood Leaches in Ceylon [notes end]

[1] Dendrochirotida, Cucumariidae, sea cucumber probably *Cladodactyla crocea* Lesson. But if it differed 'chiefly in its colour' from the sea cucumbers 586 and 594 of East Falkland Island (see p. 140), and had a diameter 1/20th inch, its length must have been 1½ and not 10½ inches.

[2] Port Low was an anchorage just south of Chiloe much used by the Scottish trader William Low, who had long experience of the waters around Patagonia, Tierra del Fuego and the south-west coast of Chile, and sometimes assisted FitzRoy as a pilot.

[3] Mentioned briefly by CD on p. 190 of *Planaria* as the only species that can crawl in both directions, but may not in fact have been a turbellarian.

[4] Hirudinea, blood sucking leeches. The genus *Hirudo* as now defined does not exist in South America, and CD's leeches attacking men and even frogs in Chile were probably the species *Mesobdella gemmata* (Blanchard, 1849) of the family Haemadipsidae.

[5] Presumably a reference to the fish leech *Pontobdella*, the development of whose ova had been studied by CD in Edinburgh in March 1827.
[6] See Reginald Heber. *Narrative of a journey through the upper provinces of India from Calcutta to Bombay 1824-25. (With notes upon Ceylon)* . . . 2 vols. London, 1828. The Ceylon leech is *Haemadipsa zeylanica*, the first of the Haemadipsidae to be described.

[**CD P. 310** continues: next 10 lines are crossed through vertically]

General Observat. Ornithology 2532 (b)

The thick forests are here tenanted by very few birds: The Barking bird[1] (a Myothera, ~~large like the Turco of Chile, of a sooty brown color, abdomen reddish)~~ & the Cheucau[2] vie with each other in uttering strange noises.— [note (b)] The Barking bird makes a noise which is frequently taken to be the whelping of a little dog.— is called by Indians of Chiloe "Guid-Guid". is very abundant with Cheucau in central forest of Chiloe: is said as well as the Cheucau to build its nest on the ground amongst sticks.— [note ends] The little black wren[3] (2502) equally inhabits the impervious mass of dying vegetable matter in the interior forest: it hops
2502 about in the same concealed manner & utters strange & loud notes: This bird I am told has been very rarely found near Valparaiso.— [note (a)]
(a) Does not the formation of legs & beak point out some affinity in structure?— The Silgaro[4] (2198), as in T. del Fuego, utters from some high tree its gentle plaintive note, which seems to proceed from no one Spot.— [note ends] The creeper of T del Fuego pursues with its harsh chirp every intruder of the solitary woods.— The thrush of T del Fuego is not |**311**| very uncommon.—

Ornithology

The little humming bird[5], uttering its very acute note, ~~dashes~~ skips from side to side, is frequent even in the damp woods, where there are no flowers.— In the stomach of one, killed at Lowes Harbor, where now in the middle of summer there are many flowers, there was a black compact mass of the refuse of small insects.— A great many of these humming birds have been shot, yet scarcely any with the shining bright head of the cock bird: yet on opening them, many appear to be of that sex. Are they
2503 moulting? In the nest there are now young ones: Specimen (2503) is a male bird, without bright head: it may be observed these have a yellow gorge, & I have seen some specimens with brown feathers on the back.— Is it a different species?

The black Furnarius[6] on the coast is exceedingly common.—

The Carrancha[7] & Chimango[8] are occasionally seen (~~NB. The genus Caracara appears to me to supply the place of our Pica, Carrion Crow, Raven & Magpie)~~.— I was surprised to find considerable numbers of the Vultur Aura[9]. They constantly attend on the herds of Seal; this (& refuse from the sea) must be their only support for there are no large quadrupeds.—

[note (b) added later] Mr Bynoe shot on Ynche Isd a large eared owl; stomach full of Marine Decapod Crustaceans [note ends]

[1] Listed in *Zoology* **3**:70-1 as *Pteroptochos Tarnii* Gray.
[2] Listed in *Zoology* **3**:73 as *Pteroptochos rubecula* Kittl.
[3] Listed in *Zoology* **3**:74 as *Scytalopus Magellicanus* Gray.
[4] Listed in *Zoology* **3**:55 as *Xolmis pyrope* Gray.
[5] Listed in *Zoology* **3**:110-11 as *Trochilus forficatus* Lath.
[6] Listed in *Zoology* **3**:68-9 as *Opetiorhynchus nigrofumosus* Gray.
[7] Listed in *Zoology* **3**:9-12 as *Polyborus Brasiliensis* Swains.
[8] Listed in *Zoology* **3**:14-15 as probably *Milvago pezoporus* in Chiloe.
[9] Listed in *Zoology* **3**:8-9 as *Cathartes aura* Illi.

[**CD P. 311** continues]

Mammalia

(a)

Many of the small islands are inhabited by mice; for instance the 3 fingered one in Lat: 45-46 where we anchored: At Lowes Harbor some of the small islands are thus inhabited (so the Chilotans told me), & others not so:— At S. Pedro we have the little deer (which is found at P. de Tres Montes), Foxes & mice: How inexplicable is the migration of quadrupeds in these Archipelagoes.— |**312**| [note (a)] On the main-land of Chiloe in the forest I see the large reddish Rat of Maldonado (& St Fe) is very common.— [note ends]

[note added later] NB. A Handkerchief which brought home the skin of the Cervus Campestris at Maldonado (V private Journal for date[1]) & has been in constant use, every time after washing gives out the smell of that animal & this is in the Jan: 1835.— [note ends]

[**CD P. 312** commences]

Otter
2529

(a)

The otter[2] (2529) is very common; the largest weighed 9 & ½ pounds; they have holes in the Forest; they roam occassionally some distance from the shore; do not live solely on Fish, for Mr Bynoes saw one in T del Fuego eating part of a Cuttle fish; & one in Lowes Harbor was shot when bringing up a large Voluta! from the bottom.— [note (a)] Both the Otter & the Seal (& birds such as Tern) feed chiefly on what sailors call Whale-food: viz. Crust: Macrou<ri> which swim in the Bays in large shoals.— Hence the dung of all the Hair Seals is quite red.— Tres Montes.— [note ends]

Nutria
2530

The Nutria[3] (of La Plata; Coypu of Chili) is also common; have burrows in the Forest; live in the Salt Water (as the Cavia Capybara in the Mouth of the R Plata is seen to do); are said to eat shell fish as well as vegetable

matter; are good to eat, flesh white.— weight 10-11 pounds: fight very boldly with the dogs.— In the females, the Mammæ are placed nearly on
Cop the back or rather high up on the sides.— On this side of S. America, Gulf of Penas appears Southern limit: on the East coast the R. Chupat:

Goats[4] On the Island Ynche, there were very many wild goats: it is not possible to know when first turned out: Their color was pretty uniform, a varying shade of reddish brown: Many had a white mark on the forehead & a few on the lower jaw: All appeared to have a singular outline of forehead. Specimen (2499).— Is the head, I thought it worth keeping as these animals clearly from uniformity in color are retrograding into their original figure & ~~color~~ kind. |**313**|

Entomology: Chiloe & Chonos Archipelagoes do not appear essentially to differ in their Entomology: Diptera & Hymenoptera are the prevailing orders: Within the thick forests, minute Staphylinidæ: (& Pselaphus) & Hymenoptera (Cynips &c &c ?) are very abundant: Anapsis is not uncommon: But generally Coleoptera are not abundant. The most characteristic genus in numbers of
2414 individuals (& indeed in species) is Lampyrus[5] (?) such as (2414). Water beetles certainly are rare.— I think the number of English genera will be curious. I could almost fancy myself collecting in England.—

Vegetation (a) At S. Pedro (SE point of Chiloe) I first noticed the Antarctic Beech of T del Fuego, but at a considerable elevation & very stunted in its form.— In Midship Bay (Chonos) Lat: 45°-46°. This tree grew to a fair size at the waters edge & formed nearly $1/5^{th}$ of the Wood.— From this point it doubtless continues to augment, till in T- del Fuego we find the woods essentially composed of it alone.— [note (a)] These remarks about the Beech must be taken with caution; for I see one of the species least common in T del Fuego is common in central forest of Chiloe [note ends] The arborescent grass which we see in Lowes Harbor (& perhaps in Lemous) is not found in this Midship Bay; Hence together with the numbers of the Beech the forest bears a different aspect from what it does in Chiloe.— Here Cryptogamic flora has reached its per-fection (V specimens). In T del Fuego I have remarked, that the forest appears to be
Vegetation too dank & cold |**314**| for even this order of plants: In this Latitude 45°S [inserted in pencil] also I see that level pieces of ground instead of
Peat supporting trees, become covered with a thick bed of peat. Trees ~~never~~ seldom grow but on a slope in T del Fuego: whereas in Chiloe the plains form the densest forest. Here the climate seems more to resemble that of T del Fuego: indeed it is remarked by old Navigators on this coast, that in the whole distance between Chiloe & C. Horn, there is no great difference of climate. The peat is here formed by the plant called in T del Fuego "Bog plant" & another, Specimen (2475): ~~It suppor~~ These socialle plants support a few tufts of coarse grass, stunted little dwarf beeches & the "Tea Plant" of the Falklands — The aspect of the Bog is precisely that of T del

Fuego.] — The Lat. 45°[6]!!—

[**CD P. 314** continues]

<u>Potatoes</u>[7]
~~(X142 Tubers)~~
(2528)

<u>Copied</u>

(1142)
in spirits
(a)

Wild plants grow in abundance on all the islands of this group: the furthest point South, where Mr Stokes saw them was at Lemous: But Mr Lowes tells me the wild Indians in the Gulf of Trinidad know them well, call them Aquina & eat them, & say they grow in that neighbourhead.— At Lowes harbor (Lat: 44°) I visited a large bed: They appear a sociable plant: in all parts they grow in a sandy-shelly soil close to the beach, where the trees are not so close together: They are now (Jan 15th) in bud & flower: the tubers are few & small, especially in the plants in the shade, with luxuriant foliage. Yet I saw one, oval with the longest diameter two inches in length. They are very watery [continued at (a) on back of **P. 315**] & shrink, when boiled: When raw have the smell of Potatoes of Europe: When cooked are rather insipid but not <u>bitter</u> or ill-tasted & may be eat with impunity (V Humboldt, New Spain Vol II P *[page no. omitted]* [)]. The stem of one plant from the ground to tip of upper leaf measured exactly 4 feet !!.— These plants are unquestionably here amongst th<ese> uninhabited Islands in their wild state (Indian<s> of South recognizing them & giving them Indian name, general occurrence on all, even very small islets &c &c).— They grow on a sandy soil, with much vegetable matter.— The Climate is very humid & little sunshine.— [later addition with different pen] The Indians of Chiloe speaking the Williche language give them a different name from Aquina, the word of ~~the~~ West Patagonia.— The potatoes has been found near Valparaiso. V. Sabine Horticultural society[7]? |**315**|

1 The deer was killed at Maldonado in June 1833 (see *Beagle Diary* p. 160), and as described in *Zoology* **2**:29-31, the specimen that smelled so strongly was eventually mounted at the Zoological Museum in London.

2 Listed in *Zoology* **2**:22-4 as *Lutra Chilensis.*

3 Listed in *Zoology* **2**:78-9 as *Myopotamus Coypus.*

4 CD's interest from the outset in the rate of change of the characteristics of an isolated population of a domesticated species should be noted.

5 CD's tentative identification of the genus may have been incorrect, since according to *Insect Notes* pp. 81-7, no specimens of Lampyridae have been found in his collections from Chiloe and the Chonos Archipelago.

6 In *Plant Notes* pp. 178-9, the Antarctic Beech is identified as *Nothofagus* spp., the "Bog Plant" is *Astelia pumila*, Specimen 2475 is *Donatia fascicularis*, and the "Tea Plant" is *Myrteola nummularia*. Lemous is CD's spelling of Lemuy Island.

7 CD's specimens are identified in *Plant Notes* p. 180 as *Solanum tuberosum* var. *vulgare* Hook. See also J. Sabine. On the native country of the wild potatoe, with an account of its culture in the garden of the Horticultural Society. *Transactions of the Horticultural Society of London* **5**:249-59 (1824).

[CD's supply of faintly lined paper ran out at this point, and **P. 315** commences with an entry on unlined paper headed 1834 July, Island of Chiloe on West Coast of S. America]

Pediculus[1] These disgusting vermin are very abundant in Chiloe: severa<l> people have assured me that they are quite different from the Lice in England: they are
1185 said to be much larger & softer (hence will not crack under the nail). they infest the body even more than the head.— I should suppose they
2561 originally come from the Indians, whose ~~race~~ blood is so ~~predominant~~ with
dry these Islanders [* inserted above deleted word refers to note on back of **P. 315**] * See my Journal under head of Chiloe for account of inhabitants[2]. few are pure bred. [note ends] I have little doubt this is the kind in common amongst the Patagonians of Gregory Bay; they are said to be ~~very~~ there also very large.— An accurate examination of these specimens will at once decide the fact of identity or difference.— Mr Martial, a surgeon of an English Whaler, assures me that the Lice of the Sandwich Islanders are blacker & different from ~~all~~ these, or any lice which he ever saw.— Several of the natives lived for months & cruized in the ship, no efforts
Climate of Chiloe temperate & very humid could free their bodies from these parasites, but he assures me as a certain fact, known to every one on board, that their lice if they strayed to the bodies of the English in 3 or 4 days died, & were found adhæring to the linen (like Pediculi from Birds or quadrupeds?). So that the Sailors who constantly slept close to the Sandwichers never were constantly infected by these vermin.— If these facts were verified their interest would be great.— Man springing from one stock according his varieties having different parasites.— ~~It leads one into many reflections.—~~ |**316**|

[1] The louse in Spirits of Wine numbered 1185 (see p. 358) has not been found; but in the Denny collection at Oxford (see *Insect Notes* p. 88), Card 2561 included a female *Pulex irritans* L. from Chiloe, and a card mount numbered 2564 carried four unidentified lice. Kenneth Smith points out in *Insect Notes* pp. 43-4 that while races of human lice have been described in the literature, not enough work has been done to substantiate the observations reported by CD, leading perhaps to his later deletion of the final sentence.
[2] See *Beagle Diary* pp. 283-5.

[**CD P. 316** commences with an entry referring back to a period when the *Beagle* was sailing northwards towards Valparaiso. The deletions and numerous corrections in the section marked **B** and enclosed within double square brackets, have been made with a different pen, and not at the original time of writing]

Infusoria[1] The sea some few leagues North of Concepcion was of a muddy color in great bands, certainly more than 1 or 2 miles long.— Again 60 miles South of Valparaiso the same appearance was very extensive; although 40 or 50 miles from the shore I though[t] it was owing to a current of muddy water brought down from the Maypo. **B**⟦ Mr Sulivan[2] however having drawn some up in a glass, thought he saw by the aid of a lens moving points.— I examined the water; — it was slightly stained as if by red dust.— & after leaving it for some time quiet, a cloud collected at the bottom: with a lens of one fourth of an inch focal distance, small hyaline spots might be seen darting about with great rapidity & frequently exploding. Examined with a much higher power *[illeg. deletion]*, their shape is oval & contracted by a ring ~~on~~ around the centre from which line on all sides ~~proceed~~ curved

little ~~bristles~~ setæ proceed & these are the organs of motion.— [see sketch in margin] One end of the body is narrower & more pointed than the other. It is very difficult to examine these animalculæ, for almost the instant motion ceases their bodies burst. Sometimes both end[s] burst at once,
(a) sometimes only one, & a quantity of coarse brownish granular matter is ejected which coheres very slightly.— [note (a)] The granular matter is contained in a thin capsular membrane, to this membrane on the ring the transparent tapering fillets or bristles are fixed.— The motion of these setæ is that of collapsing on the obtuse end.— The water only appeared as if it contained a little of the finest red dust.— [note ends]

[**CD P. 316** continues]

The ring with the setæ sometimes retains ~~life sometimes~~ its irritability, for a little while after the ends have ejected their contents, it continues a riggling uneven motion. The animal, an instant before bursting expands to half again its natural size; about ~~15~~ fifteen seconds after the rapid progressive motion has ceased, the explosion takes place.— In a few cases it was preceeded for a short interval by a rotatory motion on the longer axis. ~~Directly perhaps 2 minutes~~ ~~Very soon, perhaps~~ about two minutes
Infusoria after any number were isolated in |**317**| a drop of water, they thus perished.— The animal moves by the aid of the vibratory ciliæ with the narrow apex forwards, & generally ~~with~~ by rapid starts; ~~The setæ are rapidly vibrating around the Body~~.— The immediate bursting of the body prevented any close examination; they would sometimes explode even whilst crossing the field of vision.— They are exceedingly minute and quite invisible to the naked eye, only ~~being a trifle larger (before explosion)~~ covering a space equal to the square of .001 of an inch. Their numbers are infinite, the smallest drop of water, which I could remove, containing very many.— We passed through in one day two masses of water thus stained, ~~to day~~ of which the latter ~~of the two~~ must have been several miles in extent. ⟧**B** the edge of the blue ~~water~~ & red water was ~~quite~~ perfectly defined.— What infinite numbers of these microscopical animals! — The weather had been for some days calm & cloudy.— The color of the water as seen at some distance, was that of a river which has flowed through a red-Clay district: Looking vertically downwards on the sea in the shade, the tint was quite as deep as Chocolate.— It belongs to the family of Trichodes of Bory

St Vincent[3], but does not agree with any of his Species: The sea at this time, I fancy owing to the Calms, abounds to a wonderful degree with various animals.— This fact of sea so very extensively colored by Infusoria appears very curious.—

[note on back of **CD P. 317**] Mem: The patches of red sea in the Southern Latitudes owing to the "Whale food" or rather large red Crust. Mac: in great shoals.— [note ends]

[1] Identified in *Plant Notes* pp. 214-15 as a dinoflagellate (Pyrrophyta), probably a species of *Gymnodinium* or *Gonyaulax*.
[2] Bartholomew James Sulivan was a Lieutenant on the *Beagle*, 1831-6. Surveyed the Falkland Islands, 1838-46. Admiral, 1877.
[3] See *Dic. Class.* **16**:556.

[**CD P. 318** commences]

Spawn SE by E *[number omitted]* miles from the group of the Galapagos in the open sea, out of sight of land.— a strip of water NNE & SSW, some miles long, slightly convoluted a few yards wide, was of a very yellow mud color.— In a bucket, whole surface was covered by little nearly transparent balls in contact.— These were of two sorts.— (1st) irregular globes, the largest .2 inch in diameter: of transparent gelatinous matter, with I believe water in centre; thickly & equidistantly studded by semi-opake white little sphæres.— These are imbedded very superficially; are sphærical $1/200^{in}$ in diameter.— The (2^{d}) kind nearly the same balls rather larger reddish. The gelatinous matter divided into several distinct sphæres, united by similar substance.— The ovules rather smaller, opake red, in rather greater numbers, imbedded very superficially over whole surface, the separate sphæres & interstices.— I do not know to what animal these extraordinary
(a) numerous ovules belong.— After passing this first: There were two other similar streaks.— Of all the appearances which Sailors call "Spawn", this is the first which deserves this name.— |**319**|

[note (a) added later] Also within the Archipelago there were the same kinds: What force keeps these globules in such close order & for a length of such greatness.— Are they the eggs of fish or rather of Molluscous Pteropidous?? Capt Colnett[1] mentions much spawn as being near these Islands.— He says that the direction of bands points out that of the stream of the ocean.— Are they then Spawned in one spot (& like a river carries the brown foam from an eddy) on the edge of a Current, which sweeps them away as it encroaches on the body?— A Ball of foam in the centre from a river often becomes untwisted into a ~~river~~ ribbon?— Capt. FitzRoy remarks they now are parallel to ~~N & S~~ the direction of the Winds, viz N

& S. The difficulty of accounting for the ribbons is not much less.— Without supposing they attract each other.— [note ends]

[1] See James Colnett. *A voyage to the South Atlantic and round Cape Horn . . .* London, 1798.

[CD now inserts three pages numbered 319 to 320A without the usual margin and headings that are concerned with the red snow in the Andes. **P. 319** commences]

March 20th. 1835.= Red Snow.=

In the road from St Jago de Chili to Mendoza by the Portillo pass there are two distinct Cordilleras or ~~chains~~ lines of mountains. In both of these ~~ridges~~ on the Eastern & Western slope the road passes over ~~large~~ masses of perpetual snow.— On these I noticed much of the substance called "red Snow[1]". The elevation as calculated from Humboldt is given in Mr Caldcleugh<'s>[2] travels as 12800 ft.— Mr Miers[3] (in his account of the passage of the Andes) mentions seeing both Red & Green Snow in the *[two illeg. words del.]* ~~frequented~~ pass of ~~Uspallata~~ Uspallata or Las Cuevas: He states no particulars.— ~~I was not fortunate enough to meet with it in this~~ At the time of year I passed (April 5th) there was scarcely any snow on this road.— I first noticed the Red Snow by the color of the impression of the Mules hoofs: as if they had been slightly bloody, also in some places where the Snow was thawing very rapidly. The color is a fine rose with a tinge of brick red.— The surface of the Snow appears ~~is scattered over~~ as seen from the mules back to be scattered over with bits of dirt. My first idea was that it was the dust of the red Porphyrius, blown by the strong winds from ~~bare~~ crumbling sides of the Mountains. |**320**| The particles look as if ~~they~~ many were 1/10th of inch in size. This is an optical deception, owing to the magnifying powers of the ~~large~~ coarse crystals of Snow. Hence on being taken up the particles almost disappear. The Snow ~~being taken up &~~ crushed between the fingers or on paper communicates a red tinge, but otherwise as I have said with the exception of a few places the Snow before ~~mechanical violence~~ pressure is not coloured.— Examining it with a weak pocket lens ~~the snow on which such coarse particles appeared to have adhered~~ groups of (from 10-40) ~~minute spheres~~ most minute circular atoms were clearly visible. Each was perhaps about 2 diameters apart from the others.— These groups caused the appearance of such coarse particles.—

I placed some of them between the leaves of my Note-Book. on my return to Valparaiso, after 2 months interval, I examined the paper.— The Spots ~~where I had placed~~ were now stained of pale dirty brown (V accompanying Specimen).— The greater number of little spores had been crushed & were not to be distinguished. I ~~extracted~~ removed however some tolerably perfect.— Being placed in water they became more transparent & showed with transmitted as well as reflected light a fine Arterial Blood Red Color.— They varied in size, ~~& the outline is quite smooth~~ the largest & most perfect being exactly 1/1000th of inch in diameter. The outline is not perfectly regular or smooth. [continued on back of **CD P. 320**] The red centre is seen to have a thin ~~trans~~ nearly colourless bark: the red matter appears to be a fluid which is not miscible with water, Alcohol or Sulphuric Acid.— It would appear a fluid from being separable into variously sized perfect globules.— On applying diluted Sulp. Acid, the outer coat is either destroyed or so very soft that on the least touch falls off.

is composed of an outer most delicate tunic lined with granular matter.— This bark was ~~often~~ torn & ragged in many of those Specimens I removed from the Paper: ~~The~~ *[2 illeg. words del.]* ~~red body~~ is perfectly sphærical & smooth *[further deletions]* after remaining some minutes in the Acid is active in two very different modes: in one case, suddenly with a start the sphere enlarges to twice or three its previous diameter, the color becomes much ~~paler~~ less intense (& this continues to decrease), the whole appearing as a drop of pale red fluid, not miscible in surrounding medium. There has also fallen a cloud of equal most minute ~~circular~~ sphærical grains.— I believe they are ~~granular~~ sphærical, for they are but just Visible with my highest power 1/20th inch focal dist lens.— In other case, ~~the red ball rather contracts~~, the red fluid ~~being~~ is seen owing to its contraction to have contained in a thin colorless case & ~~contains~~ has in its middle a darker spot. In this state tolerably strong acid appears to have scarcely any further effect:— It would appear probable that in the first case, that this tunic must have suddenly [continued on extra sheet numbered **(4)**] burst & that the cloud of granules is the dark spot in middle pf red fluid,— We have then 1st colourless outer tunic, with (2^{d}) do granular lining. 3^{d} tunic of red globule. 4th. red fluid. 5th contained most minute, scarcely visible granules.— With respect to the Red fluid, ~~of course it cannot~~ its existence is only known after a short soaking in fluids: how it may exist in the dry ~~or fresh~~ specimen I do not know.— It is singular. In one case, the outer bark (= about 1/6″ focal dist.), contained two distinct red balls.— ~~The existence of this Cripto plant substance in Lat: growing at a great elevation in the Perpetual Snow appears as rather curious instance of the geographical distribution of plants.~~ I understand the late Navigators have found Red Snow in the Antarctic regions.— The existence of this Cryptogamic plant in Lat *[number omitted]* S. growing on the lower patches of Perpetual Snow is a rather interesting fact in the laws of the distribution of Vegetables. [further note in pencil] Has [been] found on many mountains in Europe and on rocks in Scotland.

[continued in pencil on **CD P. 320A** inserted after end of the voyage]
Vol IV p. 231 Greville Scottish Cryptogam Flora[4] describes — balls, fine garnet colour, exact sphærical nearly opake sited on substratum of gelatinous matter; for most nearly equal in size.= Smaller ones generally surrounded by pellucid limb, gradually becomes less as globules increase in size.— in full sized specimen internal surface appears granulated, from contained granules; granules 6-8 in number globose. capsule left floating after bursting of sclera.—

Protococcus nivalis

Decandoelle could not see granules only oily fluid[5].—

There is no notice taken of being <u>in</u> groups.—

Thinks presence owing to flowing of melted snow.— Rocks at higher level

---

Bauer states they are 1/100 of a line.—

Were my specimens going over with envelope.—

[1] For other accounts of the Red Snow see *Beagle Diary* p. 309, *Journal of Researches* **1** pp. 394-5, and *Plant Notes* pp. 207-9. The alga responsible, termed *Protococcus nivalis* by CD, is *Chlamydomonas nivalis* (Bauer) Wille.

[2] See Alexander Caldcleugh. *Travels in South America, during the years 1819 . . . 21.* 2 vols. London, 1825. CD visited Mr Caldcleugh in Santiago when he returned there on 10th April 1835.

[3] See John Miers. *Travels in Chile and La Plata . . .* 2 vols. London, 1826.

[4] Robert Kaye Greville. *Scottish Cryptogamic Flora.* Vol. 1. Edinburgh, 1825-26.

[5] See Augustin Pyramus de Candolle and Kurt Polycarp Joachim Sprengel. *Elements of the philosophy of plants.* Edinburgh, 1821.

[**CD P. 321** commences with an entry headed Chatham Is[d]]

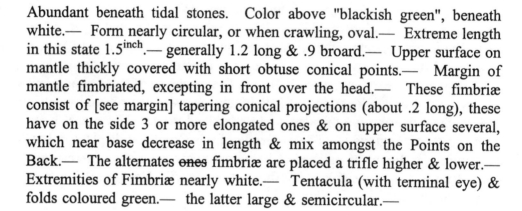

Onchidium[1]
1277 Abundant beneath tidal stones. Color above "blackish green", beneath white.— Form nearly circular, or when crawling, oval.— Extreme length in this state 1.5[inch].— generally 1.2 long & .9 broard.— Upper surface on mantle thickly covered with short obtuse conical points.— Margin of mantle fimbriated, excepting in front over the head.— These fimbriæ consist of [see margin] tapering conical projections (about .2 long), these have on the side 3 or more elongated ones & on upper surface several, which near base decrease in length & mix amongst the Points on the Back.— The alternates ~~ones~~ fimbriæ are placed a trifle higher & lower.— Extremities of Fimbriæ nearly white.— Tentacula (with terminal eye) & folds coloured green.— the latter large & semicircular.—

Actinia[2]
(1278) Body cylindrical much elongated, loosely attached to broken fragments of shells, & buried deeply in tidal sand.— When mouth is retracted, quite cylindrical, 2-5 inches long. Color uniform most beautiful ~~Carmine~~ "Lake Red".— When ~~Mouth~~ flower is expanded, body slightly conical: 3 inches long.— Flower flat, fleshy, 2 inches |**322**| in diameter.— ~~Mantle~~ Snow white.— Mouth in centre, lips pale brown.— From this to circumference, flower obscurely radiated into narrow divisions.— These stuffed with small short papillæ.— [sketch in margin] Beneath the flower there is a collar: obscurely longitudinally lobed.— This forms division from the body.— Mouth is within longitudinally folded (some of these extended with fluid). Its brown color is joined to the white of flower by serrated edge.— From the centre of the mouth there is protruded a zigzag (like section of bud) delicate membrane; precisely as happened in the Caryophillia at the C. de Verde Is[ds].— Point of attachment narrow flat rough.— Whole animal most beautiful.—

Actinia

Onchidium[3]
1285 Ordinary length 1.3[inch], breadth 1[inch].— depressed, rounded oval.— Upper surface black with tint of green, beneath pale.— Mantle far surpassing foot, studded with very short round little elevations: edges very thin & entire:

consistance cartilaginous, very tough.— Tentacula greenish .25 in length, ~~are received~~ pass edge of mantle in a little groove, .1 in length, projects beyond mantle.— |**323**| The thin membranes fold beneath tentacula.

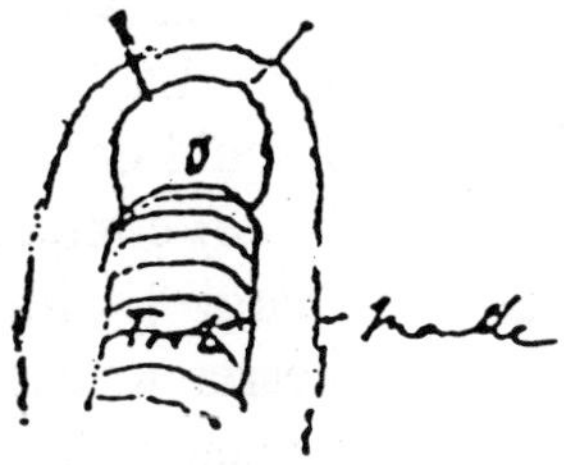

Onchidium nearly circular, broard, very large, slightly sinuated, but not bilobed.— longitudinal mouth in the lower part of surface.— Mouth can be protruded, consists of muscular tube. Respiratory tube at posterior end very long: between it & the extremity of foot are Anus & generative orifices.— I could not see the Male Organ.— When irritated emits a milky sticky fluid from under edge of mantle & from 10 to 12 longitudinal rows of pores on back or upper surface of mantle.— These rows are parallel & placed in two sets on each side of mantle. In each row there are from 16-20 pores: Pores distinct with minute round eminences. All this is well seen when animal is thrown in Spirits.— Animal inhabits tidal rocks, but different from all other species not on the under sides of loose stones: but on exposed solid surfaces. By the application of thin edge of mantle to all the small inequalities adhæeres very firmly, so as to require
(a) knife to detach them; at first [cont. at (a) on back of page] sight resembles a smooth Chiton.— Are numerous.— Crawl very slowly.— This is a very distinct species, if not subgenus of Onchidium.— |**324**|

Muricea[4] Gorgonia (3252) This Corall was taken out of 12 Fathom water: My examination imperfect owing to death of Polype & want of time.— Color dark reddish brown, described as being covered with Slime. Axis blackish brown in extremities, represented by pulpy matter: in all parts encrusted by a back of nearly same nature: When branches are thick is best seen to consist of a net work of red stony little spiculæ, round, tapering towards each end, about 5 times as long as broard, covered over irregularly with sharp points: [see sketch in margin]

These seem united by a ligamentous matter. This back rises in various points, to form the short conical cells.— The spiculæ here are placed more regularly & imbricated: On the inner, or concave side of back, there are numerous parallel narrow ridges formed of an interlacement of similar, but much more minute stony red tuberculated Spiculæ, which appear placed nearly longitudinally & actually united.— These ridges rise from amongst the coarse substance of back, & are covered by their
Muricea membrane, which lies on |**325**| the smooth horny axis. The intervals
(a) therefore between the ridges form so many distinct longitudinal vessels.— [note (a)] These channels are broarder than the ridges: in the main branches are very numerous, in the extremities few, such as 4 or 5.— In the dry back, by the shrinking of these channels the concave surface instead of being smooth is furrowed by the ridges.— [note ends] The cells seem to

lie over these at different lengths.—

The cells in the main branches have nearly a central orifice, but generally
the cell is hood shaped, the orifice being only a slit ~~up~~ under the curved
point & which looks towards extremities of branches. (This differs entirely
from Lamouroux's account[5] & must be reexamined in dead specimen).—
I could not find body of the Polypus (animal dead), but 2 or 3 (& perhaps
more) compound tentacula.— Each consisted of tapering fillet [sketch in
margin] with about 13 cylindrical ordinary papillæ on each side, those

(b) nearer to extremity thicker & longer.— [note (b)] These fillets would
appear to be contained in a thin case.— [note ends] Perhaps these fillets
are placed in pairs (2? 4? 6? or 8? in number) around mouth.— They arise
from a collar formed of parallel minute tuberculated red stony spiculæ,
Muricea precisely such as described |**326**| in the ridges only separate.— Perhaps
(a) by some process they ultimately do serve for this purpose.— [note (a)] The
Channels are distinct by air bubbles being made to travel up them.— In the
circumference of any part of branch they are more numerous than the cells
of the Polypi.— [note ends] I believe the Polypi of all those cells which
overlie any one channel must be intimately connected. The existence of
Polypi is imaginary I have seen no more than their tentacula.—

Gorgonia Growing just below low water mark: color of back fine "purplish R".—
(1306) orifice of cell lined yellow.— Structure irregularly interwoven &c &c &c:
(a) [note (a)] The expansions do not only grow in one plane but in various
directions.— [note ends] Cells very numerous, placed in main branches in
irregular rows.— Bark thick, consist<s> of numerous very small red stony
3264 dry elongated (rugose with points which are sometimes placed in concentric
specimen lines) grains, connected firmly by a ligamentous matter.— Very closely
resemble in this respect Muricæa.— In lower part of Main branches the
substance is traversed by parallel longitudinal small vessels — as in
Muricæa but not so clear. the vessels on rather clear spaces lie close to
under surface.— In the broard intermediate spaces of solid substance the
cells lie: Cells on rounded cavity, base only partially separated (by
Gorgonia membranes with the rugose red grains) |**327**| from the Horny axis.—
The vessel therefore separates rows of cells:— In extreme branches, this
structure is not visible: a mere horny fibre from the axis round which the
cells appear irregularly placed.— At very base of whole tree, the cells
disappear, the mouth becomes covered with a membrane in which the red
(a) stony grains appear by degrees.— [note (a)] It would appear that the Back
is the Living part.— is irritable & reproductive, irritable because orifice
~~closes~~.— There is little or no transmission of feeling from one to the
other.— I have omitted to state that within the saucer-shaped extremity of
tube I saw a rapid revolving motion of Particles.— V Alcyonium **P. 252** to
see resemblance in tube & ova.— [note ends]

The outer surface of horny axis appears under microscope to consist of flat brown fibres, slightly interwoven like the South Sea Islanders Cloth. The Polypiferous tube is coloured yellow, can securely be retracted within the orifice of cell.— This orifice, when tube is withdrawn, contracts into a longitudinal ellipse.— Tube short cylindrical, slightly expanded at summit.— Summit saucer shaped, octagonal: the projecting & rounded points lined on each side by by short minute Papillæ (8 or 10 in number).— Externally the collar is strengthened by few white stony striæ, which diverge from base of each point.— In centre large mouth, I believe in centre of tube the vessel is double: There is a strong resemblance |**328**|

Gorgonia

with Polypiferous tube in Alcyonium.— The Body in cell I could not examine: it is very delicate & simple: in many cells at base there were one or two oval eggs attached to side of cell.— [sketch in margin] in centre kydney shaped opake mass, point of attachment facing the concave side of this.— color pale.— Differs from Alcyonium[6] in tube being detached from external case.— But I suspect the two bear to each other somewhat same relation which Actinia does to Caryophillia.—

[1] Systellommatophora, Onchidiidae, a sluglike intertidal mollusc. *Onchidella steindachneri* is listed as endemic in the Galapagos by Yves Finet in Chapter 12, pp. 253-60, on 'Marine mollusks of the Galápagos Islands', in *Galápagos Invertebrates: Taxonomy, Biogeography and Evolution in Darwin's Islands*, edited by M.J. James. Plenum Press, New York, 1991.
[2] Sea anemone.
[3] Another of the Onchidiidae.
[4] Gorgonacea, horny octocorals. *Muricea* is often called a sea rod, and *Gorgonia* a sea fan.
[5] See *Lamouroux* pp. 36-7.
[6] Alcyonacea, a soft octocoral.

[**CD P. 328** continues]

Tortoise

This animal is, I believe, found in all the Islands of the Archipelago; certainly in the greater number.— ~~They swarm in the greatest numbers in~~ [next sentence corrected in pencil] The Tortoises frequent in preference the high & damp parts, but they occur likewise in the low & arid districts.— It is said that slight variations in the form of the shell are constant

Copy

according to the Island which they inhabit — also the average largest size appears equally to vary [pencil insert] according to the locality.— Mr Lawson[1] states he can on seeing a Tortoise pronounce with certainty [pencil insert] from which island it has been brought.— The Tortoises grow to a very large size: there are some which require 8 or 10 men to lift them:

Tortoise

|**329**| The old Males are the largest.— the females rarely grow to so great a size.— The male can readily be told from the females by the greater length of its tail.— The Tortoises which live on those Islands where

(a)

there is no water, or in dry parts of others, live chiefly on the succulent Cactus: I have seen those which live in the higher parts, eating largely of a pale green filamentous Lichen, which hangs like presses from the boughs of the trees, also various leaves & especially the ~~berrys~~ berries of a tree (called Guyavitas) which are acid & Austere.— [note (a)] The dung of the Tortoise is very large & resembles that of the S. American Ostrich. [note ends]

The Tortoise is very fond of Water, & drinks large quantities & wallows in the mud.— Even those which frequent districts far removed from the water travel occassionally to it; they stay two or three days near the Springs & then return.— My informants differed widely in the frequency of these visits.— It seems however certain that they travel far faster than at first
Tortoise would be imagined.— |**330**| They ground their opinion ~~of~~ on seeing how far ~~a~~ some marked animal has travelled in a given time. They consider they would pass over 8 miles of ground in two or three days.— One large one, I found by pacing, walked at the rate of 60 yards in 10 minutes, or 360 in the hour.— at this pace, the animal would go four miles in the day & have a short time to rest.— When thus proceeding to the Springs, they travel by broard & well-beaten tracks, which branch off to all points of the Isl^{d}.— I should have prefaced that in these Is^{ds} there are only a few watering places & these only in the highest & central parts.— When first I landed at Chatham Is^{d}; the object of these tracks was to me inexplicable.— The effect in seeing such numbers of these huge animals, meeting each other in the high-ways, ~~was some~~ the one set thirsty & the other having drunk their fill, was very curious. When the Tortoises arrive at the water, quite heedless of spectators they greedily begin to |**331**| drink: for this purpose
Tortoise they bury their heads to above their eyes in the mud & water ~~above their eyes~~ & swallow about 10 mouthfulls in the minute.— The inhabitants when very thirsty sometimes have killed these animals in order to drink the water in the Bladder, which is very capacious. I tasted some, which was only slightly bitter.— The water in the Pericardium is described as being more limpid & pure[2].—

The female Tortoise generally places her eggs in groups of four or five in number & covers them up with earth. Where the ground is rocky she ~~places~~ drops them indiscriminately.— Mr Bynoe[3] found 7 eggs laid along
(a) in a kind of crack.— The egg is quite sphærical.— [note (a)] The circumference is 7 & ⅜th inches. the eggs are white & hard.— *[line then crossed out and second version substituted in pencil]* White & hard, the circumference of one was 7 & ⅜ of inch [note ends] The young Tortoise, during its earliest life, frequently falls a prey to the Caracara, which is so common in these islands. The old ones occassionally meet their death by falling over precipices: but the inhabitants have never found one dead from Natural causes. The Males copulate with the female in the manner of a frog.— they remain joined for some hours.— During this time the Male
Tortoise utters a hoarse roar or |**332**| bellowing, which can be heard at more than 100 yards distance.— When this is heard in the woods, they know certainly

that the animals are copulating.— The male at no other time, & the female never, ~~utters~~ uses its voice.— There are now, in the beginning of October, eggs in the ground & in the belly.— The people believe they are perfectly deaf; certainly when passing ~~one~~ a tortoise, no notice is taken till ~~the animal~~ it actually sees you:— then drawing in its head & legs & uttering a deep hiss, he falls with a heavy sound on the ground, as if struck dead.—

The people employ the meat largely, eating it both fresh & salt, & it is very good.— The meat abounds with yellow fat, which is fryed down & gives a beautifully clear & good oil[4].— When an animal is caught, a slit is made in the skin near the tail to see if the fat on the dorsal plate is thick; if it is not the animal is liberated & recovers from the wound.— if it is thick it is killed by cutting open the breast plate on each side with an axe &
(a) removing ~~slicing~~ from the living animal the serviceable parts of the Meat & liver &c &c. |**333**| [note (a)] In order to secure the Tortoises, it is not sufficient to turn them like a Turtle, for they will frequently re~~sume~~ gain their proper position.— [note ends]

[1] Nicholas E. Lawson was an Englishman serving the Republic of the Equator (Ecuador) as Governor of the Galapagos Islands.
[2] In *Journal of Researches* **1**:463-4 CD says 'I believe it is well ascertained, that the bladder of the frog acts as a reservoir for the moisture necessary to its existence: such seems to be the case with the tortoise.' It would be interesting to discover what his authority was for this statement with respect to the bladder of the frog, for it does not appear to be his favourite the *Dic: Class:*, nor any other book in the *Beagle* library. But it might have been added after his his return to England.
[3] Benjamin Bynoe was the Acting Surgeon on the *Beagle*.
[4] CD very often commented on the gastronomic quality of the flesh of the animals that he encountered, and in his journal he wrote (see *Beagle Diary* p. 362): 'The Breast-plate with the meat attached to it is roasted as the Gauchos do the "Carne con cuero". It is then very good.— Young Tortoises make capital soup — otherwise the meat is but, to my taste, indifferent food.—'

[**CD P. 333** commences]

Ambly= <u>rhynchus</u>[1] (1305)

The Lizard which bears this name is said in the Blonde's[2] Voyage to have been described from a specimen brought from the shores of the Pacifick.— This animal is excessively abundant on all the Islands in the whole group.— It never proceeds many yards inland from the rocky sea beach: There, on the large fragments of black Lava, groups may be seen basking with outstreched legs.— They are hideous looking animals; stupid & sluggish in their motions. Their color is black, their general size rather more than 2 ft long. On Albermale Is[d] they appear to grow ~~very~~ much larger than in any other place, one weighed 20 £b.— I saw <u>very few</u> small ones: so that I suppose their breeding season is ~~past~~ now coming on: I could not hear ~~of~~

Ambly= Rhynchus

any particulars respecting their manner of breeding.— These animals have occassionally been seen some hundred yards at sea, swimming.— The structure of their bodies points out aquatic habits. Yet it is remarkable, that when shuffling over |**334**| the tidal rocks it is scarcely possible to drive them into the water. From this reason, it is easy to catch them by the tail, after driving them on a point.— They have no idea of biting, & only sometimes when frightened squirt a drop of fluid from each nostril[3].— Having seized a large one by the tail, I threw ~~him~~ it several times ~~into~~ a good distance into a deep pool left by the retiring tide.— Invariably ~~the Lizard~~ it returned ~~to the~~ in the same direction from which it was thrown to the spot where I stood. Its motion was rapid, swimming at the bottom of the water & occassionally helping itself by its feet on the stones.— As soon as it was near the margin, it either tried to conceal itself in the sea-weed or entered some hole or crack. As soon as it thought the danger was over it crawled out on the dry stones, & again would sooner be caught than voluntarily enter the water.— What can be the reason of this? are its habitual enimies |**335**| sharks or other <u>marine</u> animals? The manner of swimming is singular, consisting ~~entirely~~ solely in a wriggling motion of tail & body, the legs being motionless, collapsed & stretched out behind.—

Ambly <u>Rhynchus</u>

I opened the stomach (or rather duodenum) of several, it was largely distended by quantities of minced pieces of sea-weed, of that kind which grows in <u>thin</u> foliaceous expansions of a bright green & dull red color.— There was not a trace of any animal matter: Mr Bynoe, however, found a piece of a Crab in them: this might have entered accidentally, in a like manner as I have seen a Caterpillar in the stomach of the Tortoise.— I conceive the largeness of the intestine is in perfect agreement with its herbivirous appetite.— Capt Colnett[4] states they go out to sea in shoals to fish: I cannot believe this is the object, nor is it very clear what their object can be.— Does such sea-weed grow more abundantly a little way from the coast? They appear to be able to survive a long time without [continued at (a) on back of page] breathing.— One was sunk with a weight for nearly an hour, & was then very ~~lively~~ active in its motions.— Their limbs are well adapted for crawling amongst the rough & fissured rocks of Lava, & we have mentioned that with their tail & body they can swim well.— |**336**|

(B)
(a)

[notes added later in pencil] ~~Neither species were known by the inhabitants of Tahiti~~ (B) Does not the Manatee of the West Indian ocean feed on such seaweed? [notes end]

[**CD P. 336** commences]

<u>Ambly</u>
<u>Rhynchus</u>
terrestrial[5]
(1315)

This animal clearly belongs to the same genus as the last.— it being a terrestrial, whilst the other is an aquatic species. They are found only in the central division of the Islands, viz. Barrington, Indefatigable, Albermale & James Is^{d} — to the North in Charles, Hood or Chatham, & to the South in

Tower, Bindloes & Abingdon, I neither heard of, or saw one.— They frequent ~~both~~ in the above Islands both the upper, central & damp parts as well as the lower dry ~~arid~~ sterile districts: in the latter kind of soil their numbers are more especially abundant.— I cannot give a better idea of this than by stating we had difficulty in finding a piece of ground free from their burrows large enough to pitch our tents.— They are ugly animals, & from their low facial angle have a singularly stupid appearance.— Capt. FitzRoy['s] specimens will give a good idea of their size.— Their colors are, whole belly, front legs, head "Saffron Y & Dutch orange" — upper side of head nearly white.— Whole back behind the front legs, upper side of hind legs & whole tail "Hyacinth R". This in parts is duller, in others brighter passing |**337**| into "Tile R".— I have seen a few individuals, especially the younger ones, quite sooty on the whole upper side of their bodies.—

Ambly Rhynchus terrestrial

They are torpid slow animals, crawling when not frightened with their belly & tail on the ground.— frequently they doze on the parched ground, with their eyes closed & hind legs stretched outwards.— In none of their motions, is there that celerity & alertness which is so conspicuous in true Lacertas & Iguanas.— Their habits are diurnal: they seldom ~~leave~~ wander to any distance from their burrows: when frightened they rush to them with a most awkward gait: excepting going down hill their motion, from the lateral position of their legs, is not quick.— They are not timorous. When attentively watching an intruder they curl their tails, & raising themselves as if in defiance on their front legs, vertically shake their heads with a quick motion.— I have seen small Muscivorous Lizards perform the same gestures.— This gives them rather a fierce aspect, but in truth they are far the contrary. When however |**338**| being caught & plagued with a stick they will bite it severely.— Two being placed on the ground close together will fight & bite each other till blood is drawn.—

Amblyrhyncus terrestrial

As I have said they all inhabit burrows, these they make sometimes between the fragments of Lava, but more generally in the ground, composed of Volcanic Sandstone.— The burrows do not appear deep & enter at a small angle: hence when walking over the "warrens" the soil perpetually gives way.— When excavating these holes, the opposite sides of the body work alternately; one front leg scratches the earth for a short time & throws it towards the hind. this latter is well placed so as [to] heave the soil beyond the mouth of hole.— the opposite side then takes up the task.— Those individuals & they are the greater number, which inhabit the extremely arid land, ~~can~~ never drink water during nearly the whole year.— These eat much of the succulent Cactus, which is in evident high esteem. When a piece is thrown towards them, each will try to seize & carry it away as dogs do with a bone.— |**339**| They eat however deliberately, without chewing the pieces.— The Cactus is in request amongst all animals, I have seen little birds picking at the opposite end of a piece which a Lizard was eating:

Ambly-rhynchus terrestrial

& afterwards it would hop on with complete indifference on its back.— In their stomachs ~~I ha~~ vegetable fibres, leaves of different trees, especially the Mimosa were always found. In the high damp country their chief food is the berry called Guayavitas[6]; it is the same which the Tortoises eat, & has an acid astringent taste.— Here also they are said to drink water.— To obtain the leaves they climb short heights up the trees: I have frequently seen them clinging to the branches of the Mimosa.— Thus their habits are as entirely herbivorous as in the black sea-kind.— The meat when cooked is white & esteemed, by those who can bring their stomachs to such a
(B) regimen, good food.— I observe the pores on under sides of hind thighs are very large. by pressure a cylindrical organ is protruded to the length of
(a) some tenths of an inch.— |**340**|

[notes added later] (a) At this time of year (end of September & beginning of October) the females have numerous large elongated eggs.— These they lay in their burrows & the inhabitants seek for them to eat.—

Is any other genus amongst the Saurians Herbivorous? I cannot help suspecting that this genus, the species of which are so well adapted to their respective localities, is peculiar to this group of Is[ds].—

[note in pencil crossed through and incomplete] The Inhabitants of Tahiti had never seen or heard of

(B) [in pencil] Humboldt remarks that in intertropical S. America all Lizards which inhabit dry regions are esteemed as delicacies for the table.— [notes end]

[1] Listed in *Zoology* **5**:23 as *Amblyrhynchus cristatus* Bell, which is still its modern name.
[2] See George Anson Byron, 7th Baron. *Voyage of H.M.S. Blonde to the Sandwich Islands, in the years 1824-25.* London, 1826. On her way to Hawaii, the *Blonde* called at Narborough Island on 27 March 1825, where 'an innumerable host of sea-guanas' was found. A footnote on p. 92 of the book states: 'Amblyrhyncus Cristatus — described by Bell from a specimen brought to Europe by Mr Bullock among his Mexican curiosities. Mr. B. does not state the spot where it was found: probably on the Pacific shore.'
[3] The marine iguana rids itself of the excess of salt in its diet by means of a nasal salt gland, whose secretions account for this fluid.
[4] See James Colnett. *A voyage to the South Atlantic and round Cape Horn . . . .* London, 1798. On p. 56 of his book about his voyage in the Rattler, he wrote: 'The sea guana is a non descript: it is less than the land iguana and much uglier, they go in herds, a fishing, and sun themselves, on the rocks, like seals, and may be called alligators, in miniature.'
[5] Listed by Thomas Bell as *Amblyrhynchus Demarlii* Bibr. in *Zoology* **5**:22, where an engraving of the land iguana brought back by CD was published. The modern name of the species is *Conolophus subcristatus*. It exists today on Fernandina (Narborough), Isabela (Albemarle), Santa Cruz (Indefatigable), South Plaza, Baltra (South Seymour), and has been

introduced elsewhere, but in recent years has become extinct in Santiago (James) where CD saw it. A second species, *Conolophus pallidus*, not actually seen by CD, is found only in Santa Fe (Barrington).

[6] This tree is *Psidium galapageium* Hook. See *Plant Notes* p. 185.

[**CD P. 340** commences]

Ornithology I believe the collections of birds formed by Mr Bynoe, Capt. FitzRoy & myself will ~~show~~ give a nearly perfect series of the birds[1].— At this time of year (end of Septemb & beginning of Octob), from the state in which the birds appeared to be I should imagine the ~~young ones~~ last years produce had nearly attained perfect plumage.— In no female of the smaller birds, the eggs in the Ovarium were much developed.— The Ornithology is manifestly S. American.— Far the preponderant number of individuals belongs to the Finchs[2] & Gross-beaks[3].— There appears to be much difficulty in ascertaining the Species.— My series would tend to show that only the old Cocks possessed a jet black plumage: but Mr Bynoe & Fuller[4] have each a small black female bird.— Certainly the numbers of brown &
3330 blackish ones is immensely great to those perfectly black.— Species as in
3331 margin are well characterized.— I only saw them in James Is^{d} & in one place. they were there however numerous, feeding with the various other species. Mr Bynoe has a much blacker specimen.— I should state that all the Species (& doves) feed together in great numbers indiscriminately, their favourite resort being in the dry long grass in the lower & dry parts of
Ornithology Island, where in the soil many |**341**| seeds are lying dormant.— The
(a) Icterus like Finch[5] (3320...23) is distinct in its habits: its general resort is hopping & climbing about the Cactus trees, picking with its sharp beak the flowers & fruit.— not infrequently however, it alights on the ground & feeds with the flocks of other species.— Out of the <u>many</u> specimens which I have seen of this bird, the only one which was black (3320) I by good fortune procured.— Mr Bynoe has one other.— I have no doubt respecting its identity: for it was shot with the others on the Cactus: This is an illustration of the comparative rarity of the black kinds.— [note (a)] The Gross-beaks are very injurious. the[y] will strike seeds & plants when buried 6 inches beneath the ground [note ends]

The insectivorous birds are comparatively rare: they are equally found in the low dry country & high damp parts.— I was astonished to find amongst the luxuriant damp vegetation an exceeding Scarcity of insects (so much so that the fact is very remarkable) This being the case, it is no
(B) wonder that the above order of birds should be scarce.—

[note (B) added later] I neither saw or could hear of humming birds in any of the Is^{ds}. [note ends]

[**CD P. 341** continues]

3348, 3349 This birds[6] which is so closely allied to the Thenca of Chili (Callandra of
3306, 3307 B. Ayres) is singular from existing as varieties or distinct species in the
<u>Ornithology</u> different Is^{ds}.— |**342**| I have four specimens from as many Is^{ds}.— These
(a) will be found to be 2 or 3 varieties.— Each variety is constant in its own
Island.— [note (a)] The Thenca of ~~Chatham Is^{d}~~ Albermale Is^{d} is the same as that of Chatham Is^{d}.— [note ends] This is a parallel fact to the one mentioned about the Tortoises. These birds are abundant in all parts: are very tame & inquisitive: habits exactly similar to the Thenca.— <u>runs</u> fast, active, lively: sings tolerably well, is very fond of picking meat near houses, builds a simple open nest.— I believe the note or cry is different from that of Chili.—

(3374) This bird[7] was shot by Fuller on James Is^{d}: it was only one seen during our
Anthus whole residence here.— It is described as rising from the ground out of dry grass & settling again on the ground.— Showed very long wings (like a Lark) in its flight & uttered a peculiar cry.— Its structure appears very curious & interesting.— Connects Anthus & Fringilla. The body is preserved in Spirits (1309) for Dissection.

[1] This belief, also expressed in a marginal comment in pencil at the foot of **CD P. 343**, was hardly justified, for as has been pointed out by Frank Sulloway in 'Darwin and his finches: the evolution of a legend' (*Journal of the History of Biology* **15**:1-53, 1982) and other articles, CD failed to record with his usual care the exact islands on which some of his finches had been collected, and later had to borrow further specimens from FitzRoy and others, creating in Sulloway's words 'a considerable nightmare of taxonomic problems for subsequent ornithologists'.

[2] In CD's Specimen Lists (see p. 000) and *Ornithological Notes*, 21 birds later classified by Gould in *Zoology* **3**:98-106 among the Geospizinae were described as finches or "Fringilla", 4 as "Fringilla/Gross-beaks", 4 as "Icterus", and 1 as "Wren" or warbler. As explained by Sulloway (*loc. cit.*), Gould named altogether thirteen species of Darwin's finches, but four of these have since been recognized as variant forms. Those called finches by CD were the medium and small ground finches *Geospiza fortis* and *G. fuliginosa* (NHM 1855.12.19.167), the sharp-beaked ground finch *G. nebulosa*, the large and small insectivorous tree finches *Camarhynchus psittaculus* (NHM 1855.12.19.12 and .22) and *G. parvula*, and the vegetarian tree finch *C. crassirostris*.

[3] The four birds called Gross-beaks by CD were *Geospiza magnirostris*, of which Gould's *G. strenua* is a subspecies. CD claimed that the specimens with the largest beaks came from Chatham and Charles Islands, but *G. magnirostris* has not been found since on those two islands, where either the form has since become extinct, or evolution of beak size has taken place in the manner described by Peter Grant in his book *Ecology and Evolution of Darwin's Finches* (Princeton University Press, 1986).

[4] The character of the finches that most impressed CD by its variability was their degree of

blackness, now known to be dependent on age, but subsequent studies have shown that it is the size of their beaks that is most variable, though it does not change with age.

[5] The classification of this species, listed in *Zoology* **3**:104-5 as *Cactornis scandens*, and pictured by Gould feeding on a cactus, greatly puzzled CD from their resemblance not so much to a finch (Fringilla), as to another family (Icterus) that included orioles, meadowlarks and blackbirds. In the Specimen List (see p. 414) he wrote Icterus (??) against (3320...23). One of them is NHM 1855.12.19.15 type.

[6] Listed in *Zoology* **3**:62-3 as *Mimus trifasciatus* Gray, from Charles Island, *Mimus melanotis* from Chatham and James's Islands (NHM 1855.12.19.223 type), and *Mimus parvulus* from Albemarle Island (NHM 1855.12.19.92 type). It may be noted that CD had concluded at this date (October 1835) that each species was constant within its own island, leading nine months later (see *Ornithological Notes* p. 262) to his first statement of doubts as to the stability of species. However, it has been pointed out by Sulloway (*loc. cit.*) that later ornithologists have found the differences between the mocking birds in the island populations to be less distinct than was thought by CD and Gould.

[7] Listed in *Zoology* **3**:106 as *Dolichonyx oryzivorus* Swains, a migrant American bobolink. Specimen (3374) bears one of the few surviving labels written in CD's own hand. Harry Fuller was a marine in the crew of the *Beagle*, and was evidently a particularly good shot.

[**CD P. 342** continues]

3351 52: 53 (B)	This small Water Hen[1] is found in high damp central parts of Charles & James Is[d].— It frequents in numbers the damp beds of Carex & other plants, uttering loud & peculiar Crys.— There is no water in these parts, but the land is damp. is called Gallinito del Monte.— |**343**| [note (B)] Iris bright scarlet; lays from 8-12 eggs.— [note ends]
Ornithology 3356	This swallow[2] was only seen in no numbers at one point of James Is[d]. frequents bold precipices on the sea coast.—
3297 3298	Caracara[3], specimens Cock & young female. The old female is much browner on the breast: M[r] Bynoe has a specimen. (where eggs in Ovarium were very large).— As in C. Novæ Zælandæ at the Falklands the individuals with plumage like (3298) were in far preponderant numbers. I believe upwards of 30 were counted near our tents without one dark one.— Habits similar to rest of genus.— Tame, bold, sit watching on the trees when a Tortoise is killed.— Noisy, crys Very different, one Very like the C. Chimango.— Can run fast.— are carrion eaters — build in trees.— not elegant & swift on the wing. Are said to kill chickens, doves & the very young Tortoises.— They are very abundant & will eat almost anything.—

[**CD P. 343** continues. First entry inserted later in pencil]

S. America Perfect Collect. (Humming Birds)[4]

3303 There are no true Hawks — Owls are abundant. ~~Besides my species (3303) Fuller has a blackish kind.~~ With respect to the Land Birds, their extreme
(a) tameness has been described in my private Journal[5].— Little birds can be almost caught by the hand, they will [continued at (a) on back of **CD P.**
(a) **343**] alight on your person & drink water out of a basin held in your hand.— Must not this arise from the entire absence of all Cats & other similar animals & those hawks which pursue small birds? [added later in pencil] Big Tortoises.— |**344**|

Ornithology: 3299 Amongst the Marine birds we have one Duck[6] which frequents the salt Lagoons, as does a Heron[7] (3296).— There are two kinds of Bittern (3300:01)[8] on the rocks on the sea-coast.— There is one Gull[9] & one Tern both of which are common.— There is Flamingo in the Salinas.— On the shore there are several small waders[10]. Mr Bynoe & Fuller have some other species: I believe this is the most indifferent part of our collection.— At sea we have the little Mother Carys Chicken.— Procellaria (3190) & other species.— The great Pelican & common gannet as at Callao & other species of latter, beautifully white & black.—

There is also the Frigate Bird[11].— There is one part of the habits of this bird which has not been sufficiently described; it is the manner in which this bird picks up fish or bits of meat from the surface of the water without wetting even its feet.— I never saw one alight on the water.— Like an arrow the bird descends from a great height with extended head, by the aid
(a) of its tail & long wings turns with extraordinary dexterity at the moments of seizing its object with its long beak.— [note (a)] It is a noble bird seen on the wing, either when soaring in flocks at a stupendous height, or ~~when~~ as showing the perfect skill in evolutions when many are darting at the same floating morsel.— If the piece of meat sinks above 6 inches deep it is lost.— [note ends]

[1] Listed in *Zoology* **3**:132-3 as *Zapornia spilonota* Gould.
[2] Listed in *Zoology* **3**:39-40 as *Progne Modesta* Gould.
[3] Listed in *Zoology* **3**:23-5 as *Craxirex Galapagoensis* Gould.
[4] Copied in *Ornithological Notes* p. 265 as 'there are no Humming Birds'. Both these words and 'S. America Perfect Collect:' in the margin were added in pencil.
[5] See *Beagle Diary* p. 353. In *Ornithological Notes* pp. 265-6, CD's remarks about the tameness of the birds are greatly extended.
[6] Listed in *Zoology* **3**:135 as *Pœcilonitta Bahamensis* Eyton.

[7] Listed in *Zoology* **3**:128 as *Ardea herodias* Linn.
[8] One of them listed in *Zoology* **3**:128 as *Nycticorax violaceus* Bonap.
[9] Listed in *Zoology* **3**:141-2 as *Larus fuliginosus* Gould.
[10] The waders listed in *Zoology* **3**:128-32 are *Hiaticula semipalmata* Gray, *Totanus fuliginosus* Gould, *Pelidna minutilla* Gould, and *Strepsilas interpres* Ill.
[11] See also *Ornithological Notes* p. 267 for a revised version of this account of the flight of the frigate bird. But although both there and in *Zoology* **3**:146 CD questioned the usefulness of the webbing between the toes of the frigate bird, he surprisingly did not mention this species in *Journal of Researches* **1** or **2**. He nevertheless returned to the subject in the *Origin of Species* p. 185, where he said that 'no one except Audubon has seen the frigate-bird, which has all its four toes webbed, alight on the surface of the sea' (see pp. 495-502 in Vol. III of Audubon's *Ornithological Biography*, Edinburgh, 1835), and decided that 'in the frigate-bird, the deeply-scooped membrane between the toes shows that structure has begun to change'.

[**CD P. 345** commences]

Fungia[1] (1334) I kept these Specimens for a short time alive. I observe, that the tentacula are placed at the inner extremity of each ray.— The tentacula are short, are surmounted by a slightly enlarged head, are sometimes expanded with water, are very sensitive, are seated on the upper rounded part of the extremity of the ray.— from them vessels converge towards the mouth.— Each ray or plate has only one Tentaculum.— as these plates are very numerous towards the margin there the tentacula are most numerous.— None are seated on the lower side.—

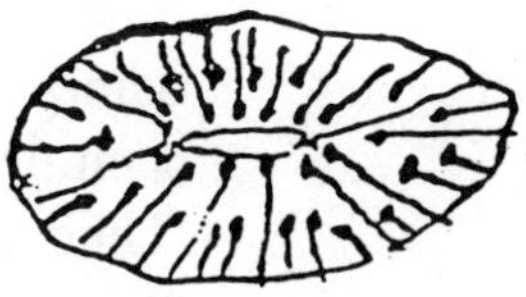

(a) The mouth has longitudinal folds on its inner lip.— [note (a)] The form of mouth & indeed whole animal identical with Actinia. [note ends] Within the cavity may be seen that delicate kind of folded drapery as described in Caryophyllia. In some points of the fleshy substance between the plate[s] which were cut & injured, similar delicate membranes were protruded.— This fleshy part, when undisturbed, projects equally with the margin of the plates.— It has considerable powers of contracting & motion.— either vertically downwards or towards the central mouth.— The Tentaculum being touched, whole of soft part contracts itself. Found in shallow water <u>within</u> the reef.— |**346**|

[1] Scleractinia, a solitary reef-inhabiting stony coral.

[Although the *Beagle*'s visit to Hobart actually took place in February 1836, CD entered the year incorrectly as 1835 in his page headings until arriving at Bahia on 1st August 1836. To avoid confusion, this mistake has been put right.]

[**CD P. 346** commences in Hobart]

Lizard[1] (1358) February.— Scales on centre of back, light greenish brown, edged on sides with black; scales on upper sides of body greyer & with less black; on lower sides reddish: belly yellow with numerous narrow irregularly waving transverse lines of black.— these lines are formed by the lower margin of some of the scales being black: Head above grey, beneath whitish. Motion of the body when crawling like a Snake.— not very active: in stomach beetles & Larvæ: common in open wood:

Lizard 1359 Two longitudinal black bands, marked with chain of yellowish white spots; upper parts of sides irregularly black with do marks: belly whitish. tail simply brown.— soles of feet pale-coloured.—

Lizard 1366 Above pale brown, with very numerous little transverse undulating irregular black narrow bands: sides richer brown,— tail same as body but paler: soles of feet black

Lizard 1361 Whole upper surface dark blackish brown, each scale with 4-6 most minute longitudinal streak.— (The black color far preponderant) Belly reddish: throat white: soles of feet black:— |**347**|

Lizard[2] 1362 Same genus as (1358).— Color — slightly dark "Wood Brown" with central longitudinal band crossed by about 5 broard very irregular bands of "Umber Br": tail with do & generally darker.— Beneath paler with most obscure undulating black lines: top of head reddish Br: Iris orange, pupil black:— Animal so torpid & sluggish a man may almost tread on it, before it will move.— I lay down close to one & touching its eye with a stick it would move its nictitating membrane & each time turn its head a little further; at last turned its whole body, when upon a blow on its tail ran away at a slow awkward pace like a thick snake, & endeavouring to hide itself in a hole in the rocks.— Appears quite inoffensive & has no idea of biting: held by the tail, collapses its front legs close to body & posteriorly.— Stomach capacious, full of pieces of a white Mushroom & few large inactive Beetles such as Curculios & Heteromerous: Hence partly Herbivorous!— not uncommon on sunny grassy hills:— Tongue colored fine dark blue.— |**348**|

Snake[3] 1363 Above colored "Hair Brown with much Liver B^{r}".— beneath mottled Grey.— The abdomen being burst in catching the animal: a small snake appeared from the disrupted egg: Hence Ovoviparous: Is not this curious in Coluber?—

Lizard 1364 Along the back a space ash coloured, which contracts over the loins; in centre of this, chain of transverse marks connected together of the richest brown: Within these marks, white spots & central pale brown line down whole back:— sides mottled with all the above colours: Belly ash, with few minute longitudinal dark streaks: Head with transverse ones of the dark brown: common: I believe also at Sydney.— |**349**|

[1] An oak skink, listed by Thomas Bell in *Zoology* **5**:30 as *Cyclodus Casuarinæ* Bibr.
[2] A blotched blue-tongued lizard, identified by F.W. and J.M. Nicholas in *Charles Darwin in Australia* (Cambridge University Press, 1989) as *Tiliqua nigrolutea*.
[3] Identified by F.W. and J.M. Nicholas (*loc. cit.*) as either a black tiger snake (*Notechis ater*) or a copperhead (*Austrelaps superba*), which are not oviparous and non-venomous colubrids, but viviparous and venomous Elapidae.

[**CD P. 349** commences]

Conferva[1] (B)

March 18th. The Ship being about 50 miles West from Cape Leeuwin, observed the sea covered with particles as if thinly scattered over with fine dust.— Some water being placed in a glass; with an ordinary lens, the particles appeared like equal sized bits of the fibres of any white wood.— On examination under higher powers, Each particle is seen to consist of from 10-15 of cylindrical fibres. These are loosely attached side by side all together; their extremities are seldom quite equal, a few projecting at each end.— The bundle was about 1/50th of inch in length, but ~~each~~ any separate fibre rather less, perhaps 1/60th.— The color, a very pale brownish green.— Each separate fibre is perfectly cylindrical & rounded off at both extremities, its diameter is as nearly as possible 2/3000 of inch; the whole is divided by transverse partitions [sketch in margin], which occur at regular intervals being about half the diameter of the fibre. ~~Between~~ Within the cells granular matter is contained; but my microscope scarcely sufficed for this.— Extremities colourless, with little or no granular matter.— The bundles must, I think, be enveloped in some adhæsive matter, because in a glass on touching the sides they almost always adhere.— The ~~number~~ ~~quantity~~ extent of sea covered by this Conferva was not very ~~extensive~~ great.— The morning was calm.— Vide similar account near the Abrolhos.— |**350**|

(a)

[notes added later] (B) On passage from Mauritius to C. of Good Hope Lat 37°30′, sea with the green flocculent tufts & sawdust during two calm day in very great quantities. Must be a most abundant Marine production.

(a) Humboldt[2] (Pers. Narr: Vol VI, P 804) mentions in the W. Indian sea, that the water was covered with a thin skin composed of fibrous particles; states is found in the Gulf Stream; channel of Bahama, & B. Ayres.— Are these fibrous particles the kind of Conferva here described? Did I not on coast of Brazil, however, myself see some real fibro-gelatinous particles[3]?—

A similar appearance is noticed by Capt King[4] on NW extremity of N. Holland. called by Capt Cooks sailor +++ "sea saw dust" a very good name.— Hawkesworth[5] Vol III, P 248.— & M. Peron (who will describe it) Voy. Vol II Chapt: 31.— +++ Cooks 1st Voy. II Vol. Chapt VII. is described as a Conferva.—

1 Identified in *Plant Notes* p. 216 as probably *Oscillatoria erythraea.*
2 See Alexander von Humboldt. *Personal narrative of travels to the equinoctial regions of the new continent . . . 1799-1804 . . . translated into English by Henrietta Maria Williams.* 7 vols. London 1814-29. In *Beagle* Library.
3 See pp. 66-8 for CD's observations at the Abrolhos Shoals in March 1832.
4 See Philip Parker King. *Narrative of a survey of the intertropical and western coasts of Australia.* 2 vols. London, 1827. In *Beagle* Library.
5 See John Hawkesworth. *An account of the voyages . . . performed by Commodore Byron, Captain Wallis, Captain Carteret, and Captain Cook . . . drawn from the journals which were kept by the several commanders . . .* 3 vols. London, 1773. In *Beagle* Library.

[**CD P. 350** commences]

Conferva[1]

(A)[During two days before arriving at the Keeling Isds in the Indian Ocean [on April 1st 1836], in many parts I saw masses of flocculent matter of an extremely pale brownish green colour floating in the sea. They varied in size from half to three or four inches square ~~in size are~~ and were quite irregular in figure. ~~& are coloured an extremely pale brownish green.~~ In an opake vessel the masses could only with difficulty can~~not~~ be distinguished; but in a glass they were ~~are very distinctly~~ clearly visible. Under the microscope the flocculent ~~masses are~~ matter is seen to consist of two kinds of Confervæ, between which I am quite ignorant whether there is any connection. Minute cylindrical bodies, conical at each extremity, are involved in vast numbers in a mass of fine threads. These threads have a diameter of about 2/3000th of an inch; they possess an internal lining; they are divided at irregular & very wide intervals by transverse septa; Their length is extreme, so that I could never certainly ascertain the form of the extremity; They are all curvilinear & resemble in position a handful of hair, coiled & squeezed together. In the midst of these threads & probably connected by some viscous fluid there are innumerable cylindrical hollow transparent bodies [sketch in margin]; each extremity of which is terminated by a cone produced into the finest point.— Their diameter is tolerably

(a)

[cont. at (a) on back of **P. 350**] constant between ~~6 and 8/1000~~ .006 and .008 of an inch. Their length varies considerably from .04 to .06 & even sometimes to .08.— Near to ~~the~~ one extremity of the cylindrical part, a green septum ~~or mass of a granular matter~~ formed of a granular matter, and thickest in the middle, may generally ~~to~~ be seen.— This I believe ~~to be~~ is the bottom of a most delicate colourless sack, composed of a ~~granular or~~ pulpy ~~matter~~ substance which lines the exterior case, but does not extend to within the extreme conical points. *[passage with similar wording erased]* In some, a small but perfect sphære of brownish granular matter supplied the place of the septum; & I observed the curious process by which these little balls are produced. A]][2]

The pulpy matter of the internal coating suddenly grouped itself into lines, so<me of> which assumed an obscure radiated position, then with irregula<r> & rapid movement the lining contracted & united itself, & in a second the whole matter was collected into the most perfect little sphære, which motionless occupied the position of the sept<um> at one end of the transparent hollow case. I can describe these motions by a simile: a bag of unequal thickness & compos<ed> of some highly elastic matter being distended by a fluid, & then such fluid being allowed to escape with some rapidity, the coats of the bag would contract & unite with similar movements.— This rapid process perhaps is a morbid one, owing to injury: certainly in many cases with such injury the process commenced.— I saw several pair of these bodies attached to each other, cone along side cone, at that end where the Septum occurs.— I do not know whether they constantly adhære in this manner when floating in the ocean.— [see sketch in margin] |**351**|

[1] Identified in *Plant Notes* pp. 216-17 as probably another blue-green alga (Cyanophyta) of indeterminate genus.
[2] The passage marked with capital A's and double brackets has been extensively revised at a period when CD was no longer using ampersands. It was reproduced with little further revision in *Journal of Researches* **1**:14-16, and was illustrated with the drawing of the method of attachment of the cones that appears above.

[**CD P. 351** commences]

<u>Meandrina</u>[1]

3605

The surface of this Coral is marked with sinuous ~~convex~~ elevated & ~~concave~~ depressed lines: a transverse section shows plates which form the lines of ridges & hollows.— Those which rim the latter are the thickest & best developed, & ~~hence~~ the mass of Coral breaks most easily in these lines. On each side there are irregular cells; & likewise on each side of those plates, which form the ridges of elevated lines, there are more regular cells. Hence we see alternate bands of different cells, on each side of different plates.— [sketch in margin] On ~~viewing~~ the external surface, the lines of ridges are seen to be composed of the plate, crossed by short plates [sketch in margin] which are either united, or nearly coincide on the line. These are the summits of the cells: but I do not understand the exact structure. At the base of the little sinuous ridges composed of the cross plates, there projects a narrow fleshy rim; this is edged by short, broard, flat, with white & rounded tips, are unequal in size, tentacula. I believe certainly they are not perforated; are soft & adhesive in their nature. They do not exactly

arise in one line. The fleshy rim, from which they arise, is united with similar substance which coats in minute folds the little cross plates, & part fills up the intervening spaces.

Directly over the plate which follows the central line of the furrows, there
Meandrina are seated the mouths: these |**352**| are cylindrical, tubular, very short,
(c) diameter $1/50^{th}$ of inch.— [note (c)] The mouths do not project so high as the line of Tentacula.— [note ends] the orifice is in folds in centre of tubular fleshy projection [see sketch below]. In the space of ½ an inch there is about 8-10 mouths, & on each side at the base of the ridges about 28-30 tentacula. This gives a proportion of from 6-7 tentacula to each

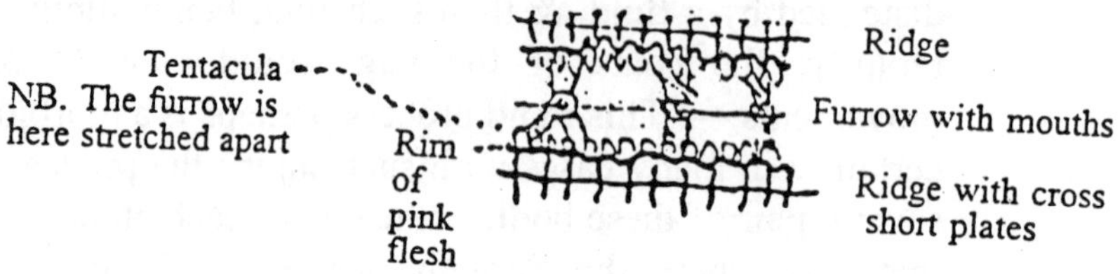

mouth, but in other parts the proportion was 8 or 10 to a mouth.— Beneath the spot of the mouth, the rays or sides of the cells unite sometimes one on each side, & sometimes more, to the central plate: On touching the tentacula, the rim is partly contracted, & together with tentacula is drawn within the folds of the little cross plates: These rims & their tentacula can likewise be partly ~~covered~~ drawn together so as to cover the mouths.— There is no division to show which tentacula belong to which mouth. The
(a) tentacula on opposite sides of the same ridge must belong to different lines of mouths.— [note (a)] The fleshy matter is united in the folds across the ridge.— [note ends] On ~~injuring~~ cutting off the mouth, a white delicate folded membrane is protruded, precisely similar to what is seen in Caryophillia.—

In splitting the coral, it generally breaks in a line with a furrow: in the two
(b) pieces, one will have the smooth plate attached to it, the other will ~~show~~ expose the cells which bordered on one side. [note (b)] The plate is crenated
Meandrina (3605) on its superior margin.— [note ends] |**353**| the principal plate.— In this longitudinal section, the living part colored pale purplish brown of the great globular mass is seen only to extend to .2 in depth, beneath this .1 deep in stained green, but dead & the whole central part entirely dead.— The upper & living row of cells are occupied, on each side of the plate, with the delicate folded membrane, which protrude out directly after splitting the coral.— As the mouth of the animal is placed directly over the plate, & where the sides of two & occasionally more cells unite, the folds depend into as many cells.— This structure is analogous to Caryophillia, where the axis of the stem is divided into many compartments. Although the greater number, not all the cells are so occupied, some being empty: this is

probably connected with the upward growth of the Coral, & agrees with the fact that the septa or bottom of the cells are not all placed in the same line.— Although the globe ~~mass~~ of Coral easily divides into sinuous layers each of which composes a ridge; yet the fleshy matter is united over the whole surface.— In its growth it has frequently surrounded & enclosed
Meandrina branches |**354**| of the stony Corals.— Such spots are centres from whence the furrows & ridges diverge. The surface of the Coral is very slimy.— The mass lies scarcely or not at all attached to a sandy bottom in the shoal lagoon.— There can be no doubt but what this animal belong[s] to the same family as Actinia & Caryophillia.—

Madrepora[2] (3560) This stony branching elegant Coral is very abundant in the shallow still waters of the Lagoon: it lives from the shoalest parts, which are always covered by water, to a depth certainly of 18 ft & perhaps more. Its color is nearly white or pale brown. The orifice of the cells is either nearly simple, or protected by a strong hood: the polypus is similar in both.— The upper extremity or mouth of the Polypus is closely attached to the edge of orifice: it cannot be protruded, nor drawn back out of sight; it consists of narrow, fleshy lip, which is divided into 12 tentacula or subdivisions of the lip. These tentacula are very short & minute [sketch in margin], are flattened vertically, are brown colored, tipped with white. The animal possesses very little irritability, on being pricked the mouth is folded or contracted into an elongated figure [sketch in margin] & partially drawn back. The body of the Polypus fills up the cell, is so excessively |**355**|

Madrepora (3560) delicate, transparent & adhæsive that I in vain tried to examine its structure. I could see a sort of abdominal sack, ~~outside of which~~ & attached to the side of this there were intestinal folds of a whitish color. These when separated from the body possessed a sort of peristaltic motion.—

3584 I examined the Madrepora (3584) also common in the lagoon & found the same sort of Polypus, & from a shorter examination I believe such will be likewise found in kinds (3612) (*[no number]*)

Madrepora[3] 3586 This foliaceous Madrepora is of a pale brownish purple color. is extremely slimy. is abundant in from 3-7 fathoms water in the lagoon. When alive, from each orifice a minute corella like lip is protruded which folds over & lies flat around the orifice. is colored "Lake red". This lip is marked by about 12 rays, but is not absolutely divided into so many tentacula. The animal possesses little irritability. the lip can be ~~with~~ drawn within, but to no depth in the orifice.— In the centre of this upper extremity, there is a small simple aperture or mouth. between this mouth & the lip there are 6 small brown points.— The whole resembled a beautiful microscopical
Madrepora 3586 flower.— The body, as well as ~~entire~~ |**356**| upper extremity is colored lake red. is likewise pulpy, adhæsive, delicate & attached to the Coral.— In the body, I could trace intestinal like compartments, & on one side ~~folds~~

of the main sack, folds of colourless intestine: each fold appears to be a loop of the intestine, the two parts joined by a granular substance. A piece of single intestine being detached revolved in the direction of the space into which it was folded.

3587 There is another species[4] (3587) found in same situation, the polypus of which is I believe of the same structure. Here however the lip is white. These foliaceous Corals appear to have an intimate connection with the stony branching Madrepores.

<u>Millepora</u>[5] I examined with considerable care two species, taking specimens from
1419 different places on successive days of this genus, & utterly failed of
<u>spirits of W:</u> perceiving anything like a Polypus, yet from the structure of the Coral I must suppose such are present.

3583 Millepora (3583) is branched & colored "Sienna yellow & Wood brown".
3609 it grows in from 0-12 Fathom water on the outer reefs. The coral consists of stony reticulations; from the axis of the branches vessels ~~cells~~ radiate through this substance to the outside. |**357**| It is divided by transverse
<u>Millepora</u> septa into cells. the orifice of the terminal one is seen outside as a white
(3583) (a) spot.—

[note (a) added later] I saw this coral at the Isle of France forming great bushes 2 ft: high.— not infrequently it coats any foreign body in place of forming distinct branches: of which specimen (3634) is an instance [note ends]

[**CD P. 357** continues]

In parts of the older branches the orifices are not quite of equal size, the smaller ones being arranged around a central larger one.— In a section of the Coral, many of these vessels will be seen to be covered up & not continued to the surface.— ~~Towards the~~ I examined a considerable number of the cells when the Coral was broken & could find nothing; but at last in one specimen, from several of the cells, with a ~~broken~~ bent needle I removed a minute portion of a gelatino-membranous substance; in this with a high power I could trace no structure. Can this be part of a Polypus? The whole Coral is of a harsh feel & not perceptibly slimy, by gently scraping the surface & particularly the whiter terminal extremities of the branches: with a high power, a most delicate transparent membrane (without any structure) will be found. beneath this is a <u>most thin</u> layer of pulpy matter in which the rounded & loosely attached extremities of the stony reticulations are imbedded. It would appear that |**358**| these
Millepora reticulations increase, by the apposition of minute stony sphæres.— In the
(3583) matter thus scraped off from several parts of <u>one specimen</u>, I found numerous little oval transparent balls, having diameter of ~~about~~ $1/1000^{th}$ of

inch. These are attached by one end to most delicate strength filaments, which commonly are divided dichotomously.— These filaments under a 1/20 lens where injured are seen to consist of a spiral fibre, which when stretched is partly unrolled.— [sketch in margin]  I at first thought it had some connection with the Coral, but now believe it to be a most minute microscopical parasitical plant or animal.—

(3610) 3611 The second series of Millepora[6] grows in strong vertical plates, which frequently intersect each other & so form a coarse honeycombed mass. In such masses the outer parts alone of the plates are alive. This Coral flourishes in the outer part of the reef where the sea violently breaks. Its color is a mixture of "Wax & Sulphur Yellow", the former being the prevailing tint in the lower & older parts, as the bright yellow is in the upper parts of the plates. The structure & superficial nature of this Coral
<u>Millepora</u> precisely resembles the foregoing |**359**| species. I could not perceive any trace even of a Polypus in the terminal cell.— I see the branches grow by concentric additions: [sketch in margin]  & that the lines of cells on vessels do not occur near to the extremities: therefore the growth of the Coral must be a process quite distinct from the agency of the Polypi, if such there be.— The orifices of the cells in this species are seen to be often very different sizes, the smaller ones being grouped around a larger central one.—

Both these species agree in having very little or rather no slimy matter on their surface; but yet possess a strong & disagreeable odor.— they likewise agree in the very remarkable property, hitherto I believe unnoticed in such productions, of producing on contact a stinging sensation. Mr Liesk[7] first observed this fact by accident in the plate kind, & I find it as strong or stronger in the branched sort.— The power appears to vary: generally speaking on pressing or rubbing a fragment on the tender skin of the face or arm, a prickling sensation will be felt after an interval of a second, & which lasts for a very short time. But on rapidly touching with the
<u>Millepora</u> specimen (3609) of the |**360**| branching kind the side of my face, the pain was instantaneous, but increased, as usual, after a very short interval; the sensation continued strong for a few minutes, but was perceptible half an hour afterwards. The sensation was as bad as that of the sting of a Nettle, but more resembled that produced by the Physa[8].— On touching the tender skin of the arm, red spots were produced, & which had the appearance, if the stimulant had been a little stronger, of producing watery pustules. With respect to the nature of these Milleporæ, I cannot help suspecting that their nature is allied to Corallina rather than to Polypiferous Corals.— I am led to this idea from not being able to discover any trace of an organized Polypus in the cells; their position with respect to extremities of branches; their size varying & their method of grouping; all which facts would better agree with the idea that the cell is the seat where the Ovum is produced.— Their manner of growth & the absence of slime is analogous to the

Corallinas.— |**361**|

[1] Scleractinia, a stony coral, but *Meandrina* is a name now used only for the corals of Atlantic reefs. CD's description leaves little doubt that this specimen belonged to the family Faviidae, and since it had a crenated plate beneath the lines of mouths was probably *Leptoria phrygia*. What CD referred to as septa would nowadays be called dissepiments, while his short plates are known as septa.
[2] The scleractinian reef coral that was most commonly referred to as *Madrepora* until the beginning of the 20th century is now called *Acropora*, and *A. corymbosa* or a related species would be consistent with CD's description of the specimen. In *The structure and distribution of coral reefs*, Smith Elder and Co., London (1842), CD identified the commonest corals as *Porites*, *Pocillopora verrucosa* and the madrepores. See also article by Brian Rosen on 'Darwin, coral reefs and global geology' in *Bioscience* **32**:519-25, 1982.
[3,4] These two madrepores would seem unlikely to have been *Acropora*, because CD describes them as 'foliaceous'. However, they may be the 'coral something like an *Explanaria*, but with stars on both surfaces, growing in thin, brittle, stony, foliaceous expansions' that he mentioned in *Coral Reefs* (*loc. cit.*), whose modern name is *Turbinaria bifrons*.
[5] Anthomedusae, *Millepora* is an athecate hydrocoral that secretes a heavy calcareous skeleton. From the colours and the polyp anatomy, specimen 3583 might be *Millepora tenella*.
[6] Another hydrocoral, probably *M. platyphylla*. Gastrozooids and surrounding defensive polyps in notched cups emerge from pores in the skeleton, and CD's "spiral fibres" may be the nematocysts that can sting quite severely.
[7] Mr Liesk was an English settler in the southern Cocos Keeling Island who had lived there for about 8 years (see *Beagle Diary* pp. 413-15).
[8] Referring to *Physalia*, the Portuguese man-of-war, a colonial coelenterate of order Siphonophora whose stinging capabilities had been experienced by CD at the beginning of the voyage (see p. 3).

[**CD P. 361** commences]

<u>Actinia</u>[1] (1415) The specimens which I found were adhæring to old shells, which were inhabited by hermit crabs: they lay beneath large stones on the outer reef.— The flower (or upper surface) has a general pale, dull "Crimson red" color. it is regularly radiated with white from the mouth to outer margin.— The tentacula are numerous & placed in many rows; the innermost are longest & they gradually decrease in length to outer row: each tentaculum tapers towards its extremity. central mouth "tile red".— where contracted, conical pins body, whole flower concealed: when expanded semitransparent & delicate. External surface colored "purplish red", irregularly mottled with white.— The mouth internally is formed by four longitudinal ribs or folds on each side, & in each corner two more smaller & obscure projections, making together 12.— About a tenth of inch above the face of the body there is a ring of 24 conical little paps which have orifices. These orifices are closed when the body is contracted. The
(a) paps or glands are white. The animal has the remarkable power when

irritated [continued at (a) on back of **CD P. 261**] of emitting from its mouth & 24 glands or pores, bunches of viscous threads. These threads are colored "Peach blossom & Aurora Red". they can be drawn out, when in contact with any object to the length of some inches, & are emitted with considerable force.— they are not at all strong. The pores near the part most irritated only at any one time ejected this substance. The animal having been kept for a day still retained this power.— Within the body in the basal parts, large quantities of these red threads are laid.— I know not ~~whether~~ their nature or use.— |**362**|

Crab[2]
1428

These monstrous crabs inhabit in numbers low strips of dry coral land; they live entirely on the fruit of the Cocoa nut tree. Mr Liesk informs me he has often seen them tearing fibre by fibre, with their strong forceps, the husk of the nut. This process they always perform at the extremity, where the three eyes are situated. By constant hammering the ~~eye~~ shell in that soft part is broken ~~in~~ & then by the aid of their narrow posterior pincers the food is extracted. I think this is as curious a piece of adaptation & instinct as I ever heard of. These Crabs are diurnal in their habits: they live in burrows, which frequently lie at the foot of trees. Within the cavity they collect a pile, sometimes as much as a large bag full, of the picked fibres of the husk & on this they ~~seat~~ rest.— At night they are said to travel to the sea: there also their young are ~~hand~~ hatched, & during the early part of their life they remain & probably feed on the beach. Their flesh is very good food: in the tail of a large one there is a lump of [continued at (a) on back of **CD P. 362**] fat, which when melted gives a bottle full of Oil. They are exceedingly strong.— The back is coloured dull brick red; the under side of body & legs is blue, but the upper side of legs clouded with dull red.— In the "Voyage par un Officier du Roi" to the Isle of France[3], there is an account of a Crab which lives on Cocoa nuts in a small Isd North of Madagascar; probably it is the same animal, but the account is very imperfect.—

Copied

(a)

NB. These Crabs are in a Cask with a black cross at one end.—

NB. Mr Liesk informs me that the Crabs with swimming plate to posterior claw employ this tool in excavating burrows in the fine sand or mud & that he has repeatedly witnessed this process.— |**363**|

[1] Actiniaria, a sea anemone.

[2] This is the Coconut Crab *Birgus latro* Linn., but the specimen has not survived (see *Oxford Collections* p. 224).

[3] See Jacques Henri Bernardin de Saint Pierre. *Voyage à l'Isle de France . . . par un officier du Roi.* Amsterdam, 1773. Copy assumed to have been in *Beagle* Library.

[For the next four pages CD reverts to some observations on terrestrial planarian worms that he had made two months earlier in Hobart. These notes are crossed through vertically like previous ones on this topic, indicating their eventual publication. **CD P. 363** commences]

Planaria[1]
3518: 19:
20: 21:

[pencil note]
[illeg. word]
Leaches
analogous
fact.—

General shape
as formerly

I found beneath a dead rotten tree in the forest a considerable number of this animal. the decaying wood was only slightly moist.— Color dirty "honey yellow", with central narrow dark brown line on back. on each side a broarder band of pale "Umber brown". beneath snow white & dotted. Both extremities pointed; the anterior one most elongated, extremity slightly turned up & furrowed on under surface. Black points situated around whole margin of foot, but much most abundant near Anterior extremity. Two orifices on inferior surface; the anterior one of them is placed in about middle of the body, & the posterior rather nearer to the first than to the extremity of the tail. Whilst the animal crawls these orifices are about .2 of an inch distant, but when at rest not above half of this.— The posterior orifice is plainest to be seen, more circular & sub. margined. the anterior one only consists of a transverse slit.— Close Before (or nearer to the head) this, the cup shaped organ is situated, its mouth is widely extensible into a conical membrane. the base of this funnel or mouth depends from |**364**| the central vessel, which run[s] towards anterior extremity. Animal placed in weak spirits of wine, this organ is protruded.— On each side of whole length of body that opake branching structure is visible, which has formerly been described.—

length { A good sized individual crawling was 1.5 of an inch long, but when at rest only .8.— Manner of crawling &c similar to what has formerly been described.— I kept some specimens alive in a saucer with rotten wood from Feby 7th to April 1st, when apparently from the excessive heat of the latitude which we then entered, they gradually sickened & died.— during this period ~~they~~ some increased in size; the most perfect one the day before its death, I found with the skin on its back ruptured & the cup shaped organ partly protruded through the hole. I observe they have a particular dislike & immediate apprehension of the light, directly crawling to the under side of bits of wood.— Having neglected to put any water with the rotten wood, it became one day perfectly dry. the largest & only perfect |**365**| specimen ~~contin~~ did not suffer any injury.—

Planaria

Feb. 10th

Cut an individual into two pieces, without attending to where the section was made (possibly it [was] at the anterior orifice). On the 16th both ends quite lively, wounds healing; one orifice manifest in posterior half, but more in the anterior. March 6th. Posterior half quite lively, the posterior orifice visible, wound unhealed, crawls ~~with~~ in the proper direction: Anterior half with its truncated end quite healed & pointed, slightly pink.— I can see no orifice on the lower surface.— These specimens were lost by neglect.—

March 6th

On same day (6th) cut another specimen into nearly equal halves, one having the two orifices & the other none.—

20th

The posterior half had become a perfect animal, the wound quite obliterated; the new anterior extremity was rather suddenly pointed & of a slight pink color. The anterior ~~extremity~~ half resembled in figure to a perfect animal.

31st
Planaria

The posterior half in no way to be distinguished |**366**| from any other individual.— The anterior ~~extremity~~ half had increased considerably in length since the 20th.— Near to its tail, by the aid of transmitted light, a pear-shaped clear space was most distinctly visible. [sketch in margin] it was united to a short clear vessel or space which lead to the tail.— Within the pear-shaped clear space, an opake cup-shaped could be obscurely distinguished. But by no mean could I discover a trace of any orifice on the corresponding part of the lower surface or foot.— It is impossible to doubt, if the hot weather had not killed all the specimens, that in time the cup shaped organ & its orifice would have been produced & the animal completed, similar in every respect to the one produced by the Posterior half.— Thus we see these 25 days sufficed to complete one animal in every respect & another in its external form & partly in its internal structure.— In the first case, the fact of the wound in the posterior half not being healed after 26 days may perhaps be accounted for by supposing the section was made at the [continued at (a) on back of page] point of the anterior orifice.— |**367**|

Their death was hastened by this last examination

(a)

[there follow on the back of **CD P. 366** several entries made at various later dates ignoring the margin, and with different pens or in pencil]

I must here mention that at New Zealand I saw a species of this genus, but lost it in bringing it home.— We thus see that in the Southern hemisphære, America, New Zealand & Van Diemens land all possess this curious family of terrestrial animals.

[different pen] In the Isle of France I also saw a small specimen beneath a stone in the mountain of La Puce.— May, 1836.—

[written vertically in pencil up the lefthand side of the page] Anyone accustomed to *[illeg. word]* Planaria is surprised at being terrestrial.=
Reproduction of cut body same in terrestrial & aquatic species.

[written in pencil at an angle across the page]

Duges[2]

p. 12 movement well described.
p. 14 In Derostoma œsophagus apparently not essential.
p. 15 P. tremellaris marine spec. *[illeg. word]* folded in zigzag.
p. 28 aquatic species allow water to act on their lower surface, apparently for aeration. Has observed tenacity of life in trunk.
p. 29 softening of body not fluid.— dissolution of body.
p. 30 cicatrica Diaphane.
p. 34 In P. tremellaire "2 pores genitaux rounded".—

p. 35 eggs placed between branches of Intestinal case.

[1] Described by CD in *Planaria* p. 188 as *Planaria Tasmaniana*. Is now the type of the genus *Tasmanoplana* in sub-family Geoplaninae. CD gives excellent accounts of the regeneration of bisected specimens and of the death of a terrestrial flatworm, and notes their photophobia.
[2] See article by Ant. Dugès on 'Recherches sur l'organisation et les mœurs des Planariées' in *Annales des Sciences Naturelles* **15**:139-83, October 1828, to which CD's notes written partly in French refer.

[**CD P. 367** commences with a final entry written in Bahia when the *Beagle* anchored there from 1st to 5th August 1836[1]]

<u>Corallina</u>[2]
1463
<u>Spirits</u>
3854:55:
56

This species is very common encrusting the smooth surfaces of the granitic rocks in the tidal pools.— Its colour in the under surfaces is rather paler than that of Corallina officinalis, but generally it is cream-coloured, with a tinge of flesh-red.— The extremities of the short rigid branches (in 1463) are either rounded & white or acuminated into a cone. In this latter case the summit is surmounted by a perfectly circular minute orifice, which leads by a short cylindrical tube into a circular cavity occupying the base of the cone.— [sketch in margin] ~~The structure of the branch shows rectangular intersections of concentric with vertical plates, & the cavity does not appear to lie conformably with these plates.—~~ On breaking off the terminal cone, the cavity is seen to be occupied by a white mass, which from the disturbance ~~has~~ appears like an intestinal mass. It is found to consist of from 20-40 separate cylindrical bodies attached by the lower extremities & embedded in a pulpy matter; ~~which~~ they are placed in a vertical & nearly parallel position.— These occur, & in the same cavity, <u>in several states</u>; some consist of a simple elongated sack with a little granular matter, which presently assumes one or two obscure ~~artic~~ circular contractions.— But the greater number & most perfect ones are in dimensions 3/500in long & 1/500 broard; under the |**368**| microscope they are seen to consist of an envelope ~~containing~~ full of a pale brown granular matter.

<u>Corallina</u>

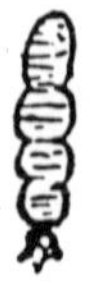

The envelope has a necklace form, owing to three ring-like contractions, partially dividing the little cylinder into four beads.— [sketch in margin] These contractions do not appear to form true articulations, for they are far from separating the internal granular matter:— At the lower end, the terminal ~~arti~~ lobe has a point or navel of attached flocculent granular matter: the superior lobe is generally rather larger & more elongated than the others.— These articulate-like contractions in the most perfect bodies amounted to three, but in the less developed were two & even one; & lastly as I have said, an obscure sack can alone be distinguished in the enveloping matters.— I conceive these are the gemmules or seeds.— In one case I saw one of their cones placed on the side, but near the summit of a branch.— ~~Many of~~ The greater number of the ~~terminal points~~ extremities of the branches are white & rounded. ~~Are these~~ Have the cones been removed from these? I am inclined in some cases to think so, from marks of a slight depression & a scaling structure, which appear general manner of healing.— I saw in section of some branches the trace of an obliterated cavity.—

[1] See *Beagle Diary* pp. 433-4.
[2] Identified as the coralline alga *Melobesia mamillaris* by William Henry Harvey in *Nereis australis*. London, 1847. Specimen 3857 collected on the same occasion was *Melobesia scabiosa*. See *Plant Notes* pp. 200-2.

Specimen Lists

As has been explained by Duncan Porter in his article on 'The *Beagle* Collector and his Collections'[1], the lists of zoological and botanical specimens collected on the *Beagle* by CD were entered in ink in six small notebooks, 7 by 4½ inches in size, now kept at Down House. For purposes of reference these are catalogued at the Cambridge University Library as Down House Notebooks 63.1-6. Notebooks 1-3 were used for specimens preserved in spirits in jars, and labelled with metal tags on which numbers were stamped. Notebooks 4-6 were used for dried specimens, which had coloured paper tags with numbers printed on them. Tags from the same sets were also used for numbering CD's geological specimens listed in his four Geology Specimen Notebooks[2], thus accounting for numbers missing from the list of Animals not in Spirits.

The lists of specimens were drawn up on the right hand pages of each of the notebooks. The pages in the notebooks were not numbered, but at the head of each page CD generally entered the year, month and place at which the specimens had been collected. The specimen numbers were entered in the margin, followed by a capital letter in pencil to indicate to which of the lists drawn up by Syms Covington towards the end of the voyage they should be assigned. Sometimes there were instructions to Covington such as 'Copy' or 'Come to me'. These letters were: A for animal (*i.e.* a mammal), B for bird, C for crustacean, F for fish, I for insect, P for plant, R for reptile or amphibian, and S for shell. Against quite a few of the specimens CD has also written X or XX, apparently as reminders to himself to look at them again. On the left hand pages of the notebooks, additional and sometimes quite lengthy notes about some of the specimens on the opposite page were entered. There are frequent cross references to the entries in the *Zoology Notes* concerning the specimen in question, CD's own pagination being shown here as before in heavy type. For the specimens in spirits a ringed letter or sometimes another symbol in the margin indicated the particular jar in which they were stored. But although this labelling of the jars was essential for CD in subsequent handling of the specimens, it no longer conveys any information of interest, and the ringed letters have therefore been omitted from the transcript.

The fishes stored in spirits were identified by Leonard Jenyns, with descriptions of a number of new species. CD's specimens were presented to the Cambridge Philosophical Society and later passed on to the Museum of the University Department of Zoology in Cambridge, from which a substantial number were transferred to the British Museum (Natural History) in 1917. The list of 'Fish in spirits of wine' now held in the Cambridge University Library as MS DAR 29.1 was copied out by Syms Covington, and the names of the species were added to it by Jenyns. His written identifications and comments on the condition of some of the specimens have been included here, together with a page reference where applicable to *Zoology* **4**, and those specimens that still remain in the Cambridge University Zoology Museum have been marked with an asterisk.

The reptiles, amphibians and crustaceans were entrusted for identification to Thomas Bell, who through procrastination and poor health was responsible for the late appearance of *Zoology* **5**. However, he had evidently taken on a larger task than he could easily handle, and although he had obtained useful help with some of CD's reptiles from the eminent herpetologist Gabriel Bibron in Paris, it was unfortunate that Bibron then died without completing his examination of CD's snakes. There is nevertheless no excuse for Bell's subsequent failure to take any action on the crustaceans. The notes on 'Reptiles in Spirits of Wine', mainly in Covington's hand with some corrections by CD and preliminary notes by Bell, are in the Zoology Library of the Natural History Museum (MS 89FD), together with Bell's identifications of the lizards and frogs, and comments on the specimens in Spirits of Wine that have been transcribed here. There is also a list in the hand of John Edward Gray of 32 specimens received by the British Museum (Natural History) from Bell in 1845; but the whereabouts of the other specimens are unknown. The notes on 'Crustacea in Spirits of Wine' have not been preserved, but a fair number of CD's specimens in class Malacostraca have been located in the Bell Collection at the Oxford University Museum, and those well enough preserved to be assigned to orders, suborders or genera in *Oxford Collections* have been marked with asterisks here. It was Thomas Bell, incidentally, who presided over the reading of the Darwin-Wallace papers at the Linnean Society on 1 July 1858, disapproved of what he heard, and wrote in his annual presidential report that the year had not 'been marked by any of those striking discoveries which at once revolutionize, so to speak, the department of science on which they bear'!

A collection of CD's *Beagle* invertebrates stored in spirits was presented to the Zoology Museum of Cambridge University in 1870 by Francis Darwin, and was catalogued by S. F. Harmer in 1901. Harmer's identifications are noted here, and the surviving specimens are again asterisked. Others of Darwin's polyzoa not stored in spirits were identified by George Busk[3] at the Natural History Museum in London. The identifications entered by Busk against CD's bryozoans now held at the Natural History Museum in the Busk Collection are noted here, and the specimens are asterisked.

The list of birds was copied by CD himself from the *Zoology Notes* with a number of important additions. It has been transcribed by Nora Barlow, and published with annotations as 'Darwin's Ornithological Notes[4]'. The skins had been presented to the Zoological Society of London on 4 January 1837, and when the Society's Museum was closed in 1855, the British Museum was given first choice and accepted about 50 of them. Its Assistant in the Natural History Department, George Robert Gray, who had helped CD to complete *Zoology* **3**, acquired some types. CD presented others to the British Museum in 1856, Gould sold more in 1857, and the rest of his collection was sold to the British Museum after his death in 1881. The acquisition numbers of those birds added in 1855 and 1856 to the Natural History Museum's collection now at Tring have been listed here, and together with some specimens from other sources to which CD's numbers, sometimes in his own hand, were attached the skins still surviving from the *Beagle* have been asterisked.

The list of insects was transcribed by Kenneth Smith as 'Darwin's Insects'[5]. Reference should be made to these *Insect Notes* for the present location of CD's specimens.

The list of plants was transcribed by Duncan Porter as 'Darwin's notes on *Beagle* plants'[6]. CD's manuscript notes are preserved in the Herbarium of the Department of Plant Sciences in Cambridge. The vascular plants themselves had been given to Henslow for identification, and dried specimens are now to be found in the herbaria of the Natural History Museum in London, Cambridge University, the Royal Botanic Gardens at Kew, and elsewhere. The coralline algae were given by CD in the 1840s to the Irish botanist William Henry Harvey at Trinity College, Dublin, in whose herbarium they remain. Those of CD's specimens that Duncan Porter was able to examine have been noted here, together with his identifications.

The list of mammals was drawn up as the unpublished notes on Animals in CUL MS DAR

29.1. The first page is headed 'Gt. Malbro' in Syms Covington's hand, so that the notes were presumably written some time shortly after 13 March 1837, when CD and Covington moved into furnished rooms at 36 Great Marlborough Street. After half a page, the remainder of the notes were written by CD. An edition of *Darwin's Mammals*, accompanied by a classification of those of the skins that can be located in the Natural History Museum, is in preparation.

Some pages headed 'Mr Darwin's Shells' were copied in numerical order from entries in the Specimen lists, with additional information taken from the Zoology Notes, and are mainly in Syms Covington's hand, but with some additions by CD. There are 8 pages on 'Shells in Spirits of Wine' in CUL MS DAR 29.1, and 8 more on 'Shells' in DAR 29.3. Written wholly in Syms Covington's hand, apart from a note by CD at the start saying 'NB The shells which I want out are marked with a cross || about 100 ||' there is also in DAR 29.3 a partially identified and numbered list of shells in alphabetical order, with the localities at which they were collected. The identifications can probably be attributed to William John Broderip, for CD wrote to Henslow on 1 November 1836[7] 'I also heard that M[r] Broderip would be glad to look over the S. American shells'. But the later fate of the collection of shells is not known.

CD's first consignment of specimens was dispatched to Henslow from Monte Video by the Emulous Packet in August 1832[8], and Henslow reported on it in January 1833[9]. The second consignment, including the fossil bones discovered at Punta Alta on 23 September 1832[10], was sent back with the Duke of York Packet a month later, though it reached Henslow only in August 1833[11]. On 18 July 1833[12] and 12 November 1833[13] CD packed off from Monte Video boxes containing several hundred skins of birds and mammals, 'an immense box of bones and geological specimens', a bundle of seeds, and casks containing insects and bottles of fish, on which Henslow duly reported in July 1834[14]. Writing from his sick-bed in Valparaiso on 4 October 1834[15] CD told Henslow that H.M.S. *Samarang* was sailing for Portsmouth with 2 casks containing 'very valuable specimens'. In February 1835 H.M.S. *Challenger* sailed from Valparaiso with more specimens that were probably transferred to another ship at Rio de Janeiro[16]. CD's last letter to Henslow[17] 'from the shores of America' was written on 12 August 1835 announcing the dispatch of two boxes of specimens on board H.M.S. *Conway* that would be followed by those carried by the *Beagle* herself. Although several of CD's consignments of specimens were considerably delayed in transit, all of them eventually arrived safely in England.

Endnotes to Specimen Lists

1 See *Darwinian Heritage* pp. 973-1019.
2 See CUL MS DAR 236.
3 See George Busk *Catalogue of Marine Polyzoa in the Collection of the British Museum, Cheilostomata Parts I (1852) and II (1854), and Cyclostomatous Polyzoa (1875).* The Trustees of the British Museum, London.
4 See *Bull. Br. Mus. nat. Hist.* (hist. Ser.) 2:201-278 (1963).
5 See *Bull. Br. Mus. nat. Hist.* (hist. Ser.) 14:1-143 (1987).
6 See *Bull. Br. Mus. nat. Hist.* (hist. Ser.) 14:145-233 (1987).
7 See CD to Henslow *Correspondence* **1**:515-6.
8 See CD to Henslow *Correspondence* **1**:250-3.

9 See Henslow to CD *Correspondence* **1**:292-5.
10 See *Beagle Diary* p. 107.
11 See Henslow to CD *Correspondence* **1**:327-8.
12 See CD to Henslow *Correspondence* **1**:321-3.
13 See CD to Henslow *Correspondence* **1**:351-3.
14 See Henslow to CD *Correspondence* **1**:394-6.
15 See *Beagle Diary* p. 263, CD to Henslow *Correspondence* **1**:410.
16 See CD to Henslow *Correspondence* **1**:420.
17 See CD to Henslow *Correspondence* **1**:461-3.

Specimens in Spirits of Wine

Down House Notebook 63.1

From Jany 1832, to June 1833
Catalogue for Animals
in Spirits of Wine.
No.^{n}: 1 to 660.
C.Darwin

V. **2**(c)
means Vide Page **2** of note books
& article (c) in it

V. Pl: 1 Fig 1
means Vide Plates one accompanying
note book figure one

1832

Specimens preserved in spirits & with tin Labels

1		Lat 22°N. Jan 10th. Chiefly Pteropodous animals, viz Cleodora, Limacina, Atlanta porrinii, Hyalena, Orthoceræ (?) or Creseis, Rang
2		Jan. 10, Lat. 22 N, Biphora, V. **2**(c)
3		Radiata V. **1**(c) Velella scaphidia?
4		Medusariæ V. **1**(d)
5		Tunicata
6		Small dark blue animal allied to Physalia
All Jan. 10th Lat. 21°N		
7		Tunicata
8	C	Chiefly minute crustaceæ Lat 20°N
9		Physalia V. **3**(e)
10		Medusaires V. **3**(c)
11		Velella, different from (3) Lost description
12	C	Jan. 17 & 18: 5 small Crustaceæ from Quail Island *
13	R	Gecko (Hemodaelites Cuv:) [*Platydmetylus Darwinii*, very near *Delalandii* TB]
14		Aplysia V. **14**(c)

1832

15		Asterias "arterial blood red"
16	S	Patellæ
17	F	Porto Praya, caught by hook [*Serranus*??, Exd. LJ] *
18	F	Hab: do. Vermilion, with streaks of iridescent blue [*Upeneus Prayensis* Cuv. et Val., *Zoology* **4**:26-7, Exd. LJ] *
19	F	Fish, Quail Island: they bite very severely; having driven teeth through Mr Sullivans finger *
20	F	Fish, do. [*Salarias atlanticus* Cuv. et Val., *Zoology* **4**:86-7, Exd. LJ] *
21	F	Fish, do. [*Salarias vomerinus* Cuv. et Val.?, *Zoology* **4**:88-9, Exd. LJ] *
22		Echinus, in profusion. Quail Island.
23	S	Patellæ. Archa. Quail Island.
24	R	Lizard: Porto Praya

25	R	do
28		29, 30, 31. Aplysia. V **14**(c) [*Aplysia* SFH] *
32		Nereis Quail Island [4 Amphinomidae SFH] *
33		Lobularia [Zoanthid SFH] *
34	R	Gecko. Red hill [*Hemidactylus* (also 219) *Hemidactylus Mabonia* young TB]
35		Crustaceæ: Centipedes with blue long legs. Red hill (700 high} Blattæ.—
Jan. 22^{d}		1832 Porto Praya
36		Actinia. tentacula & base scarlet red, body dark arterial blood red: common [1 Actinian SFH] *
37	R	Gecko. with mended tail [*Platydactylus Darwinii* (13) TB]
38		Lobularia. Actinia [Zoanthid, Actinians SFH] * Quail Island
39		Mouse
40		Centipedes Arachnidæ. NE of Port Praya
41	C	Two upper Crustaceæ from Praya.— & others taken at sea between this & Canary.—
42		Terebellæ, small pongiform masses. White & brown with cylindrical tube Jan.25th [1 Sabellidae SFH] *
43	F	Fish dark greenish black above, beneath lighter. Sides marked with light emerald green. tips of Anal, Caudal & hinder part of Dorsal tipped Saffron yellow.— Tip of pectoral, orpiment orange. Quail Island. [*Serranus aspersus* Jen., *Zoology* **4**:6-7. Exd. LJ]
Janu		1832 Porto Praya
		Jan. 25th Quail Island
44	F	Fish. Do.— [*Blennius palmicornis* *Zoology* **4**:83. LJ] *
45		Fish. do [*Stegastes imbricatus* Jen., *Zoology* **4**:63-5. LJ]
46		Fish. do [*Muraena* ——? Zoology **4**:145. Exd. LJ] *
47	C	Centipede, & fresh water Crustaceæ. St Martin
48	S	Lymnea & Physa. St Martin
49		Alcyonium
50		Octopus V. (**5**) [*Octopus* SFH] *
51		V **7**(a):
52		V **6**(a):
53		V **6**(b)
54 & 55		V **6**(c). Doris
56		Cavolina: V: **6**(d)
57	S	Bulla: V: **7**(c)
58		Worm V. **7**(d) [*Gephyrea* SFH] *
59	S	Containing Crustaceæ, Echuria, Sepunculus & white animal allied to it: Actineæ, Fissurella, Chiton. W of Quail Island [1 Calcareous sponge, 2 Actinians, 1 Eunicidae, 2 ?*Aspidosiphon*, 1 ?Echiuridae, 1 Ophiurid, 1 Holothurian SFH] *
60		Fistulariæ. V. **7**(e)
61		Fistulariæ. V. **8**(a)
62		63: 64: Do [2 Holothurians SFH] *
65		Black spongiform substance with large apertures
Jan.		1832 Porto Praya
66		Orange coloured alcyoniums
67		Lobularia (?) polypi with 15 inner & thicker tentacula & 15 outer & intermediate ones [Zoanthid SFH] *
68		(Tunicata Lam:) with a very strong & bad smell.— All animals from 51 caught Jan. 28th

		W of Quail Island [1 ?*Halisarea* SFH] *
69		Fistularia. V. **8**. (b) [1 Holothurian SFH] *
70		Actinia V **9**. (a) [1 Actinian SFH] *
71		Aplysia V **8** (c)
72	S	Cypræa, with ova (?)
73		Octopus, same as (50) [*Octopus* SFH] *
74		75 Fistularia, same as (61) [2 Holothurians SFH] *
76	S	Fissurella & Patella
77		Spongia & Corallina
78		Echiura: snow white asterias Several Sipunculus. Chitons [1 Phyllodocidae, 1 Cirratulidae,
	S	5 Sipunculidae, 1 Asterid, 2 Ophiurids SFH] *
Jan		1832 St. Jago
79		Bulla, same as (57) Nitidula
	S	Doris, as 51 & 53. Worm as 58.— Ascidea: aggregate tunicata dirty orange:
80		Peronia (Blain:) V. **9**(b)
81		Actinea V. **9**(c) [1 Actinian SFH] *
82		Physalia (escaped), described V **3**(c)
83		Alcyonia V. **9**(d) [1 Compound Ascidian SFH] *
		[note opposite] Vide infrà Feb. 3^{d}
97	S	Bulla: nitidula [Bulla SFH] *
98		Caryophyllia orange, with young polypi! ~~on side~~
		Pyrgoma
100		102: Do, without young
99	S	101: Caryophyllia yellow.— 99 with ~~young polypi~~
		Pyrgoma [(Dry, all in one box) *Coenopsammia* + *Pyrgoma* SFH] *
84		Planaria V **15**(a)
85		Cavolina V **6**(d)
86		Doris V **9**(e)
87		Doris V **9**(f)
88	S	Bulla. animal. fine red with edges greenish white
		1832 St Jago
89		Fistularia small, same as (61) [1 Holothurian SFH] *
90		Sipunculus (?) [1 Holothurian SFH] *
91		Crustaceæ. Chiton. Bulla. Doris *
	S	Fistularia. Echiura. Doris, same as 51: 52: 54 & 55.
92		Doris, same as 52
93		Pleurobranchus. orange: (description lost)
94		Actinia V. **10**(a) [1 Actinian SFH] *
95		Tunicata, aggregate [1 Compound Ascidian SFH] *
96		Alcyonium V **10**(b)
		[note opposite] Vide suprà Feb 5th
103		Doris **10**(c)
104	S	Bulla nitidula &c: Chiton fine orange colour: Worm same as
		58: Cavolina, same 56 : Annelidæ
105		Planariæ: (one same as 84) [note opposite] V. **15**(a) & **18**(a & b) & **17**(a)
106		Onchidium 80: Doris 86 & 54 Terebelleæ.—
107	S	2 species of Creseis, V. **19** & **18** Limacina (violet) V **19**(a)

1832		St Jago to Fernando Noronha
108		Porpita. V **19**(b) *
	C	Crustacea. ~~Biphora~~ Salpa *
109		2 species of Dyphyes & one Salpa Feb 17. 1°30′ S.—
110		Octopus. same as (50) St. Jago
111	C	Crustacea. St. Jago
112		Pyrosoma. St Jago [2 Pyrosoma SFH] *
113		Do
114	F	Sucking fish off a shark near St Pauls.— [*Echeneis Remora* Linn., *Zoology* **4**:142, LJ]
115		Centipedes Fernando Noronha
116	S	Patellas (very flat)
117	C	:118 Crustaceæ St Jago
119	C	Do & a large centipede
120		Spider St Jago.— under stone at sea side at low-water mark Most certainly overflowed by tide [note opposite, not in CD's or Covington's hand] little glass tube unfortunately uncorked so that it may be lost
121		Sabella. St Jago.— [1 Terebellidae SFH] *
1832		Feb.— March
122		Octopus St Jago same as (50) *
123	R	& 124. Lizard. Fernando Noronha
125	S	Lepas & crab: St Jago
126	F	Fishes Do
127	S	128 K 129 K Caryophillæ & upon them Pyrgoma. St Jago [128 and 129 *Coenopsammia* + *Pyrgoma* SFH] *
		[note opposite] 130 to .... 143 all taken at Bahia from Feb 29th ... March 17th.—
130		Centipedes &c. Bahia
131		Arachnidæ do
132	F	Diodon V P **22** [*Antennatus* ?, *Zoology* **4**:151. LJ] *
133	C	Land Crab
134		Vespertilio V P **24**
135		Echinus
136	P	Fungus
137	F	Fish [thrown away: bad LJ]
138	F	Fish [*Syngnathus crinitus* Jen., *Zoology* **4**:148-9. LJ]
139	R	Lizard Bahia Brazil
140	S	Murex (for dissecting)
141	S	Fissurella & Ostrea
142		Echina
143		do
1832		March Bahia
		[note opposite] 144, 145, 146, 148, 150, 151, 152, 153, caught at Bahia from Feb 29th to March 17th.—
144	C	Crab *
145	S	Shells, Crustacea & fish *
146	R	Hyla (Laurent's) Shot running up a lofty palm. [note added to List of Reptiles: (is a lizard, *Paraguira smithii*) CD] [*Ecphymotis torquatus* TB]
147	F	149. Fish caught on the 21st Lat. 14°20′ S. Long. 38°8′ West about 65 miles from nearest land; became of a pinker colour from spirits of Wine [*Balistes Vetula* Bl. young, *Zoology* **4**:155, LJ]*

148		Echiura
150		Caterpillar
151	R	Snake, both last given to me by Mr Wilkin of Samarang
152	R	Frog (Barnetti Cuvier?)
		[*Bufo ~~semi~~cinctus* (Prince de Neuwied) compare with *B. chilensis* TB]
153	R	Lacerta
154	F	Fish (very small) Lat 17°12′ S: Long 36°23′ W.
155	S	Atlanta (pinkish shell) Lat & Long do [note opposite] 154 & 155 March 23^{d}
156	F	157. Fish V P **30** Copy [*Psenes* ——?, *Zoology* **4**:73-4. LJ] *
158	S	Janthina. Crustacea. small fish Lat 18°6′ S: 36°6′ W.
1832		March
159	C	Minute crustacea & animal described at Page **2**.— the former found in great abundance 20 Fathoms few miles W of Abrolhos.— *
		[note opposite] 160....177 Abrolhos March 29th
160	C	Crab
161	C	do
162	R	163. Ecphimotes (?) Cuvier [*Ecphymotis torquatus* TB]
164	R	165. Lizards [D^{o} TB]
166	R	Agama
167	R	Gecko
168		Arachnidiæ
169	S	Phasianella
170	S	Minute Multilocular shell
171	C X	Crustaceæ, taken by one of the Sailors out of the *[illeg. word]* of a large eatable fish
		[note opposite] So reported to me.— (is it Cyanum? June 12th)
172	C	Crustaceæ *
173	S	Coronula. on rock in profusion high water mark.— March 29th. Abrolhos
1832		April Rio de Jan.
174		Onchidium or Peronia on Abrolhos.— 29th March
175		Tubipores (?). Abrolhos, do. V **33** [*Idmonea milneana* D'Orb. SFH] *
176	C	Crustaceous animal almost buried in the body Exocætus Communio [?]. April 2nd. 120 E of Rio. [notes opposite] (Cyamus? June 12th) Socêgo is on the Rio Macaè & all the places lie on the road to it.—
177	C	Crustaceæ. Bahia
178	R	Snake. Socêgo. called Corral snake.
179	F	Fish. sallt water lake Lagoa de Boacica [*Gerres Gula* Cuv. et Val. ?, *Zoology* **4**:58-9. Exd. LJ] *
180	F	Fish. running brook. Socêgo not common; pectoral fin causes painful pricks [*Pimelodus gracilis* D'Orb., *Zoology* **4**:110-11. LJ] *
181	F	Another species from same site *
182	F	do do [*Tetragonopterus taeniatus* Jen., *Zoology* **4**:126-7. LJ] *
183	R	Lizard Socêgo
184	R	Frog do
185	R X	Ceratophis (Cuvi:) V **36** Copy [horned toad TB]
		[note opposite] Ceratophis Vide **36**. From 178 to 187 all taken middle of April.
186	R	Frog. Campos Novos [*Cystignathus ocellatus* TB]
187	S	Land shell. Socêgo

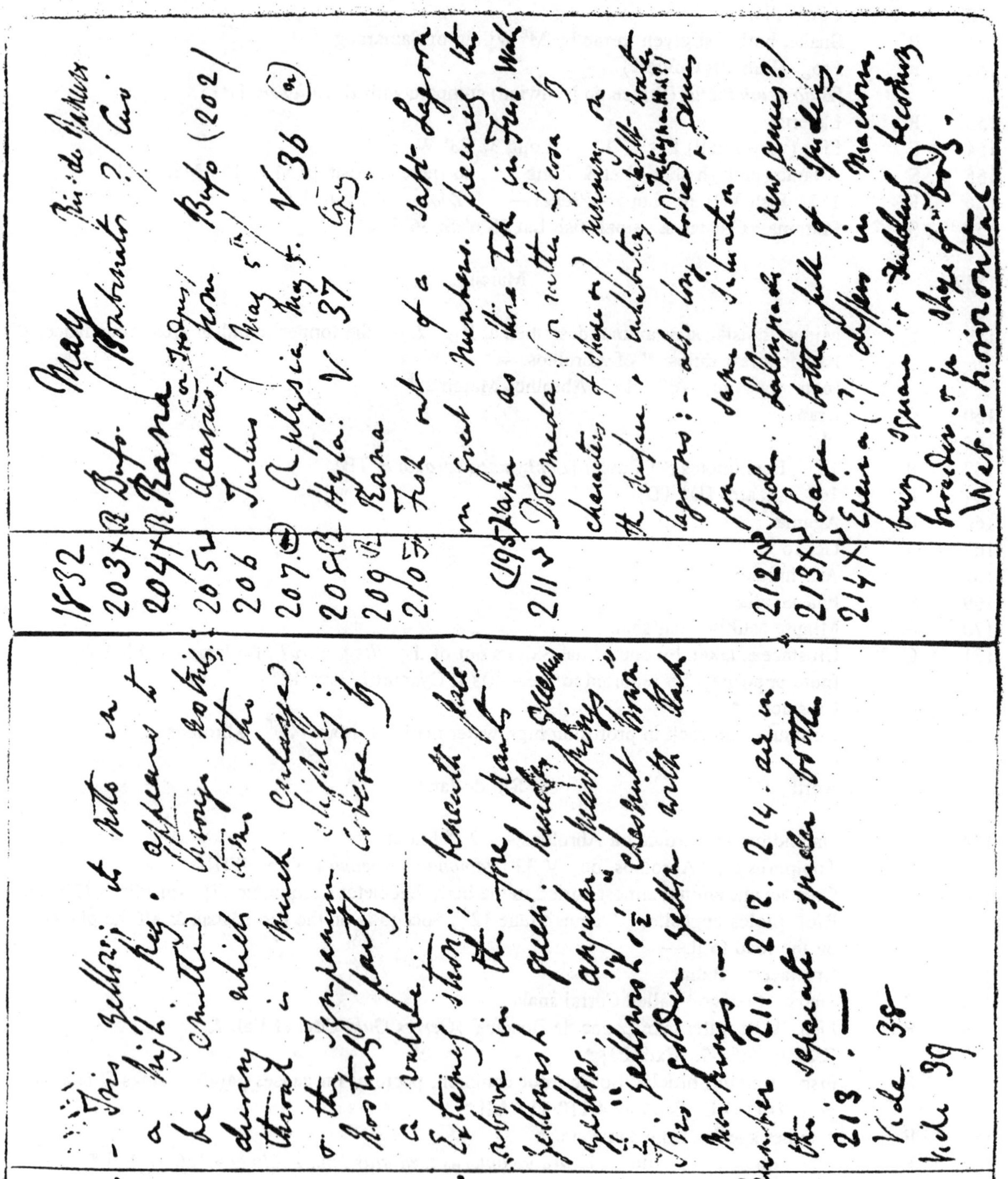

A typical page from the list of Specimens in Spirits of Wine

1832 April Rio de Janeiro

[note opposite] Socêgo & dates, Vide Suprà
188 S Fresh water shell Socêgo
189 S Two species fresh water shell Campos Novos
190 R Snake Botofogo
191 R Hyla (Lam?) do.— colour greenish.— stomach yellowish [*Hyla albomarginata* (Spix) TB]
192 R Frog Botofogo.— Iris yellow with black mark [*Bufo agua* TB]
193 R Lizard Botofogo [*Ecphymotis torquatus* TB]
194 P X Fungus [note opposite] Growing on a wet plank in a darkish outhouse. uniform colour rather light "reddish brown".—
All numbers 190....202 (both inclusive) taken at Botofogo
195 F F Water fish, in great numbers in a small ditch [*Pœcilia unimaculata* Val., *Zoology* **4**:114-15. Exd. LJ]
196 C Fresh W. Crab
197 R Frog [Amph (Bib) *Hyla* ? very young TB]
198 Parmacella (Cuv) V **34**
199 S Physa
200 Arachnidæ from ~~Botofogo~~ Rio de Janeiro & the bottom ones Rio Macaè
201 R Coluber (Cerberus ?) May 5th.— Cuvier
202 R X Bufo (Bombinator?) Cuvier [note opposite] Colour on back dirty "Lemon yellow". *[tous parlientary [?]]* inner face. one darker brown. Iris black spotted with golden yellow.— [*Bufo agua* TB]
[note opposite] All numbers 190...202 (both inclusive) taken at Botofogo

1832 May Rio de Janeiro

203 R X Bufo (Bombinator ?) Cuv [note opposite] Iris yellow; its note in a high key. appears to be emitted through nostrils, during which time the throat is much enlarged, & the Tympanum slightly: Nostrils partly covered by a valve.— [*Bufo ~~semi~~cinctus* 152 TB]
204 R X Rana [note opposite] Extremely strong; beneath pale, above in the fore parts yellowish green, hinder greenish yellow; angular markings ½ "yellowish" & ½ "chesnut brown". Iris golden yellow with black markings.— [*Cystignathus ocellatus* TB]
205 Acarus or Trodpes from Bufo (202)
206 Julus (May 5th) *
207 Aplysia May 5. V **36**
208 R Hyla. V **37** Copy
209 R Rana [*Cystignatus ocellatus* TB]
210 F Fish out of a salt Lagoon in great numbers, precisely the same (195) as those taken in Fresh Water [same as 195, LJ] *
211 Dolemeda (or rather Lycosa by characters of shape &c) running on the surface & inhabiting salt water lagoons:— also long bodied Tetragnantha (?) genus from same situation
212 X Spider Latengrade (new genus ?) [note opposite] Vide **38**
213 X Large bottle full of spiders
214 X Epeira (?) differs in machoires being square & suddenly becoming broarder & in shape of body. Web.—.horizontal [notes opposite] Vide **39**. Number 211, 212, 214 are in the separate spider bottle 213.—

1832 May Rio de Janeiro

215 R X Bufo [note opposite] Is very like (203) differs in not having pale dorsal line between darker ones; & in having dark transverse marks on the legs & obscure angular ones behind head.— [*Bufo ~~semi~~cinctus* (152) TB]

216	P	Fungus. colour pale dirty yellow [*Agaricus salebrosus*, see *Plant Notes*, p. 219]
217	R	Coluber (called the Corall snake}
218	R	Steltion [?] [*Ecphymotis torquatus* (193) TB]
219	R	Gecko (Hemidactyles) [*Hemidactylus* (34) *H. Mabonia* TB]
220		Acari from young or larva of an Orthopterous insect
221	X	Acari buried in skin of a Rana [note opposite] These appear to be able at least as well to move in water, as on a solid.—
222	C	Salt water Crab; Julus; Lepisma; Worms. Wood lice. Acari *
223	S	Cyclostoma, roots of trees on the wooded hills.—
224	P	Cryptogamous plants; when shaken let fall fine yellow powder. growing on rough bark of Palm.—
225		Fungus "scarlet red" on turf
226	C	F Water crab
227	R	Hyla. Palm tree [listed by TB as *Hyla Vauterii* in *Zoology* **5**:45-6]
228	F	Fish. same (195) F Water [same as 195, LJ] *
229		Parmacella. same as (198)

1832	May	Rio de Janeiro

230		Acarus from Phalangium
231	X	Spider Saltigrade [note opposite] in the tube numbered if it is not Tessacrisso [?] & new genus: 6 eyes. skin brass & coloured conaceous punctured: mouth little developed.—
232	P	Lycoperdon on turf. colour white
233		Two species of Acari
234		Julus, Polydemus
235	X	Spider. orbiteles new genus [note opposite] Leucauge Darw: Vide **39**
236	S X	Chiefly fresh water shells as Planorbis. Chondrus Physa Sucunea, Cyclades & Bulimus [note opposite] Chondrus in water in great numbers.
237		Arachnidæ in a tube
238	X	Epeira, web much inclined [note opposite] generally among the Yuccas (?) on sandy plains; rests on the rock with 2 front legs approximate & stretched out before & the hinder ones behind.—
239	C	Fresh W Crab. Caucovado
240	S	Bulimus (with animal V **39**) copy
241		Ixodes, adhering fast to a Bufo
242	R	Bufo. colour intense "Dutch orange" wooded summit of Caucovado [*Brachycephalus ephippifer*, Brachycephalus is Fitzinger's name, Ephippifer being Cocteau's TB]
243		Leucauge (Darwin) Web horizontal species differs from any in (214) or (235) in bottle (252)
244	S X	Ampullaria [note opposite] The animal when kept in a basin is continually absorbing & expelling with noise air.— Can live very well several days out of water & all probability is buried in the mud when the small ditches in which it now abounds are dried up. When first taken, forces quantity of air out: as the animal retracts itself

1832	June	Rio de Janeiro

245	P Copy	Hymenophallus the specimen is mainly in fragments from having no method of carrying V 43
246	P X	Fungus [note opposite] Growing on other trees in the forest.— Çolour "Amber & Chesnut brown" Cup orbicular, regular & most elegant, margin folded down
247	P	Fungi on rotten wood, in forest [*Laschia infundibuliformis*, see *Plant Notes*, p. 220]
248	C	Julus; Polydemus; Ants from the forest; fresh water ~~Malaco~~ Entomostraces : fresh W crab: Cloporta: fresh water leeches &c &c. V No. 1486
249		Hemiptera. (water). covered with ova
250	S	Ampullaria, same as (244)

251	R	Lizard [*Ecphymotis torquatus* (163) TB]
252	XX	Bottle of Arachnidæ [notes opposite] contains 243: 258. V **49**(a & b & c)
253	C	Marine crustaceæ
254	C	Fresh W Do
255		Orthopterous insects
256		Vaginulus n , 2 specimens V **44-45**
257	S	Land shell, with animal V **44** Copy
258	X	Dolimeda; living on the large stones in the middle of [cont. opposite] very rapid brooks, where it may be seen standing motionless, with its legs fully stretched out; in spider bottle (252)

1832 June Rio de Janeiro

259	R X	:260. Rana, brooks in forest. Acari buried in the skin. [note opposite] hind toes edged with membrane, & curious subdivide at extremities [New genus (palatine teeth in corner) listed by TB as *Limnocharis fuscus* Mihi [= Bell] in *Zoology* **5**:33]
261	R	:262. Rana: from the forest [*Cystignathus ocellatus* TB]
263	C X	Crab. (~~Felumnus??~~) fresh water brook in the forest under large stones.— [note opposite] I may mention I saw in brook a Decapode Macrourus: but could not catch it
264	X	Nudibranches allied to Doris. Botofogo bay V **46**.
265	X	Coralline V **47** [note opposite] Sertularia Lamark, but widely different from any I have seen.
266		Arachnidæ living on web of Epeira. V **47**
267	P	268. Cryptogam plants in the forest on bits of stick.
269	F X	Fish, swimming surface. Rio bay [cont. opposite] Above & scales olive brown with red spots & Mark. beneath silvery white; edges of pectoral fin Prussian blue.— Emitted a sound like a croak. [*Prionotus punctatus* Cuv. et Val., *Zoology* **4**:28-9. Exd. LJ] *
270	R X	Leposternum (Spira) [note opposite] Taken in the ground, with other specimens, whilst digging in the garden.— When placed on turf made no attempt to escape, but on soft mould soon (like a worm) forced its way into it.— [TB noted *Leposternum* (to be well figured especially the teeth, Spix's fig. being bad) but the animal was all the same not included in *Zoology* **5**]
271	R	Rana (forest) [Amphib. in too bad a state to be determined TB]
272	R	Lacerta, (do), iridescent with blue
273		Polydernus Tulus Cloporta &c
274		Scolopendra Tulus.—

1832 June Rio de Janeiro

275	X	Acari from a Passalus [note opposite] Riciniæ Lat: but do not agree with any genus.—
276		277. Peneus (+ Dic Class) very abundant. Salt water.—
278		Planaria; under bark V **50**
279		Aplysia V **53**
280	S X	Bulla [note opposite] Animal speckled all over with green & orange spots
281		Aphrodita (Sigalion [?] And: & Edn:)
282	X	Sertularia Tubularia & Amphiroa 1525 growing in great plenty. [notes opposite] Could perceive in Amphiroa no trace of Polypus. Amphiroa V **56**
283		Comatula. V **45**
284	C X	Læmodipodes. Caprella.— Isopods Ligia.— Amphipods Dec. Macromers &c [note opposite] Caprella, in enormous numbers crawling on a Fucus.—
285	C	Decapodes (notopodes)
286	P	Fucus Botofogo
287		Polydemus. Tulus. Scolopendra.
288	F	Fish fresh W. same as (195) [*Tetragonopterus scabripinnis* Jen., *Zoology* **4**:125-6. LJ]
289	R	Bufo (Bombinator) V **53** Copy [~~*Bufo semicinctus*~~ (152) *Bufo melanotis* Bibr. TB]

290 Planaria V **53**
291 Vaginulus V **45**(d)
292 R Leposternum same as (270)

1832 June Rio de Janeiro

293 C Palemon, reported by fishermen to be from fresh Water.
294 Comatula, differs from (283) in its colour & pinnæ on arms
295 Comatula same as (283)
296 X Asterias, back brownish black [cont. opposite] with irregular markings of "Hyacinth red"
297 Sertularia (?) same as (265) [Sertularian (with gonangia) on which were found *Membranipora ornata* Busk and *Schizoporella hyalina* SFH]*
298 Tubularia n
299 ~~Tethys~~ V **55** Cavolina ?
300 C Arachnidæ; Ligia from the Lagoa & minute Crusts. from Bay.
301 F Fish. Rio Harbor [*Gymnothorax ocellatus* Spix & Agass., not in *Zoology* **4**. LJ] *
302 C Pilumnus, with 9 lateral spines, alternating longest.— terminal one not remarkable; common Botofogo Bay
303 C Pilumnus
304 & 305: 306: 307. Loligo, Lamarck n [note opposite] division Calmar of Cuvier. sold in the market for eating. common [1 *Sepiola* SFH] *
308 R Gecko (Hemidactyles) [this number was without any specimen in the bottle TB]
309 F Fish. Rio Harbor.—

1832 Monte Video (July)

310 X Biphora. Dianœa. July 100 miles off Rio Plata [note opposite] Vide **68**, **69**, **70** (pages)
311 C Cancer. on the rocks, Rat island: water but slightly salt.
312 C Pilumnus: Habitat do
313 C Cancer do do
314 R Bipeda (Cuv) Hysteropa. Dic. Class. under a stone. Rat Island
315 Dysdera. Hab. same as last
316 C Cloportæ do do
317 R Bufo. open plain, with horny plate on hind feet [~~*Zonopterna Delalandii*~~ listed by TB as *Pyxicephalus Americanus* Bibr. in *Zoology* **5**:40-1]
318 C Cancer. fresh running water
319 C Cloportæ. Scolopendra. Tulus. fresh water Crust: Amphipodes
320 Scorpio, under a stone on the mount.—
321 Several species of Lycosa: Mygalus. Segestaria: Gonoleptes under stones: the latter in families
322 S X Land shell. July 29th. hybernating [cont. opposite] in chinks of rock; with pellucid membrane in mouth of shell: animal pale nankeen colour.—

1832 Monte Video. August

323 S X Creusia & Mollas: Bivalve living in fresh running brook [note opposite] In case of Bivalve it is not perhaps impossible the water may occasionally be brackish: & from situation of Creusia not improbable: the water was at the time perfectly fresh; & at no time can be very salt as the partial communication at high water was only with Rio Plata & that is brackish.— The river was not at the time low, & the fall & rise of tides is very little.— The Creusia being brought home next day were placed in fresh water & for some hours expanded & retracted regularly their plumose cirrhi.— This fact is curious, showing change of habits.—
324 Lycosa (2 species) & do of Phlodorus [?]

325		Cloporta. Scolopendra Amphipode. Rat Island
326	C	Graspus, with pincers coloured "purplish red" & Plagusia (+ Dic. Class.) with pincers white: tail, lower joint of pieds machoires & base of joints in legs, coloured dark "peach blossom red". both these live in numbers under stones at Rat Island where the water is only brackish.—
327	C	do do do
328	X	Minute Larva (?) congregated in groups of countless numbers on the puddles near the river. Rat Island. [note opposite] I have seen them in several places on the surface of the water, in such numbers as to be quite black
1832		August Monte Video
329		Lycosa. (Spiders same as (668 & 9) printed numbers) & Gonoleptes
330		Vaginulus V. **71**
331		Planaria V. **71**
332		Planaria V. **72** & in [jar o] !
333		Worm
334	F	Fish; little pools near the river
335	X	Plagusia (2 species) [note opposite] Differs from Plagusia (326) in not having tail &c coloured pink
336		Scorpio (2 old & 2 young specimens) under stones on the Mount.
337	S X	Chondrus, Helix, Bulimus, under stones on Mount [note opposite] Chondrus, animal "wood brown" colour
338	R X	Snakes (2 species) [note opposite] Back with black dorsal band; on each side is one of a paler "tile red"; then a black: then primrose yellow & then the black central abdominal one.— Other species is above dark "yellowish brown", beneath pearly white.—
339		Amphipodes, Gonoleptes, Cloporta (2), Epeira (2), Mygalus, Lycosa (2), Sallicus, Philodromus (?). Mount
340	R X	Monitor (Ameiva?) [note opposite] Living in a hole, not near any water: very thin & torpid
341	R	342: 343: Lacerta, (palatine teeth small) under stones. Mount [TB noted these as *Ameiva*, young specimens (Bibron to send the adult with the name)]
1832		August
344	R X	Rana. Mount.— [note opposite] When frightened, puffed itself up with air.— [*Cystignathus ocellatus* TB]
345	R	Coluber. Mount.—
346	C	Cyclops. length 1/30 of inch, in the ocean between Point St Antonio & Corrientes:— motions rapid:—
347	F	Fish. Coast of Patagonia Latit. 38°20' August 26th.— Sounding, 14 fathoms. Caught by hook & line V **77** Copy [*Percophis Braziliensis* Cuv., *Zoology* **4**:23-4. Exd. LJ] *
348	F	Fish. Habitat same as last V **77** Copy [*Plectropoma Patachonica* Jen., *Zoology* **4**:11-12. Exd. LJ]
349	C	Pilumnus, out of stomach of fish (348). Colour purplish red.
350	C	Isopod. Cymothoudes on fish (348)
351	C	Isopod. (Bopyrus?) on fish: & ~~curious Decapod~~ Porcellana. can swim tail first: & Amphipode &c: ~~Habitats on Corallina same as (347)~~ 14 Fathoms. Coast of Patagonia *
352	X	2 species Cellaria. Habitat, same.— [note opposite] One of them grows in rigid funnell-shaped pieces; the spines on the cells are of two sorts, one simple; the other long flexible with distant notches only visible with lens ¼ focal distance. ~~Perhaps this belongs to Flustraceæ.~~ [*Caberea rostrata* Busk, *Scrupocellaria*, in pencil *Menipea*?? SFH] *
353	X	Corallina. Habitat, same.— [note opposite] Colour pale. (Hornera Lamouroux?)
354	F	Fish. Habitat same as last V **77** Copy [*Pinguipes fasciatus* Jen., *Zoology* **4**:20-1. Exd. LJ]

1832		Coast of Patagonia
355	X	Flustra (new genus) V **78** [note opposite] 355:356: Habitats &c same as (347) [*Bugula* SFH] *
356	X	Cellepora (?) V **77** [*Cellepora eatonensis* Busk SFH] *
357	C	Porcellana. same (351)
358	X	Fish. Habitat same (347) [note opposite] Colour above salmon coloured [*Plectropoma Patachonica* Jen., *Zoology* **4**:11-12. Exd. LJ]
359	F	Squalus. V **81** Copy [Great Shark (bad) LJ]
360		Mollus: Tunicata. V **82**
361	C	Erichthus (new species) V **88**: & Mysis (new species) V **89** & a new genus in Amphipod Heteropodes V **90**: & Cyclops:
362	C X	Crustacea (pelagic) taken between Rio de Janeiro & Monte Video.— [note opposite] The lower Amphipode taken from anchor, Rio de Janeiro harbor: upper (in tube) coloured purple.
363	X	Loligo V **90** [note opposite] Abundant Baia Blanca
364		Pelagia V **91**
365		Mollusc: Tunicata, different from (360) V **91**
366	C	Crustaceæ: Schizopodes V **96**: Amphipode, Heterom same as (361) & Macrourus (new genus) V **97**:
1832		Septemb:
367	F X	Fish. Lat 39° Long 61 W [note opposite] Body semitransparent colourless: with a bright silver band on each side: also so marked about the head; taken some miles from the land [*Atherina incisa* Jen., *Zoology* **4**:79-80. Exd. LJ]
368		Loligo, same as (363 V)
369	C	Crust: Macrour. (new genus) perfect specimen V **97**
370	C	Isopod. Cymothoudes. V **98**
		Bahia Blanca
371	X	Fish. [note opposite] Body silvery, excepting back greenish blue.— [*Clupea arcuata* Jen., *Zoology* **4**:134. N.S. LJ]
372	X	Mygalus; Epeira. 2 Lycosa. B. Blanca [note opposite] Small Lycosa; body pale, with abdomen with purplish marks: inhabits short tubes in sand near to the sea.—
373	R	Iguaniens; approximates to Quetzpales (Cuvi). sand hillocks [*Proctotretus* n.s. (386) Listed by TB as *P. Weigmannii* in *Zoology* **5**:15-16]
374	R	Lacerta; on sides 2 dark red streaks: tail red: [*Ameiva* TB] [listed by TB in *Zoology* **5**:28-9 as *Ameiva longicauda* Mihi.]
375	XX	Armadillo (~~Encoubert~~ Cuv:) Pichiz. I have also found roots in their stomachs. [notes opposite] Live in the sandy hillocks in very great numbers near the sea & pampas: do not make attempt to escape but try to hide themselves. In stomach were Larva & pupa & perfect insects of several sorts of Coleopt. insects, & an Amphisbœna chiefly the Pupa: which live underground V **204**
376		Pulex from the hairy under sides of the (Encoubert): also curious (vagabond) Riciniæ
377	R	Bufo V **99** [in pencil] Come to me [listed by TB as *Phryniscus nigricans* in *Zoology* **5**:49-50 with the comment that it is not figured in Bib. VIII p. 465]
378	R	Lizard same as (373) [*Proctotretus* n.s. 373 *P. Weigmannii* TB]
379		Scolopendra
380	X	Dynamenæ (with its ovaries) n [note opposite] Polype with 14 arms ? [see also Specimen 297, to which SHF's note could equally well apply] *
381	X	Polyclinum. (Sigillina ?) [note opposite] Mouth of *[illeg.]* n reddish orange. body pale do: the stem appeared to have power of solidifying & relaxing its body.— in plenty 10 fathoms

water.— [see also p. 190 in *Zoology Notes*]

1832 Septemb: Baia [*sic*] Blanca

382 X Sigillina (same as 381) [note opposite] In the fish barrell:
383 R Coluber: Heterodon V **99** [in pencil] Come to me
384 R Amphisbena; in sandy hillocks near the sea; same colour as "earth *[illeg. word]*".
385 R Iguaniens, same as (373) [*Proctotretus* n.s. TB]
386 R X :387: Same genus as (385) but different species.— [note opposite] Differs in having orange coloured gorge, faint lateral stripes of blue; & general markings [*Proctotretus Weigmannii* same as 373 doubtful TB]
388 Vaginalis V **99** (same as 710 not spirits) very bad specimen:
389 R X Bufo.— Marshes, the Fort [note opposite] Body "oil green" with spots & bands of pale blue.— [*Bufo semicinctus* (152) TB; but not listed in *Zoology* **5**]
390 F Fish V **100** Copy [*Clupea (Alosa) pectinata* Jen., *Zoology* **4**:135-6. N.S. LJ]
391 F Fish do.
392 F Fish do. [*Umbrina arenata* C. & V., *Zoology* **4**:44-5. Exd. LJ] *
393 F Fish do. [*Mugil Liza*. & V., dried & in bad condition (Thrown away) LJ]
394 F Fish do. [*Platessa Orbignyana* Val., *Zoology* **4**:137-8. LJ]
395 F Fish do. [*Rhombus* ——?, *Zoology* **4**:139, dry & bad, LJ]
396 F Fish do. [Young Ray (bad) LJ]
397 R Agama (?) V **100** Copy [Listed by TB as *Proctotretus pectinatus* in *Zoology* **5**:18-19] [to be well figured (there are two males & one female) see 443.686 TB]
398 R Lacerta (same as 374) [*Ameiva* TB]
399 R Lizard (variety of 373?) [new, with notched scaling on the side of neck & abdomen like Pr. nigromac. but with two or three series of inferior labial plates TB] [Listed by TB in *Zoology* **5**:14-15 as *Proctotretus Darwinii* (same as 445 not 421)]
400 E Intestinal worm taken out of the stomach of an Ostrich
401 X Virgularia V **106** also some specimens loose in Jar (H) [note opposite] In the fish barrell, in a bottle:—

1832 Sept: Baia Blanca

402 F X Fish; cast up on the beach [note opposite] Above purple-coppery; sides pearly; beneath yellowish with silver dots in regular figures; iris coppery.— not uncommon [*Batrachus porosissimus* Val.?, *Zoology* **4**:99-100. LJ]
403 Dasypus tricinatus; not nearly as abundant as Dasypus (375)
404 C Crust: Isopod: body very flat.— crawling on sand beach at lowest ebb:
405 Annelides (2 ~~species~~ genera)
406 Asterias; beneath "orpiment", above "brownish orange"
407 C Pagurus in a Buccinum *
408 X Ascidia; with outer tunic dissected off: on the beach [note opposite] Orifices bright orange: respiratory one fringed, the other plain: fæces from one to two inches long; diameter $1/40^{th}$ brownish;
409 X : 410: 411.— Ascidia, same species; were very much larger when distended with water:— [note opposite] In the fish barrell.—
412 S Buccinum. (with ovules) V **100** Copy
413 Actinia V **101**
414 C Plagusia; body pale *
415 F Squalus (very small specimen) [Young Shark (bad) LJ]
416 F Fish; back blue, belly silvery [*Clupea arcuata* Jen., *Zoology* **4**:134. N.S. LJ] *
417 S Mya: dug out of the mud on arenaceous clay-bank; 6 inches within; in numbers

1832 Sept: Ba[h]ia Blanca

418 Tringa; shot out of large flock:—

419 C Crust: Cymothoudes; from fish:

420 C X Crust: Isopod; closely allied to (404). tibiæ & tarsi coloured orange [note opposite] I do not think any of Latreilles families agrees with this specimen

421 R Lizard Rio Negro [*Proctotretus* (new species) 373 *Weigmannii* TB]

422 ~~Diodon (given to me) Pacific ocean~~ (spoiled)

423 R Lizard [*Proctotretus* n.s. (386) TB]

424 X Mygalus; Plagusiæ: [note opposite] I observe in Mygalus, the claws are covered by two moveable (at will) organs

425 X Hirudo, from sea fishes mouth [note opposite] Colour dark green; posterior sucker very large [now preserved at the Natural History Museum]

426 C Plagusia. Arms & mouth rose-coloured.

427 R Lacerta (different species from 374) [*Ameiva* (341) TB]

428 Tetragnanth, on the beach.—

429 S Crepidula V **102** Copy

430 S Chiton

431 F Fish; above reddish lead colour.— [Great Conger? Eel at Trinity (bad) LJ]

432 R X Lizard (Iguaniens propus) [note opposite] Above mottled brown & yellow, gorge faint yellow: on the beach.— [*Proctotretus multimaculatus* or very near it TB]

433 R X ~~Heterodon (diff: species 383)~~ [note opposite] Trigonocephalus ~~same as (439) V 99(b)~~ [in pencil] (Come to me) [Snake TB]

434 R X : 435. Lizard (Galeotes ?) [note opposite] Above "liver brown", with latero-dorsal pale streak: thighs of hinder legs pale yellow:— [*Proctotretus* n.s. see 1061 *P. cyanogaster* TB]

436 C Plagusia (two species) *

437 Clytia V **103**

1832 Octob: Bahia Blanca

438 X Clytia. V **103** [note opposite] On one specimen there were numerous Crustacæ: Ostracodes.— These specimens are not so good as those in (477):—

439 R Trigonocephalus ~~V 99(b)~~ Same as (433) [Snake TB]

440 R X Coluber [note opposite] Belly plates yellowish; dorsal scales, with central band greenish, tip black, sides pale;— back mottled greenish:

441 Acari (Riciniæ) from the Felis *[illeg. word]* [CD recorded in *Zoology* **2**:18-19 that he killed
a half-grown specimen of *Felis pajeros* at Bahia Blanca in August, but no specimen is listed]

442 R Trigonocephalus (same as 439)

443 R Lizard same as (439) [*Proctotretus pectinatus* (397) TB]

444 R Lacerta same as (374) [*Ameiva* (374) TB]

445 R Lizard [*Proctotretus* n.s. 399 *P. Darwinii* TB]

446 R Trigonocephalus (same as 439) [Snake TB]

447 X Spider (Latengrade) [note opposite] does not exactly agree with any of Lat: genera.—

448 C Small Crustaceæ, from Corallinas &c i.e. not pelagic.— *

449 X Fish; uniform bright silvery [note opposite] ridge of back blueish; black patch on gill-cover, & another under pectoral fin [*Paropsis signata* Jen. (N. gen), *Zoology* **4**:66-7. LJ]

450 X Fish; Scales silvery iridescent, [cont. opposite] back especially greenish; caudal fins yellow: remarkable from circular dark green patch behind Gill-cover [same as 390.— dry & bad LJ]

451 R Coluber same as (440)

452 R Lacerta; Punta Alta [*Ameiva* (341) TB]

453 R Lizard (Quatrapodes? Having palatine teeth) [*Proctotretus* (n.s.) 421 TB]

454 R :455. Lizard. V **113** Copy [*Proctotretus multimaculatus* ? very young 432 TB]

456 Ophiura [1 Ophiurid SFH] *

457 C Entomostræ (Lophyropes) V **115**

1832 Nov: Monte Video

458 F Fish. Monte Video [*Corvina adusta* Agass., dry, *Zoology* **4**:42-3. LJ] *
459 F Octob: 29th.— [*Mugil Liza* ? Dry & bad LJ]
460 X Gonoleptes. Lycosa. Salticus. Epeira under stones. M Video Octob 29th [note opposite] Epeira: hinder part of thorax & under sides; under side of abdomen & marks on the upper "Lake red":— the rest of body black & cream coloured:—
461 R X Lizard (?) with palatine teeth: [note opposite] Gorge orange colour: abdomen & 2 dorsal longitud: bands pale do: [*Proctotretus Weigmannii* to be figured. try 399 for length of tail TB]
462 Gossamer spider, Rio Plata Octob 31. V **117**
463 R :464. Bufo, in the marshes B. Ayres: belly dirty yellow, with do dorsal line: note very high: odour singular & fetid.— [*Bufo D'orbignii* adult 652 TB]
465 F :466. Fish. M. Video
467 F Fish do do
468 R X Ameiva M Video [note opposite] Back emerald green with black patches & white lines.— lateral ventral scales bright blue [*Acrantus viridis [illeg.]* (D'Azara) figured by D'orbigny under the name of *Ameiva cœlestina* TB]
469 R Lacerta (common on Mount) [*Ameiva* (341) TB]
470 F Fish. fresh water [*Lebias multidentata* Jen., *Zoology* **4**:117. N.S. LJ]
471 Vaginulus V **71**
472 C Fresh water Crust Amphi: & Cloporta
473 Scorpio & Buthus (latter new species?) under stones Mount.—

1832 November: & Dec: Monte Video

474 R Amphisbœna. under stones in ground
475 Poly-desmus, dusky red: & Sentigera [?]; found in ship (British ?) *
476 P Chara. V **119** Copy
477 X Epeira. Salticus. Tetragnatha [note opposite] Epeira.— body orange & black. very common amongst the Agaves.
478 C Crust: Branch. (pelagic, some leagues S. of C. Corrientes V. **122**).—
479 F Fish. San Blas [Unkn.— Dry & in bad order LJ]
480 F do : do [*Achirus (Plagusia)* ?, *Zoology* **4**:139-40, Dry & in bad state.— LJ]
481 R Coluber : do
482 Octopus
[479-482 bracketed as:] The Schooners coast of Patagonia
483 C Decapod Notopod V **125**
484 X Acetabulum G [note opposite] Colour fine blueish purple growing mud banks. Brightones [?] Bay.— The Schooners.— [*Renilla* (much macerated) SFH] *
485 C X Zoea & Erichthus, showing the transition V **131**
486 C Zoea V **128**
487 C Zoea. (486 & 485) Notopod (483) Polypi (Page **2**) & other Crustaceous animals.— San Blas: This includes everything caught in the net:—

1832 Dec:

488 C Cyclops (most minute) V **134**:—
489 Clytia V **135**: Clytia V **136**; growing on Fucus & Flustra
490 S Anatifa. Lat 45°S.— on same Fucus:—
491 C Crust. Mac V **98**(b): (its young?) & Crust. Amphipod — off San Blas. Dec. 4th: *
492 X Crust. Amphipod & Cyclops. pelagic. off the Cape Fairweather about 70 miles: Decr 13th.—

[note opposite] Crust. Amphipod: V **138**
493 Animal described Page **96**(b)
494 F X Fish; in Coral 35 fathoms, about 30 miles off Northern Terra del Dec 15th.—
[note opposite] Head coloured purple (color abruptly truncate posteriorly) with a white line over the nose.— belly purplish: rest of body dirty yellow:—
495 Fistularia. V **141**:—
496 Worm: fresh water alpine.
Crust. amphipod: Cloporta:
Dec 20th Epeira & Arachnidæ from under stones on the mountains around Good Success Bay
497 C X Sphæroma, in great numbers ~~both species~~ rocky pools on coast of Bay of Good Success
[note opposite] The Sphæromidæ are most exceeding abundant: Desmarest mentions numbers existing at Tristan d'Acunha; is it the presence of the Fucus giganteus?—

1832 Decr 20th Good Success Bay:

498 C Crab: orbicular: back purple
499 C Crab. triangular
500 C Crustaceæ living on surface of water. G. S. Bay
501 S Various shells. ~~Asterias & Crustaceæ~~. 30 fathoms. 53° S. Decr 15th.
[note opposite] V. **149** [?]
502 X Arachnidæ. (Lycosa) & scarlet Acarus: very summit Kater Peak
[note opposite] K P. Hermit Island height 1700 feet:—
503 Cryptogamic plant ~~V 145~~ same as 980 not Spirits

[note opposite] January 1833 from No^{r} 504

504 F X Fish: very common in the Kelp [notes opposite] Coppery orange, with dark brown marking: Pectorals & Ventrals reddish orange.
505 F X Squalus.— Goree Sound: same Hab: as last [note opposite] Above with white & dark spots & transverse marks, do breast, & pectoral & ventral fin clouded with "scarlet red".
506 C X Porcellana, swimming on surface outside Wigwam Cove: coloured red:
[note opposite] swims rapidly backwards & uses its small legs in cleaning its body
507 C Crust: (Spræromulæ) very fine <u>on</u> Fucus. Hermit Island *
508 X Trachea of common goose [note opposite] May be known by Male white, female with breast banded black & white
509 S Crepidula with animal
510 C Pagurus. Goree Sound *
511 F X Fish. with irregular bands of pale reddish brown; the [cont. opposite] pale parts with a most beautiful metallic violet coloured glitter along the sides.— Grows to be one foot long:—

1833 Jan: Goree Sound

512 Coralline V **146** & Clytia creeping, same as (P **163** ?)
513 F X Fish. dusky orange red above obscure. Kelp
514 F X Fish. coppery orange above obscure. common in the kelp [note opposite] Fish 504, 514, 520 & others form chief subsistence to the Fuegians
515 F : 516 Fish V **146** Copy [*Myxine australis* Jen., *Zoology* 4:159. LJ] *
517 F Fish, above curiously marked with reddish purple, grey & black
519 F Fish, uniform yellow
520 F X Fish, Anal & Ventral fins black, pectoral orange: 3 orange stripes on the side
[note opposite] All these fish caught by hook in the Kelp
521 Worms on scales of fish (517)
522 X Sipunculus Holuthuria H [note opposite] Holuthuria salmon-colour, tentacula round the

mouth very long, irregularly branched & sparingly, also the branches irregularly branched, truly shrub like. papillæ numerous long:

523 F Fish, colour "crimson red": Kelp

524 F Fish } Caught in Kelp

525 F Fish

526 F X Fish (3 specimens) these were [cont. opposite] caught in the mouth of a fresh-water stream: the water was quite fresh. upon being placed in salt water, they immediately died: [*Aplochiton taeniatus* Jen., *Zoology* **4**:132-3. N.S. LJ]

1833 Tierra del Fuego

527 X Arachnidæ.— Ponsonby Sound [note opp.] the Epeira with abdomen bright green.— & eggs in bag enveloped by brown silk:—

528 P Yellow excrescences of the Fagus antarcticus. esculent: V **147** [*Cyttaria darwinii*, see *Plant Notes* p. 221]

529 P Lycopodium (?) on do:—

530 C Cancer. Amphipod & Sphæromidæ under stones. Beagle Channell

531 F Fish.— Beagle Channell

532: P 533: 534. The junction of the parasitical plant (977) with the Fagus

533 F Fish. abdomen with a fine red:

536 F Alpine fresh-water fish in lake: Hardy Peninsula: [*Mesites alpinus* Jen., *Zoology* **4**:121. N.S. LJ]

537 Arachnidæ & Scolopendra: do: [note opposite] NB In the same cask as the Skate there is a Gadus, caught in G.S. Bay & common about Cape Fairweather: it leaves the coast in March

538 F Skate. 2 specimens both sexes Good Success B.— Colour "Broccoli brown" & marked (like binding of book) with rings & lines of "Chocolate red".— Iris silvery Grey; upper part depending fringed; sometimes almost concealing the pupil.—

539 C XXX Crustaceæ. parasitical on a large, slimy Raia

540 C Galathea (?) in the sea off Hardy Penins. in great numbers, colouring the sea red

541 C Sphæromidæ. Hardy P. under stones

542 F Fish. in pools left by the tide

543 F Fish. fresh water brook Hardy Pen: [*Mesites maculatus* Jen., *Zoology* **4**:119-20. N.S. LJ]

544 Arachnidæ. Hardy Peninsula

545 X Trachea of a Goose [note opposite] The goose called Steamer or Logger head.—

1833 March Tierra del Fuego

546 F X Merlus; caught in G.S. Bay by hook & line: colours reddish brown & white, variously marked [note opposite] Eye with singular fleshy appendage: to the fish are sewn another pair from another specimen:—

547 C X Crust: Isopod & Macrouri. stomach of a Gadoid: G Success Bay [note opposite] The fish seem exclusively to feed on Crustaceæ:—

548 Hirudo (Pontobdella) adhering to a Raia: body pale brown, H with darker circular rings & pale longitudinal stripes.—

549 F 550. Fish, in the rocks on coast. Good Success Bay

551 C Sphæromida from stones & a Crust. Macrouri from stomach of a Gadus.— G S Bay *

552 C A parasitical animal buried in the tail of a Gadoid of a red colour: & Crust. from a Raia:—

553: F X 554: 555: Fish from fresh-water lake (Silurus?): dull leaden [cont. opposite] color: good eating grow about half as large again: common: Falkland Islands.— This lake is not far from the sea & connected by a brook:— [*Aplochiton Zebra* Jen., *Zoology* **4**:131-2, (Three specs.) N.S. LJ]

1833 March Falkland Islands

556 Lithobius[?]: ~~Gonoleptes~~: Arachnidæ: Oniscus: Lumbricus: Falkland Islands *
557 P Hepatica (Marcantia?) damp shaded rocks: Falkland
558 F Fish; under stones; sea coast
559 Gasteropod V **149** Sigaretus
560 C Sphæromida & genus closely allied to Atylus; all under stones; Falkland Is^{ds}:— *
[note opposite, circled] Bottle not in H
561 Mouse, caught near a Wreck in Falkland Islands: stated to be common in the Island: (European?)
562 F Fish (2 species). Rocks. Sea coast
563 Asterias, Cidarites. Bipappelaria.—
564 Ascidia (Bolteria)
565 Box XX do (?) (?) (perhaps mistake in Numbers)
566 X : 567 Box XX : 568.— Ascidia (Phallusia): Polyclinum dirty green colour: [note opposite] Phallusia in enormous numbers on the coast, thrown up by a Gale: This whole order seems abundant.
569 Doris V **151**
570 Gasteropod (same genus 559) V **150**
571 Doris (same as 569).—

1833 March. Falkland Islands

572 C Crust. Decapod: Sphærom: Amphipod:
573 Nereis:—
574 S Creusia V **159**
575 S Patellæ, 2 beautiful species; & Fissurella; & Chiton &c
576 X : 577 Trachea of the male & female Upland goose. [note opposite] Male nearly white & female with much brown lives in the upland plains & swamps.—
578 F : 579: 580: 581.— Small fish in rocky pools; at low water
582 X Asterias (2 species) H [note opposite] One rich "chesnut brown" above, found eating a fish: the other pale salmon colour:—
583 Sigaretus same as (570)
584 do (different species?) V **150**(b):
585 Corallina V **161**
586 Holuthuria; Sipunculus V **163**
587 Young Rat: Europæan or native ??
588 Acarus on common Snipes breast, back yellowish brown; legs, head & spot behind it black
589 C Crustaceæ; crawling on Corallines: the long one is of a very curious structure:— *

1833 E. Falkland Is^{d}

590 S Balanus (Linn) V **160**(a)
591 S Balanus (Linn) V **167**
592 F Fish, fresh water, embourchure of brook.
593 Sigaretus
594 Holuthuria V **163**(a)
595 Synoicum V **167**
596 Holuthuria V **173**
597 X 2 species of Obelia V **173** & 2 Flustra, one gelatino-membranous.— [note opposite] On tentacula of gelatinous Flustra I saw a vibratory motion, as if produced by minute ciliæ:
598 F :599 Fish caught amongst Kelp [*Phucocœtes latitans* Jen., *Zoology* **4**:168-9. N.S. LJ]
600 S Patelliform shells on Fucus gigant:

601	C	Cancer *
602		Nereidès & Tubicolès
603		:2 Macrouri. Amphipod. Sphæroma *
604	C	Isopod, caught by dredging sandy bottom: adhering together stomach to stomach. Male & female, edges of dorsal plates tinged with red: a most curious genus.—
1833		May Maldonado
605	R	Frog, above prettily marked with dark olive green above & greenish white; on hind thighs & base of body a little red.— [~~*Adenopleura oculata*~~ *Pleurodema Darwinii* 665 676 TB]
606	R	Hyla V **187** Copy [Listed in *Zoology* **5**:46-7 as *Hyla agrestis* (figured in Spix or Prince de Neuwied) TB]
607	R	Rana V **187** Copy [*Cystignathus ocellatus* TB]
608	R	Bipes V **176** Copy [*Carococca* of Spix TB]
609	R	Quetzpales.— sand dunes [*Proctotretus* n.s. (~~386~~) 373 *Weigmanni* TB]
610		Insecta (?) V **191**
611	C	Fresh W Crust. Amphipod.— colour coppery & metallic lustre *
612		Scorpio. Scolopendra. Arachnidæ
613	R	Bufo (diaboliens!) ~~V 191~~ same as (377) [Listed by TB in *Zoology* **5**:49-50 as *Phryniscus nigricans* Weigm.]
614		Limas V **177**
615	R	Ameiva or Lacerta (allied to Centropiæ) R. Marmagaya. Bando Oriental. [*Ameiva* (341) TB]
616	R	Amphisbœna in ants nest, under stones
617		Saltobus & Julus. Sierra las Animas *
618		Arachnidæ. Saltigrade & Latengrade
619	X	Arachnidæ [note opposite] the largest Latengrade has a brown coloured body: the next in size green: the other, of which there are 5 or 6 specimens, with body lead-coloured, legs red with black bands.—
1833		May Maldonado
620		Tongue of Picus (1237)
621	R	Hyla. (same as 606?) but body silvery white instead of green [*Hyla agrestis* (606) TB]
622		Priscus. Lithobius. Scolopendra. rocky hills.— *
~~623~~		~~Aphodius, one of the rare instances of finding one under horse dung: the dung was old.~~
623	R	: 624 Coluber V **176** Copy
625	C	Grapsus. in holes in mud in water but very little brackish.
626	XX	Galeodes [note opposite] Caught whilst running fast over stones on sandy hill. upper part of abdomen with three longitudinal brown bands.
627	X	Planaria (terrestrial) 2 species V **192** [note opposite] one dark brown with narrow central & pale dorsal line: the other cream-coloured with two broard longitudinal brown bands.— in imperfect examination could only see one orifice on under surface
628	P X	Lycopodium growing in sand dunes [note opposite] Are not uncommon on bare sand from the size of pea to that of specimen: surface rough with pointed pyramids: colour nearly pure white, internal mass of larger specimen becoming soft & brown
629		Zoophite (?) on beach [? *Flustra* in pencil SFH] *
630		Certhia same as (1228) only with a tail
631	R	Rana V **187** — Copy [*Pseudys paradoxa* Wagler. TB]
632		Limas same as (614)

1833		May Maldonado
633		Intestinal worm from the duodenum of a Cavia Capybara, floating amidst the green digesting mass.
634	S	Patellæ (?) fresh water grassy pool with stream of water.—
635		Acarus (Limnochares?) swimming in water; colour "arterial blood red": 2 species of Hirudo: Habitat do:
636	F	Fish. Hab. do Fresh Water
637	X	Gordius, muddy pool, [note opposite] One extremity of body bifid, the other truncate, coloured white surrounded by dark brown. [Nematoda etc. 1 *Gordius* SFH] *
638		Pediculi, very minute, but curious from head of Certhia (1248)
639	R	Coluber V **176**
640	C	Crust Amphipod: one isopod & Ostracodes, grassy pool with stream.
641	C X	Crust: Macrouri. Hab do [note opposite] Body transparent. colourless. in great numbers.— (Pontonia ?)—
642		Lithobius. Scolopendra
643		Planaria [note opposite] Body when fully stretched, two inches long & .3 broard, breadth uniform: but tapering suddenly at anterior extremity; tail abruptly terminates by point: body much depressed: colour above uniform blackish brown, beneath pale: found under rotten bark.—
644	R	Coluber V **187**
1833		May, June Maldonado
645		Small worm from mouth of Coluber (644) V **187**(a)
646	X	Pediculi from the Aperea [note opposite] As this animal is supposed to be the wild Guinea-pig, it would be interesting to compare these parasites with those inhabiting an Europæan individual to observe whether they have been altered by transportation & domestication: It would be curious to make analogous observation with respect to various tribes of men.—
647	P	Lycoperdium or rather Phallus V **189**.— Copy [*Clathrus Crispus*, see *Plant Notes* pp. 224-5]
648:	R X	649 Lacerta—Ameiva V **190** Copy [note opposite] Are these 2 specimens the same. differ in form of scales of Head [648 is *Ameiva longicauda* (374); 649 is *Acrantus viridis* (341) TB]
650		Parus same as (1257) preserved because the beak of skinned one is broken
651	R	Bufo same as (613) [*Phryniscus nigricans* (377) not figured TB]
652	R	Hyla V **190** Copy [*Hyla Vauterii* (227) TB]
653	R	Rana V **190** Copy [*Upoderonotor* [?] n.s. TB]
654	R	Bufo V **190** Copy [*Bufo D'Orbignii* Bibron 463 not yet described or figured TB]
655	R	Amphisbœna. Always under stones:
656		Julus. Lithobius Lentigera: Hab do.—
657		Scorpio (2 specimens Lycosa. Mygalus
658		Pediculi from Toco Toco (1267)
659		Toco Toco (same as 1267) for dissection: appears to be blind:
660	F X	Fish. F Water lake [note opposite] Lake left dry by breaking of bank: Lake sometimes a little brackish: above greenish black, sides paler, slightly iridescent.— [*Chromis facetus* Jen., *Zoology* 4:104-5. Exd. LJ]

Down House Notebook 63.2

Catalogue for Specimens
in Spirits of Wine.—
$\mathrm{N^r}$ 661-----1346

C.Darwin

1833		June Maldonado
661	F	Fish. F Water lake; blueish silvery.— [*Hydrocyon hepsetus* Cuv., *Zoology* **4**:128-9. LJ] *
662	C	Crust: Macrouri, grassy bank, colour blackish when alive.—
663	R	Coluber V **190** Copy
664	P	Lycoperdium V **190** Copy
665	R	Bufo V **194** Copy [Listed as *Pleurodema Darwinii* Mihi. in *Zoology* **5**:36-7 (605, 676) TB]
666	R X	Amphisbœna (2 species) [note opposite] one has vertical ridge on nose: differ in proportion of tail. [*Anops Kingii* Bell TB]
667		Julus (2 species) Lithobius *
668		Scorpio (2 species) Gonoleptes
669	F X	5 species of fish from a lake which was suddenly drained [note opposite] The fish with beard I have seen 8 or 9 inches long.— The smallest fish with black spots on the side I think is full grown — I have taken them so repeatedly in brooks &c of the same size. [*Pœcilia decem-maculata* Jen., *Lebias lineata* Jen., and *Tetragonopterus interruptus* Jen., *Zoology* **4**:115-17 and **4**:127-8. N.S. LJ]
670		Mygalus.— Arachnidæ
671	C	F W Macrouri; terrestrial onisci
672		?: By mistake, I know not what: [*Proctotretus* n.s. TB]
673	R	: 674: 675 Coluber V **194**
676	R	Bufo same as (665) [*Pleurodema Darwinii* (605) (665) TB]
677	R X	Amphisbœna; larger one paler coloured [note opposite] It appears there are now three species.—
678	A	Head of mouse (1288) to show teeth.—
1833		
679	R X	Coluber. From Port St Antonio [note opposite] 679....to 692 Specimens collected on coast of Patagonia by the small schooners.— There is some mistake in the numbering by (692) occurring twice & 688 being omitted:—
680	C	Macrouri. pelagic. Port Desire
681	R	Agama V **195** Copy [*Diplolœmus Bibronii* new genus TB]
682:	R	683: 684: 685.— different species & specimens of Agama, for descriptions of colour V **195**. Copy [*D. Bibronii* & *D. Darwinii* TB]
686	R	Lizard V **195** Copy [*Proctotretus pectinatus* (397) TB]
687	R	Lizard V **195** do [*Proctotretus* n.s. listed as *P. gracilis* Bell in *Zoology* **5**:4]
688		(?) [*Adenopleura* n.s. 778 *Pleurodema bufonium* TB]
689	R	Bufo V **195** Copy [*Bufo Agua* from Bibr. TB]
690	R	Bufo V **195** do
691		Aphrodita [Polychaeta, 1 Aphrodite SFH] *
692	F	Fish. [*Percophis Brasilianus* Cuv., *Zoology* **4**:23-4. Exd. LJ]

May, June Maldonado.—

692 F Fish mottled with red, beneath beautiful white: common. good eating
693 F Fish. Silvery, with silver lateral band. above blueish grey: very common also in brackish water [*Atherina argentinensis*? Cuv. & Val., *Zoology* **4**:77-8. Exd. LJ]
694 F Fish silvery white, above iridescent with violet purple & blue.— [*Otolithus Guatucupa* Cuv. & Val.—, *Zoology* **4**:41-2. Exd. LJ]] *

1833 June.— Maldonado

695 F Above more coppery, with irregular transverse bars of brown; beautifully iridescent with violet. [*Corvina adusta* Agass. *Zoology* **4**:42-3. Exd. LJ] *
696 F X Sides with numerous waving longitudinal [cont. opposite] lines of brownish red; intermediate spaces greenish-silvery, so figured as to look mottled; head marked with lines of dull red & green. Ventral & Anal fins coloured dark greenish blue.— The above 5 fish Maldonado Bay: [*Dules Auriga* Cuv., *Zoology* **4**:16. Exd. LJ] *
697 Arachnidæ
698 Head of Mus (1287 ns) to show shape
699 Cavia Cobaya. (for dissection & comparison with domesticated variety. same as (1266 ns)
700 Mus. same as (1289 ns) to show teeth &c
701 Rhycops same as (1264 ns) for dissection
702 R Coluber same as (639) large specimen
703: 704. Vespertilio, very common in the town
705 R Coluber same as (639)
706 R Amphisbœna
707 Perdrix—Scolopax. same as (1224 ns). dissection
708 R Amphisbœna
709 Body of Didelphis (1283 ns). dissection
710 F X Plectropoma. caught in 40 fathom water off the mouth of R. Plata [note opposite] Closely allied to Brazilianum [*Plectroploma Patachonica* Jen., *Zoology* **4**:11-12. N.S. LJ]
711 Head of Phocœna V **174**
712 P X Phallus growing on sand dumes [note opposite] Head with much greenish black fluid; smell but little offensive.—

1833 June Maldonado

713 R Bipes
714 F Fish. colour bluish silvery. fins darker: [*Umbrina arenata* Cuv. & Val., *Zoology* **4**:44-5. LJ]*
715 R Lizard. ash grey, with dark brown marks & specks of orange on them [*Proctotretus* n.s. 373 *Weigmannii* TB]
716 R Bipes
717 (Number destroyed)
718 R Lizard [*Proctotretus* n.s. 373 *Weigmannii* TB]
719 C Cancer. Guritti Island
720 R X Coluber— [note opposite] Ventral plate "greyish black": sides pale primrose yellow: back with 3 black bands & two intermediate "tile red" ones. behind head with collar of pale yellow. Round anus collar of black.—
721 Furnarius (?) same as (1222) for dissection
722 do same as (1260) do
723 F Diodon, picked up on beach [*Diodon rivulatus* Cuv., *Zoology* **4**:150-1. LJ] *
724 C Ligia (?). extremities of legs scarlet: on rocks. Guritti Island:—
725 Latengrade spider; colour *[illeg. word]* brown
726 S Patelliform & Balanidæ. Guritti I.

727	C	Daphnia & Ostracodes from pot of fresh water amongst Arthridiiès V **202** [it is recorded in the Natural History Notes of Sir John Lubbock for 2 June 1853, Royal Society LUA 1, that these specimens were among those recovered from Thomas Bell by CD in order to instruct his young friend on the comparative anatomy of crustaceans]

1833

728		Long billed Causaria (1467) R. Negro
729	R	Snake. R. Negro
730	R X	Viroia [?] B. Ayres [note opposite] Colour ash grey with regular corresponding marks of rich brown edged with black.—
731	R	732 Toads B. Ayres [*Bufo Chilensis* ? to be examined by Bibron TB]
733	R	Snake do
734:	R	735: 736: Toads B. Ayres [*Bufo Chilensis* (731) TB]
737	C	Crab. caught in dry hole in one of the low islands of the R. Parana, above Rosario:— *
738	F	.....to 745 Fish bought in market of B: Ayres & all edible:— [note opposite] 741 is a fish excessively abundant high up the R. Parana.— In like manner is (745) the Armado.— This fish is peculiar by the very loud harsh grating noise which it can make.— heard even before hawled out of water: Is able to seize very firm hold of any object with the serrated pectoral bones & dorsal fins.— [739 is *Achirus lineatus* D'Orbig., *Zoology* **4**:139-40. LJ]
746	F	Fish. High up the Parana. V **203** Copy
747	F	Locality do [*Tetragonopterus Abramis* Jen., *Zoology* **4**:123-4. N.S. LJ] *
748	F	do do [*Tetragonopterus rutilus* Jen., *Zoology* **4**:125. N.S. LJ]
749	F	do do
750	F	do do
751	F	Jar with the above & B Ayrean fish, also a Cod & one sex of a Ray from Good Success bay, Tierra del Fuego:

1833

752		Corallina. Coast at mouth of R. Negro
753	C	Crustaceæ inhabiting the above coral, mouth of R. Negro Coral *
754	P	Lycoperdium growing on the most dry part of camp R. Negro [*Bovina cervina*, see *Plant Notes* p. 225]
755		Arachnidæ Goritti Island.—
756		Leeches. Onisci n. spirits to the Bajada
757	X	Arachnidæ. Bajada [note opposite] October. One of these is an Epeira. is black with ruby coloured marks on its back: they live in societies of some hundreds.— are of same size, not old & young ones.— all the vertical webs are connected with very strong main lines to which the separate webs are attached about 2 feet apart. The higher parts of large bushes are thus fairly fortified with webs.—
758		Fleas. Bajada, for comparison.—
759	X	Bottle of salt, from Salinas up the ~~river~~ R. Negro; for analysis. [note opposite] This salt since being in my possession has been stained by contact with Iron. All these salts ~~more~~ properly belong to Geological Book, placed here on account of sort of Ticket
760		Muddy sand from bottom of Salina in which crystals of Gypsum were imbedded in & Sulp of Soda lying on.—
761		Saltpetre shaken off roots, twigs &c bottom of shallow muddy pools when dry; very abundant North of Punta Alta, Bahia Blanca
762		Salt from Salinas chicitas, 12 leagues SW of Bahia Blanca [note opposite] Given me at B. Blanca

763 Sulp: of Soda, crystallized in cross spiculæ. Salina up R. Negro

1834 Jan Port Desire

764 to 768 Lizards various species taken in numbers dry sterile plains.— V **209** Copied
764 R [Listed in *Zoology* **5**:21-2 as *Diplolœmus Bibronii* (681) TB]
765 R [~~*Ceratophrys marmorata*~~. *Proctotretus* ~~*Fitzingerii*~~ *Kingii* as listed in *Zoology* **5**:13-14; to be figured, also 773 TB]
766 [*Proctotretus* new sp. *[several alternatives discarded] Kingii* examine TB]
767 [*Proctotretus* n.s. *Bibronii* as listed in *Zoology* **5**:6-7 TB]
768 [*Diplolœmus* ~~*Darwinii*~~ *Bibronii* Mihi. TB]
769-771 R Gecko V **209** (new genus?) do [*Gymnodactylus* n.s. (*Gaudichaudii*) to be figured TB] [as listed in *Zoology* **5**:26-7]
772 R Lizard do do
[*Proctotretus Fitzingerii* as listed in *Zoology* **5**:11-12 TB]
773 R Lizard V **210** do [probably same; to be figured TB]
774 R Rana V **210** do
[~~*Pleurodema*~~ *Leiuperus salarius* n.s. as listed by TB in *Zoology* **5**:39-40]
775 X Fish, in rocky pools of salt water [note opposite] Beneath dirty white; back with olive brown, darker in the middle.—
776 F Fish left among the mud banks [*Aphritis porosus* Jen., *Zoology* **4**:162. N.S. Exd. LJ]
777 X Scorpios, under stones [note opposite] The two tied together, the one was eating the other.— I found several in the stomach of an Ibis!
778 R Rana, back greenish brown, with pale medial line & sides do.—
[*Adenopleura* n.s. (688) *Pleurodema bufonium* Bell as listed in *Zoology* **5**:39]
779 C Small Crust.— from fresh water, yet rather brackish: drunk in the Beagle
780 Spider.—
781 X Water very brackish from a small pool 50 or 60 feet above the level of the sea, resting on porphyry but draining great North Plain, for Analysis. [correction opposite] resting on Porphyry, but in all probability draining the Plain.— for analysis.

1834 Jan: 8th.—

782 Various marine productions 4 or 5 miles from shore; 19 fathoms: Lat 48°56′
783 X Asterias (caput medusæ). site do [note opposite] Colour above & beneath "orange coloured white", disc between the ridges "red lilac purple". exquisitely beautiful & delicate.
784 X Alcyonium, pale "flesh red". site do. & fragment of Eschara.— [note opposite] Mass slightly branched, turned transparent. Polypi when protruded .2 long, with eight shortly fimbriated arms: sensation does not appear to be communicated from one Polypus to another.—

Port St. Julian

785 F Fish. rocky pools
786 C Crustaceæ found with (782) *
787 Arachnidæ, under gravel above high water mark
788 F Fish, whole body silvery, upper part of back iridescent blue, lower greenish, spotted with coppery-lead circular patches, common size.
[*Stromateus maculatus* Cuv. & Val.?, *Zoology* **4**:74-5. In bad condition & thrown away. LJ]
789 F X Fish. Back blackish; centre of each scale greenish white; reach to 1 or 2 feet long. [note opposite] At Port Famine, common. one was $2^{ft}4^{in}$ in length.— A Pescado <*illeg.*> was there likewise, 20 inches long & wonderfully numerous
790 C Pale mud-coloured Crab & Crust: Isopod.
791 F Small fish

792 Sepia, upper parts (chiefly) with small circular spots of dark red varying to pink; upper part of Sclerotica dark green

793 R Lizard. V **210** Copied [*Diplolæmus Darwinii* (683) TB]

1834 Jan:

794 P Cactus & plant growing near the Salinas.— Port St Julian [note opposite] The Cactus from Port Desire.— The stamens when touched collapsed rapidly & with force on the Pistol; as also did the Petals, but in a less sudden manner.—

795 S Balanus (for dissection) Hab: as (782)

796 Sponge & Ascidia. P. St. Julians

797 P Cellaria, very pale "Vermilion red": Sea-weed (~~same color~~ Vermilion Red, best seen extremities of branches: Corallina Habiting do dark "crimson red" (with flat articulations).— 1529 [in margin] Port Desire

798 Ascidia (compound) dirty "scarlet red". *[illeg. word]* Port Desire.—

799 C Onisci. Arach: &c &c under stones P. Desire

800 C Crust. sea beach. do *

801 C X Crustaceæ. pelagic. Watchman Cape. L. 48°18′ [notes opposite] Caught at night, could not catch any by day under similar circumstances: small white Entom. creeping numerous. Small white Ento: with long antennæ very numerous at night Lat 51°53′ Long. 68°11′— *

802 C X Crust. Isopod.— I believe certainly was on the body of a large dog fish. color above mottled greenish grey & tile red: edge dark brown [note opposite] Same Locality as last.— *

803 R Gecko. same as (769). P. Desire [*Gymnodactylus* n.s. (769) TB]

804 X Salt from Salina near Thirsty Hill. Port St Julian [note opposite] A great mass as white as snow adhæred to a root.—

805 X Worms from the branchiæ of a Lota (Gadus) C. Fairweather bank.— [note opposite] Head very slightly globular. tail bluntly pointed (sides of body fimbriated?) protruded eggs, I know not from where: sphærical, granular, semi-opake matter in transparent arm diameter $1/500^{th}$ of inch.— [Nematoda SFH] *

1834 Jan: St^{s} Magellan.—

806 X ~~Holuthuria~~ entrance of St^{s} Magellan. [note opposite] Body thickly covered on one side with Papillæ, on the other placed in rows.— color "orpiment & reddish orange".— skin where no papillæ pale blackish grey: papillæ very ~~thick~~ numerous & protruded round one orifice, did not see true arms.— length 3 inches. caught accidentally by fishing hook.—

807 P Sea-weed V **211** Copy

808 P Sea-weed; 1st Inarticulate, "Hyacinth with little arterial Blood red": 2^{d} with capsules or ovules on sides of branches pale "Hyacinth red". main stems with much green: 3^{d} finely pinnate rather more "Art. blood red" than in 1st: Conferva. bright "sage green".—

809 P 1st Very finely pinnate, "cochineal with "Hyacinth red": 2^{d} inarticulate brownish "sulphur yellow": 3rd with necklace-like stem, brownish "wax yellow": 4th a Coralline. This & former St Gregory Bay, St^{s} of Magellan 15 fathoms

810 C 3 Crustaceæ; amphipod with three spines, mottled pink & white; Hab: do *

811 P X Plants. Elizabeth Island: Diandriæ (?) plants "dutch orange"; hinder surface ~~mottled~~ shaded with "brownish orange" Beneath snow white lip, space [cont. opposite] mottled with the richest "brownish orange".— curious appearance. [This is identified in *Plant Notes* p. 227 as *Calceolaria darwinii*, and was painted by Conrad Martens (see *Beagle Record* p. 184)] Orchis, 5 outside petals veined with "duck green": head of stamens (?) on anterior petals (?), green on yellow margin: two holes in centre of flower surrounded by space of fine yellows——

1834 St^s of Magellan

812 R Lizard. Port Desire [*Diplolæmus Darwinii* (683) as listed by TB in *Zoology* **5**:20-1]

813 X Medusa; "arterial blood with little hyacinth red" [note opposite] Edges of the umbrella transparent viewed like old tree with do colour.— This & next three specimens near Elizabeth: 5 fathoms.—

814 C Crab, white, above "tile red" with pimples of "art: blood do." eggs color of yoke of egg.— *

815 C Crab, mud colour: eggs bright "scarlet red" *

816 P Sea-weed, dark "olive green"

817 Wind pipe & worms from Stomach of Diomedra escula. St^s of Magellan

818 F Fish above coppery yellow. with 5 or 6 transverse brown bands: hook & line, P. Famine

819 F Dog-fish yellowish brown clouded with "cochineal red". P. Famine

820 C Crab. Macr: do *

821 P Esculent parasitical balls on the Beeches: do

822 C Crustaceæ — Cape Negro *

823 P X Orchis. Petals all white, 2 central & anterior ones spotted with purple [note opposite] The Orchis inhabits the darkest forests: Lichen on rocks common pale green (Lichen colour) <*illeg.*> beautiful "vermilion & Arterial blood red" Port Famine [*Codonorchis lessonii*, see *Plant Notes* p. 227]

824 S Balani. adhering to wood picked up on beach. P Famine. same as not spirits ( ) [no number entered]

1834 Port Famine

825 P X Orchis. very shady damp wood no leaves [cont. opposite] white: 2 central & interior petals white spotted with purple.— [*Codonorchis lessonii*, see *Plant Notes* p. 227]

826 (lost)

827 Gossamer spider: about 60 miles off the Plata in the spring of 1833.—

828 Body of Puffinus (1816 not spirits)

829 F X Fish pale yellowish brown, with figure of S (or muscles) on sides [cont. opposite] pale coppery: about mouth branchial covering tips of pectorals & ventrals, reddish orange; caught by hook, uncommon

830 C Crab, back "brownish orange with purple", legs mottled "orpiment orange" *

St^s of Magellan

831 S Small shells. 20 fathoms.

832 Sigillina V **214**

833 C X Crust: Schizopod. St Sebastian Bay. Vast numbers. snow white except black eyes. 12 Fathoms [note opposite] Caught at night:— Here there were very many Whales: 5 miles out at sea.—

834 C X Crustaceæ. S of C. Penas. 11 Fathoms, 3 miles out at sea: caught at night. [note opposite] Largest & most abundant specimen color pale red, like half boiled crab; excessively numerous: (833) & 2nd sized Amphipod (with dark blue eyes & back) also very numerous: *

835 F Fish. Hab do: beautifully silvery with raised lateral line: upper parts of back, pale "Auricular purple" most beautiful

836 Parasitical worm from under Branchial covering of above Fish.—

1834 Feb: T. del Fuego

837 C Crustaceæ, 13 Fath. 2 miles from shore, caught at night.— C. Ines. Feb. 19th.—

838 F Fish caught with above.— [*Clupea Fuegensis* Jen., *Zoology* **4**:133-4. LJ] *

839 C X Crust.— some miles to South of (837) under similar circumstances. (Ship at anchor at night) [note opposite] The Amphipod (largest & most *[word missing]* specimen) excessively

numerous; in different places, different sorts appear predominant. Viz this species here.— a Shizopod near St Sebastian (833, 834) & others at C. Penas *

840 F X Dog Fish. Color pale "Lavender purple" with cupreous gloss.— sides silvery do.— above with regular quadruple chain of circular & oblong snow white spots. tip [cont. opposite] of dorsal & caudal blackish. under part of caudal reddish; iris pearly white.— length of old specimen tip to tip 2^{ft}3^{in}. breadth from tip of pectoral to tip of other 8 inches. Young specimen out of belly. with it is posterior spine of old specimen.

841 C Crust: one mile from shore, caught by night. East of Woollaston Is^{d}. *

842 C Crust: on Corallines. low water mark on Corallines. Wollaston Is$^{d.}$ *

843 Holuthuria V **215**

844 X Asterias. Wollaston Is^{d}.— [note opposite] Above "purplish & Cochineal red", beneath "Dutch orange".—

845 S Balanidæ. Hab do.— abundant

846 X Sigillina. handsome "Aurora & little vermilion red". Apertura of animals "scarlet red". Tadpole-like [cont. opposite] ovules free (!) in water "scarlet red": also an encrusting, cream-coloured Sigillina. structure of animal ?. Hab: as above

847 F X Fish. Above greenish black: beneath yellowish white; sides iridescent where [cont. opposite] the dark back shades away.— NB. Bough[t] of & cleaned by the Fuegians. Kelp Fish. East entrance of Beagle Channel.

<u>1834</u> Feb: Tierra del Fuego

848 F X Fish. Pectoral. Ventral. red orange: Anal Caudal. Dorsal blackish: back & sides [cont. opposite] mottled reddish & greenish — blackish. Kelp. E. entrance of Beagle Channel.

849 F X Fish: Pectoral. Dorsal & Caudal, "Tile & vermilion red" side of head, 4 or 5 very [cont. opposite] irregular rows do color.— Anal, Ventral & Branchial covering dark blue black.— Hab: as above.—

850 C X Crust: 1 mile from shore: 16 F. caught at night. NE. end of Navarin Is^{d} [note opposite] Is the small & most numerous specimen with rudimentary legs the young of (834) *

851 C Crust: from sea-weed &c &c at bottom, 16 Fathom. NE. end of Navarin Is^{d} *

852 X 3 sorts of Spongia. Hab. do [note opposite] One with tube V **217**: one with net-like bag.— white.—

853 Animal ? V **217**

854 Doris-like Holuthuria V **215** in 871

855 Encrusting & other Corallines. &c &c. Hab: do

856 Sinoicum (lost): Gasteropod. Spongea: Hab: do.

857 Optiura. Planaria. Hab: do:

858 S Shells.— Hab: do:

859 X 2 Coralline. Flesh color: Hab: do L [note opposite] The branched one with simple Polypi 14 or 16 arms, on common retractile stem: the sphærical extraordinary one has an orange mass ½ way down the cell, pursed up in the centre, from which I once saw some arms protruded. So that these are drawn within the body & not as commonly extended with the footstalk. Alliance with Actinia-like coralls.—

860 C Crust. Mac. V **217** Copy *

861 Pleurobranchus V **218**

862 Octopus V **218**

863 do. Young of former

1834 March. Tierra del Fuego

864 S Cupidula. V **218** Copy

865 X 2. Ascidiæ & Spongia: 10 Fath: roots of Fucus Giganteus. East end of Beagle Ch^{l}. [Note opposite] Ascidia with mamillated hairy surface: has narrow edge of orifices "Vermilion red":

Ascidia with long footstalk & Sponge dirty "flesh color".—

866 F X Fish: very active: roots of Fucus. Hab. do [note opposite] Sides transverse bar of "chocolate & brownish red" separated by narrow grey spaces.
[*Conger punctus* Jen., *Zoology* **4**:143. N.S. LJ]

867 C Crust. Brachyu & Macrou: Hab: do: *

868 C Crust. Amphi. & Isopod: Hab: do: *

869 C Crust. Isopod: Hab: do: Brow*[n]*ish purple red" with darker spots of same.

870 F Fish. Hab: do: (Young of 866)

871 Asterias "deep reddish orange" brilliant: & (853 & 854)

872 S Shells. Hab: do.—

873 C X Crust. Brac: above "crimson & purplish red". beneath do but paler [note opposite] Some of those caught by the Fuegians were a yard wide from tip to tip of legs.—

874 Flustra (with Capsules) V **219** [*Schizoporella hyalina*, var (= *Escharina brongniartiana* D'Orb.), *Beania magellicana* Busk, *Tubulipora organisans* D'Orb. SFH] *

875 Various encrusting Corallines. East Entrance of Beagle Ch: 10 Fathoms. leaves & roots of Fucus

876 F X Kelp. Fish. Beagle Channel [note opposite] Back mottled with dirty red & green; fins with orange. eyes coppery

877 F Gadus. Back "Yellowish & Chesnut brown". dorsal fins "Liver brown"

1834 March. Tierra del Fuego

878 Flustra (encrusting) V **223**.—

879 S X Terebratula; deepish water; Ponsonby Sound [note opposite] for Dissection. I imagine the depth to be between 20 & 50 fathoms

880 Flustra V **224**(a).—

881 Polype (?) V **224**

882 F Dog Fish: upper part coppery "Brownish purple & Cochineal red" with small white spots & large blackish ones: Ponsonby Sound.—

883 F X Kelp Fish. Hab. do [note opposite] Mottled with orange: Pectoral & part of Caudal do color: Anal. Ventral. Dorsal blackish green.

884 X ~~Loricaria V 226 Flustra same as (880) Ponsonby Sound~~. [note opposite] Flustra Box XX same as (874). Ponsonby Sound:

885 Loricaria V **226**

886 do(?) V do.— [*Menipea* or *Cellularia* (in pencil) SFH] *

887 Flustra. same as (878)

888 Asterias. above "Imperial & Auricular purple". Ponsonby Sound: Fucus roots:

889 Corallines on Fucus leaves. Ponsonby Sound.

890 Obelia V **173**(a)

891 X Polypus ??? in tufts on Fucus leaves. Ponsonby Sound. [note opposite] V 234(b).—

892 C X Crustacea. roots of Kelp. Beagle Channel [note opposite] 3 specimens of Crust Mac. were excessively abundant, swimming in deep water in the channel: surface appearing as if raining

893 Nereides. Kelp roots do

1834 March. E. Falkland Is[d].

894 Clytia. V **227**

895 Flustra V **229**

896 F : 897: 898: Fish roots of Kelp. Berkeley Sound

899 Asterias. dirty orange. Hab do

900 Corallines &c &c. Hab. do

901 Holuthuria (same as 596?). Hab do

902 Sigaretus Hab. do.—

903	C	Crab. South coast of E. Falkland Is^{d}.— *
904		Trachea of an Upland Goose.
905	F	Fish. General color "Gallstone & Honey Yellow" browner on its back.
906	F X	Fish. More brown on back; same general color: [cont. opposite] with small irregular patches on sides of body. head, branchial covering pale silvery blue.
907	F X	Fish. [note opposite] Pectoral, Ventral, Caudal fins mottled with orange: body with brown — black: Much more tenacious of life than latter two: All caught in Kelp. *
908	S X	Bivalve. dirty yellow: caught in deep mud [cont. opposite] in 3 fathom water: Body dirty yellow, one end with long siphon: other with oval foot. fringed on edges folding up longitudinally & seated on foot stalk.—
909		Pleurobranchus.
910	X	Spongia. color "Saffron with little Gallstone Grey" [note opposite] Grows in large masses: Main tubular orifice. Must be nearly ½ an inch in diameter.— roots of Kelp.—
911		Ovules of shells (?). Large sort of common Trochus??
912	X	Flustra encrusting "Paletta deep reddish & brownish orange": those on a Univalve dark obscure reddish color [cont. opposite] & 3^{d} sort "Tile & Flesh red"
1834		March. E. Falkland Is^{d}
913	X	Flustra encrusting. reddish orange. [note opposite] Body essentially the same as of Flustra P **223**: the part where revolving organ lies is more globular & larger & more distinctly separated from the Cæcum & Liver: (the revolving organ seems to lie in cylindrical vessel in globular enlargement ? Arms 16.—
914	X	Loricaria. Polypi in the smaller tuft with 14 arms: if same species with (885) the latter is wrongly described with 16.— [note opposite] The difficulty of counting the arms is great.
915		Cellaria. V **230**
916	C	Crustaciæ. crawling on scales of fish (906 & 7)
917	C	Crust: NB caught as all these marine productions by pulling up roots of Kelp *
918		Annelidæ
919		Spongiæ Box XX
920		Flustra (with Vulture heads) V **230** [*Beania costata* Busk SFH] *
921		Arachnidæ. under stones on high hill 960 feet above sea.—
922		Gonoleptes. Arachnidæ. *[3 words illeg.]*
923		Earth worm. under stones: generally hills:
924	X	Spongia. color "crimson red". in pools, low water. [note opposite] Currents very evident: slight galvanic actions appeared to stop this: spiculæ. cylindrical. reticulated.— growing loosely on shells, on a vertical surface or in corners.
925		Sigillina. 2 or 3 small transparent species: Kelp leaves
926		2 Ascidiæ: one on creeping base, dirty yellow colour. with red orifices: the other entirely colourless & transparent.— Kelp leaves
1834		April E. Falkland Is^{d}.
927		Flustraceæ (species of). Stony. V **232**
928		Sigillina "Hyacinth & Arterial Bl. R"
929		Sponge. dirty yellow — all Kelp roots
930	X	Hippothoæ (Lam.) allied to last cells connected by long curved brackets [cont. opposite] in irregular patches. Polypus same general structure as in the Flustraceæ: growing on the smooth leaves. tree-like stem, Fucus [*Hippothoa* in pencil SFH] *
931		Flustraceæ V **233**
932		Flustra (with capsule 4th species) V **233** [*Micropora uncifera* Busk SFH] *
933	X	Cellepora. cells most minute connected by strong net work [note opposite] Very abundant on both sorts of Kelp: in circular patches. I believe the connecting links are perforated, for a cell

		broken from the rest & containing air, when placed under water sent out air bubbles through the broken net work
934		Various minute Corallines from Kelp leaves such as above species.—
935		Spongia "dutch orange". Sigillina "Sulphur & gamboge yellow". Ascidia milk white, transparent, orifices like a coronet of leaves; inner tunic with a scarlet spot corresponding to each indentation [note in margin: Bowerbia (lost)]
936	C	Crustaceæ
937	X	Flustraceæ species of V **234** L [note opposite] Specimen very poor: there is also a very minute Coralline allied to Clytia on same Fucus leaves.
938		Encrusting sponge on Kelp. Color "Wood & yellowish brown"
1834		April E. Falkland Isl[d].—
939		Flustra same as (874)
940		Polyclinium[?] V **244**.—
941		Globular bodies (Compound ascidiæ ?) "Scarlet & Art. Blood R" & Sponge on Ascidea
942		Ann [?]: Tubicol: 10 Fathoms. entrance of S. Cruz
		Santa Cruz
943		Tubularia.— Clytia. V **245**.—
944	S	Balanidæ. Ships bottom Santa Cruz
945		Acari, in ear of wild Guinea pig
946	X	Salt occurring with gypsum [cont. opposite] in veins in soft earthy sandstone: V Geological notes: [further note] number destroyed, replaced by 954
947	F X	Fish, found dead, high up river of Santa Cruz; pale yellowish brown [cont. opposite] with black mottlings [*Perca lævis* Jen., *Zoology* **4**:1-3. N.S. Exd. LJ]
948	R X	Bufo, high up river Santa Cruz. Beneath white, above "yellowish, with some chesnut B": black punctures lumbar glands, rather coppery with large "Liver B" marks, upper part of thigh, faintly as glands, under, flesh color [note opposite] I daresay, nearly the southern limit for this genus.— [*Adenopleura* n.s. (688) *Pleurodema bufonium* as listed by TB in *Zoology* **5**:39]
949	R X	Lizard, central Patagonia up river [note opposite] Blackish grey, with pairs of square black marks: 4 irregular, longitudinal yellow lines: Beneath do grey with waving black lines [*Proctotretus Fitzingeri* (765) TB]
950	R X	Lizard [note opposite] Above do do, but with white longi: lines, & belly orange [certainly the same (see 766) TB]
951	R X	Lizard [note opposite] Above with three broarder whitish lines, pairs of irregular square brown marks; intervals dirty orange: altogether much paler coloured: beneath as before. [*Proctotretus* n.s. ??? surely *Fitzingeri* (at least the same as 765) TB]
1834		May Santa Cruz
952	F X	Fish, numerous in streamlets & creeks high up river; pale greenish brown with small irregular transverse bars ~~marks~~ of black [note opposite] Belly, snow-white.— [*Mesites maculatus* Jen., *Zoology* **4**:119-20. N.Gen. LJ]
953		Arachnideæ, high up river central Patagonia
954		in place of 946 (Salt) [note opposite] (946 destroyed)
955	C	Crustaceæ, caught by night off C. Virgins 10 Fathom.— 3 miles from shore.
956	X	Sigillina & Spongia. C. Virgins [note opposite] Sigillina. dead, opake white, growing abundantly on shingle in pear-shaped & globular masses: 9 Fathom water
957	F X	Fish found dead on beach. C. Virgins [note opposite] Saw many skeletons in estuary of S. Cruz.— [One of the Gadida; but in very bad condition, & thrown away.— LJ]
958	X	Spatangus. Echiura & Crust Iso. 16 Fathoms muddy bottom off C. Espiritu Santo [note

		opposite] Spatangus "Sulphur Y." with crown shaped mark of dark red.
959		Sertularia & Clytia V **248**
960		do 2^{d} species V **250** [Sertularians SFH] *
961		do 3^{d} species V **249**
962		Alcyonium V **252**
963		Escara V **250**
964	P X	Fucus (?). "Blackish Br", excepting near root "Yellowish Br", this [cont. opposite] curious substance was abundant in 8 Fathoms, on rocks off St^{s} of Magellan.—
965	S	Shell: Hab as above.
<u>1834</u>		May
966		Polyclinum, pale "Brownish Orange"
967	C	Crust (from sea weed).
968		Sertularia = Clytia & Dynamena V **255** [Sertularians SFH] *
969	P X	Sea-weed, color same as common red, delicate sea-weeds: all these come from rocky bottom 8 Fathoms off Sts of Magellan.— [note opposite] There are some small Ascidia, the greater half of which are coloured pale "Vermilion R".—
970		Crisia V. **255**.— [*Caberea minima* Busk (?) SFH] *
971	X	Corallines. 10 Fathom, St^{s} of Magellan: [note opposite] The Obelia has a polypus with the structure of the Flustraceæ:
972		Sertularia — Flustra V. **255**.—
		Port Famine
973	P X	Fungus on Beech trees, [cont. opposite] Cup shaped: bright "Dutch with little Orpiment O" internal surface with down like green-gage
974	X	Loligo (Orychothentes[?]) cast up on beach: mottled in patches, chiefly on fin, with "Hyacinth & little Vermi R" [cont. opposite]— Very large: ink dark, greenish brown.—
975		Ho*[l]*uthuria V. **261**
976		Spongia.— faint yellow.—
977	C	Crab — mud-color.—
978	X	Gerbillus: Tierra del Fuego: Is it same with (2032)? Is it an old one? — for Dissection.— [note opposite] The fur can easily be dried by Heat.— Sweepstakes Foreland.— This animal is I believe certainly gregarious in its habits.—
<u>1834</u>		June <u>Port Famine</u>.—
979	X	Eyes of the Vulpus Antarcticus from Falkland Is^{d}.— [note opposite] For dissection, to know whether it is Fox or Wolf
980	C	Crustaceæ. Kelp Roots *
981		Nereidous animals & Ophiura: do: [1 Ophiurid (the worms were thrown away) SFH] *
982	S X	Shells. 10 to 20 Fathoms. [note opposite] The number is loose in the bottle.—
983		Flustra V **262**.— [*Membranipora membranacea* L. SFH] *
984	P	Fungi (esculent) V **147**(a).—
985		Earth-worms: Spiders: Acari found under stones at near high water
		Chiloe
986	R	Frog. above "<u>pale</u> Umber-brown" sides with angular spots of dark "chesnut Br" behind eye : & others paler on sides & on thighs uper part of eye golden: throat brown. [New genus see 1170 & 1176 TB]
987		Actinia V **264**
988	S	Shell-fish for dissection, curious internal plates &c &c

989 C Cancer Brach:
990 C X Crust. Parasit: said to have been taken from the gills [cont. opposite] of a Sun-fish on the NW coast of N. America.—

1834 July. Chiloe

991 C Crust: Parasit, said to be taken from skin of ~~Spermaceti~~ the Black or Right Whale
992 Limas, very abundant; I believe the same as in La Plata
993 C Crust: Scolopendræ &c &c *
994 C X Crust, scarlet red, in clusters on roots of Kelp;—— [note opposite] Could adhere pretty firmly even to glass.—
995 Clytia, pinkish red stems; structure of Polypus true Clytia. growing on stones on which Kelp adhered.
996 Gasteropterus V **264**
997 C X Crab: Brachy: both species tinged with dark crimson red [note opposite] Kelp roots *
998 S Shell fish for dissect: in soft stones 3 or 4 fathom water.—
999 C Crust. Kelp roots *
1000 } C Squilla: often caught when fishing with nets here; given me. *
1001 }
1002 F Fish said to come from the Japan sea.— a Whaler.— [*Cheilioramosus* Jen., *Zoology* **4**:102-4. N.S. Exd. LJ]

1834 July. Chiloe

1003 C Crustaceæ. Kelp
1004 C Crab: Brachy: trigon: back "Wax G" with black spots; beneath "Sulphur Y". clinging to the Kelp.
1005 C X Small Crust: at sea off Valparaiso, taken chiefly 3 or 4 feet beneath surface: day time [note opposite] Some minute ones are curious *
1006 Animal V **73**. Hab —— do.—
1007 F X Fish. Valparaiso [note opposite] Above dirty "Gallstone & Honey G", posterior half of body becoming reddish.— White spots on side & smaller ones above head.—

Valparaiso

1008 } F Various fish bought in market
1018 } V **269** Copy
[the descriptions of these eleven fishes were duly copied by Syms Covington from CD P. **269** in the *Zoology Notes* (see p. 272), but only four were classified as follows by Leonard Jenyns]
1011 *Heliases Crusma* Cuv. & Val., *Zoology* **4**:54-5.
1012 *Pinguipes Chilensis* Val., *Zoology* **4**:22-3.
1014 *Sebastes oculata* Val.?, *Zoology* **4**:37-8.
1017 *Latilus jugularis* Cuv. & Val. (Young), *Zoology* **4**:51-2.
[CD's list continues]
1019 C Crust. Mac. V **269** Copy
1020 Spiders. Scolopendræ under stones
1021 R Lizard. "Yellowish R, with dark brown markings [*Proctotretus* n.s. (434) TB]
1022 R do. Brown nearly Chesnut [note opposite] This is very common size & color. Is it young of last?? [D^{o} TB]
1023 R Bufo V **271** Copy [Amph. Bibr. *Cystignethus nodonis* ? TB]
1024 R do.
[~~*Adenopleura* n.s.~~ *Pleurodema obscurum* 1088.1168.1213 not yet decr^{d} & fig^{d} TB; but listed as *Pleurodema elegans* in *Zoology* **5**:37-8]

1025	R	do. [*Bufo Chilensis* Bibr. n.s. TB]
1026	F	Fish. above "Wood & yellow br" with white & dark brown spots; grow another foot long
1027		Vaginulus V **272**
1028	C	Crust: adhering in numbers on under side of Asterias (1031) *

<u>1834</u> <u>Aug.</u> Valparaiso

1029	C	Crab. above dark "Cochi R" legs Hyacinth & tile R", front pincers purplish *
1030	C	Crab, whole body mottled with "Carm & *[illeg.]* R" *
1031		Asterias V **273**
1032	S X	Animal of the Conchilepas [note opposite] This is a littoral shell; after being once detached, does not willingly reattach itself. [further note dated Decem. 1834] Lives chiefly in the pools left by the tides & not on uncovered rocks: does not adhære with more force than a Buccinum of equal size: siphon projects a short distance: foot trails after the shell for a little length: & this is covered by the Operculum, which may be considered to complete the covering of that extreme part of foot which the shell does not.— This shell is common all over peninsula of Tres Montes & is said to be so in West part of St^{s} of Magellan.
1033	S	Chiton
1034		Sertularia V **273**
1035		Actinia V do
1036		Limax, common under stones, same as at Chiloe & M. Video
1037		Body of Bird (2174)
1038		~~Body of~~ Didelphis (2204)
1039	X	Turdus — *[illeg.]* large species: dissect: [note opposite] Myothera (2172)
1040	X	Rat — dormouse. dissect: [note opposite] Degu (2202)
1041	R	Snake, above "Wood B" with 3 bands of "Umber B". ventral scales, with double arches of black, the rest colored <u>pale</u> yellowish green.
1042	C	Squilla. dull red; back lead-colored
1043		Emberiza, with toothed bill for dissection. "Rara"

<u>1834</u> August Valparaiso

1044		Arachnidæ
1045	C	Crab: admirably adapted for its habitation under surface of round stones, & Asterias, central part of disc purple, edges red.—
1046	C	Crust. Mac: fresh-water brook
1047		Vaginulus same as (1027)
1048		Head of Rat (2312)
1049		Didelphis. female same as (1038)
1050		Large humming bird (2179)
1051		Spider under stone & Oniscus
1052		Myothera. "Tapacola" same as (2173)
1053	R X	Snake: 4 longitudinal bands of [cont. opposite] dark brown, 2 central ones broardest, elegantly connected transversely.
1054	R	Snake. one central broard, dark brown band
1055		Mygalus
1056	F	Fresh-water fish [*Atherina microlepidota* Jen., *Zoology* **4**:78-9. N.S. Exd. LJ] *
1057	C X	Fresh-water Crust. Mac.— [note opposite] Mentioned by Molina as the builder. the mud which it brings up in making its barriers is placed, so as to form a circular wall, several inches high, round the edge of mouth of burrow.— Burrow in marshy field, generally near a brook.
1058	X	2 Scorpions. Gonolepta. Scolopendra. Julus under stones. Mountains. [note opposite] The largest Julus emits much yellow fluid with very pungent smell like mustard.— *
1059		Arachnidæ. Hab do do do

1834		Sept. Oct. Valparaiso
1060	R	Bufo. Bright yellow lines & patches & red punctures [*Bufo D'Orbignii* young TB]
1061:	R	62. Lizards.— [*Proctotretus* n.s. (434) *P. cyanogaster* TB]
1063	R X	Lizards [note opposite] Above blackish, posterior half of body with many scales mottled with brilliant blue: anterior half with do greenish blue: These brilliant colors are nearly absent (excepting few scales on head) by degrees in different individuals, in some individuals as (1063), which is simply brownish black, with transverse black bars: basking in sun on rocks in mountains. Nancagua.— [*Proctotretus tenuis* ? Bibron IV p. 279; as listed in *Zoology* **5**:7-8; to be figured TB]
1064		[*Proctotretus pictus* (same as mine from Capt. King) TB; as listed in *Zoology* **5**:5-6]
1065	P	Edible Fungus on the Roble. V **281**
1066	F X	F.W. Fish. (most extraordinary) [note opposite] Tadpole!!
1067		F Water Leach vide 1486 & Crust Amphipod
1068		Arachnidæ: procured by sweeping.—
1069		Arachnidæ: do do do
1070		Body of Cururo (2315)
1071		Arachnidæ sweeping
1072	R	Young Frog. like Fish ~~same as (101~~
1073	X	Leaches, young ones adhering to centre of belly [note opposite] Smooth specimen (with young) yellowish brown: other greenish.—
1074		Arachnidæ sweeping
1075	F	Fish.——— Valparaiso [*Syngnathus acicularis* Jen., *Zoology* **4**:147]
		Archipel of Chiloe
1076	R	Frog V *[no page entered]* [new genus *Rhinoderma Darwinii* listed by TB in *Zoology* **5**:48]
1077	F	Fish. Lemuy
1834		Nov. & Dec. Archip: of Chiloe
1078	C	Crab *
1079		Octopus
1080	F	Blennius under stones [*Iluocœtes fimbriatus* Jen., *Zoology* **4**:166-7. N.S. LJ]
1081	F	Sucking fish.— do.— [*Gobiesox marmoratus* Jen., *Zoology* **4**:140-1. N.S. LJ]
1082	R	Lizard [*Proctotretus pictus* (1064) TB]
1083	F	Fish
1084	C	Crabs in the greatest numbers under stones *
1085	X	Long extensible worm [note opposite] Dark black-blue, with numerous fine circular yellow lines
1086	R	Frog. V **285** Copy
1087	R X	do. on back a longitudinal [cont. opposite] narrow bright yellow line; above blackish brown.—
1088	R	Frog, in forest.— [~~*Adenopleura* n.s.~~ *Pleurodema obscurum* (1024)(1168)(1213) TB]
1089		Nereidrus. Tubuliferæ
1090	R	Frog. forest. bright dorsal longit. line of yellow; beneath orange tint [new genus TB]
1091		Doris & Cavolina V **284**
1092		Peronia V **284**
1093		Onisci, F. Water Leaches &c &c *
1094	X	Land Leaches (V **309**) G [note opposite] These in summer are excessively numerous in the forests. they crawl about the grass & low plants & thus crawl on to the legs of any person walking: they will bite through stockings, the pain is said to be very annoying.— every person suffers much from this pest.
1095	X	Antipathis. 16 Fathoms. Is^{d} Huafo. Color "Orpiment & Buff orange". [notes opposite] A

small piece of this Corall came up with lead in Cockburn Channel T. del Fuego.— This ~~piece~~ specimen has been kept in water till immersed in spirits.— The two basal parts are preserved dry (2428).—

<u>1834</u> Arch of Chiloe

1096 Vaginulus V **272**

1097 Holuthuria (Doris-like) V **283**.— [1 ? *Psolus* SFH] *

Decemb Chonos Archipel.

1098 C X Crust: amphipod; burrows & feeds in the leaves of a Fucus [note opposite] Growing like the Durvillæa utilis. Dic. Class.— Plates.— *

1099 Holuthuria. V **288**

1100 Sponge, "Gallstone yellow" encrusting under surfaces.— brackish water.

1101 Lumbricus. Oniscus. Scolopendra. C. Tres Montes, Cone Harbor *

1102 P Necklace-like bright green Conferva: do:

1103 C Crab F Water stream do *

1104 C X Crust pelagic colourless — do *
[note opposite] Caught in day time in harbor several yards beneath the surface.—

1105 X Holuthuria like Doris. 13 Fathoms. do [note opposite] I do not believe it is quite same (color same) as that caught in T. del Fuego, because hyaline spots are collected in groups on under surface.— [1 ? *Psolus* SFH] *

1106 Nudibranch. Molluscous Animal V **291**

1107 Tubularia V **294**

1108 Doris V **293**

1109 C Crustaceæ pelagic; night time; do: do: *

1110 C Crabs: The Amphipods, red coloured, under putrid Kelp.— *

<u>1834</u> <u>Decemb.</u> <u>Is^{d} of Inchy, North part of Tres Montes</u>

1111 P Fungus, disc bright scarlet, bordering hairs black: growing on wood in the most rotten state in forest.—

1112 Animal V **297**

1113 X Whitish Holuthuria & 2 Sponges m [note opposite] One is yellow: the other finely reticulated is quite white

1114 C Crustaceæ (littoral). *

1115 Arachnidæ. Oniscus. Scolopendra

1116 F Fish. tidal rocks

1117 R Frog V **299** — Copy [Amph. Bib. Dorso impresso punctato *Cystignathus roseus* TB]

1118: R 19. ~~Bufo~~ Natter Jack V **298** — do [*Bufo Chilensis* young TB]

1120 R Frog V **299** — do
[Amph. Bibr. new genus *Ellipticeps (alsodes) monticola* TB]

1121 Arachnidæ. Oniscus. Scolopendra. same Local: as Bufo (1118).—

1122 Echinoderm &c &c V **301**

1123 F X Fish: Hab: same; caught in middle of Cove [note opposite] Coloured pale "Reddish Orange" with black spots on the fins: & a dusky shade on back:— strange appearance with its bony face.— [*Agriopus hispidus* Jen., *Zoology* **4**:38-40 and 163. N.S. Exd. LJ] *

1124 C Crustaceæ, by night: pelagic in harbor several feet beneath surface: do: *

1125 Holuthuria V **303**

1126 ———————— (number lost)

1127 C X Crust. Mac: caught in middle of Cove, beneath surface, at night, [cont. opposite] must be excessively numerous, because both great herds of seal & flocks of Tern appear to live on them: Above light purplish black; mouth, joints, rings of abdomen & all thin places, fine pinkish red:— Anna Pinks Harbor. Jan 4th.— [see *Beagle Diary* p. 277] *

<u>1835</u> Jan: Chonos Archipel

1128 S Tubinicellæ from Whale (Spermacetti?) Lat 44°30′.—
1129 C Crustaceæ from do. do. between the Balanidæ.—
1130 Two Echini. P. Tres Montes
1131: S 32. Balanidæ V **305**
1133 Nereidous animals, from centre of shell of Concholepas Peruviana
1134 Ascidias, compound, with viscera brilliant scarlet (Huxley) *[inserted above with different pen]*: simple, transparent white: [the name 'Huxley' was probably added in April 1853 when CD learnt that T.H. Huxley was cataloguing the British Museum collection of the Ascidiacea, and wrote (see *Correspondence* **5**:130-1) offering to send him 12-15 specimens collected on the *Beagle*.]
1135 Holuthuria V **308**
1136 R X Frog: above fine "grass gr" mottled all over with Copper color: which nearly forms two longitudinal bands: beneath entirely [cont. opposite] a lurid reddish lead-color.— Iris brown, eyes very prominent large: mouth very much rounded: jumps; inhabits thick forest:— [~~probably~~ New genus of Hylidæ *Hylorina hilendia*, listed by TB as *Hylorina sylvatica* Mihi in *Zoology* **5**:44-5]
1137 Hirudo V **309** [6 Hirudinea SFH] *
1138 F Two sorts of marine Fish: Lowes Harbor [see *Beagle Diary* p. 279] [*Aphritis undulatus* Jen., *Zoology* **4**:160-2. N.S. Exd. LJ]
1139 F Fish, tidal pool; pale lead-color coarsely reticulated with brown [*Gobius ophicephalus* Jen., *Zoology* **4**:97-8. N.S. Exd. LJ]
1140 C Littoral crustaceæ.—
1141 F Fish; Silvery, bright: back blue. Lowes Harbor.—

1835 Jan^{y}

1142 P X Potatoes (wild) Lowes Harbor Chonos A [note opposite] V **314**
1143 Salt, encrusting in great quantities inside of dry cavern of Huafo.—
1144 Arachnidæ. Lowes Harbor. Chonos

<u>Chiloe</u>

1145 F Cabora del Cavallo. Fish same as at Chonos
1146 F Fish, silvery blue above with regular circular leaden spots.— [*Stromateus maculatus* Cuv. & Val.?, *Zoology* **4**:74-5. Exd. LJ] *
1147 F All silvery
1148 F Above with fine tint of purple
1149 F Mottled Reddish above, beneath white
1150 F do ~~——~~ do
1151 F Silvery, irregular leaden coloured marks
1152 C Crab. above uniform dull red *
1153 F Fish. above dusky. [*Aspidophorus Chiloensis* Jen., *Zoology* **4**:30-3. Exd. LJ]
1154 F Fish.—— [*Syngnathus* — ?. In very bad condition, & thrown away LJ]
1155 Balanæ, buried in a stick & ~~a Crust Macrouri~~
1156 Sepia: All the above caught with the Net: Jan 20th.—

<u>1835</u> Jany Chiloe

1157 X Barking bird: Interesting to dissect throat of this genus [note opposite] Following 6 Specimens collected at East Coast of Chiloe by Mr Sulivans party.— [see *Beagle Diary* p. 280)

1158 Bat

1159 X Worms from the stomach of a Porpoise.— [note opposite] In the stomach there were the beaks of Cuttle Fish.

1160 X Vaginulus I believe same as (1096) [note opposite] Estero de Guildad, N of Caylen

1161 F Fish under stones on sea beach Chauques Is^{d}

1162 X Ascidia called Peure, colored reddish grey almost buried in Sand Chauques Is^{d}
Huxley [note opposite] vide (1165 infrà) [see also 1134 above. The only red ascidian found in Chiloe is *Pyura chilensis*, which appears to have been especially well known for its culinary qualities, a characteristic that always appealed to CD.]

1163 R Lizard — Pale, with dark brown spots & two lines along back of a blueish color.— [*Proctotretus* n.s. (434) *P. cyanogaster* TB]

1164 C Amphipod: Crust: feeding on dead crab on sand-beach at Cucao Chiloe * [see *Beagle Diary* p. 283]

1165 X Ascidia, called Peure do do [note opposite] Am told adhære to stones on the Sand beach — many adhære together; hence shape is rather angular; orifices red: sides with curious hard sandy horns: <u>body</u> within outer tunic orange. are good to eat & are much esteemed in Chiloe. [1 Simple Ascidian SFH] *

1166 C Hermit Crab: Chonos Archipel. *

1167 S Marine shells for dissect: — do —

1168 R X Frog. Valdivia: above ash-grey with blackish [cont. opposite] brown marks oblong on back.— [~~*Adenopleura*~~ *Pleurodema obscurum.* (1024) 1000.1213 TB]

1169 Arachnidæ. Scolopendra. Valdivia

1170 R X Frog, above reddish brown, with [cont. opposite] blackish marks on back.— [~~new genus *Cyclorhamphus* n.s.?~~ (986) *Borborocœtes Bibronii* n.g. as listed by TB in *Zoology* **5**:35]

<u>1835</u> Feb. Valdivia

1171 R Lizard. pale chesnut brown, with 2 cream colored longitudinal bands on back: emerald spots on sides — belly greenish [*Proctotretus* n.s. 434 *P. cyanogaster* TB]

1172 R Lizard. (do?) without emerald spots. [D^{o} TB]

1173 R X Lizard. Sides, head & tail black [cont. opposite] with brownish obscure marks. down back row of pair of small do colored marks.— centre of back obscure mixture of colors, edged with dirty green:— belly yellow, throat greenish, both freckled with dark spots.— [*Proctotretus* n.s. (434) TB]

1174 S X Great Balanus (for dissection) Chiloe [note opposite] Are esteemed very good eating. grow to 5 or 6 times the size of specimen; sometimes at the lowest spring tides can be seen, generally grow in deeper water.—

1175 Arachnidæ. Julus. Scolopen:— & Cancer from nearly Fresh Water.—

1176 R Frog: forest: above ash grey, posteriorly with Chesnut B. marks, sides & under surface, white & pale brown mottled. [New genus Borbocætes Grayii same genus as 986 & 1170 but difft species TB; listed as *Borborocœtes Grayii* Mihi in *Zoology* **5**:36]

1177 R X Frog. above bright rusty red, beneath [cont. opposite] fuscous. band over eye along side & lumbar gland blackish brown: forest: [Amph. (Bib.) *Cystignathus roseus* TB]

1178: 79. Frog. ~~V 283(a).~~ [note opposite] See account of Frog (285) [~~New genus~~ *Rhinoderma Darwinii* (1076) TB; listed as *Rhinoderma Darwinii* Bibr. in *Zoology* **5**:48]

1180 Vaginulus V **272**(a). *

1181		Arachnidæ
1182	C	Crab. fresh water Brook
1183		Acari (red) skin of Lizard.—
1184		Vaginulus V **272**(a).
1185		Pediculi V **315**.
1186	C	Crustacea. open sea, a degree S. of Concepcion.—

1835

1187	R X	Snake. sand dunes Concepcion
1188	F	Fish. disc of body yellowish brown with minute spots; 4 transverse bands in front part, & superior convex edge most beautiful cobalt blue. body generally dark yellowish brown.—
1189:	R	90. Frog same as (1178) Valdivia [note opposite] Above light brown with three narrow ash-colored bands bordered with a darker shade of brown: belly very pale with a row in centre of triangular black marks; on cheeks two oblique lines of do.— [*Rhinoderma Darwinii* TB]
1191	R	do. Valdivia [Amph. Bib. n.g. listed by TB as *Batrachyla leptopus* Mihi. in *Zoology* **5**:43-4]

[specimens 1192-1234 mostly collected April-June 1835]

1192	R	Lizard. Guasco. deformed double tail. [Listed by TB as *Proctotretus Chilensis* in *Zoology* **5**:2-3]
1193	C X	Fresh W Crab. Coquimbo [note opposite] Caught in gr numbers excellent food
1194	R	Lizard, nearly black, common on sea beach Copiapò [*Microlophus Lessonii* young TB; but not listed in *Zoology* **5**]
1195	C	Crab — Coquimbo *
1196:	R	97: 98. Lizards Concepcion
1199	R	Snake Is^{d} St. Mary do [see *Beagle Diary* p. 298]
1200	R	Frog Concepcion [? listed as *Litoria glandulosa* Mihi in *Zoology* **5**:42]
1201:	F	202. Fish, marine do [*Blennechis fasciatus* Jen., *Zoology* **4**:84-5. N.S. LJ]
1203	C	Squilla Coquimbo *
1204	F	Fish do [*Clinus crinitus* Jen., *Zoology* **4**:90-1. N.S. LJ]

1835

Coquimbo

1205	C	: 206. Crabs
1207	R	Lizard [*Proctotretus* ~~*Fitzingeri* (765)~~ *nigromaculatus* as listed by TB in *Zoology* **5**:30]
1208	R	Lizard [*Proctotretus* n.s. (767) TB]
1209	C	: 210. Crabs *
1211	F	Fish &c &c [*Blennechis ornatus* Jen., *Zoology* **4**:85-6. N.S. LJ]
1212		Scolopendræ. Julus

Valparaiso

1213	R	Frog [*Pleurodema obscuranum* (1024)1088)1168) TB]
1214		Scolopendræ Julus Valparaiso *
1215	F	Sucking fish —— R. Maule
1216	F	: 217. Fish Coquimbo
1218	F	: 220. Fish do [*Umbrina ophicephala* Jen., *Zoology* **4**:45-6. (Bad) N.S. LJ]
1223	C	224 Crabs do *
1225	C	Crab. Horcon (North of Valparaiso)
~~1226~~		~~Asterias Coquimbo~~
1227		Nitrate of Soda (V number 3054)

1228 C Crab Iquique Peru
1229 F Fish do — [*Engraulis ringens* Jen., *Zoology* **4**:136-7. N.S. LJ]
1230 R Lizard do — [*Microlophus Lessonii* (1194) TB]
1231 X Salt, edible V. Geology of Iquique [note opposite] from Poste largo , Copiapò (Specimen lost)
1232 F X Flying fish. Lat: 18° July [note opposite] Whole upper part of body & fins beautiful dark violet-blue. beneath snow white [*Exocœtus exsiliens* Bl., *Zoology* **4**:122-3. LJ]
1233 R X Snake. Traversia S. of Mendoza.— [note opposite] Color primrose yellow with broard jet black barbs which contain bright scarlet red square marks. belly black, excepting beyond tail, where rings of black & scarlet are continued all around. Scarlet brightest near head.— young one.—
1234 X Spiders, Los Arenales. (7000 ft?). East valley of Andes, S. of Mendoza. [note opposite] Web vertical: habitation of Spider in centre, strong lines proceed in all directions, between which there is a regular interlacing so as only to form a segment of a circle instead of a complete one.— [see sketch in margin] Spiders & webs very numerous in the Bushes. Autumn:

1835 July. Lima. S. Lorenzo

1235 Echinus. dark colored
---36 : 37 do. black.— Box XX
1238 F Fish dull lead color, with pale transverse bands. Pectoral, Ventral, Anal, Caudal Fins pale Vermilion.— [Unkn. Dry & in bad order.— LJ]
1239 F Fish. Dull colored but with numerous red spots like a trout. Pectoral & Caudal, orange
1240 F X Fish Pale greyish blue with black specks [cont. opposite] & clouds of do. Tips of fins pinkish.
1241 F : 42. Small silvery fish
1243 S Bulla — body yellowish —
1244 F X Fish, back iridescent green, belly [cont. opposite] white. All Fins & Barbillons reddish-purple. [One of *Siluridœ*.— Very bad & thrown away.— LJ]
1245 Spider & Hermit Crab
1246 F Fish. above nearly black
1247 C Crab, above purple, legs speckled *
1248 (Number lost)
1249 C Crab. white. back with purple punctures & legs marked with paler do
1250 C X Crab. beautifully marked with dark Lilac purple in regular forms, color brighter on legs [note opposite] Crab, very common. when taken draws all its legs close to its body & shams death.— *

1835 August Lima

1251: F Fish do [*Clupea sagax* Jen., *Zoology* **4**:134-5. N.S. LJ]
1252: F 53: 54: 55. Fish do
1256 C Crabs. Coquimbo. Mr King. *
1257 C Crab. Callao.—
1258 F Fish. Coquimbo.— 2
1259 C Crab. Mac. do. *
1260 F Fish.— Callao [*Otolithus analis* Jen., *Zoology* **4**:164. N.S. LJ]
1261 F Fish do
1262 C Decapod Notapod.— do *
1263 R Little Snakes 2 do
1264 Cryst. transparent salt; covered by layer of other saline substance (& spec: of do): sufficient covering of country.— Arica

Chatham Is^{d}. Galapagos

1265 F X Fish [note opposite] Above dull green, base of Pectorals & Dorsals black: a white patch beneath the pectorals: inflatable [*Tetrodon angusticeps* Jen., *Zoology* **4**:154. N.S. LJ] *

1266 F Fish. bluish-Silvery [*Pristipoma cantharinum* Jen., *Zoology* **4**:49-51. N.S. Exd. LJ]

1267 F X Fish. Beneath silvery white, above [cont. opposite] mottled brilliant "tile Red".— [*Prionotus miles* Jen., *Zoology* **4**:29-30. N.S. Exd. LJ]

1268 F XX Above & fins obscure greenish, [cont. opposite] sides obscure coppery, passing on belly into salmon color.— Pectoral fins edged with pale blue: Iris yellowish brown.— pupil black-blue [*Latilus princeps* Jen., *Zoology* **4**:52-4. N.S. Exd. LJ]]

1269 F Centre of each scale pale [cont. opposite] "Verm Red" Lower jaw quite white; large irregular patch above the Pectorals bright yellow. Iris red & pupil blue black [*Cossyphus Darwini* Jen., *Zoology* **4**:100-2. N.S. Exd. LJ] [In Natural History Museum, accession number 1918.1.31.11]

1835 Septemb. Galapagos Chatham Is^{d}

1270 C Crust. Parasite on Fish (1269) *

1271 R X Lizard. above brown obscure [cont. opposite] coppery, black; mottled with dirty white, so as to form two irregular white bands on each side.— Crest coppery: Before arm jet-black fold. Head & throat very dirty brown. Belly do yellowish, under side of do white.— [*Holotrepis* n.s. — listed by TB as *Leiocephalus Grayii* Gray in *Zoology* **5**:24-5]

1272 Fine orange. Holuthuria-formed animal: ½ smooth ½ with long Papillæ; former half most brilliant. [Veretillidae ? *Cavernularia* SFH]

1273 F X Fish. Mottled brown-yellow, black & white [cont. opposite] upper & lower edge of tail, edges of Ventral & dorsal (art & purplish Red). [*Serranus labriformis* Jen., *Zoology* **4**:8-9. N.S. Exd. LJ]

1274 F X do White, with four dark brown much [cont. opposite] interrupted bands, giving mottled appearance: do colored about head: top of do, ridge of back: edges of dorsal; tail & ventral fins tinted with fine "azure blue".— [*Chrysophrys taurina* Jen., *Zoology* **4**:56-7. N.S. Exd. LJ]

1275 F do Common large mottled brown fish [*Serranus olfax* Jen., *Zoology* **4**:9-11. N.S. Exd. LJ]

1276 Asterias. very dark "Art blood R". surface of suckers & tips of spines "Scarlet R"

1277 Onchidium V **321**

1278 Actinia V **321** [1 Actinian SFH] *

1279 R Lizard. above cream coloured & pale brown.— [*Holotrepis* n.s. allied to microlophus. Listed in *Zoology* **5**:24-5]

1280 R Lizards. both common.— [*Holotrepis* n.s. (1271) very young TB]

1281 R do. certainly distinct species: above dusty olive green, mottled with blackish brown; belly cream-color, with band on each side of orange: patch of do. beneath throat.— Sand dunes [D^{o} 1271 TB]

1835 Septr. Galapagos Chatham Is^{d}.—

1282: F X 83. Whole body "scarlet Red". fins [cont. opposite] rather paler: small irregularly shaped light black spots: very extraordinary: Are they distinct? [*Scorpaena histrio* Jen., *Zoology* **4**:35-7. N.S. Exd. LJ]

1284 F Pale yellowish brown, with numerous transverse bars, of which upper parts reddish black: lower "Verm R", Gills covers. Head & fins tinted with do. Fish.— [*Prionodes fasciatus* Jen., *Zoology* **4**:47-9. N.S. Exd. LJ]

1285 Onchidium V **322**

1286 F X Eel. tidal pools. color dark [cont. opposite] reddish purple brown, with pale or whitish brown spots. Eyes blueish.— [*Murœna lentiginosa* Jen., *Zoology* **4**:143-4. (Same as 1299) LJ]

1287 F Fish.— tidal pools.— [*Gobius lineatus* Jen., *Zoology* **4**:95-7. N.S. LJ]

1288 F Fish. do [*Gobiesox pœcilophthalmos* Jen., *Zoology* **4**:141. N.S. LJ]

1289 S X Balanus. one common on the [cont. opposite] rocks.— the other on the green Turtle. (Capt. FitzRoy sends specimen of this animal). for Dissection:

1290 C X Crust. tidal rocks: the Decapod [cont. opposite] is black: runs extraordinarily fast like Spiders: inhabits the hole in cliffs above high water mark: the Crust. allied to Squilla was bright green:—

1291 C do. do: except a shrimp from 50 Fathoms deep.—

1292 C do. tidal rocks: big claw from holes. mud-bank. Lagune.—

1293 F X Diodon. Beneath snow white. Above dark brownish black. this color is placed in broard rings one within the other on the back, so that on the side they form oblique ones which point both [cont. opposite] ways: Whole upper surface spotted with darker black circular spots.— Pectoral & dorsal fins yellowish brown. Iris inner edge clouded with orange. Pupil dark green-blue. Make a loud grating noise
[listed as *Tetrodon annulatus* Jen., *Zoology* **4**:153-4. N.S. LJ]

<u>1835</u>: Septr Galapagos. <u>Charles Is^{d}</u>

1294 R Snake: 2 dark "Liver B" stripes separated by lighter brown, on each side of these, pale yellowish B. stripes edged again by the "Liver B", which shades beneath into pale "Clove B".—

1295 R X Snake: sides "clove B" [cont. opposite] back do tinged with "yellowish B" marked with circular patches of "Blackish B" forming a double band

1296 R X Lizards. Upper part "Clove B" [cont. opposite] passing into "Black B" with black spots, sides tinted lightly with orange; some scales of crest near head white: belly whitish, whole throat before front legs glossy black.— This is commonest variety; <u>black</u> spots on scales not infrequently are arranged in sinuous transverse bars & sometimes longitudinally: vary in numbers much [*Holotrepis* n.s. (1271) TB]

1297 R X Lizard. above "Hair B" mottled with "yellowish B" (sometimes arranged in longitudinal bands) throat light black: on each side intense orange mark.— [D° 1271 TB]

1298 Scolopendra. dark reddish brown. grow to 14 inches long.— *
[see *Oxford Collections* p. 222]

1299 F Fish fine dark <u>purplish</u> brown, with yellow circular spots.
[*Murœna lentiginosa* Jen., *Zoology* **4**:143-4. N.S. LJ]

1300 R X Lizard, perhaps variety of (1296). above greenish sooty brown, mottled with yellowish brown. clouded & spotted with jet black: fold before front leg [cont. opposite] do.— belly yellowish.— some hair side of face colored orange [note on RSW list: 1300-1271-1281-1303-1296-1280-1297- *Holotrepis* of Bibron new species (*Leiocephalus* Genus of Gray founded upon bad characters says Bibr.); D° 1271 TB]

<u>1835</u> Galapagos Is^{ds}

1301 R Snake: sides "Clove B", belly pale do. ridge of back narrow band of do: bordered by sinuous margin of "Umber B".— Charles.—

1302 F X Fish. silvery. above, shaded with [cont. opposite] brown & iridescent with blue: fins & iris sometimes edged with blackish brown.— Flap of gill-cover edged with black.— [*Pristipoma cantharinum* Jen., *Zoology* **4**:49-51, (Young). N.S. LJ]

1303 R Lizard.— Above blackish brown, beneath yellowish edged with orange, throat black

[*Holotrepis* n.s. (1271) TB]

1304 F X Fish, varys much in color [cont. opposite] Above pale blackish green: belly white: Fins, Gill covers & parts of sides dirty reddish orange: on side of back 6 or 7 good sized snow white spots, not very regular outline. In some specimens the blackish green above dark, separated by straight line from the paler under parts: Again others colored dirty "Reddish orange & Gallstone yellow", upper parts only rather darker: but in all, white spots clear.— (5 or 6 in one row & one placed above). Sometimes fins banded with orange & the black green lengthways.— [*Serranus albo-maculatus* Jen., *Zoology* **4**:3-5. N.S. LJ]

1305 R Black Lizard (for dissect) V **333** [*Amblyrynchus cristatus* young TB]

1306 Gorgonia V **326** [Gorgonian (purple) SFH] *

<u>James Is^{d}.—</u>

1307 Arachnidæ (sweeping)

1308 R Snake: on back 2 yellowish brown bands, between which & on each side dark umber brown: belly pale

1309 Body of bird (3374)

1310 R Snake; centre of back, one dark Umber Br. band, edged with black each side Clove brown shading into the whitish belly

<u>1835</u> Galapagos

1311 Salt water from Salina in Crater at James Is^{d}.— V Geology

1312 In place of 3156

1313 do of 3157

1314 S Balanidæ. bottom of Yawl

1315 R Lizard V **336** [*Amblyrynchus Demarlii* Bibron IV. p. 197; listed by TB in *Zoology* **5**:22]

1316 C Crust. Mac^{r} from <u>fresh water pool</u> near Sea Beach Chatham Isl.

<u>Otahiti</u>

1317 F Fish. splendid Verditer blue & green
[*Acanthurus humeralis* Cuv. & Val., *Zoology* **4**:76-7. Exd. LJ] *

1318 F Fish. [*Scarus lepidus* Jen., *Zoology* **4**:108-9. N.S. LJ]

1319 F Fish. [*Scarus globiceps* C. & V., *Zoology* **4**:106-8. Exd. LJ] *

1320 F Fish. [*Upeneus trifasciatus* ?, *Zoology* **4**:25-6. LJ] *

1321 F Fish. [*Caranx torvus* Jen., *Zoology* **4**:69-71. N.S. LJ]

132~~2~~3 P Sea weed (& minute club-head Corallina) growing on the reef. greenish brown.

1322 Scorpion. Oniscus &c Mountain of interior

1324 F Extraordinary Fish.—

1325: F 26 —— Fish.— [*Ostracion Meleagris* Shaw = *O. punctatus* Schn., *Zoology* **4**:158. LJ] *

1327 F Fish. [*Murœna* LJ] *

1328 F Fish. [*Syngnathus conspicillatus* Jen., *Zoology* **4**:147-8. LJ]

1329 F Fish. [*Balistes aculeatus* Bl., *Zoology* **4**:155-6. LJ] *

1330: F 31 Fresh Water Fish [*Dules Leuciscus* Jen., *Zoology* **4**:17-8. N.S. Exd. LJ] *

1332 C Fresh water Prawns *

1835 (November) Tahiti

1333 C Fresh water Shrimp & one Marine *

1334 Fungia V **345**

December Bay of Is^{ds}.— New Zealand

1335	F	F. W. Fish
1336		Julus & Scolopendræ. Arachnidæ
1337	F	F Water Eel [*Anguilla australis* Rich., *Zoology* **4**:142, bad spec. LJ] *
1338		Land Snails, color blackish. back slightly carinated
1339	F	F Water Fish [*Eliotris gobioides* Val., *Zoology* **4**:98-9. LJ] *
1340	F	do [*Mesites attenuatus* Jen., *Zoology* **4**:121-2. N.S. LJ]
1341	F	Fish. whole body bright red [*Trigla kumu* Less. & Gurn., *Zoology* **4**:27-8. Exd. LJ] *
1342	R	Lizard. whole body fine green: lives in trees: is said to make a laughing noise.— [*Platydactylus* of Bibron - *Gehira elegans* of Gray in Zool. Soc. not described nor figured TB; but listed as *Naultinus Grayii* in *Zoology* **5**:27-8]
1343	F	Fish [*Trypterygion Capito* Jen., *Zoology* **4**:94-5. N.S. Exd. LJ]
1344	F	Fish tidal rocks [*Acanthoclinus fuscus* Jen., *Zoology* **4**:92-3. N.S. Exd. LJ]
1345	F	Fish do
1346	C	Crabs —— do —— *

[some illegible pencil notes on last pages]

Down House Notebook 63.3

Catalogue for Specimens
in Spirits of Wine
1347|————|1529

Jan^y.		1836	New S. Wales
1347	C	F. W. Crab & Tulidæ	
1348	R	Snake	
1349	R	Frogs [*Cystignathus ocellatus* TB]	
1350		Frog [*Hyla Peronii* TB]	
1351		Frog [Amph. Bib. *Hyla fusca* ??? too bad to determine TB]	
1352		Frog [Amph. Bib. *Cystignathus Georgianus* genus, listed by TB in *Zoology* **5**:33-4]	
1353:	R	Lizards [*Grammatophorus muricatus* TB]	
1354:		Lizard [*Tiliqua* [?] (Cocteau) *T. tæniolata* Gray TB]	
1355		Lizard [*Grammatophorus muricatus* TB]	
1356	S X	Oyster: small pools; muddy [cont. opposite] almost separated from the Sea.—	

		February:	Van Diemen's. Hobart town
1357	S	3 species of Balanus	
1358	R	Lizard V **346** [*Tiliqua Casuarini*, listed by TB as *Cyclodus Casuarinæ* Bibr. in *Zoology* **5**:30]	
1359	R	60: 61. Lizards [*Tiliqua* ? named by Bibron but not fig^d TB]	
1362	R	Lizard V **347**	
1363	R	Snake V **348**	
1364	R	Lizard. V do [*Grammatophorus barbatus* young (1353) TB]	

		March	King George's Sound
1365	F	.....1386. Various fish caught by net in Princess Royal Harbor	

[Those classified by Leonard Jenyns are listed below]

1365, 1366	*Caranx Georgianus* Cuv. & Val. *Zoology* **4**:71-2. *
1368	*Platycephalus inops* Jen. *Zoology* **4**:33-5.
1371	*Arripis Georgianus*. *Zoology* **4**:14-15. *
1372	*Helotes octolineatus* Jen. *Zoology* **4**:18-20.
1374	*Caranx declivis* Jen. *Zoology* **4**:68-9.
1375	*Upeneus?* Dry.
1378	*Dajaus Diemensis* Rich. Dry. *Zoology* **4**:82-3. *
1380	*Aleuteres velutinus* Jen. *Zoology* **4**:157-8. *
1381	*Platessa ——?* Dry & bad. *Zoology* **4**:138.
1386	*Apistus?* Dry. *Zoology* **4**:163. *

[CD's list continues]

1387 S X Balanus (for dissection). [note opposite] Some Specimens have numerous eggs or Larvæ, each of which is when immature a sharply pointed oval, afterwards a small animal with six (?) legs furnished with setæ

1388 R Lizard, caught in trap baited with cheese; frequents rocks near coast
1389 F Skate, above muddy cream color
1390 F Fish. Above varied dull green, with pale do, beneath snow white
1391 F Fish. pale copperish brown [cont. on next page] with water marks of a fine darker brown.— [*Loucate* ? new genus. Dry & in bad order. LJ] *

March 1836 K. George's Sound

1392 F F. very pale brown, fins pale orange. [*Aleuteres velutinus* Jen., *Zoology* **4**:157-8. N.S. LJ] *
1393 F F. mottled with pale blackish green, leaving white spots [*Aleuteres maculosus* Rich. ?, *Zoology* **4**:156-7. LJ] *
1394 F F. sides fine dark green, & pale silvery green, fins tipped with red. Iris fine green. handsome fish
1395 F F. silvery, ~~above~~ back dark greenish do
1396 C X Crust: colourless. inhabiting [cont. opposite] a most enormous cavity in proportion to size of parts, which not more than 5 or 8 times larger than the parasites. Cavity in flesh on sides & behind the gills & extending apparently to membrane of abdominal viscera.— whole animal, excepting tip of tail included in cavity.— Fish class of Diodon, but with many teeth.—
1397 R Frog, centre of back rich brown, lateral bands of pale orange, under which a narrow irregular line of the brown; orange brightest on flanks, hinder thigh & legs.— [Amph. (Bib.) *Cystignathus georgianus* Bibr. (*Crinia Georgiana* Tschudi) as listed by TB in *Zoology* **5**:33-4]
1398 R Tree frog, above with irregular marks of bright green, margined with copper, intermediate [cont. on next page] spaces, pale silvery brown. [*Hyla Jacksonianus* Bibr. TB]

March 1836 K. George's Sound

1399 R Rana, above bright green, brown streak along the eyes, hinder thighs orange. [*Hyla furca* ??? too bad to determine TB]
1400 Arachnidæ; sweeping caught by; neighbourhead of Sydney.—
1401 C X Fresh water Crust: Mac. above jet black, beneath & basal joints of legs with tinge of dark blue. [note opposite] Inhabits holes in soft mud in swampy places
1402 F Fish. blue. silvery.—

April Keeling Islands

1403 C X Crabs; the Decapod is nearly white. runs like lightning with erected eyes on the white sand beaches: Hermit Crab, colored bright scarlet, frequent a particular univalve & swarms on the coast & in all parts of the dry land far from water: Another Hermit Crab is likewise found inland [note opposite] its front legs form a most perfect operculum to the shell.— *

[as noted on p. 299 in the *Zoology Notes* CD wrote 1835 in place of 1836 for the next few months, and made the same mistake in the *Specimen Lists*. This error has been corrected without comment wherever it appears.]

1836 April Keeling Is[d]

1404 C Large brown crab with scarlet eyes.
1405 X Black Holuthuria } Both excessively
1406 Brown do. } abundant on the beach
[note opposite] These species afford a very inferior kind of Trepang

1407 F X Fish: whole body dull "imperial purple". Dorsal fin intense yellow, tail banded white & black: face from snout to behind the head jet black, throat white. [note opposite] Colors imperfectly described

1408 F Whole body most beautiful "Verditer blue", said to eat coral; Fish common on the outside reefs in small shoals

1409 F X Eel: cream coloured with rows of large black rings, which send off coarse reticulation of the same color [note opposite] a very fierce fish; immediately on seeing any person opens its mouth & will even spring out of water; is very strong & has great power in its jaw: lives on crabs, the hard & large shells of which are broken with ease: frequent very shallow places.—

1410 F X Fish: dirty white, with 3 longitudinal bands of pink & some transverse lines [cont. opposite] on its sides.—

1836 April Keeling Is^{ds}.—

1411 X Holuthuria. Very pale brown [cont. opposite] with numerous irregular rings of dark rich brown: mouth with 16 (?) long tentacula. bordered on each side with simple papillæ. surface of body covered with extraordinary adhæsive points: length about 3 ft, but body excessively extensible & of a soft flabby nature so as to be easily broken.

1412: F 13: Fish Coral reefs.— [*Acanthurus triostegus* Bl. Schn., *Zoology* **4**:75-6. LJ]

1414 F Fish Coral reefs [*Chætodon setifer* Bl., *Zoology* **4**:61-2 LJ]

1415 Actinia. V **361**

1416 P 2 species of sea-weed from holes in the reef. The one with a reticulate structure grows in square pyramidal leaves or masses in little tufts, is colored with the same red, as common to Fuci: the other species, common, is of a reddish Salmon color:

1417 Nereis, under stones; on the slightest touch the hairs remain fast in a persons skin & thus cause considerable pain. [1 Amphinomidae SFH] *

1418 C Various Crabs beneath stones *

1419 Millepora V **356**

1420 F X Fish: dirty metallic olive green, with white circular spots. belly white with streaks [cont. opposite] of same color with back.— [*Tetrodon implutus* Jen., *Zoology* **4**:152-3. N.S. LJ]

1836 April Keeling Is^{ds}.

1421 P Sea weed. pale green: a fragment of a kind growing like a lichen.

1422 C Crab. commonly inhabiting Fresh Water wells.—

1423 F Fish. Band on side "Azure blue". above a duller greenish blue; beneath two greenish metallic stripes: lower half of body snow white
[*Seriola bipinnulata* Qu. & Gaim., *Zoology* **4**:72-3. LJ] *

1424 F Fine Verditer Blue, with some yellow stripes about head & fins
[*Scarus chlorodon* Jen., *Zoology* **4**:105. N.S. LJ]

1425 F Most beautiful silvery white, dorsal fin blueish, upper part of body with a beautiful shade of "siskin green". on mid side a row of few irregular spots of "gamboge yellow": eye jet black.

1426 S Bivalve adhering to branching stony Corals in the Lagoon. when young, bright green:

1836 April Keeling Is^{ds}.—

1427 F Beautiful white, with a yellow tinge on back: on each side form bands of a pale blueish green: the edges of these being darkest give the appearance of having been sirled.—

1428 C Crab. V **362** Copy

1429 F Fish. dull red transverse lines [*Salarias quadricornis* Cuv. & Val., *Zoology* **4**:87-8. LJ]

1430 F do. body dull reddish & greenish colors, blended & mottled; fins banded lengthway with "Vermilion R." head with waving bright green lines.
[*Scarus* ———?, *Zoology* **4**:109. Exd. LJ] *

1431 F do. body pale with narrow dark straight lines which form network; across eye black band; posterior half of body bright orange. upper part of prolongation of dorsal fin edged with black & round patch of do [*Chaetodon setifer* Bloch., *Zoology* **4**:61-2. Exd. LJ] *
1432 F Whole fish silvery.— [*Gerres Oyena* Cuv. & Val. ?, *Zoology* **4**:59-60. Exd. LJ] *

1836 April Keeling Is[ds].

1433 F Upper part pale lead color, pectoral fins yellow; ventral & anal orange: sides very pale yellow [*Diacope marginata* Cuv., *Zoology* **4**:12-13. Exd. LJ] *

[note opposite] NB. There is a Cask with a Cross marked with ink, in which are the great land crabs & some small fish.— in another cask there is the large Coral eating Fish in the dung of which I could perceive bits of Coral. These fish colored beautiful "Verditer blue" are found within the lagoon. live on the delicate branching stony Madrepores & Seriatopora.

1434 S X Oysters for dissection [note opposite] NB Mr Liesk informs me that Cypræa hatches its young, if removed will return to its eggs: that there is considerable difference between certain male & female shells: that certain kinds always go in pairs [see *Beagle Diary* pp. 413-19]
1435: C 36. two Crabs
1437 Ophiura, curious spines. [1 Ophiurid SFH] *
1438 Actinia (allied to). the individuals are so closely packed together as to form a cushion on the outer part of outer reef: color dull leaden & reddish, centre of each animal bright green.—
1439 F Fish, coloured in circles red, white & dull brown.—
1440 F Dull silvery fish [*Mugil*———? Dry & in bad condition.— *Zoology* **4**:81-2. LJ]
1441 F Do. with yellow stripe on side [*Upeneus flavolineatus* Cuv., *Zoology* **4**:24-5. Exd. LJ]
1442 Spiders, caught by sweeping. King George's Sound.
1443 A X Mouse. south of Concepcion, Chili. [note opposite] Given me by Capt FitzRoy one of a complete pest which overruns the country.

1836 May Mauritius.—

1444 Scolopendra. Julus. Scorpio. Mountain of La Puce.—
1445: R 2 46: 47. Frogs: swampy places near the sea: extraordinary high jumps.— [listed by TB as *Rana Mascariensis* Bibr. in *Zoology* **5**:32]
1448 X Limax. on summit of La Puce. 2600 ft. [note opposite] Body nearly transparent, slight tinge of yellow: superior tentacula black line, inferior colourless: upper surface very slightly mottled with most minute black points in patches: anterior margin of shell black, shell itself & skin which covers it with brown & white.—
1449 F Fish —
1450 C Crab.— *

June Cape of Good Hope

1451 Asterias
1452 R Lizard [*Cordylies griseus [?]* TB]
1453 Lizard [*Scincus travittatus [?]* TB]
1454 Lizard [*Gerrhosaurus sepiformis* not fig[d] TB]
1455 R Frog [Bibron to write about TB]
1456 Frog [listed by TB as *Rana Delalandii* in *Zoology* **5**:31]
1460 C Fresh water Crust. Amphipod: *
1461 Arachnidæ. sweeping.—

1462 Nereidous animal: Ascension [note opposite] Case (3818) consists of nearly parallel tubes of cemented particles of shells: Mouth one side always highest forming sort of hood.— Is said to produce large masses within short space of week.—
1463 2 species of Corallina: Bahia Brazil. One inarticulata with corals V **367**
1464 Halimeda. Bahia. tidal pools.— at extremities of branches coarse vascular structure.—

1836

1465 P Flower & leaves of a low shrubby tree growing on the hills: flower singular dirty white. petals fleshy.—

August end Cape de Verd. St Jago.

1466 S Balanus. on sandstone reef. Pernambuco. Brazil
1467 A Bat. common in old Lime Kiln. Hab. do— [listed as *Phyllostoma grayi* in *Zoology* **2**:3-4]
1468 Associated Actiniæ. very abundant, forming layer on sandstone reef (also on coral reef in Pacific) in small cavities & where most exposed.— Pernambuco [Zoanthid SFH] *
1469 S Shell embedded in recent calcareous agglomerate, in reef of Pernambuco.
1470 A Mouse. Bahia Blanca; caught in grass far from houses. given me by Mr Bynoe

[final entries evidently added after the end of the voyage]

1836 Decemb.

1474 (same 78) Chiton. Porto Praya
1475 (same 59) Shells. Porto Praya
1476 (—— 79) Bulla. do
1477 (91) Crust St Jago *
1478 104. Shells St do
1479 125. Crab do
1480 159 animal described at P 2 [sev. Chaetognaths SFH] *
1481 158 Crust. Lat 18°6′ S 66. W [?]
1482 (145) Shells. Bahia
1483 (145) Fish do
1484 300 Crust, from Lagoa, Rio
1485 Arachnidæ. <u>Rio (?)</u>
1486 (248) F. W. Leaches. Rio
1487 (339) Arachnidæ M. Video
1488 424 Spider B. Blanca
1489 number lost Crust. Falkland Isl[ds] East of S. America *
1490 Crust (501) 30 Fath. 53° S
1491 Asterias (do) do do
1492 Shell. Fissurella. T. del Fuego

[last page of notebook]

1493 (no lost) Lycoperdium Maldonado?
1494 (1058) Scorpions Valparaiso
1495 (958) Crust. 16 F. C. Espiritu Santo
1496 (1067) F.W. Leach Valparaiso
1497 (797) Sea weed. Port Desire
1498 Crab. Keeling Isl[d] (?) *

1499	(1155) Crab. —— Chiloe.— *			
1500	501. 502. 503. 504. Crust Brachyuri. Mauritius Dr Page *			
1505	Crust. Mac.	do	do	
1506	Crust.	do		
1507	Hermit Crabs	do	do *	
1508	Asteria Echinus	do	do	
1509	Crust. Keeling Isl[d] (?)			
1525	for 282 Nullipora			
1529	for 797 do			

Specimens not in Spirits

Down House Notebook 63.4

C.Darwin

~~H.M.S. Beagle~~

<u>12 Upper Gower Street</u>

North Latitude

<u>1832</u>		<u>Paper number in white</u>
1	S	Spirula Peronii. 6th Jan. off Santa Cruz. stomach of Larus fuscus. [in margin] <u>Cabinet</u>
2	I	Taken on board Jan. 10th, Lat. 21-2 [for specimens 2-5 see *Insect Notes* p. 45]
3	I	Acrydium. owing to prevailing winds must have come from C. Blanco in Africa, 370 miles distant Jan. 13th V. Kirby Vol. 1, P 224 [see William Kirby and William Spence. *An introduction to entomology*. 4 vols. London, 1815-26. In *Beagle* library.]
4	I X	Jan. 14th — 10 miles at sea from St Jago insect
5		Jan. 12th Lat: 19° insect
6		Jan. 11th Lat: 21-2 Crustaceæ
7	S	Atlanta peronii (fuscus Dic: Sc), judging from shell only: differs from figure in Blainville (~~by the whorls being separatedby keel produced:~~) by having more keel: & the oblique markings not so strong: taken in plenty, Lat. 21-2. Jan. 11th — also some in Spirits. A. (1)
8	S	Atlanta Madraunii (Dic. Sc:) from the animal. Lat. 19: — Jan. 13th.—

1832 St Jago

9	S X	Creseis (Rang) or ~~Orthyra~~ in profusion. Lat: 21-2. Jan. 11th [note opposite] Length 35 near the acute point at slight contraction: with high power, transverse bands were visible
10	S X	Creseis agrice rotundo. Lat 19. Jan 13th [note opposite] Vide P. **3**(a) Copy
11		Dust fallen on board, Jan 16th 10 miles W of St Jago.— V. **3**(f)
80	S	Serpula. Quail Island [note opposite] Jan. 18th
63	S	Univalve shells (chiefly). Quail Island [note opposite] Jan 17th
94	S	Coralls ~~Shells from Quail Island, interesting to be compared with those found on the upraised shore~~
95	S	Do
96		Corallina. Quail Island.
97		Do
98		Do
99		Do
100		Operculum & horny pieces out of stomach of two Aplysias.—
138	S X	Limnea: turreted Limnea: Physa in profusion: Planorbis: [note opposite] In a little pool Jan 26th in the valley of St Martin, W of Praya

1832 St Jago

139		Echinus. very common [note opposite] Quail Island
140		Asterias. tops of prickles scarlet red, & other parts tile red
141		Asterias

142		Astræa. reddish coloured
143		do common
145		Astræa greenish.—
185	B	These birds were shot in neighbourhead of Porto Praya from 16th of January to 7th of Feb. Gull.
186	A	A mouse very common [listed as *Mus musculus*
187	A	on Quail Island in *Zoology* **2**:38]
188	B	Cock bird common in the interior on the table land. Pyrrhalauda in small flocks: females of the same colour as back of bird.— runs like a Lark.— [listed as *Pyrrhalauda nigriceps* Gould in *Zoology* **3**:87-8]
189	B	Uncommon [identified as *Passer hispaniolensis* G.R.Gray in *Zoology* **3**:95; skin at NHM carries CD's own label] *
190	B X	Sparrow. The commonest bird in the island.— generally in small flocks, both round the houses & in wild desolate spots [listed as *Passer Jagoensis* Gould in *Zoology* **3**:95]
191	X	Alcedo. [Senegal kingfisher, listed as *Halcyon erythrorhyncha* Gould in *Zoology* **3**:41-2] Very frequent in the valleys where there is no water.— but still more abundant near water as at St Domingo. Their stomach[s] contain wings &c of Orthopterous insects, & ~~Mr Maccormick~~ one was caught ~~one~~ with a lizard in its craw.— It is a solitary tame bird & has not the swift flight of the European species.— It is the only brilliantly coloured bird in the Island.— [specimen labelled 0192D at NHM] *
		[note opposite] N.B. The Island abounds with hawks & a small raven.— with Quails & Guinea fowl. On the coast the beautiful Tropic bird builds.

1832 St. Jago

193	S	Shells. [original entry crossed out] Buccinum. Murser[?] includes, Porcelina, Fissurella.— Quail Island, interesting to compare with those in the former coast
196	P	Fruit from the great Adansonia, NE of Port Praya [see *Beagle* Diary, pp. 26-8, and *Plant Notes*, p. 155]
197		~~Corallina~~ & other Corals 4 Box
198		Sertularia. Dinamena.
199		Jania. V. **14** (a) [*Jania micrarthrodia*, see *Plant Notes* p. 187]
200	P	2 sorts of Fucus. V. **14** (b) [see *Plant Notes*, p. 155]
201	I	202. Harpalidæ Quail Island [for specimens 201-231 see *Insect Notes* pp. 45-7]
203	I	Allied to Crypticus
204	I	do. These two insect<s> are found in the greatest profusion under stones, all over St Jago
205	I	Allied to Trechus
206	I	Bee. Common, making nest in the rocks
207	Lost	Spider, out of Cathedral at Ribera grande.—

1832 St. Jago

208	I	209. Hygrotus, stream at St Martin. [note opposite] is not this genus generally confined to colder climates? [next sentence later crossed through] Is it not interesting finding fresh water animals in the islands, supposed to be part of Atlantis?—
210	I	Corixa. do.
211	I	Lice from head of gull (185). I observed they continue alive on bird many days after its death
212	I	Blatta St Domingo
213	I	214. — Gyrinus, allied to Dineutes. MacLeay (?). [note opposite, later deleted] Hab. same as 216 &c Did not recognise them as Gyrini in the water [see *Insect Notes*, p. 46]
215	I X	Gyrinus. Do (?) [note opposite] Solitary habits like some European species.

216	I	217. 218. Hydrobius. stream near St Domingo
219	I	Hydrobius & Gerris. Hab. do
220		221. Spider. in water. St Domingo
223		Mucor Linn: V **20** (a)
225	I	226. 227. Ornithruga Latr: /Feronia Leach/ from the Booby. frequent Feb 16th St Pauls
1832		Feb St Jago
228	I	Moth St Pauls
229	I	Staphylinus do: Birds dung
230		Spider do. common
231	I	Oniscus do
232		233. 234. Tic[k]s. do

		[note opposite] N.B. Following plants collected at St Jago from 16th to Feb 8, 1832
269	P	The commonest tree in the island. growing in the valleys: the juice abounds with Gallic Acid, making all iron things directly black
270	P	near stream at St Domingo [*Christella dentata*. See *Plant Notes* p. 156]
271	P	272. Plants
273	P	Ribera grande
274	P	275.— Plants [*Chenopodium murale* L. See *Plant Notes* p. 157]
276	P	277 Water cress & other plants from St Domingo [See *Plant Notes* p. 157]
278	P	279 Plants [*Campanula jacobaea*. See *Plant Notes* p. 157]
280	P	St Domingo. damp place. [*Tagetes patula* L. See *Plant Notes* p. 157]
281	P	Do [*Achyranthes aspera* sp. See *Plant Notes* p. 157]
1832		Feb St Jago
282	P	St Jago [note opposite] Plants continued
283	P	Dry places
284	P	St Martin
285	P	
286	P	the two lowest plants are the commonest on desert places.— the bush smells sweet.—
287	P	288 [*Dalechampia senegalensis*. See *Plant Notes* p. 158]
289	P	Quail Island
290	P	291. Do
292	P	Plant with stalk on rocks near sea.
293	P	294 St Domingo [note opposite] ~~Altogether about 70 species~~
295	P	296. 297. 298 [See *Plant Notes* p. 158]
299	P	300.— St Martin
301	P	302. 303. St Jago
304	I	Termites. Fernando Noronha.
305	I	part of their nest.— Vide **1** Geological Notes.—
1832		Feb
306	S Lost	Bulima. roots of trees Fernando Noronha
307	S	Murex Bucanum Mytilus Arca Turribella &c &c St Jago
308	P	Rhynchites. Seeds of the Tamarind. St Jago
309	P X	Lichen from the highest peak of Fernando Noronha
325	I	Numerous small Coleoptera & Hemiptera from Bahia [see *Insect Notes* p. 47]

345 346 Corallines Fernando Noronha
347 P Fuci. do
[added opposite] 348...353 Taken at Bahia from Feb 29th to March 17th
348 I X 349 Numerous Coleoptera from Bahia. part of a couple of hours collecting [see *Insect Notes* pp. 47-8]
350 S Three species of land shells, the smallest on a Parasitical Orchis

[for specimens 351-368 see *Insect Notes* p. 48]

351 I Onthophilus. perceiving the smell of human dung with singular quickness
352 I Elater Noctilucus: Vide **25**:
353 I Cimex. drove its proboscis deep into my finger

1832 Feb. 29th March 17th Bahia

354 I Geotrupes
355 I Acarus from do
356 I Louse from Vespertilio [see *Insect Notes*, p. 48]
357 I X 358. Specimens from an enormous migration of ants [note opposite] (V Page **28**)
359 I 360. 361. 362. 363. 364. a very common species of ant: the winged ones were flying in numbers from the nest
365 I 366 Feb Hymenopterous insects
367 I Nest of do.— When large & complete is globular
368 I X Curious habitation of some insect on a root in a sand bank.—
[note opposite] May 31st have found out it belongs to Hymenoptera
384 P A leafless tree bearing beautiful pink flowers at Fernando Noronha, an essential character in landscape

1832 March

385 S Atlanta
386 I Mantis ~~V. 29~~ (Copy) [for specimens 386-389 see *Insect Notes* pp. 48-9]
387 I Butterfly. very common. on main island of Abrolhos March 29th
388 I Helops do
389 I Ornithomya.— nearly all the birds in island were Totipalmes: yet this insect I think differs from those taken at St Pauls from the bodies of a Sula.— Abrolhos.— 29th.—
390 391. Oscillaria V **31**.
392 P Conferva. V **32**
393 S Coronula. Abrolhos
394 S Balanidæ.— do in corals. Shells &c Bahia
395 X Halimeda.— [?] [*Halimeda opuntia*, see *Plant Notes*, p. 187]
2 Eschara, 20 fathoms off Abrolhos.— [note b opposite] One of the Eschara had opening of cell this shape: body of polypus doubled up behind cell.— [sketch in margin] Cell covered with pores [Busk Collection: *Membranipora magnilabris*] [*Stegenoporella magnilabris*, Type by exchange with B.M.(N.H.) SFH] *

1832 April

412 S Phasianellas. Abrolhos. March 29th
413 B Eggs of the Booby, & a smaller. Another one of the Noddy.
St Pauls.—

414	I X	Coleoptera from the neighbourhead of the Rio Macao [note opposite] From 414 - - 437 all taken middle of April. All the places on road to Rio Macaè from Rio de Janeiro [for specimens 414-538 see *Insect Notes* pp. 49-56]
415	I	do Rio de Janeiro
416	I	417. Cicindela from the woods, Socêgo.—
418	I	Carabidous.— from Rio Frade
420	I	421. Colymbites, small puddles Socêgo
422		423. Diptera R. Macaè
424	I	425. 426. Blattæ under bark of rotten tree at Socêgo
427	I X	428. 429. Blaps do [note opposite] Emitted a musky, together with the usual disagreeable smell.— Stained my fingers for some days of a purplish red colour
430	I	Erotylus. Socêgo
1832		April & May Rio de Janeiro
431	I	Cimex [note opposite] 431-457 for dates & places Vide Suprà
432	I	433. Gyrini. Campos Novos
434	I	Diptera. Mandetiba.
435	S	Lymnœa (?) Vide **36** (a)
436	S	2 Species, fresh water Shell, Campos Novos
437		Coralline. Mouth of Macaè [*Amphiroa variabilis*, see *Plant Notes*, p. 187]
438	I	Coleoptera. Botofogo
439	I	Diptera do causing intolerable itching
440	I	Lampyrus V **41**
441	I	Do; both highly luminous
442	I	Females of this insect. Larva do
443	I	Do. — luminous V. P. **42**
444	I	Lopha (?) taken in great number on sand walk. at night.
445	I	Coleoptera
446	I	Freshwater Coleoptera, ~~including Minute Hydroporus Hygrotula Hydrobius & some Hydrophili~~
1832		May. Rio de Janeiro
447	I	Hydrobius inhabiting salt water lagoon (road to Botanic Gardens)
448	I	Hydrophilus, together with the last.—
449	I X	Ants found in (I do not know whether making) a nest like (368) found at Bahia [note opposite] Ants do not make it.— I found ones somewhat similar, filled with half dead Spiders, evidently collected by some Hymenopterous insect. It is the case, V No 536
450	I	Ricinus from a pretty, but common yellow Certhia
451	I	Ricinus, do do. (another species)
452	S	Helix V. **38** (Copy)
453	I	Insect ~~Santerella~~, colour changed by boiling water from grass green into a yellow
454	I	do do do do
455	B X	Krotophagus [note opposite] In the stomach were numerous remains of various Orthopterous & some Coleopterous insects. [listed as *Crotophaga ani* in *Zoology* **3**:114]
456	I X	Lampyrus [note opposite] different species from (440) shine nearly as brightly; uncommon; caught in web of a small Epeira.—
457	I	458. Geotrupes; collect human dung into balls & push it along with hind legs

1832		May. Rio de Janeiro
459	I X	Acarus. from a Passalus [note opposite] 459 & 460.— in very moist rotten wood.—
460	I X	Circulio nearly covered with Acari. Uropodes Lat^{r}.
461		Spider: web horizontal
462	I	Hymenoptera. the most common species, in great numbers.—
463	P	Lichens. mosses &c on trees chiefly oranges old trees.
464	P	Lichens [*Pycroporus sanguineus* L. See *Plant Notes* p. 160]
465		Spongia (?) skeleton irregular siliceous particles
466		Ova of some marine animal both the latter. Botofogo Bay.
476	I	477. Curculio with Acari
478	I	Numerous Coleoptera
479	I	Beetle, exceedingly numerous on sandy plain near the sea
480	I	do. Sandy plain
481	S	Helix Vide **38** (c). Copy
482	I X	Hymenop: Was carrying off a large Mygalus; they seem to prey on & kill large spiders [note opposite] N.B. The only two Mygalus I have yet caught were in the jaws of this insect
1832		May. Rio de Janeiro
483	I	Bee, the most frequent sort.—
484	I	Diptera. vibrates its wings as its congeners do in England
485	I	Diptera. runs swiftly laterally.
486	I	Cicindela, woods on Caucovado
487	I X	Capsida. Caucovado
488	I X	489. 490. Larva [note opposite] As the Capsida was found on the Larvæ, they most probably belong to it.— The Larvæ were curiously placed in two groups heads to heads round a stick.— They adhered by the remains of a capsule & each groupe was thickly imbricate.
491	I	Coleoptera
492	I	Cerambyx. with Acari, by the friction of the thorax it made a most extraordinary noise.
493	I	Diptera. very summit Caucovado
494	I	Diptera. hovered over sand-bank like a Bombylius
501	I X	Diptera, This is the insect called [note opposite] Sand fly, & notorious even at Ansons voyage, from the painful bite which causes a swelling that lasts for many days; in centre a circular red mark is visible: the pain is half itching & half aching.—
502	I	Xenos (??) Sandy plain; sweeping.
503	I X	Libellula, I observed this insect [continued opposite] as it proceeded along the edge of a pool, strike the water violently with its curved tail, so as to throw some several inches on the bank; is this connected with Oviposition? [see *Insect Notes* p. 55]
504	I	505. Cicindela, habits precisely the same as Cicin: hybrida
1832		May. Rio de Janeiro
506	I	507. The Larva or female of Lampyrus. V **42**
508	I	do; another species: all luminous
509	I	510: 511: 512: 513: Coleoptera from the very summit Caucovado
514	I	Coleoptera. habit do
515	I	516: Hemiptera habits do
517	I	518: 519: 520. Diptera. habits do
529	I	Coleoptera, living in the water or caught in my water net:
530	I X	531.— ~~Leptidactyles Latreil~~ Insects [note opposite] New genus, habits the same as Elmis, living under stones in running water: differs remarkably from that genus in shape of body &

palpi (& in spear to sternum?) [named *Psephanus darwinii* by Waterhouse. See *Insect Notes* p. 55]

532 I 533 Diptera. plague the horses terribly

534 I X Hymenoptera. Pompilus (?) [note opposite] This family runs very quickly amongst the herbage, continually at the same time vibrating its wing.— Excavates cylindrical holes in a trodden path.—

535 I Hymenop. caught killing spiders. V **39**

536 I X Hymenop.

537 I X Cell made by the latter for its larva [note opposite] I observed this insect carrying a large green caterpillar, & watched it to the cell (537): Where with its mandibles by degrees it forced the caterpillar inside.— The rim of the cell is broken: This is the same as (368) found at Bahia

538 I Orthopterous with Acari

550 I Leiodes from Hymenophallus V **43** (a)

551 I Beetle from the dense forest

1832 June. Rio de Janeiro

552 I Cicindela. from the forest.

553 I Forficula. from do. (forceps curious)

554 I : 555: Gyrini, rapid brook in the forest; emit an odour like G. natator.—

564 I Larva of Lampyrus, highly luminous

565 I Aphodius, the only species I have yet seen in Brazil

566 I Agrion from the forest: Common

567 I Frigania do

568 I Geotrupes do

569 I Diptera. common do

570 I X Diptera. called sand fly [note opposite] caught whilst inflicting its painful bite on the knuckle, its favourite place

571 I Curculio. covered with yellow down, when first taken

572 I X Onthophilus [note opposite] Inhabits the forest in plenty & does not I suppose feed on dung

573 I Gyrinus, brooks in the forest

574 I Coleoptera

575 P Cryptogamous plant, like a hollow horse hair on a dead tree in the forest [see *Plant Notes*, p. 160]

1832 June. Rio de Janeiro

580 I X Tricoptera (Stephens) allied to [note opposite] in Fungus in forest.— the smallest beetle I have seen in Tropics [later named *Trichopteryx darwinii* Matthews. See *Insect Notes* p. 57.]

581 P Fern, hanging from tree [*Asplenium mucronatum*. See *Plant Notes* p. 160]

582 P X 583 Leaves & flower of Palm tree [note opp.] 582....584. Tree height 9 ft: circumference at bottom $3\frac{1}{3}$ inches; at top $2\frac{2}{3}$. on the trunk there were 305 rings.— Do these mark the year? Shaded forest on hill. [see *Plant Notes* pp. 160-1 and *Beagle Diary* pp. 75-6]

584 P X Stem of do

585 P Cryptogam: Cascade Tijeuka [*Selaginella jungermannioides*. See *Plant Notes* p. 161]

586 P X 587: 588. Crypto: plants on Caucovado, about 2000 above sea;
[cont. opposite] Clouds generally resting on it, the dampness produces innumerable Cryptogamous plants.— These were procured May 30th.—

589 P Tea-tree. Botanic Garden [*Camellia sinensis* L. See *Plant Notes* p. 162 and *Beagle Diary* pp. 67-8]

590 P Cryptogam: plants. Caucovado

591	P	Crypt: growing in number on the old trees. on arid planes, near the sea; giving a most fantastic appearance to them.
592	I X	Bee. (social) [note opposite] Burrows its nest in the ground in forest. projecting tube. with folding edges leading to it.
593	I	Lampyrus, abdominal rings shining
594	I X	Curculio (diamond) feigns [cont. opposite] death to a most remarkable degree; is this to compensate for greater danger brought on by brilliancy of colours
595	X	Cellaria. Sertularia. Plumularia. Amphiroa. Tubularia. Fucus [note opposite] Amphiroa. V **56** [*Amphiroa exilis*, syntype specimen. See Fig. 1 in *Plant Notes* p. 188] [Busk Collection: *Nichtina tuberculata*]
596	P	: 597: 598: 599: 600 Cryptog: plants, chiefly on rotten trees in forest. [see *Plant Notes*, p. 162]
1832		June. Rio de Janeiro
601	S	Ampullaria. fresh W & land shell
602	X	Pilumnus [?]. Botofogo Bay. [note opposite] 9 lateral spines. extreme one not biggest: alternating layer.— bad specimen.—
614		Bottle of Arachnidæ; also labelled with Tin (252) in case of losing one
615	I	Butterfly V **55**
616	S	Land shells
617	S	Water do
618	I	Coleoptera [see *Insect Notes* p. 58, and *Journal of Researches* **1**:38]
619	S X	Marine shells & Turbo (Linnæ) [note opposite] This small Turbo in the greatest numbers in the Lagoa.— where the water is not quite so salt as in the sea.—
629		Amphiroa. [*A. exilis*. See Fig. 2 in *Plant Notes* p. 189] Cellaria. Plumularia. [Busk Collection: *Membranipora rozieri*]
630	I X	Coleoptera. taken in Beagle between [cont. opposite] Rio de Janeiro and M. Video
631	I	Cloporta [sic]. Beagle
632	I X	Meligethes. Beagle. common [note opposite] ~~Appear to~~ Come from the ripe fruit of the banana.
633	I	Acrydium. Rio de Janeiro
634	I	Lampyrus do do
635	I	Diptera. Beagle
636	I	Lepidoptera. diurna } Rio
637	I	638.— Moths } Rio
639	X	Tubularia (2 species) [cont. opposite] Growing in great abundance on the Anchor in Rio harbor. The Iron was fairly hidden.— The anchor had been down exactly one month, so that the quickness of growth may be seen from these specimens.—
1832		July Monte Video
640	I	Colymbetes. taken on board, must have at least flown 45 miles from Cape St Mary
641	I	642: 643: 644: Gnats, in same situation as last in great numbers
645	I	Pediculus from the petrel called Cape-pidgeon, in the open ocean
		August Monte Video
646	I	Diptera. Rat Island, M. Video
647	I	Blatta do do
663	X	Segestaria [note opposite] Inhabiting tubes in the chinks of rock. about 1 & ½ inch long. inferiorly wider, or bag shaped. spider frequent.

664	I	Pediculus, from a Tringa (Peewit)
665	I	Curculio, on sandy hillocks near the sea.—
666	I X	Cillenum? (Leach) under stones in mud. Rat Island. Water brackish. August.
667	I	Agonum ? allied to; elytra singularly sculptured; Habitat do.—
668	X	669. Spider (allied to Clotho or Philodromus) ? Lost [note opposite] Living under stones. with small irregular web. in damp places: Rat Island. pretends death: is a rectigrade.—
1832		August Monte Video.
(670		Crab Rio de Janeiro harbor)
671	I	Diptera, very common here
672	I	Acarus from Cavia Capybara (Linn:)
673	I	Ricinus from Rhynchops
674	I	Moth common on the Mount
675	I	Beetle. found in the middle of an ants nest.
676	I	Carabidous beetle. common under the drift of the tide.
677	I X	[entries 677 and 678 later crossed through] Heterom: 4: Poecilus. Dermestes. Necrobia. Haltica. Galeruca. Coccinella. Forficula. Harpales. Omarus. Pterostichus. Trechus. Peryphus. 2 Curculio. Forficula. Corixa. 2 Harpalus. Noloptes. Capsida. Colymbetes. Formia. Pentatoma. Silpha. Hygrotus. Hister. 2 Crysomela. [note opposite] The greater number found under stones & sticks.— Hybernating on the Mount.—
678	I X	7 Lamellicorna. 2 Heteromi 2 Curculio. 9 Carabidous insects
1832		
683	B	Fringilla. Monte Video [specimen numbered 0683D in NHM data bank, listed as *Emberizoides poliocephalus* G.R.Gray in *Zoology* **3**:98; see also specimen 1207] *
684	B	Numenius, in habits like a Jack Snipe: swamps
685	B X	Alauda [note opposite] This bird flies upwards & then suddenly falls with its wings expanded like some Titlarks in England in Spring time.—
686	P X	Lichen, growing on stones near summit of Mount.— [note opposite] The Mount is 450 feet high: [*Usnea densirostra*. See *Plant Notes*, p. 162]
687		Lycosa. under stone. Mount
688	X	Lycosa with bag of eggs [note opposite] These inhabit a hole about an inch in diameter & 8 in depth. Made quite smooth & lined with web.— Over the mouth the blades of grass were connected by meshes forming a ball the size of turkeys eggs. There was no opening, so that the spider which sits at the bottom with eggs must cut its way through every time it goes out.—
689	S X	Fresh water shells.— Turbo on stones in running brook.— [note opposite] The spiral univalve is closely allied to a Turbo taken at Rat Island: was found same site as ~~Creusia~~ Bivalve (323 Spirits)
690	S	Floating in the open Atlantic
691	I	Harpalidæ. (one of). Mount
692	I	Cerambyx. buildings M Video

Baia Blanca

694	I	Harpalidæ. (one of) Baia Blanca
695	I	Meloe: elytra with bright yellow spots. sides of abdomen red. emitted yellow fluid:—
696	I	: 697: 698. Trox (3 species)
699	I X	: 700: 701: 702. [above in pencil] given to me by Mr Clare. 4 species of melasomes [notes on next page] 699. Tolerably abundant, in sand hillocks. 700. The commonest insect in the

place. on a hot day runs very actively on the sand.

703 I X Scarabiidæ: All these beetles in=habit sandy hillocks near sea [note on next page] This beetle seems to live on the dung of Ostriches.— I saw one busily employed in pushing along a large piece with its pointed horns.— Sept. 19th.

1832 Sept: Bahia Blanca

704 X Sertularia; washed up on beach: [note opposite] Dynamena (Lamouroux)

705: I 706: 707: Heteromerous insects. Sandy plains.—

708 I Staphylinus

709 I Insects

710 B Vaginalis V **99** (Stop)

711 B the tail of the latter, taken from another specimen.—

712 B Charadrius [listed as *Hiaticula trifasciatus* in *Zoology* **3**:127]

713 B X Podiceps; Iris "scarlet red" [note opposite] Live in flocks in the salt marshes.— [listed as *Podiceps kalipareus* in *Zoology* **3**:136]

714 X Dynamena (same as 704) [note opposite] The specimen has the 3 ovaries.—

715 Flustra. encrusting thrown up on the beach.—

717 I Harpalidous. sandy plain

718 I X Meloe. hillocks [note opposite] Hind legs very long. forehead angular; sides of abdomen bluish

719 I Lamellicorn. (Hoplia?) copulating in great number. sandy plain: Sept. 19th.

720 I Lamellicorn:

721 I Coccinella

722 I Coccinella (allied to).—

1832 Sept. Baïa Blanca

723 A Stone (Bezoar) said to be from the ~~stomach~~ Guanaco

724 I Coleop. Heterom: Rio Negro

725 I Colymbetes

726 I Carabidous beetle, from the mud banks

745 B Sterna [listed as *Viralva aranea* in *Zoology* **3**:145]

746 B XX Hirundo (not common) [cont. opposite] building in cliffs in holes near sea. I saw at the cliffs at M: Hermoso a flock of these birds pursuing each other & screaming much in the same manner as the English Swift: in its characters it seems also to approximate to it: ~~How frequently does structure & even trifling habits go together~~:— [listed as *Progne purpurea* in *Zoology* **3**:38-9; numbered 0746? in NHM data bank] *

747 B X Psittacus living in flocks, on cliffs near the sea, where there are no trees [cont. opposite] & I have no doubt, breed in holes; I found a nest with rather small white egg & several parrots feathers:— It is so: [listed as *Conurus patachonicus* in *Zoology* **3**:113]

748 B Larus. common [listed as *L. dominicanus* in *Zoology* **3**:142]

749 B Motacilla; runs on the beach of the bay & looks like a lark; its note is high & is repeated like a young hawk.—

750 B Fringilla

751 B X Sylvia; concealing itself in low thicket. [note opposite] In habits like a kitty wren.—

752 I Carabidous beetle, inhabiting sand hillocks

753 I Crysom. on a flower

761 P X Succulent plant, covering large tracts of pampas & looking at a distance like our heaths: [note opposite] Grows chiefly in salt plains overflowed occasionally by the sea.— [*Allenrolfia patagonica*. See *Plant Notes* p. 163]

1832		Bahia Blanca
762	P	A very abundant grass growing in tufts & on sandy plains [*Poa ligularis*. See *Plant Notes* p. 163]
763	P X	Oxalis in great quantities [note opposite] Flowers bright pink [*Oxalis floribunda*. See *Plant Notes* p. 163]
764	P X	Bush. very common; growing in tufts. like our Gorss banks.— [*Discaria longispina*. See *Plant Notes* p. 164] [note opposite] NB. Sept; 23^{d} All these plants were in full flower [further notes heavily deleted]
765	I	Lamellicorn
766	I	Crysomela. near the sea Monte Hermoso
767	I	Harpalus B. Blanca
768	I	Elater
777	A	Dipus (Gme:) or Gerboise D. Class V **103** copy [listed as *Mus elegans* in *Zoology* **2**:41-2]
778	I	Bruchus from the Calavances
779	B X	Sylvia. [note opp.] Lies concealed in the thickets like the Kitty Wren } M. Hermoso
780	B X	Alauda. [note opp.] In small flocks running on the sea-beach.— } B. Blanca
786	I	Curculio
787	I	Lamellicorn
788	I	Amara; sandy hillocks
789	I	Clavipalpes. Heterom. Lat: living at roots of grass: sandy hillocks
790	I	Pulex from the Armadillo (375)
791	P X	Clover. very common.— [note opposite] This plant characterises all the low & more fertile spots; mingled with grasses & the Geranium (792) it forms a thick mass of herbage, in places nearly a yard deep: Sept. 15 to Octob 1st It is said the cattle do not eat it:— [*Melilotus indica* L. See *Plant Notes* p. 164]
1832		Octob: Bahia Blanca
792	P	Geranium, very abundant, in flower middle of Septemb. [*Erodium cicutarium* L. See *Plant Notes* p. 164]
793	P	A low bush common near the sea. [*Ephedra ochreata*. See *Plant Notes* p. 164]
794	P	do. flowers smelling sweet growing near the sea.— [*Lycium chilense*. See *Plant Notes* p. 164]
795	I	Carab: sand hillocks; beautiful comb of spines over the Tarsi
796	I	Silpha. in number feeding on carrion with Trox & Dermestes.
797	I X	Lamellicorn:— [note opposite] I think this number has been used twice: once for a large bush, bearing very sweet flowers & no leaves.—
814	B	Egg of Struthio rhea. V ~~122~~ account
815	A X	Cervus (campestris?) [in pencil] (Stop) [note opposite] Common inhabiting the sandy plain, often in small herds: they are very curious & if the sportsman remains in a crouching position they will approach close to reconnoitre him:— Many of the Does have kidded, Octob 10th:— They weighed from 60 to 70 £b [further note written in small hand between these lines] It is curious to observe how much more afraid of a man on horseback than on foot: every person in this country riding: so totally the reverse of what happens amongst English deer: the crawling position will also attract Guanaco, & they will advance neighing to reconnoitre a person in that position: [listed as *Cervus campestris* in *Zoology* **2**:29-31]
816	A	Antlers. belonging to same animal
817	A	Cavia patagonica V **112**
818	A	Fox. not uncommon [listed as *Canis Azaræ* in *Zoology* **2**:14-16]
819		Janthina. Atlantic, between Bahia & Rio Janeiro
820	I	Harpalus. Mon: Hermoso
823		Virgularia & stony axis (same as 401) V **106**

824	S X	:825. Shells, living on the sand banks. [note opposite] The Mya lived about 5 inches within the mud: a small blackish Buccinum on the mud. The Voluta had no operculum.— Shells to be compared with fossil ones from P. Alta.—
826	S	:827. Shells & Coralls, on beach for comparing with those at P. Alta. 811-813 [CD's shells: *Mactra elegans*]

[new page, no heading]

828	B	Sylvia. in thickets: Bahia Blanca
829	I	Fly, just killed a gnat:—
839	I	Saperda on the trunk of the Phytocalla: (a large tree). B. Ayres
840	I	841 Diptera. on flowers.— Buenos Ayres
842	I	Coleoptera B Ayres
843	I	Coleoptera M: Video
844		845. Spiders. B Ayres [note opposite dated December 1833] I sent home a skin of a large lizard Iguana.— I know not its number. I saw it one day catch & kill a green lizard 7 or 8 inches long, & shake it like a dog:—
846	A X	Rodentia. B Ayres [note opposite] This animal by some is said to be the Chinchilla; by others as a young Viscacha.— I bought it as the latter:
847	B	Muscicapa. B. Ayres.

Monte Video

848	I	Heterom: common under stones
849	I X	Nest of Bee, under stones [note opposite] Contained leaden blue, slightly sweet honey: mouth closed by a sepal of a flower.—
850	I	Heterom, feeding on Compositæ & when touched, like Meloe emitting yellow fluid
851	I	Heterom: habits. do do.—
856	S	Shells. Rat Island
857	S	Fresh water shells.—
858	I	Coleoptera, The Mount [for specimens 858-884 see *Insect Notes* pp. 66-9]

1832

859	S	Land shells. Monte Video
860	I	Meloe. San Blas: Bay of Patagonia
861	I	Belostomus. in water. Rat Island. M: Video
862	I X	Calosoma: flew on board when we were about 10 miles from the shore: Bay of San Blas [note opposite] Others were found on shore: Bay of San Blas North of R. Negro:
863	I	Lamellicorn
864	I	Heterom San Blas
865		Mygalus
866	I	Moths, flying about the ship, the crysalis were in the fire wood
867	I	:868: 869. Carabidous beetle. dead in the sea. 40 miles off the St^{s} of Magellan.—
870	I	871. 872. Butterflies V **138**
873	I	Libellula. M: Video
874	I	Cimex. San Blas
875	I X	Fresh-water & carabidous beetles found <u>alive</u> in the sea. S of Cape Corrientes. flown off the shore?— [cont. opposite] I was very much surprised to <see> how perfectly alive & active the fresh-water beetles were (Colymbetes, Hydroporus, Hydrobius &c; & there were other

insects which I by accident lost). This may be a very instrumental means in peopling islands with insects: I cannot help suspecting they were washed down from the Plata; although 250 miles distant from fresh-water.— I think this from the numbers of living & dead ones floating in the sea.— The distance from the nearest shore was 17 miles, off Cape Corrientes: Capt Cook saw numerous insects blown off near St Georges bay: & formerly in last voyage this fact was frequently noticed: it must be owing to flat country, without trees ∴ no shelter; insect once in air cannot stop:— [see *Insect Notes*, p. 67]

1832 Decem. 20th Good Success Bay

879 Succinea. common in Wooded hills & in Navarin Island [cont. opposite] feeding on plants close to water

880 I X Carabus, damp forest: [note opposite] Carabus does not ascend the mountains.—

881 I X Harpalidous; found flying in numbers about sea coast in evening.— [note opposite] These insects live amongst the soft yellow balls which are excrescences ~~produced by some (?) insect~~ [corrected later by CD to 'or rather fungi growing', see *Insect Notes*, p. 67] on the Fagus antarcticus & which are eaten by the Fuegians.—

882 I Harpalidous; the most abundant insect under stones &c. damp forest

883 I X Harpalidous.— the only insect which I found inhabiting the very bare summits of the mountains.— [note opposite] &c &c. These woods are all more or less ~~lofty~~ above the sea.— [see *Insect Notes* p. 69]

884 I Lamellicorn, common in the forest.—

885 Flustra encrusting Fucus giganteus; picked up Lat 45°S at sea.— (came from the South)

886 S Ampullaria, very abundant in marshes near R. Plata. B: Ayres.—

887 S Its eggs, color "scarlet red", on rushes few inches above the water.— in great numbers; so as to be beautiful

1832 Decemb: 20th ~~Good Success Bay~~

[note opposite] All specimens from 888 to 900 much injured by the gale of Jan[y] 13th.— & Numbers 894....900 changed into 931...937 [see *Beagle Diary* pp. 131-2]

888 X Celleporaria (?) V **143** [Busk Collection: *Adeonella atlantica; A. fuegensis*]

889 do do V **142**

890 Favosites V **144**

~~891 Corall (?) V 144~~ (now 926)

892 do do V **144**

893 Retepora, fine Salmon colored

~~894 : 895 : 896 : 897 : 898 :~~

~~899 The above & various~~ other Coralls all collected at 30 fathoms, 53°S. Dec 15th

900 Echini (Ciclants?) Habitat do

901 B Tringa, inhabiting in small flocks the bare stony summits of highest mountains, Good Success Bay [classified as *Squatarola cincta* Jard. & Selby in *Zoology* **3**:126]

902 B Fringilla. Mountain ~~forests~~ summits; about the turf bogs

903 B Alauda

904 B Fringilla. Mountain forest [listed as *F. formosa*, *Zoology* **3**:93-4]

905 S Marine Shells, on coast as above, Good Success Bay.—

1832		Tierra del Fuego
906	I	Coleoptera: wooded hills, Good Success Bay. Dec 20th.— [for specimens 906-914 see *Insect Notes* pp. 69-70]
907		2 species of Corallina. G Success Bay:
908	I X	Coleoptera from the very summit under stones: Katers Peak [note opposite] Katers Peak. abrupt cone of greenstone. 1700 feet high. in Hermit Island near Wigwam Cove, not far from Cape Horn
909	I	Carab: very abundant. Hab: do
910	I	Carab: under stones sea beach. Wigwam Cove, also in hills Navarin island
911	I	Carab: (same as 883?) very abundant. summit Katers P.—
912	I	: 913. Heterom. common very summit Katers P.—
914	I	Curculio on Fagus Antarcticus
915	S	Helix. very summit of Katers peak.—
923	I	Ricinus from Albatross. C. Horn. Jan:
924	P	Lichen from very summit of Mount. M: Video [*Parmelia fistulosa*. See *Plant Notes* p. 164]
925	I	Libellula. Navarin Island
926		instead of (891)
1833		Jan. Tierra del Fuego
929	S	Marine shells. Good Success bay
930	I	Harpal: Navarin Island.
931	1	....937 instead of 894....900. original having been lost
	X	[note opposite] 937 Specimen destroyed
932		Coralls 30 fathoms [Busk Collection: *Schizoporella harmeroides*; *Cellarinella dubia*; *Idmonea*]
934		Coralls 30 fathoms [Busk Collection: *Eschara fuegensis*; *Adeonella atlantica*; *A. fuegensis*]
967	I	Hymenopt: Ponsonby Sound:
968	I	Lucanus in rotten Beech: do:
969	I	Hemip, in great numbers under rotten bark. Ponsonby Sound
970	B	Tringa. in flocks on the beach Goree Sound: [cont. opposite] & in the Falkland Islands [*Squatarola cincta*, *Zoology* **3**:126]
971	B	Sylvia: do:
972	B	Fundus: do:
973	S X	Marine shells. adhæring to the Kelp & stones about the roots [note opposite] the brown bivalve in great numbers on the Kelp:
974		The White & red paints of the Fuegians V **148**
		[note opposite in Covington's hand] Stop

976	P	Plant, chief origin of the peat bogs. V **155** [*Astelia pumila*. See *Plant Notes* p. 164]
977	P	Parasite plant on the Beach [*i.e.* Beech] [*Myzodendron brachystachyum*. See *Plant Notes* p. 166]
1833		Feb. Tierra del Fuego
978	P X	The infusion makes a pleasant drink, much used by the Sealers instead of tea: on the hills: [note opposite] Bear a pale pink berry: with a fine sweet Juniper flavor: the plant is said by the Sealers to be diuretic [*Myrteola nummularia*. See *Plant Notes* p. 166]
979	P	Cryptog: when alive partly enveloped in gelatinous matter:
980	P	(same as 503 in spirits) V **145** Copy

981 P Growing generally near the Wigwams [*Epilobium ciliatum*. See *Plant Notes* p. 981]
982 P Plant very alpine [*Senecio darwinii*. See *Plant Notes* p. 167]
983 P Pretty pink flower growing near to a Cascade
984 P X Scurvy grass (very good) growing near the Wigwams [note opposite] V **155**
985 P X Generally growing near Wigwams [note opposite] Fruit. pulpy. sweet pleasant. as large as a red current ripe. Feb 25th.— [*Senecio acanthifolius*. See *Plant Notes* p. 167]
985(bis) Currant bush. generally near to the Wigwams [*Ribes magellicanus*. See *Plant Notes* p. 168]
986 P Lichen universal on rocky summit of mountains
987 P Lichen. mountain. G. S. Bay: [note opposite, later crossed through] N.B. The red color on which number is printed is supposed to be equal to <u>1000</u>
1001 B Fringilla. Goree Sound
1002 A Mouse. on the hills in Hardy Peninsula.— [listed as *Mus xanthorhinus* in *Zoology* **2**:53-4]

1833 Feb: Tierra del Fuego

1003 B Emberiza. on the Mountains, Hardy Peninsula:— [listed as *Chlorospiza* ? *xanthogramma* G.R.Gray in *Zoology* **3**:96-7; numbered 1003?D in NHM data bank] *
1004 S Helix. under stones. summits of mountains. Hardy Peninsula
1005 I Alpine Colymbetes : do : [for specimens 1005-1012 see *Insect Notes* pp. 70-1]
1006 I & 1007: Heterom. Mountain. H. Penin.
1008 I Byrridæ: Mountain : do :
1009 I Carab: do : do :
1010 I Carab: do : do :
1011 I Cimex. do : do :
1012 I Haltica do : do :
1013 P X Beech, foliage yellowish green [*Nothofagus betuloides*. See *Plant Notes* p. 168]
1014 P X Bright green (Beech) [note opposite] These Beech trees are the only ones which grow on the mountains in this district (Hardy Peninsula) The first is by far the most general, almost universal, & grows to a larger size: the other follows the course of a rivulet on more sheltered nook.— the contrast of the two greens is at all times striking:— [*Nothofagus antarcticus*. See *Plant Notes* p. 168]
1021 I 1022: 1023:1024: Heteromerous insects V **149**
1025 I Alpine Bembididous insect:
1026 B Motacilla, <u>common</u> on the mountain } [listed as *Opetiorhynchus*
1027 B do: on the Mountains } *vulgaris*, *Zoology* **3**:66-7]
1028 B Falco.— Hardy Peninsula.—
[listed as *Milvago pezoporus* in *Zoology* **3**:13-14 and *Ornithological Notes* p. 213]
1043 I Heterom: under stone at just above high-water mark (Vide 1021)

1833 March Tierra del Fuego

1044 I Ricinus from the Falco (1028) [see *Insect Notes* p. 71]
1045 P X Plant in habits much resembling the common rush in England
[note opposite] ~~V 156~~ [*Marsippospermum grandiflorum* L. See *Plant Notes* p. 168]

<u>March</u> Falkland Islands

1046 B X Emberiza. Falkland Islands [note opposite] I have seen these two constantly in the same flock.— They are by far the commonest land-bird in the Island.— [listed as *Chlorospiza* ? melanodera in *Zoology* **3**:95-6, NHM 1855.12.19.50, numbered 1046? in data bank] *
1047 B do (not shot with the last, but perhaps it is the male)
1048 B Scolopax Falkland [listed as *Limosa hudsonica*, *Zoology* **3**:129; or might be *Scolopax*

Magellicanus, *Zoology* **3**:131]

1049 I Coleoptera. Tierra del F, chiefly Hardy Peninsula

1050 I Harpalidæ. Falkland Island

1051 I Ricinus from Scolopax (1048)

1052 P Lichen common in mountain on the rocks. Tierra del F.

1053 B X Sylvia Falkland Islands [note opposite] Beak & legs large in proportion. lives in the coarse herbage, close to the ground:— [with different pen] I never saw a bird so difficult to make to fly after marking it down within a few yards in open plain it could never *[illeg.]* [listed as *Troglodytes platensis* Gmel. in *Zoology* **3**:75, NHM 1856.3.15.20, and see *Ornithological Notes* p. 213] *

1054 B Falco

1055 P Excrescences or Fungi; edible; on the Beech same as in spirits (528) [*Cyttaria darwinii*. See *Plant Notes* p. 168]

1056 P Junctions of Parasite bush with the Beech of Tierra del F. same as in spirits (532-534)

1833 March

1057 I Moth. on leaf of Black Currant bush. G. Success B.

1060 I X Harpal: (Sphodrus?) Falkland Island [note opposite] Was this insect imported or is it an original inhabitant

1061 I Harpal; abundant near coast. Falk: Isl.

1069 S Marine Shells. Wollaston Island & G Success Bay: the Balanus with crenated sections coats <u>all the</u> rocks at low water

1070 S Marine shells; good Success Bay [Busk Collection: *Crisa edwardsiana*]

1071 I Fly.— Falkland Island

1072 Cancer. Wollaston Island:—

1073 P A square piece cut out of the peat whilst forming.— Tierra del Fuego V **156**

1074 P X Tierra del Fuego A very abundant bush in T. del F. [note opposite] Does not reach above 4 or 500 feet up the mountain: bears a very pleasant but bitter berry: colour & size varies, from white to dark red: I eat great numbers of them.— [*Pernettya mucronata* L. See *Plant Notes* p. 170]

1075 P Bog Plant. same as (976) [*Astelia pumila* L. See *Plant Notes* p. 170]

1076 P Celery. generally growing near the Wigwams: very good flavor when boiled in Soups &c [*Apium australe*. See *Plant Notes* p. 170]

1077 P Plant growing in the Peat & closely resembling in general habits & tint our heaths. [*Empetrum rubrum*. See *Plant Notes* p. 170]

1833 March E Falkland Island

1086 I X Harpalidous insect Falkland Islands

1087 I 1088. Heterom. near coast Falkland Islands
[note opposite] Both insect<s> are common to Tierra del Fuego

1137 I Gonoleptes

1138 Cancer

1139 X Cellaria & stony dichotomous Coralline [note opposite] The latter growing on the leaves of Fucus giganteus [Busk Collection: *Salicornaria malvinensis*]

1140 Spongia. colour "gamboge yellow"

1141 [Busk Collection: *Membranipora galeata*]

1142 I Cellaria on the beach; Halimeda; Serpula. Tierra del Fuego
[Busk Collection: *Tricellaria aculeata*; *Menipea patagonica*]

1143 Corallina (true) V **165** [*Corallina officinalis* L. See Fig. 3 in *Plant Notes* p. 191]

1144 B Bird common in Tierra del F & Falkland Islands.—

1145 B Tringa. upland marshes

1146	B X	Sturnus, I believe same as that at M. Video, Bahia Blanca [note opposite] Sturnus ruber. at St Fe Bajada.— most common at R Negro & coast of Patagonia [*Sturnella loyca*, listed as *Sturnella militaris* Vieill. in *Zoology* **3**:110]
1147: 1148	B	Scolopax. feeding flocks on the mud banks on sea-coast.— [listed as *Limosa Hudsonica* in *Zoology* **3**:129, No. 1147 is now in Smithsonian Institution, Washington, Reg. No. 8074] *
1149	S	: 1150 Marine Testaceæ
1151	I	Coleoptera [see *Insect Notes* p. 72]
1153		Coralline (inarticulata) V **161**: **164**: **164** 3 species.—

1833 Falkland Island

1154	P	Lichen growing near the Sea. very common. Falkland Island:
1155	P	Parasitic plant on Beech. Tierra del F [note opposite] all plants from South part of Tierra del Fuego
1156	P	Grass. Wollaston Island & other unfrequented places
1157	P X	Syngenesia plant. on sand dunes Wollaston Isl: also Falkland Island.— [*Senecio candidans*. See *Plant Notes* p. 171]
1158	P X	Alga. Wollaston Island.—
1159	A	Rat. Falkland Is? evidently not Europæan.— (bare hind legs &c)
1160	B	Falco, probably the male of (1054): as these are the only sorts common in Falk Island.— [listed in *Zoology* **3**:30-1 as *Circus cinerius* Vieill]
1161	7	Coralline on Fucus giganteus. chiefly Obelia same as (597 spirits) [Busk Collection: *Porella margaritifera*; *Tubulipora phalangea*]
1162	P X	The common grass which so universally covers the whole island [note opposite] Growing on the peat. [*Cortaderia pilosa*. See *Plant Notes* p. 171]
1163	P	This is largest <u>tree</u>, sometimes growing 2 & 3 feet high [*Chiliotrichum diffusum*. See *Plant Notes* pp. 171-2]
1164	P	Common low shrub [*Berberis magellicanus*. See *Plant Notes* p. 172]
1165	P	Plant. very abundant. resembling in habits our heaths.— [*Empetrum rubrum*. See *Plant Notes* p. 172]

1833 March E Falkland Island

1166	P X	Lichen. particularly abundant on the level country
1167	P	Lichen abundant on hills.— [note opposite] All the lichens are very abundant in this island. The same lichen (986) which is so common in Tierra del is found here:
1168		Asterias. superior surface smooth colour "leek green" & blackish green
1169		Asterias. sup: surface smooth colour "leek green"
1170		Asterias. "tile red"
1171		Aster: Pale, most beautiful "Auricula purple"
1172		Asterias. "deep orange brown"
1173		Asterias. beautiful "vermilion red"
1174	S	Marine shells
1175		:1176 Cancer
1177		Ophreiza
1178		Coralline (found dead)
1179	S	Marine shells:—
1180	I	Diptera. Hardy Peninsula

Maldonado

1181 I X Scarabaus [note opposite] Feeding on horse dung & throwing up the sand like Gastrupes.— sand dumes.

1182 I Coleop: feeding on Lycoperdium & Fungi

1183 I Notonecta

1833 Maldonado

1200...1224 B Various birds skinned in the month of May.— for particulars V **177, 178, 179, 180**:—

1225 I X Aphodius [note opposite] One of the rare instances of finding these insects in this country under horse dung; it was not fresh.—

1226...1236 Various birds. for particulars V **180**;

1237 A X Mouse common in the houses in the town. Europæan?

1238...1264 B Various birds. for particulars V **180, 181, 182, 183**,

1253 [note opposite] Coleoptera in Lycoperdium (1346)

1254 [note opposite] Brachinus. emits loud & visible explosions, lives in families under stones in open camp.—

1265 A Mus, very abundant in gardens & hedges, not near houses; easily caught by trap bated with cheese or rind. [listed as *Mus obscurus* in *Zoology* **2**:52-3]

1266 A Aperea. Cavia Cobaya V **187** Copy

1267 A Bathiergus ?? Toco Toco. V **188** Copy

1268...1277 B Various birds for particulars V **P 184**.—

1278 A Gulo, called here 'Huron' or thief Weighs 1 £b: 8 oz: (Imperial wt:) [listed as *Gallictis vittata* in *Zoology* **2**:21]

1279 A Felis. killed in a rocky mountain. [note opposite] Whether this is a distinct species or domestic cat run wild I know not.— it was much larger & stronger & more regularly coloured: It would be interesting to compare it with the aboriginal of the domestic, if they are the same species: [Listed as *Felis domestica* in *Zoology* **2**:20, it was mounted by Sowerby, and is preserved at the Natural History Museum]

1833 June. Maldonado

1280 A Mus V **196** Copy [*Mus (decumanus) maurus* in *Zoology* **2**:33]

1281 A X Didelphis. called in this country Comadreja. inhabits burrows:very offensive smell: common. nocturnal. steals poultry [note opposite] This whole genus is called Comadreja, which properly Weasel.— [listed as *Didelphis azarae* Auct. in *Zoology* **2**:93]

1282 A X Didelphis, tail prehensile. weighs 14 & ¼ oz. Imperial weight.— [note opposite] Abdomen possessed the bones articulated to Pelvis:—
[listed as *Didelphis crassicaudata* in *Zoology* **2**:94-5]

1283 A X Didelphis.— [note opposite] Intestine full of remains of insects chiefly ants & some Hemipterous; caught by digging.— [listed as *Didelphis brachyura* in *Zoology* **2**:97]

1284 A X Gerbillus (?) Eyes & ears very large. looked like a small rabbit. [note opp.] Caught in open camp by trap bated with cheese. V ~~188 (a)~~ account of Toco Toco [further note] Having seen small specimen at B. Blanca, the tail is not tufted
[listed as *Reithrodon typicus* in *Zoology* **2**:71-2]

1285 A Mus. caught in so wet a place & so surrounded by water, I should certainly think it aquatic.— [listed as *Mus tumidus* in *Zoology* **2**:57-8]

1286 A X Mus. closely resembling (1265) [note opp.] Untill I had seen many of the latter & two of these, I did not believe they were different.— But the latter are so constantly of a size larger & of a lighter colour, I think they must be different species.—

1287 A X Mus. nose much acuminated. from nostrils to curve in lip ¾ of inch. Caught in open camp.

trap bated with ~~mouse~~ bird: Head in Spirits (698) [listed as *Mus nasutus* in *Zoology* **2**:56-7]
1288 A Mus. in camp caught by bird: inhabiting especially sand dumes. Head in Spirits (678).— [listed as *Mus arenicola* in *Zoology* **2**:48-9]

1833 June Maldonado

1289 A X Mus, most beautiful; I imagine lives in families: Spirits (700) [listed as *Mus bimaculatus* in *Zoology* **2**:43-4]
1290 A Mus.
1291 I X Brachinus. Explosion very loud & visible: the skin of my fingers was for many days afterwards stained brown: [note opposite] At the instant of explosion a sensation of warmth was felt: taste very acrid even when diluted:— [*Brachinus* sp., see *Insect Notes* p. 73]
1292 A Cervus V **196**
1293...1297 B Various birds for particulars V P. **185**.—
1298 I Hymenoptera. Bahia Blanca
1299. I 1300. Hymenoptera. Bay San Blas
1301 I Lepidop taken 60 miles from nearest land, but much further in direction of wind. Mouth of R. Plata
1302 I 1303. Coleoptera do.
1304 I Brachinus. Maldonado
1305 I Hydrous
1306 I Hemiptera
1307 I 1308: 1309. Hymenoptera.
1310 I Coleoptera [see *Insect Notes* pp. 73-4]
1311 A Head of Toco Toco same as (1267)

1833 June Maldonado

1312 Shells. from bed of mud beneath Fresh W lake V Geological notes
1313 S Shells from a Fresh Water lake now having become brackish by inroad of sea.— [note in margin] Cabinet
1314 I Fresh Water Coleoptera [see *Insect Notes* p. 74]
1315 S Fresh Water Shells
1316 I Coleoptera [see *Insect Notes* p. 74]
1317 S Fresh Water Shells.—
1318 A Head of Cavia Cobaya. same as (1266)
1319 S Land shells
1320 First maxillary tooth of Coluber (624). V **176**
1321: I 1322: 1323. Coleoptera [for specimens 1321-1332 see *Insect Notes* pp. 74-5]
1324. I 1325. Leionotus
1326: I 1327: 1328: Lamellicorns
1329 I : 1330 Orthoptera
1331: I 1332 Hemiptera
1333 S Fresh water shell from the banks of R. Negro. Patagonia.

1833 June Maldonado

1334 P X Gum-resin from the bosses of the Hydrocotile gummifera. Spoilt [note opposite] Much oozes out naturally, but if the plant is cut vast quantities of this milky fluid flows, which in a few days hardens: said to be good for cuts:—
1335 B Procellaria from Bay of St Matthias caught by fishing line: [listed as *Procellaria glacialoïdes* A.Smith in *Zoology* **3**:140]

1336 I Pediculi from do.— Procellaria (1335) [see *Insect Notes* p. 75]

1337 A 1338. Horns from a deer same as (1292)

1339 S Ampullaria & Helix. very common:

~~1340 X Pecten from bay of St. Joseph. animals were in them.~~

1340 B Palomba ~~V 185~~

1341 A X Mus.— It was caught in trap in house in town: but surely it is one of the country: the common grey rat is in the town: [note opp.] Rat is very abundant in out houses ~~near~~ at St Fe. Bajada: runs about the hedges & climbs well: The female has 6 Tits on each side.— the 3^{d} is as far distant from the 4th as 1st from the 3^{d}.— [see sketch in margin] [later note] In forest of Chiloe, I saw this same rat

1342 A Mus: very common at R. Negro

1343 S X Pecten. Bay of St Josephs.— Animals were in them: [note opposite] Interesting to compare with those found in the cliffs:—

1345 P Fungus (2 species) the flat kind growing on under side of timber

1833 July Maldonado

1346 P X Lycopodium & Lichen [note opposite] The Lycopodium often grows in open camp [i.e. 'campo' or country] to three or four times the size of this one; but always in same singular shape.— The lichen grows on damp indurated bare, not very pure sand near the dumes.— It has a very singular appearance, where there is much of it: [see *Plant Notes* p. 173]

1347 S Ampullaria. Helix. same specimen by mistake as (1339).—

1348 Fuegian colours from the West Coast. Cap Fitz Roy

1349 B Thalassidromus ~~V 185.~~ [listed as *Thalassidroma oceanica* Bonap. in *Zoology* **3**:141]

1374 Fuegian white paint. V **148**

1375 :1376. Siliceous tubes, supposed to be produced by lightning V **197**

1377 S Paletta (?). Falkland Islands

1378 B Egg of Perdrix (1223)

1379 I X Forficula near sand dunes: [note opp.] There is another species in the houses: they are held in extreme dread: it is curious this prejudice against a harmless insect, being so general. [see *Insect Notes* p. 75]

1380 I Coleoptera (chiefly Carabidous) under stones Guritti Island. Maldonado

1381 I X Excrescences, containing larvæ; aperture most beautifully constructed: are found in a particular valley near M: Video [note opposite] It is said that a large fly which bites horses is produced [see *Insect Notes* p. 75]

1382...5 B Various birds. for more particulars, V **185**.—

1833 July Maldonado

1386 Specimens of a substance resembling Peat.— the heavy black sort was undermost. V **200**(a):—

1387 : 1388. Dorsal fin & tail of Phocæna. V **174**

1389 S X Land shells. Guritti Island [note opposite] The largest specimen (Helix?) inhabits in great numbers the sand dumes.—

1390 B Larus V **185**(bis)

1391 P Grass. Cape Blanco: plant from R. Chupat; root eat for liquorice:—

1392 Salmacis V **201**

1393 Arthrodiès V **201**

1394 I Phalangium

1395 I Pediculi from Falco (1396) [see *Insect Notes* p. 75]

1396 B Falco V **185**(bis)

1397 I Coleoptera for (1380 number destroyed) [see *Insect Notes* p. 76]
1398 S Shells. fresh water pools Guritti: Animals coloured in bands black & yellow
1399 Fossil shells from new Bay: Patagonia:
1400 A X Lutra: bought after being quite skinned & therefore spoilt: 1401 Head cut from the carcase:— [note opposite] Killed by the dogs on a peninsula which projects ~~from~~ into the salt water: & far from fresh water (ie 3 or 4 miles). I do not however feel sure it is a marine species: Mr Sorrell (an old sealer) says he thinks it is same as the Marine one of Tierra del Fuego.— & that the Males are much redder coloured: if so the lower incisives are six in number: [listed as *Lutra platensis* in *Zoology* **2**:21-2]

1833

1402 B X Ptarmigan inhabiting <u>summit</u> of Katers peak: Hermit Island: not uncommon on bare summits of most Southern mountains. [note opposite] Katers peak is 1700 feet high. bird lay close & tame: were shot also on Hardy Peninsula amongst the mountains. These three birds by oversight were not numbered at the time.
1403 B Tringa } Falklands Islands
1404 B Sylvia } Berkeley Sound
1412 S Fresh water shells, found dead banks of the Pacana.—
1413 A Molita (sort of Armadillo) V **204** Copy
1414 B 1415 1416.— Birds St. Fe. Bajada:
1417 B Sparrow (appears different from common species)
1418 B Icterus. Bajada. was also found by Fuller at Maldonado

<u>Buenos Ayres</u>

1419 B Duck. Buenos Ayres
1420 B Long legged Plover
1421 B Duck
1422 B Charadrius. common in small flocks. planes of Buenos Ayres
1423 B in small flocks: inland
1424 B Shot in vessel on R. Plata
1425 B Bird. Marsh. inland.—

[a list of specimens given to Mr Owen for shipment follows]

Down House Notebook 63.5

C.Darwin

H.M.S. Beagle

Printed numbers
N^{n}. 1426......3342

Red = 1000 + &c
Green = 2000 + &c
Yellow = 3000 + &c &c

A animal
B bird
I insect
S shell
P plant

mem: double cross : (Copy beginning) : Ask me :

mem: General observations at Port Desire & St Julians P **210**
Before Falkland. General observations
——— S. Cruz do P **260**
——— Chiloe P **265**
Write myself Valparaiso P **274**
———

Chonos Gen Obser, P **310**, Introduce it before (2479)
——
Galapagos P **340**.— do before No 3296

1833		October.	Buenos Ayres
1426	B	Icterus	
1427	B	Small flocks. very noisy chattering bird.—	
1428	B	Woodpecker	
1429	B	Grebe. fresh-water	
1430	B	1431. Birds	
1432	B	Specimen of female. was shot at Maldonado.—	
1433	B	Charadrius. Rio Plata	
1434	B	Bird	
1435	B	Arenaria. banks of the Plata	
1436	B	Duck	
1437	B	Bird female of (1439) [listed as *Pyrocephalus parvirostris* Gould in *Zoology* **3**:44-5, labelled 1437D in NHM data bank] *	
1438	A	Rat. common in houses & in the camp	
1439	B	Not present in the winter at Maldonado. Now common there therefore migratory.— labelled 1439? in NHM data bank] *	
1440	A	1441. Horns of the common deer at Bahia Blanca	

1833 Bahia Blanca

1442 A Biscatcha B Ayres V **205**
1443 A Gato pajero, lives amongst the thick straw at Bahia Blanca, also found in Banda Oriental.
1444 B Bird lives near the beach [probably the specimen labelled 1443D at NHM]
1445 B Swallow. nests in holes of the Barranca [listed as *Hirundo cyanoleuca* in *Zoology* **3**:41. Labelled 1445D at NHM] *
1446 B Owl. [listed as *Strix flammea* Linn. in *Zoology* **3**:34, and labelled 1446D at NHM] *
1447 B X Perdrix. (same species?) [note opposite] Inhabits sand dumes & barren very dry country
1448 B Sylvia
1449 B Charadrius

Buenos Ayres

1450 B Bird, same as at Maldonado
1451 B X Fuller shot this bird at Maldonado. it is common at St Fe Bajada [note opposite] This belongs to the tribe of birds allied to Certhia of which I shot so many at Maldonado
1452 B Bird [listed as *Pachyrhamphus albescens* in *Zoology* **3**:50, and labelled 1452D at NHM] *
1453 B Shot on board the Beagle:

R. Plata

1454 B Duck. Bahia Blanca
1455 B X Gull do [note opposite] Beak "saffron yellow" lower mandible of at base brownish orange; legs yellow not so bright as the beak.— These birds often flies 50 or 60 miles inland: frequent slaughtering places, & make a noise like the common English Gull when its breeding place is disturbed. [listed as *Larus dominicanus* Licht. in *Zoology* **3**:142]
1456 B Falco Bahia Blanca

1833

1457 A Fresh water rat, lives in the Streams at Bahia Blanca. Hinder feet demi=palmated. [listed as *Mus Braziliensis* in *Zoology* **2**:58-60]

August Rio Negro

1458 B Ibis very common in large flocks in the Great Plains between B Blanca & B. Ayres. Flight soaring. very graceful [listed as *Ibis (falcinellus) Ordi* Bonap. in *Zoology* **3**:129]
1459 B X Bird. cry loud, singular, single.— [note opposite] Is very remarkable from the great activity. it runs along at the bottom of the hedges, resembling some animal; flies very unwillingly [listed as *Rhinomya lanceolata* in *Zoology* **3**:70, NHM 1855.12.19.169] *
1460 B Thrush [both *Turdus rufiventer* and *T. Falklandicus* were collected at the Rio Negro. See also specimen 1470]
1461 B X Bird called the Callandra. V **179** (d) [note opposite] V account of Bird (N° 1213) [listed as *Mimus Patagonicus* G.R.Gray in *Zoology* **3**:60-1]
1462 B Sylvia
1463 B Dove
1464 B Falco. I have seen it at Maldonado
1465 B Fringilla

1466 B Fringilla, is found in small [cont. opposite] flocks, inhabiting the most desert parts of the passage between Rio Negro & Colorado.— [listed as *Fringilla carbonaria* in *Zoology* **3**:94]
1467 B X Long billed Causaria [note opposite] Inhabiting the dry plains & Bahia Blanca; flies quietly about & hops very quickly along the ground picks the dry pieces of dung.— [listed as *Uppucerthia dumetoria* in Zoology **3**:66, and see specimen 2827]
1468 B X Fringilla
1469 B X Little long pointed tail bird [note opposite] Hops about bushes very like a Parus, runs on the ground, very quickly, uttering harsh shrill quickly reiterated cry: Found at B. Blanca [listed as *Serpophaga parulus* Gould in *Zoology* **3**:49]
1470 B Turdus [see specimen 1460]
1471 A X Aperia but decidedly different from the Maldonado one [note opposite] V **187** (d)
1472 A Rat. dry camp. ears very large & delicate

1833

1472 X 1473. Wood silicified. R. Uruguay.
[note opposite] By mistake Number (1472) is used twice.—
1474 S Shells, sold in the market at B. Ayres for spoons.—
1487 A Skeleton of a Biscatcha
1488 I : 1489 : 1490 Coleoptera. R. Colorado [see *Insect Notes* p. 76]
1491 I Copris. Bahia blanca V **200**(b)
1492 I X Aphodius, flying by thousands [note opposite] but not alighting on plentiful horse dung: 10 leagues North of Sierra de la Ventana. V **200**(b)
1493 I Hemipte: <u>very</u> abundant in herbage. Bahia Blanca
1494 S Fresh water shells. very abundant on banks of rapid brook, R Sauce:
1495 I Coleoptera Bahia Blanca [see *Insect Notes* p. 76]
1496 I Carabus Bajada St Fe
1497 I Brachinus Gorodonà. R Parana
1498 I Heterom:— St Fe Bajada
1500 I : 501: 502: 503 Coleoptera Bajada
1504 I Heterom. Rozario
1505 I Coleoptera Bajada [see *Insect Notes* p. 76]
1506 S Land shells Bajada
1507 I : 1508: 1509: Onthophagi [note opposite] Caught crawling in a ditch Buenos Ayres

1833

1510 Fishes teeth. Liniston Bajada
1511 Pectoral bone from the Armado. Fish
1512 Rattle from a snake killed at the Bajada.—
1513 A X Head of the fresh water rat; brooks Bahia Blanca
1514 S X Land shells, in the most extraordinary numbers on the arid plains N. of B. Blanca. [note opposite] All picked up without animals.—
1587 A Head of Cavia (1471) shot at R. Negro
1588 A Bezoar stone out of stomach of some animal. Sold by the Indians. R. Negro
1589 Part of Concha; perforated by some animal allied to Cliona celata of Grant
1590 P X Sort of Lichen growing on the dry sandstone plains of R. Negro.— [note opposite] The patches are circular from size of shilling to half a crown; the ground is blistered, that is the patches are convex & partly hollow underneath.— It is abundant.
1591 S Pecten. Port St Antonio
1592 B X 6 eggs. M: Video: 3 spotted with brown of the Sparrow: 2 spotted with red, the Anthus (1202) [note opposite] one larger egg, also spotted with red, was with some of the Sparrow eggs in

its nest on the ground.— I had previously heard that a bird called the Cusco lays its eggs in sparrow or other birds nests

1833

1593	P	Bearded wheat, injured by the Pulvilho.— V **208** [see *Plant Notes* pp. 174-5]
1596	I	Cerambyx. Maldonado
1597	I	Moth flew on board in <u>wonderful</u> numbers. Mouth of R. Plata [see *Insect Notes* p. 77]
1598	I	Flew on board in considerable numbers in Lat. of R. Negro.—

November Monte Video

1600	B X	Lanius; common about St Fe Bajada [note opposite] This for a land bird is most singularly white; it is most beautiful.— rather shy.— [listed as *Fluvicola irupero* Gray in *Zoology* **3**:53]
1601	B	Sylvia
1602	B	Vanellus or Pteru Pteru V ~~186~~ additional notes at Maldonado [listed as *Philomachus cayanus* in *Zoology* **3**:127]
1603	B	Tringa
1604	B	Muscicapa (same as at Maldonado?)
1605	B	Fringilla
1606	B	Charadrius in large flocks
1607	B	: 1608. 2 species of Tringa
1609	B	Swallow (most common sort) [listed as *Hirundo frontalis* in *Zoology* **3**:40]
1610	B	Trochilus (not very abundant) [listed as *Trochilus flavifrons* in *Zoology* **3**:110]
1611	B	1612 Fringilla
1613	B	Muscicapa

<u>1833</u> November M: Video.

1614	B	Fringilla (same as at Maldonado)
1615	B X	Fringilla (common sparrow) same as (683) [note opposite] V ~~186~~ [in pencil] add: notes at Maldonado V: No 1592 eggs. [listed as *Zonotrichia matutina* in *Zoology* **3**:91]
1616	B	: 1617. Fringilla. Cock & Hen, shot together.
1618	B	Swallow
1619	B	Oven bird ?
1620	B	Callandra ? (same as (1213)) [listed as *Mimus orpheus* G.R.Gray in *Zoology* **3**:60]
1621	B X	1622. Cock & Hen Scissor tail [note opposite] So called by the Spaniards, & well deserves its name, as it flies from a bough it pursues insects & turning short, opens & shuts its tail vertically & laterally exactly like scissors.— [listed as *Muscivora Tyrannus* in *Zoology* **3**:43-4]
1623	B X	Caprimulgus. [note opposite] Not uncommon at St Fe Bajada, rises from the ground like European: alights on a rope rather diagonally. [listed as *Caprimulgus parvulus* Gould in *Zoology* **3**:37-8, NHM 1855.12.19.241 type, labelled 1623D in NHM data bank]

Rio Plata

1624	B X	Procellaria, shot Lat.42°.20′ S. also seen little S of R. Plata [note opposite] In stomach beak of cuttle fish. [listed as *Procellaria gigantea* Gmel. in *Zoology* **3**:139-40]

1834		Janu:	Port Desire

1641 P Lichen, common on pebbles

1661 B Duck, 20 miles up the creek.—

1693 A X 1694.— Large eared mouse excessively abundant in all situations [note opposite] Caught with cheese & biscuit. [listed as *Mus xanthopygus* in *Zoology* **2**:63-4]

1695 A Gerbillus? very like Maldonado one

1696 A Mouse very common in long dry grass.— [listed as *Mus canescens* in *Zoology* **2**:54-5]

1697 A XX Taturia Pichiz. [note opposite] This differs in general appearance from B. Bahia one.— form of head, shape of scales, number of foreward rings; & yet the difference is but little; is not abundant.— Parts of the case were seen at S. Cruz, Lat 50°.—

1698 B Long billed Casaia. utters reiterated shrill cry; tolerably common in the most dry & desart places.

1834 Jan. Port Desire

1699 B : 1700.— 2 species of Lanius, both shy scarce solitary; wild vallies with thorny bushes. [listed as *Agriornis micropterus* and a nestling *A. striatus* in *Zoology* **3**:57, NHM 1855.12.19.253 and 1855.12.19.298]

1701 B Fringilla not very uncommon in the vallies

1702 B Furnarius; same habits as (1222). More in bushes.— rare.— [listed as *Eremobius phœnicurus* Gould in *Zoology* **3**:69-70, labelled 1702D at NHM]

1703 B Sylvia (?) in bushes near sea-coast.

1704 B Sparrow. Apparently same as (1615) but the egg is decidedly different, & I do not believe there is any mistake in either case.— [note opposite] The most common bird in desart plain, rocks & bushy vallies.— Eggs No (1710)

1705 B Certhia, actively flying about bushes. apparently same genus with (1250) [listed as *Synallaxis brunnea* Gould in *Zoology* **3**:78-9, NHM 1855.12.19.99 type]

1706 B Hawk. Nest in low bush. eggs (1710) [listed as *Falco femoralis* Temm. in *Zoology* **3**:28]

1707 S X 1708.— The commonest shells now existing on the rocks.—
[deleted note opposite] For Comparison with those from North plains.—

1709 X Obelia & Cellaria. latter pale "hyacinth red" setæ *[illeg.]* 6bis Lat 48°.56′, 4 miles from land; 19 Fath; [note opposite] Globular shaped. Polypus protruding from aperture at the foot of spine; very simple, with 12-- or 16--- arms; highly retractile.— Coralline pale "peach blossom red".— [Busk Collection: *Flustramorpha flabellaris* var. *patagonica*]

1710 B Eggs of Hawk (1706) & sparrow (1704)

1712 I Cicindela (2 specimens) taken on dry mud bank, encrusted with salt. habits like Hybrida. Port St Julian.

1834 Jan.— Port St. Julian

1713 I Truncatipennis, under salt-loving plant just above high water.

1714 I Hab: do (Young specimen) [see *Insect Notes* p. 77]

1715 I Colymbites. nearly drowned in salt water head of Harbour; proving fresh water although we could find none.— [see *Insect Notes* p. 77]

1716 I X Diptera, very numerous, bite very badly [note opposite] What animals did Nature intend for them? they are out of all proportion too numerous for Guanacoe & scarcely any other large animal exists here: [for specimens 1716-1717 see *Insect Notes* pp. 77-8]

1717 I Heterom. (found dead)

1718 S Small bivalves, under smooth round stones, on mud, in little pools near high water.— [note opposite] When alive tinged with Carmine.

1734 Corallines (found with animals 782 spirits)

1735		Corallines, port of St. Julians.—
1747	I	Cells of Bee (1748) adhering to round stones; Hills. plain cylinders applied side to side.— [note opposite] Honey thick, yellow, very little sweet.— (chiefly pollen)
1748	I	Bee (nest above)
1749	I	Diptera
1750	I	Curculio, sterile plain
1751	I	Heterom. do do.— [see *Insect Notes* p. 78]
1834		Jan. Port St Julians
1752	B	Lanius. (amongst bushes) [listed as *Agriornis micropterus* Gould in *Zoology* **3**:57, NHM 1855.12.19.253 type, labelled 1752?]
1753	B	Sylvia do
1754	B	Furnarius same as (1702). Coleoptera in stomach. [listed as *Eremobius phœnicurus* Gould in *Zoology* **3**:69-70. Specimen at NHM carries CD's own label 1754]
1755	A X	Gerbillus (?). Weighs 579 grains. [note opposite] Killed by falling down a cliff.—
1756	B XX	[on opposite page] Cormorant: skin round eyes "Campanula blue" cockles at base of upper mandible "saffron & gamboge yellow".— Mark between eyes & corner of mouth "orpiment orange". I saw this bird in the Falkland Islands catch a fish, let it go & catch it again 8 times successively as an otter does a fish or Cat a mouse.— [listed as *Phalocrocorax carunculatus* Stephens in *Zoology* **3**:145-6]
1757	B	Larus; Beak coloured palish "arterial blood red", legs "vermilion red". [listed as *Larus hœmatorhynchus* King in *Zoology* **3**:142]
1758	B	Hawk. iris light brown. legs gamboge yellow.—

1759	S	Shells, adhering to Corallines &c at Lat: 48°.56′: 19 fathoms: 5 miles from shore
1760	I	Coleoptera. Port Desire [see *Insect Notes* p. 78]
1761	S XX	1762 Balanus & its valves. Hab same as (1759) [note opposite] Horns of valves fine pink.—
1763	S	Solen. Port St. Julian
1770		Halimeda & Cellaria. Port Desire [note opposite] Halimeda V **211** [CD's *Halimeda* was a green alga (Chlorophyta) and is shown in Fig. 4 of *Plant Notes* pp. 186-217 as the syntype specimen of *Amphiroa orbigniana* Harvey ex Decaisne. The specimen of *Cellaria* bottled with it was identified in the Busk Collection as *Menipea patagonica*]
1771	B	Sparrow same as (1704)
1772	B	Hawk. iris dark brown. legs blueish
1773	B	Ibis V ~~210~~ Notes P. Desire
1774	A X	Aperea or Cavia. old female [notes opposite] Old male weighed 3530 grains. live amongst ruins & have burrows under bushes; very regular tracks from bush to bush.— [note opposite] 1774 to 1775 all Port Desire. The Gregory Bay Indians had mantles for children of this animal. Falkner says the tribe has its name from this cause.— [listed as *Cavia Cobaia* in *Zoology* **2**:89]
1775	S	Shells. Port Desire & St Julian
1777		Old shells (one partly perfect) with encrusting Corallines. 5 Fathoms. Elisabeth Island, St of Magellan
1834		St^{s} of Magellan
1778	B }	Two species of ducks; fresh water,
1779	}	Cape Negro
1780	B	Grebe. iris scarlet red: Hab: fresh water [listed as *Podiceps rollandii* in *Zoology* **3**:137]
1781	B	Hawk; iris brown; male: Hab do

1782 B XX Petrel; legs "flax flower blue". in stomach remains of Crust: Maia. Hab: do [notes opposite] This bird is a complete diver in its habits. In the evening often flying in direct lines from place to place: P. Famine. frequents quite deep inland seas.— flight direct rapid drops from the air & instantly like a stone dives; far & long; rises to the surface & will then instantly take to the wing; this is when frightened; generally quietly swimming & diving after its prey:— Common in the Beagle Channel. I saw this bird between the Falklands & Patagonia [listed in *Zoology* **3**:138-9 as *Pelecanoides Berardi* Gray]

1783 B Gull; legs & base of bill brownish cream yellow: Hab: do

1784 B X Icterus: (also common Port Famine): Hab do [note opposite] Small flocks. runs on ground. noisy chattering bird like a starling

1793 I Heterom. Cape Negro: (it is here that the features of Patagonia & T. del F. are united)

1794 I Carab: Hab do [for specimens 1793 and 1794 see *Insect Notes* p. 78]

1795 A Head of Toco Toco. Common on the burrowed ground of C. Negro V **188**(a)

1814 B X Egg of ostrich, Port Desire [note opposite] I believe Avestruz petise. V **212**

1816 B Puffinus V **213** Copy

Port Famine (beginning of Feb:)

1817 B Tringa. sea coast

1818 B Fringilla. common in outskirts of the Wood

1819 B X 1820.— Muscicapa, inhabits the [cont. opposite] gloomiest recesses of the forest, generally high up amongst the trees.— Constantly uttering plaintive whistle in same tone.— very difficult to be seen or found as the noise seems to come from no particular spot or place or distance.— [listed as *Xolmis pyrope* in *Zoology* **3**:55]

1834 Feb. Port Famine

1821 B Muscicapa, not uncommon in outskirts of forest. sits on dead branch:

1822 B X Furnarius. same as (1260) of Maldonado? [deleted note opposite] This is not common, & I think certainly different from (1823). This latter bird has same reiterated shrill cry: walks — always on the beach.— [*Opetiorhynchus patagonicus* or *O. vulgaris*]

1823 B Furnarius, very abundant, feeding at high water mark: very tame. [deleted note opposite] Feeds entirely on marine substances even on the floating kelp: (1822) up in the hills & beach. wilder.— I was surprised to see (1823) in central Patagonia April, banks of the S. Cruz: [*O. patagonicus* or possibly *Uppucerthia dumetoria* in *Zoology* **3**:66-7]

1824 B Muscicapa. gloomy forest. tame, quiet, very rare; Specimen shattered.— [listed as *Myiobius parvirostris* in *Zoology* **3**:48, labelled 1824D at NHM] *

1825 B Creeper. (rare) actively hopping about bushes. shrill rapid note.—

1826 B Sparrow. (not uncommon) [listed as *Zonotrichia canicapilla* Gould in *Zoology* **3**:91-2, labelled 1826D at NHM] *

1827 B Swallow. builds in cliffs (do) [listed as *Hirundo Leucopygia* Licht. in *Zoology* **3**:40] *

1828 B X Wren, very curious loud cry, frequents [cont. opposite] bottom of stumps; outskirts of forest: hard to see or put up. [listed as *Scytalopus Magellicanus* in *Zoology* **3**:74]

1829 B do. Shot in deep forest. (Cock of last?)

1830 B Fringilla. active. tops of beeches deep forest: wild.—

1831 B X Wren. very abundant outskirts of forest. [note opposite] This & the common creeper the two most abundant birds in the woods.—

1832 B XX 1833. Feathers of Ostrich V **212** [note opposite] Ask me

1834 B Head of do (P Desire) V do See account of

1835 B 1836. Legs of do V do small species

1837 B Feathers. Gregory Bay V do of Ostrich

1838 B Hide do V do
[listed as *Rhea Darwinii* Gould in *Zoology* **3**:123-5]

<u>1834</u> Feb: Port Famine

1839 I Coleoptera, under bark.— [for specimens 1839-1843 see *Insect Notes* p. 79]

1840 Coralline, articulata (Jania) & *[illeg.]* on rocks.—

1841 I Fly Port Famine except

1842 I Lepidop. Lepidop. Cape

1843 I Bee Negro —

1844 S Shells (some of them old)

1846 Coralline on Balanus. "orpiment or reddish orange", could not see Polypi: 2. Sertularia & their encrusting small Corallines. 11 Fathoms: East entrance of St^{s} of Magellan.

1852 X Corallines, encrusting rock 40 Fathoms off C. Deceit & Barnevelts [notes opposite] The one lapped up in paper has a delicate Polypus with 12 arms.— Color pale red.— One in blotting Paper was found in 75 Fathoms between Falklands & Santa Cruz.— [Busk Collection: *Mucronella tricuspis*; *Discoporella fimbriata*; *Lepralia (Pentapora) monoceros*; *L. bicristata*]

1871 Sertularia. Dynamena. East Coast of Tierra del Fuego. the Former with purple or ovular capsules.

1872 Corallines: low water mark Wollaston Is^{d}:

1874 Flustra. V **219**

1875 B Owl, bought from the Fuegians. Ponsonby Sound [listed as *Ulula rufipes* in *Zoology* **3**:34]

1876 X Corallines: 54 Fathoms, some miles from Staten land going to Falklands [notes opposite] White branching one V **227**. There were same species in water from 60 to 70 Fathoms.

1877 X Obelia & Corallines on Fucus leaves. Ponsonby.— [note opposite] Obelia is I believe same as the one of P **173**: angular Polypus has same essential characters.
[Busk Collection: *Porella margaritifera*; *Diastopora tubuliporide*]

<u>1834</u> March

1878 S Pecten: Fucus G. leaves. Ponsonby Sound.

East Falkland Island

<u>Shells!!!</u> 20 numbers

1879 B Fringilla: ~~very~~ abundant in large flocks in all parts of Island; very tame.
[listed as *Chlorospiza ? melanodera* in *Zoology* **3**:95-6, NHM 1855.12.19.50] *

1880 B Tringa. in flocks on sea-beach. [listed as *Limosa Hudsonica* in *Zoology* **3**:129]

1881 B Hawk.— [listed as *Circus cinerius* Vieill. in *Zoology* **3**:30-1, labelled 1881D at NHM] *

1882 B X Carrion Vulture V **238** [note opposite] Legs "ash grey".
[listed as *Milvago leucurus* in *Zoology* **3**:15-18, labelled 1882D at NHM] *

~~1883 Lepus Magellicanus~~

1883 S Land shells. I do not know where from.— I believe M: Video

1884 Corallines. encrusting roots of Fucus G [Busk Collection: *Microporella malusii*; *M. personata*; *Lepralia (Pentapora) personata*; *L. labiosa*]

1885 A Lepus Magellicanus V **236**

1898 B Lark. not uncommon [listed as *Anthus correndera* in *Zoology* **3**:85]

1899 B Muscicapa. inhabits chiefly the dryer & more stony hills & sea-coast

1900 B Thrush. Hab do & houses: tame, inquisitive like English thrush; silent, cry peculiar [listed as *Turdus Falklandicus* Quoy et Gaim. in *Zoology* **3**:59, labelled 1900? at NHM]

1901 B Owl. [listed as *Otus palustris* Gould in *Zoology* **3**:33. Specimen at NHM has CD's own field label 1901]

1902 A Head of Lepus Magell: specimen was differently marked from the description of Lesson

1834 March E. Falkland Is^{d}.

1910 I Sphodrus. with four indistinct orange spots on elytra: under dead bird sea-coast [for specimens 1910-1912 see *Insect Notes* p. 79]

1911 I Catops. under old dead calf: far in country

1912 I Curculio. in berry of Tea-plant.—

1913 Flustra (same as 920 spirits) V **250**

1914 XX [note opposite] Minute Berenica on Cellipora (933 spirits) cells. oral punctured. mouth surrounded by long diverging spines.— Ask me

1915 B XX Vultur (aura?): female: skin of head "Scarlet & Cock" red; flight soaring, elegant; shy: Iris dark coloured.— tolerably numerous [note opposite] Is said to occur in Tierra del Fuego & I believe in La Plata [listed as *Cathartes aura* Illi. in *Zoology* **3**:8-9]

1916 B Hawk: female (dissection): chiefly lives on rabbits: [listed as *Milvago leucurus* in *Zoology* **3**:15-18]

1917 B X Grebe: female: lives in the far inland tranquil arms of sea.— Iris dark red [note opposite] Male is exactly same colour.—
[listed as *Podiceps rollandii* Quoy et Gaim in *Zoology* **3**:137]

1918 B Grebe: only seen in one fresh-water lake. female: legs same color as back. iris "scarlet & carmine r": pupil dark: [listed as *Podiceps kalipareus* Quoy et Gaim in *Zoology* **3**:136]

1919 B X 1920. Emberiza. shot in the hills in the same large scattered flock [note opposite] This Emberiza is commonly shot on the lower land & may be seen with (1879). I do not believe (1920) is at all common.— [listed as *Chlorospiza* ? *xanthogramma* in *Zoology* **3**:96-7]

1921 A Mouse. caught far from houses ¾ of mile in grassy bank: English? if so curious change from Ship to such a country as this: [listed as *Mus decumanus* in *Zoology* **2**:31-3]

1922 B Emberiza: female shot with (1919)

1923 B X do: shot in plains: same or different? [note opposite] If different is possibly young cock of (1829): both were shot in plains.—

1834 March E. Falkland Is^{d}.—

1924 Flustraceæ. V **233**

1925 Flustraceæ, little circular orange patches, on Kelp leaves: Polypus 16 arms: ova <*illeg.*> in length. oval with kidney shaped internal mass

1926 B X Hawk: male: iris "honey yellow" [note opposite] M. Bynoe has female of this: larger: secondaries more various. legs & skin above blue, beak bright yellow: feeding on bits Carrion & by necessity? — [listed as *Milvago leucurus* in *Zoology* **3**:15-18]

1927 : 1928. Small encrusting Corallines from the Kelp: specimens appear small, but are very interesting. [Busk Collection: *Salicornaria malvinensis*]

1929 S Shells. Kelp: excepting small thin bivalve on the beach

1930 P X Gum-resin, exuded from the bosses of the Hydrocotile gummifer [note opposite] Chemical analysis.—

1931 B Furnarius (1823). (Male). V **241**(a)

1932 B X Caracara (female) V **238**(a) [note opposite] Specimen unfortunately burnt, tail quite spoiled.—

1933 B do (sex?) V do

1934 A Teeth of rat out of stomach of a Hawk shot in the country.—

1935 Corallines encrusting stems & leaves of Fucus G. 4 to 10 Fathoms. [Busk Collection: *Porella margaritifera*; *Lepralia (Pentapora) Margaritifera*; *L. discreta*; *Chorizopora discreta*; *C. bougainvillei*]

1936 X do do on stones, Kelp, roots [note opposite] The inarticulate encrusting Corallina was in about 4 fathom water. This is curious in Corallina. [Busk Collection: *Micropora stenostoma*; *Smittina landsborovii*]

1834 April. E. Falkland Is[d].

1937 S Shells (chiefly from Beach).

Port Famine

1938 S Balanidæ growing on old Beech trunk washed up on beach in profusion.

1947 X Pebbles of Porphyry with delicate encrusting Corallines. Santa Cruz. The one in paper, tubular Coralline, has Polypus with structure of body as in the Flustraceæ. [note opposite] Interesting to Geologists; 10 Fathom water, 3 miles from shore where most rapid tides, yet living Corallines possessed most delicate spines, showing how little pebbles are moved at the bottom.

1998 S Balanidae: Beagles copper.

1999 I Fly, under dead birds, sea-beach, from Falkland Islands

2000 S X Shells, excessively abundant in every streamlet which enters the S. Cruz high up: [note opposite] Is it not same as found in R. Negro & Sauce.

2002 I X Coleoptera, high up S. Cruz river. All the Carabidous & Staphylini under stones on the beach [note opposite] Green colored Paper signifies +2000:— [see *Insect Notes* p. 80]

2003 S X Terebratulæ, abundant on large shingle (this appears a common [cont. opposite] locality for this genus) in 7 Fathoms, where tide runs 3 & 4 knots & in the ebb is brackish! S. Cruz

1834 May S. Cruz

2004 B Feathers of Ostrich, I suppose of Petise; found high up river of S. Cruz.—

2005 Sertularia = Clytia, same as (959) V **248**

2006 Sertularia = Flustra V **246**

2007 Escara V **254** [Busk Collection: *Eschara gigantea*; *Aspidostoma giganteum*]

2008 Encrusting Corallines; 48 Fathoms [Busk Collection: *Mucronella tricuspis*; *M. ventricosa*]

2009 X Corallines. St. Cruz [note opposite] This is another Sertularia, besides those described P **248-250**: & the Tubularia = Clytia P **245** [Busk Collection: *Tubulipora labellaris*]

2010 Favosites V **144**(a).— [Busk Collection: *Fasciculipora ramosa*]

April Birds: S. Cruz

2011 B Lanius called Callandra, sings very prettily amongst the spiny bushes not uncommon.— [note opposite] Female [listed as *Mimus Patagonicus* in *Zoology* **3**:60-1, NHM 1855.12.19.221 and .311] *

2012 B X Lanius, rare, interior of the country also found in; chases insects very quickly half flying, half running [note opposite] The Lanius with white tail I hear belongs to the Andes of Chili. Don Pedro Renous told me. excellent authority.— [added later] Yes. I saw it there. [listed as *Agriornis maritimus* Gray in *Zoology* **3**:57, NHM 1855.12.19.251] *

2013 B Lanius rare

2014 B X Hawk: female: flutters stationary over one spot like the Kestrel [listed as *Tinnunculus Sparverius* Vieill. in *Zoology* **3**:29]

2015	B X	Fringilla: Cock? — Cordillera [note opposite] In small flocks, 6 to 10, in the bushy valleys: not common; also seen at Port St Julian, & no where else; uttered a very peculiar, pleasing note, & with a peculiar soaring flight from bush to bush; here I never noticed this; but heard the noise. [listed as *Fringilla fruticeti* Kittl. in *Zoology* **3**:94]
1834		April S. Cruz
2016	B	do: Female:
2017	B	Fringilla: cock: abundant in the valleys & whole way up into the interior [listed as *Fringilla Gayi* in *Zoology* **3**:93, NHM 1855.12.19.42] *
2018	B	do: Female:
2019	B	Fringilla : cock : rare [listed as *Emberiza luteoventris* in *Zoology* **3**:89]
2020	B	Red throated creeper: Male [listed as *Synallaxis rufogularis* Gould in *Zoology* **3**:77, NHM 1855.12.19.104 type, and .171] *
2021	B	do : not uncommon amongst the thickets [labelled as 2021D at NHM]
2022	B	2023. Long-tailed Creeper: male: habits do [listed as *Synallaxis ægithaloides* Kittl. in *Zoology* **3**:79, labelled as 2022D at NHM] *
2024	B X	Creeper: Male.— [note opposite] All these are called Males because Ovarium did not appear (even with aid of lens) granulated. [listed as *Synallaxis flavogularis* Gould in *Zoology* **3**:78, labelled as 2024D at NHM] *
2025	B	Furnarius: flies about under the bushes: & cocks up its tail: not common: reiterated shrill cry: female: [listed as *Eremobius phœnicurus* Gould in *Zoology* **3**:69-70, NHM 1855.12.19.73 type, labelled 2025D] *
2026	B	Wren. female; harsh chirp [listed as *Troglodytes Magellanicus* Gould in *Zoology* **3**:74]
2027	B	Parus: Female; by 3ˢ and 4ˢ together [listed as *Synallaxis ægithaloides* in *Zoology* **3**:79, NHM 1856.3.15.11] *
2028	B	Caracara: Cock: (Carrancha?) [listed as *Polyborus Brasiliensis* in *Zoology* **3**:9-12]
2029	B	Caracara: female: new: [listed as *Milvago albogularis* in *Zoology* **3**:18-21]
2030	B	Hawk. legs pale yellow: bill blueish black. [listed as *Buteo ventralis* Gould in *Zoology* **3**:27-8, NHM 1855.12.19.204] *
1834		April. S. Cruz
2031	B	Owl.— [listed as *Otus palustris* in *Zoology* **3**:33]
2032	A X	Mouse with grooved teeth; very abundant: caught by cheese, bread, flesh &c &c: every-where up the country: weight 1336 grˢ [note opposite] 1336 grain. Apoth: weight [listed as *Reithrodon cuniculoïdes* in *Zoology* **2**:69-71]
2033	A	Mouse: common: caught same manner
2034	A	Caught up the river.—
2035	A	Mouse. ears, feet, tail, nose, dusky orange. extraordinarily numerous up country & every-where: weight 329 grains.— [listed as *Mus xanthopygus* in *Zoology* **3**:63-4]
2036	A	Cat; in a bushy valley: did not run away: but hissed: [listed as *Felis Pajeros* in *Zoology* **2**:18-19]
2037	A	Head of Mouse (2032).—
2038	A	Head of Mouse (2035).—

Plants

2039	P X	Very sweet smelling plant with a rather biting aromatic taste: used for making tea by the seamen.— [*Satureja darwinii* Benth. See *Plant Notes* pp. 175-6]

[note opposite] (NB Write this at Beginning) As all these plants were collected during end of April & beginning of May, they are late autumnal plants.— I collected every one in flower: as indeed I have done every-where in Patagonia.— Country same dry sterile shingle bed as before.—

1834		April.	S. Cruz

2040 P Plant on the dry banks: (flowers minute?): High up the river interior: [*Euphorbia portulacoides* L. See *Plant Notes* p. 176.]

2041 P X : 2042. Plants, 140 miles up the river: character of country same as at coast: as these plants, I never saw to the Coast, are they not Cordilleras plants crawling downwards [note opposite] Grows rather near river: [*Quinchamalium chilense* and *Oreopolis glacialis*. See *Plant Notes* p. 176]

2043 P Very adhæsive, abundant about the Lava cliffs, 8 or 900 feet above sea in the interior: very adhesive: [*Senecio tricuspidatus*. See *Plant Notes* p. 176]

2044 P Same locality; shady nooks amongst the rocks [*Sisymbrium magellicanum*. See *Plant Notes* p. 176]

2045 P Plant.— interior.— [*Galium richardianum*. See *Plant Notes* p. 176]

2046 P Grass; this characterizes all the arid plains of S Patagonia [*Stipa speciosa*. See *Plant Notes* p. 177]

2047 P Plant.— interior.— [*Descurainia appendiculata*. See *Plant Notes* p. 177]

2048 P Plant on the wet shingle: river side [*Arenaria lanuginosa*. See *Plant Notes* p. 177]

2049 I Curculio lying dead by thousands on all parts of plains: interior far up & Coast [for specimens 2049-2055 see *Insect Notes* p. 80]

1834		April	S. Cruz

2050 I : 51: 52. Curious Heteromerous insects. 50: & 51: far up the country ∴ quite original

2053 I Lamellicorn: lying dead in great numbers: interior probably feed on Guanaco dung.

2054 I Galeruca: a tribe very rare in such countries

2055 I Fly: feeding on a Phallus.—

2066 A X Rat — Choiseul Bay — E. Falkland Is^{d} — (same as last year?) [note opposite] these 3 specimens brought in the Adventure.— [listed as *Mus decumanus* in *Zoology* **2**:31-3]

2067 A Head of a rat, W Point W Falkland, probably same as above.—

2068 A Rats Head. Port Egmont, W Falkland English Rat?.—

2069 Crisia V **255**

2070 X Coralline & shells. 10 to 20 Fathom St. Magellan [note opposite] The Tubularia.— Clytia from Ships bottom.— [Busk Collection: *Mucronella tricuspis*; *Arachnopusia monoceros*; *Membranipora umbonata*]

2071 X Holuthuria - doris V **215** [note opposite] The obscurely-colored ones were put in Spirits

2072 S Balanus, 10 to 20 Fathoms St of Magellan. (common) [note opposite] 2072: 73: Both rather remarkable in their Hab. as belonging to tribes generally found on tidal rocks.— Large stones from the above depth are oftentimes entirely coated with the Corallina.—

2073 X Corallina. Hab. do: the layers of cavities for the ovules here well seen

1834 June. P. Famine

2080	B		Procellaria gigantea. V196(c) Copy [see *Zoology* **3**:139]
2081	B		Tyrannus. (not uncommon) [listed as *Xolmis pyrope* in *Zoology* **3**:55]
2082	B		Fringilla, in small flocks feeding near the beach: I never saw any before there.— I do not believe the cocks (if this is not one) are brighter coloured
2083	B		Tyrannus [listed as *Myiobius parvirostris* in *Zoology* **3**:48, labelled 2083D at NHM] *
2084	B	X	Certhia, excessively abundant [note opposite] Inhabits all parts high up & low-down of the beech forests: very tame, not at all shy: follows with apparent curiosity every person who enters these silent forests: continually utters a small harsh cry: will approach to within a few feet of a person['s] face: does not run up & down the trees but very little or never.— [Listed as *Oxyurus tupinieri* Gould in *Zoology* **3**:81-2]

July Chiloe

~~2101~~			Geotrupes } [in pencil] 2107:8 (number lost)
2102	I		Earth-balls } V **264** [see *Insect Notes* pp. 80-1]
2103	S		Fresh-water muscles [sic]
2104	S	X	the above & various shells.— [note opposite] 2 species of simple Patelliform shells, littoral as is the Balanus & Nerita: the Patelliforms with internal plates are Oysters: all the other on or in soft sandstone in 2 or 3 Fathom
2105	S		Terebratulæ with internal processes 17 Fathoms back of Chiloe: & a helix from the Forest.—
2106			lost
2107:8	I		Geotrupes
2109	I		Carab. Bemb: in moss.—

1834 July.— Chiloe

2110:111	I		The great curious Lucanus [stag beetle]: given me by Mr R.Williams: caught when flying about in summer. [Note opposite] The male insect is said to make a very loud clacking noise with its horns when molested or even approached: is not very uncommon: is found abundantly on Mainland near Valdivia.— [different pen] In end of Jany, Chiloe 1835, I found 3 females flying about during the day: when touched stood on four hind legs & raised their head, as in battle: very strong:— Caught male at Valdivia, fought most boldly, turning round to face enemy: the noise alluded to is not very loud and produced by friction of abdomen, when even frightened, but not touched: jaws not so strong as [to] produce pain to fingers. [Further note inserted on separate page] Mr Douglass sent me 12 specimens of this fine insect & the following account.— 'I found them in the crutch of an Atenihue tree, thirty ft above the ground, in a nest of moss.— I was led to the spot by following one of them morning & evening for several days & always lost sight of it near this tree. I at last climbed up the tree & discovered them as mentioned.— This is in the Isle of Conahue.—' XX Copy this [Lucanidae, *Chiasognathus grantii*. See *Insect Notes* p. 81.]
2112			Corallines. 30 Fathom. Cockburn Channel
2113	S		The Succinea & brown Helix on land, the others in fresh-water brooks.—
2122	B	X	Kingfisher: female [Note opposite] in Stomach a Cancer, Brachyus & small fish.— fishes in the quiet waters of the creeks & coves.— Is it common for a Kingfisher thus to live on salt-water produce. [listed as *Ceryle torquata* in *Zoology* **3**:42]
2123	B		Tringa: inland fields: large flocks [listed as *Limosa Hudsonica* in *Zoology* **3**:129]
2124	B	X	Lanius; iris scarlet, ~~tolerably~~ very common [Note opposite] generally perched on top of branch looking out for its winged prey [listed as *Xolmis pyrope* Gray in *Zoology* **3**:55]
2125	B	X	Thrush: Male, in stomach seeds & berrys [note opposite] Common

			[listed as *Turdus Falklandicus* in *Zoology* **3**:59]
2126	B		Furnarius. Male(?)
2127	B		Red-breast: female: (Cheucau) [listed as *Pteroptochos rubecula* Kittl. in *Zoology* **3**:73]
2128	B		Muscicapa. always on the beach, expands its tail
2129	B		Creeper. female: Coleoptera in stomach [listed as *Dendrodramus leucosternus* Gould in *Zoology* **3**:82-3]
2130	B	X	Creeper same as in T. del Fuego [note opposite] female.— [listed as *Oxyurus tupinieri* Gould in *Zoology* **3**:81]
2131	B		Fringilla
2132	B		Common Fringilla, very
2133	B		abundant: Cock & Hen!
2134	B		Humming Bird [listed as *Trochilus forficatus* in *Zoology* **3**:110-11]
2135	B	X	Humming Bird: female of last? [note opposite] I believe is a female by Dissection
1834			July
2136	B		Hawk: female: Chiloe [listed as *Buteo erythronotus* in *Zoology* **3**:26]
2137:	I		2138: Heterom: Coleopt: P. St Julian
2139	I		Cicada, very abundant, uttering shrill cry on the plains of Patagonia.— P. Desire &c
2147	B	X	:48 Bones supposed to belong to the small ostrich. at P. St Julian [note opposite] Because the so far South, we know in St^{s} of Magellan there are no others

			August Valparaiso
2151			Corallina, abundant V 279 [*Corallina chilensis*, see *Plant Notes* p. 195]
2152	I		Pulex from Didelphis (2204)
2153	I	X	Ricinus from a Condor [note opposite] V ~~281(a)~~ account of Condor [see *Insect Notes* pp. 81-2]
2158	I		Coleopt. Onthoph: under stones, not dung feeder, rolls up like Armadillo

		X	Birds shot during August & September
2159	B		Partridge male [note opposite] Sexes distinguished by dissection by S. Covington. For many particulars respecting habits &c &c V **274**...**278** [see pp. 277-83]
2160	B		Pidgeon (large sort) female [listed as *Columba Fitzroyi* in *Zoology* **3**:114]
2161	B		Woodpecker male
2162	B		Owl male
2163	B		Small dove female [listed as *Zenaida Boliviana* or *Z. aurita* in *Zoology* **3**:115-16]
2164	B	X	Water Hen — male [note opposite] Bill "Grass & Emerald green", iris scarlet.— [listed as *Rallus sanguinolentus* Swains or *Gallinula crassirostris* J.E.Gray in *Zoology* **3**:133]
<u>1834</u>			August. Sept. Valparaiso
2165	B		Water Hen — female [see above]
2166	B		Plover, male.— middle claw serrated
2167	B		Lanius female [listed as *Agriornis gutturalis* in *Zoology* **3**:56, NHM 1855.12.19.344] *
2168	B		Snipe female [listed as *Scolopax (Telmatias) Paraguaiæ* in *Zoology* **3**:131]
2169	B		Callandra female } Lanius [listed as *Mimus thenca* G.R.Gray in *Zoology* **3**:61,
2170	B		do male } NHM 1855.12.19.230, labelled 2169D] *

2171 B Caprimulgus male
[listed as *Caprimulgus bifasciatus* Gould in *Zoology* **3**:36-7, NHM 1855.12.19.241] *
2172 B Myothera. "Turco". female [listed as *Pteroptochus megapodius* Kittl. in *Zoology* **3**:71-2]
2173 B do "Tapacola" female [listed as *Pteroptochus albicollis* Kittl. in *Zoology* **3**:72]
2174 B do do
2175 B Rara — male.—
2176 B do female
2177 B Blue sparrow — male
2178 B do do
2179 B Large Trochilus [listed as *Trochilus gigas* Vieill. in *Zoology* **3**:111-12]
2180 B do — male.—
2181 B :82 Larks — both males
2183 B Water Hen. male
2184 B Bittern — female
2185 B Woodpecker "Carpintero" female. [listed as *Picus kingii* G.R.Gray in *Zoology* **3**:113, NHM 1855.12.19.88, carries CD's own field label number 2185] *

<u>1834</u> Aug. Sept. — Valparaiso

2186 B Icterus (yellow spots on wing) Male
2187 B do female [listed as *Xanthornus chrysopterus* G.R.Gray in *Zoology* **3**:106, carries CD's own label numbered 2187] *
2188 B Arenaria — male.— [listed as *Hiaticula Azaræ* G.R.Gray in *Zoology* **3**:127, labelled 2188D] *
2189 B Fringilla — male — (yellow beak)
2190 B Red-throated Certhia — male [listed as *Synallaxis rufogularis* Gould in *Zoology* **3**:77, NHM 1855.12.19.170] *
2191 B Certhia (red wing) femaĺe [? listed as *Synallaxis humicola* Kittl. in *Zoology* **3**:75]
2192 B do male
2193 B Long-tailed tit — male.—
2194 B Wren — female [listed as *Troglodytes magellanicus* Gould in *Zoology* **3**:74, labelled 2194D] *
2195 B Black, head yellow Fringilla — male.— [listed as *Chrysometris campestris* Gould in *Zoology* **3**:89, NHM 1856.3.15.5, labelled 2195D] *
2196 B Emberiza — male
2197 B Muscicapa (grey bird) female.—
2198 B White tuft, Muscicapa of T. del Fuego (1819). female.—
2199 B Muscicapa of T. del F. — female
2200 B Swallow — male [listed with 2201 as *Hirundo leucopygia* Licht. or *H. cyanoleuca* Vieill. in *Zoology* **3**:40-1]
2201 B Swallow, other species — male [see above]
2202 A 2203. Rat with tuft on tail. Degu of Molina.— V **279** [listed as *Octodon Cumingii* in *Zoology* **2**:82-3]
2204 A X 2205. Opossum, excessively abundant; caught in traps by meat or cheese [note opposite] Inhabits the dry hills amongst the thickets. Can run up trees, but not very well.— does not use its tail much. Could distinguish in stomach Larvæ of Beetles (1038)
[listed as *Didelphis elegans* in *Zoology* **2**:95-6]

1834 Aug. Sept.— Valparaiso

2206 A Mus

2207 A X Large rat. Aconcagua, foot of Andes; as largest [notes opposite] Norway rat, its great ears gave it a very singular appearance.— I saw some of these climbing up Mimosa trees. [listed as *Abrocoma Bennettii* in *Zoology* **2**:85-6]

2208 B Red-back Muscicapa.—

2209..2213 I Coleoptera, under stones. on Mountains, valley of Aconcagua

2214 I X Serica, flying about in evening gr. numbers, 4000 ft. elevation [cont. opposite] Campana of Quillota, 6000 ft

2215 I Dromius, under dead bark, foot of Andes

2216 I Harpal. — Hab. do

2217 S Septaira, under stones, brook, valley of Cauquenes.— high up.—

2218 I Colymbetes, rapid brook. Hab. do

2219 I Coleopt. flying about in evening: 4000 ft. elevation.— Campana of Quillota

2220 B Dove — female.—

2296 B Myothera (Turco) (2172) [listed as *Pterooptochus megapodius* Kittl. in *Zoology* **3**:71-2]

2297 B Furnarius, same as La Plata 1222 [listed as *Furnarius cunicularius* G.R.Gray in *Zoology* **3**:65-6]

2298 B Tufted tit of T. del Fuego

2299 B Common Sparrow

1834 Aug. Sept. Valparaiso

2300 B Fringilla — female

2301 Planaria, white V **270**

2302 Planaria. "Chocolate R" central, longit. space cream colour: with two faint fine lines of the R: minutely spotted with white beneath white.—

2303.. 2308 I Coleoptera, Diptera &c: all the latter & most of others taken by sweeping in month of October.— [see *Insect Notes* p. 82]

2309 S F. water shells

2310 S Marine, tidal shells

2311 S F. W Unio

2312 A Rat, common about houses. (same as at S. Fe & La Plata?)

2313 A :14 Rat, caught in the country

2315 A X "Cururo", makes extensive [cont. opposite] burrowings, like the Toco Toco of La Plata, into which horses sink: not generally common, but abundant in some places (chiefly hills?) [listed as *Poephagomys ater* in *Zoology* **2**:82]

2316 ~~Fox~~ Lost

2317 I Hister, under dry human dung. abundant [note opposite] Red spots were much brighter

2318 I X Gonoleptes. certainly from [cont. opposite] West coast of S. America but I cannot find out what part.— given to me.—

1834 Sept & Novemb. Valparaiso

2319 B Nest of large Humming bird

2320 B Nest & eggs of Fringilla diuca, Molina [listed as *Fringilla diuca* in *Zoology* **3**:93]

2321 B Penguin, coast near Valparaiso [listed as *Spheniscus Humboldtii* Meyen in *Zoology* **3**:137]

2322 A Bat. Valparaiso:

2323 I X Curculio, first appears in November [note opposite] Very abundant, injurious to young shoots of plums & peaches: This is time of year when many Lamellicorn beetles first appear: [see *Insect Notes* p. 83]

2324		Stone with encrusting Corallines. 27 Fathoms. Tea. Island, W of W. Falkland showing little motion.—
2325	I	Lamellicorn [note opposite] Flying in numbers round the young peach trees. first appeared in first week in November.—

		December Archipelago of Chiloe
2326	I	Coleopt. in Fungus
2327	I X	Blue Carabus, under logs of wood in the forest: Is[d] of Lemuy [note opposite] I notice all the Blue ones are males & coppery ones females, yet surely they are diff. species: Do not Carabi abound in one sex at one period.— Emit a powerful acrid fluid & smell like some of the Heteromerous insects very disagreeable & powerful.— [see *Insect Notes* p. 83]
2328	I	Carabus: far more common. same Hab: & local.
2329	I	Brighter Variety (?) diff. locality
2330	I	Carab: Harpal. same Habitat & Locality
2331	I	Heterom. rotten Wood
2332	I	do , under stones, near beach
2333	I	Carab. Harpal. very abundant
<u>1834</u>		Nov. & Decem. Archi: of Chiloe
2334		Coralline (?). adhering in the back of a large Asterias
2335	S	Land shells collected by sweeping bushes
2336	S	Bulimus
2337	S	Fresh W shells
2338	I X	Elmis. small stream, under stone [note opposite] Various parts, East coast of Chiloe
2362	S	Balanus, adhering to a stick
2363	S	Marine shells
2364	S	do. same Hab & Locality with Peronia. (P **284**)
2365		Corallines encrusting very abundant
2366	S	F: Water shell.—
2367.. 2372	I X	Coleopt. Diptera &c &c collected by sweeping the bushes & some from a Fungus.— [note opposite] The whole country is one great forest [see *Insect Notes* pp. 84-5]
2373:74	S	Marine Shells.—
2375	B	Egg of Lanius (2124)
2376	I •	Elater from considerable height S. Pedro
2377	P	Lichen — Confirm V **287** .— Copy
2389	S	Land shell: within forest on hill. Cone Harbor. Peninsula de Tres Montes
<u>1834</u>		Decemb: Arch of Chonos
2414	I X	Lampyrus? the genus to which this insect belongs, is in number of <u>individuals</u> & [cont. opposite] species the most abundant kind in Chiloe & Chonos.
2415	I	Curculio (of T del Fuego). St Andrews Harb. Cape Tres Montes.
2416	I	Locality do: Carab: in rotten wood, high up on hilly forest.
2417	I	Curculio. Local & Hab, same
2418	I	Harpal: under log of wood. Local: do [see *Insect Notes* p. 85]
2419	I	Bee. Midship Bay, Chonos
2420	I	Libellula. East coast of Chiloe
2421		Peronia V **284**
2423	I	Corallinas. Midship Bay. Chonos

[*Corallina officinalis*, *Amphiroa orbigniana* and *Amphiroa darwinii*, see *Plant Notes* pp. 196-7]

2424 Coleoptera, thick forest: do, do [note opposite] In the very thick (cryptogamic flora) damp forest. Pselaphidæ & small Staphylidæ the most abundant insects

2425 B X Nest & eggs of Trochilus (2134) [note opposite] Is^{d} S. Pedro S. extreme of Chiloe

2426 B Egg of black Furnarius of coast. Midship Bay, Chonos

2427 B Egg of Partridge (2159) Valparaiso

2428:29 The two basal parts of Antipathis of which the 3^{d} piece is preserved in Spirits (1095) [in margin] Cabinet

1834 Decemb. C. Tres Montes

2430 Corall. encrusting in spots stones 13 Fathoms C. Tres Montes.— [Busk Collection: *Discoporella* sp.]

2431 A X Fox. (Blue fox of Molina) a not very common animal. SE point of Chiloe. [note opposite] killed by blow from my geological hammer, on the rocks on sea-beach. [listed as *Canis fulvipes* in *Zoology* **2**:12-13]

2432 A X Mus: Midship Bay, Chonos Archipelago [cont. opposite] on a small island!

2433 A Mus: do. East coast of Chiloe

2434 B Godwit. in large flocks. E. coast of Chiloe

2435 B Grebe do

2436 B Myothera — rare — do [note opposite] Called by inhabitants Cheuqui, the similarity of which word to Cheucau (2127) shows what is the case, the general resemblance in plumage & habits.— [listed as *Pteroptochus paradoxus* in *Zoology* **3**:73-4, NHM 1855.12.19.159] *

2438 I X Fly: bred from the soft putrid kelp on coast of Tres Montes [note opposite] I never saw such immense numbers in clusters under side of stones

2439 X ~~Crust: pelagic, night lines. do do~~ [note opposite] "Soundings" 30 Fathom a few miles off shore of C. Tres Montes; curious small white bodies

2440 X Planaria V **289** [note opposite] In drying. a broard crack was formed on the under surface, which is thus shown from the upper.—

2444.. 2455 I Insects, from under stones at an elevation of 2500 ft, bare granite mountain. "Patch Cove" North part of Tres Montes. 2444:46 Curious Hemipterous insects: it may be remarked there are three species of Curculio [cont. opposite] The Elater in numbers are far most abundant: this is good example for Alpine Entomology: for I sedulously turned up the stones: Libellula 2455 from base of Mountain.— [see *Insect Notes* pp. 85-6]

1834 Decemb: C. Tres Montes

2456 Land shells under logs of wood & on Fungi Ynche Is^{d}.— The smooth one abundant. [note opposite, crossed out] N.B. Mr Bynoe shot on Ynche Is^{d} large eared owl, stomach full of marine Decapod Crustaceæ.—

2457 Planaria V **295**: only fragments; ova can be seen: anterior part with eyes: & two pieces (right of number) of the saucer-shaped organ

2458 X Planaria; "cup" protruded: V **298**. [note opposite] The box also contains Specimen (2457).—

2462 I Carab. Trechus. Ynche Is^{d}. Ynche Is^{d}. Forest

2463 I Curculio —— do —— do

1835 Jan^{y}.

2474 Coronula. from whale. Chonos Archipel

2475 P X Little plant, very abundant. on hills Midship Bay, Chonos Arch: L. 45°.19′ [note opposite] This & the "Bog plant" of T del Fuego (& grass:), here form great beds of peat & the Latitude

45°!! December. 1834 V **314** [*Donatia fascicularis*. See *Plant Notes* p. 178]

2476 P X Cryptogam: (all ensuing ones) do do [note opposite] All these Cryptogam: were gathered in 5 minutes & within a space of 10 yards square.— a most wonderful profusion.— [see *Plant Notes* pp. 179-80 for a list of the liverworts and mosses collected by CD in the Chonos Archipelago that are in the herbaria of the British Museum (Natural History) and of Cambridge and Manchester Universities.]

2477 X Coralline: 40 Fathoms: Chonos Archipel. [note opposite] Has the structure of the Flustraceæ: Tentacula 14 or 16.— [Busk Collection: *Idmonea milneana*]

2478 Pebble from beach, showing manner of abundant growth of Corallina: now bleached.—Chonos Archipel:

2479 B 2480. Male & Female Woodpecker: high mountain. P. Tres Montes [listed as *Picus kingii* G.R.Gray in *Zoology* **3**:113, NHM 1855.12.19.101, 2480 carries CD's own field label] *

2481 B Dove do.— [listed as *Columba Fitzroyii* King in *Zoology* **3**:114]

<u>1835</u> Jany Chonos Archipel:—

2482 I :83 :84 : Coleoptera from B. Blanca, Patagonia [see *Insect Notes* p. 86]

2485 I Acari (black) under stones & on putrid vegetable matter; on beach in immense numbers.—

2486 I Fly (biting my flesh).—

2494 Pebble. 57 Fathoms encrusted with Corallines.—

2495 S Shell of living Concholepas, with cavities of Balanus (V **305**) & a small Pholus which inhabited the thicker part.

2496 X Planaria, closely allied habits, manner of crawling I hope to do (2457) [note opposite] Color of a uniform pale brown, had not time to examine its structure.— Lowes Harbor

2497 I Fly.— on Coast.— Lowes Harbor

2499 A Goats Head. Ynche Is^{d} V **312**

2500 Bits of Corall. shoal water. Lowes Harbor.—

2501 B X Curlew: Chonos: this bird is very [cont. opposite] abundant on all the mud-banks which surround Chiloe: are in large flocks: as they rise in flight utter shrill note. Specimen male.— [listed as *Numenius hudsonicus* in *Zoology* **3**:129]

2502 B Wren: male ~~V General observ.~~

2503 B Humming bird. Chonos Archipel. Male.— [see also 2134]

<u>1835</u> Jany 15th Chonos Archipel:

2504 S Voluta. Lowes Harbor

2505: I 06: Coleoptera in dense forest

2507 I Cicada

2508 I Carab: young

2509 I X Diptera. Hymenopt. Coleoptera: all above insects, taken on borders of wood by sweeping. Lowes Harbor.— [note opposite] There are two <u>excessively</u> minute Atomariæ.— [see *Insect Notes* pp. 86-7]

2510 Planaria V **308**

2520 I X Carabus.— Centre of Chiloe, in forest level of water: all three under one log of wood [note opposite] It is remarkable that the same variety (2329) is also a female & was equally found low down: Is it distinct species? — [for specimens 2520-2525 see *Insect Notes* p. 87]

2521 I Glow-worm. Centre of Chiloe.—

2523 I Insects, sweeping. Chiloe

2524 I 2525. Flys which bite both man & horses the first especially abundant: Chiloe

2526 Corallines. 96 Fathoms, inland sea East of Chiloe (& bit of a Pecten) [Busk Collection: *Discoporella fimbriata*]

2527 S Marine shells. Chonos Archipel

2528	P	Wild Potatoe V **314** Copy [*Solanum tuberosum* var. *vulgare*. See *Plant Notes* p. 180]
2529	A	Otter V **312** [listed as *Lutra chilensis* in *Zoology* **2**:22-4]
<u>1835</u>		Jan^{y}
2530	A	Nutria V **312**
2531	B	Barking bird (male). Chiloe [listed as *Pteroptochos tarnii* G.R.Gray in *Zoology* **3**:70-1, labelled 2531D] *
2532	S	Marine shells. Chonos Archip:
2544	I	2 Beetles from either Cocao or Sugar on board
2545	I	Insects from S. Carlos de Chiloe
2546	I	Meloe. common: crawling about grass & flying about. Cudico S. of Valdivia [note opposite] The Padre told me that the Indians use this as a Poison & likewise apply it as a Caustic
2554		Planaria V **298**(a).—
2555	B X	:56 2 Myothera. Baldivia [note opposite] 2556 is a male bird: stomach almost full of large seeds & remnants of few insects. [listed as *Pteroptochos paradoxus* G.R.Gray in *Zoology* **3**:73-4, NHM 1855.12.19.159] *
2557	I	:58/59 Insects, sweeping, in & on borders of forest. Baldivia.— [see *Insect Notes* pp. 85-6]
2560	S X	Fresh W Shells.— do —— [note opposite] The large specimens abound in water very slightly brackish.—
2561	I X	Pediculi V **315** & Pulex [note opposite] The Fleas may be compared with some of those I collected at St Fe.— [see *Insect Notes* p. 88]
2569	I	Fly which together with (2524:25) torments man & horse in forest of Chiloe.—
2594	S	Venus Concepcion
2595		Beautiful rose-color Sigillina (?) Moche Is^{d}.—
2596	I	:97 Heterom: Sand-dunes. Concepcion.—
2764	I	.......72 Small insects from Concepcion S. Covington [see *Insect Notes* p. 88]
1835		Ask me [in pencil]
2773	I X	:74:75. Small insects. Coquimbo, S. Cruz [cont. opposite] &c Insects of Coquimbo & Valparaiso taken in the Winter.— those of Concepcion in the Autumn:— [for specimens 2773-2837 see *Insect Notes* pp. 88-9]
2776		:2836:2837: do Valparaiso do
2838	I	Lamellicorn. Is^{d} of S. Maria
2839	I	40. Insects. Copiapò.—
2841	I X	Insects. Mendoza. Cicindela. Elmis [note opposite] The Cicindela comes from the saline mud-banks of R. Estacado: The Elmis & Colymbetes from the tepid & slightly mineral waters of Villa Vicencio in Cordilleras. The Cryptocephalus is Chilian insect. [see *Insect Notes* p. 89]
2842	S	Land shell. Cordilleras of Uspallata.
2843	S	do Valparaiso

Birds &c Coquimbo

2821	B	Coot — Concepcion
2822	B	Hawk male
2823	B X	Partridge Cordilleras of Coquimbo [note opposite] I saw this bird amongst snow of Cordilleras at Copiapò; covey of 5 rose together; uttered much noise fly like Grouse. wild.— very lofty, never descend. N.B. All these birds, shot by S. Covington in Winter [listed as *Attagis gayii* Less. in *Zoology* **3**:117]
2824	B	Turco 2172 [*Pteroptochos megapodius*]

2825	B	Like the Tapacola, first met this a little N of Illapel half way between Valparaiso & Coquimbo: habits nearly similar. does not so much erect tail. extends to Copiapò.—
2826	B	Furnarius on beach ? same (1823?) as in T del F: [listed as *Opetiorhynchus nigrofumosus* Gray in *Zoology* **3**:68-9, NHM 1855.12.19.244, labelled 2826D] *
2827	B	R. Negro Furnarius 1467 [listed as *Uppucerthia dumetoria* Geoffr. & D'Orb. in *Zoology* **3**:66, NHM 1855.12.19.75, but the skin is now in Edinburgh Museum, Reg. No. 1931.76.10] *
2828	B	Grey bird, very common in Traversias Male

<u>1835</u> Coquimbo

2829	B X	Port Julian. Fringilla [note opposite] S.Covington also saw the <u>little</u> bird of T. del Fuego & Patagonia with crest
2830	A	Mouse with large ears
2831	A	:32:33 Bat: a Vampire caught while sucking blood from back of a horse [listed as *Desmodus D'Orbingyi* in *Zoology* **2**:1-3]
2913	I	Bug mentioned by all authors [cont. opposite] as so great a pest near Mendoza; in the Traversias; sucks very much blood, frequents houses, but this was caught in sandy ravine of Cordilleras of Copiapò; called Benchuca, caught in my bed [see *Beagle Diary* p. 315 and *Insect Notes* p. 89]
2914	I	2915 Insects Valdivia
2916	I	Heterom. high valleys of East Cordilleras & Traversia of Mendoza
2917	I	Lamellicorn, abundant do Traversias
2918	S X	Balanus Coquimbo [note opposite] On a chain cable, this had been only six months under water
2919	S	Land shells do
2920	S	do do Concepcion
2921	S	Pecten Coquimbo
2922	S	:23 Marine shells in the neighbourhead of Coquimbo.—
3055	S.	Marine Shells, from rocks Iquique
3056	P X	Lichen, lying, without any adhæsion [cont. opposite] on the bare sand, at Iquique elevation 2-3000 ft. (where clouds often hang) sufficiently abundant in patches to give a green tint to sand seen from distant: I saw one other species of minute yellow Lichen on old Bones, & a Cactus on lofty rocks on coast. Besides these three, there is neither Cryptogam or Phanergam on coast or for 14 leagues inland: & this specimen are only seen on the coast mountains. [see *Plant Notes* pp. 180-1]

1835 August Lima

3057	S	Shell 12 fathom Iquique
3152	I	Locust V Private Journal P *[number omitted]* Mendoza [see *Beagle Diary* pp. 314-15]
3153	P	: 154: 155. Lichens. San Loronzo. Lima — 1000 ft. region of winter clouds.—
3187	A X	Large Fox Copiapò [note opposite] This animal was caught in a trap after having destroyed 200 fowls; is a bold animal, wanders about by day: runs very fast, & barks so precisely like a dog, when chased, that for some time I could not tell whether this sound came from the dog or fox.— The pupils in a dead animal appear round & very large, & this is the Culpen of Molina, whose account of habits I have heard repeated: inhabits whole of Chili, but have never seen it in the Traversias. [notes in margin] Fox of Chiloe 2431. I believe T. del Fuego but not Patagonia. [listed as *Canis Magellicanus* in *Zoology* **2**:10-12]
3188	A X	Small Fox. do. [Copiapò] [note opposite] This small & quite distinct Fox is found in the most desert Traversias (this was killed in the Despoblado) is common, as also in the plains of Patagonia. Is very abundant at Concepcion.— does not run fast, must live on Rodentia. Is

		called Chilla of the English Translation of Molina) is recognised as quite distinct animal: The blue fox of Chiloe is sent home; the common fox is the same as the specimens from B. Blanca: Thus I have all four:— [listed as *Canis azaræ* in *Zoology* **2**:14-16]
3189	B	Petrel. Callao Bay [listed as *Daption Capensis* in *Zoology* **3**:140-1, labelled in error 1389D]*
3190	B	Petrel Iquique Peru [listed as *Pelecanoides Garnotii* G.R.Gray in *Zoology* **3**:139]
3191	B	Sand Plover do. do.—
3192	P	Cryptogam. plants. Same locality with numbers 3153—155.—
3193	S	Land shells. North part of Chili
3194	S	Sea shells. Coquimbo
3195	I	Insect. (interesting) from the <u>country</u> near Callao [for specimens 3195-3201 see *Insect Notes* p. 90]
3196	I	& 97 Male & female Crysomela, <u>about</u> 1400 ft elevation, Lima lower limit of winter vegetation
3198		Crab. Callao
3199	I	Prionus. Valparaiso (interior country)
3200	I	Pulex (I believe irritans). Callao
3201	I	Insects. sweeping. Callao

<u>1835</u> Lima

3202	S	Shells. Lima, upper layer land.— second with Bulla from brackish lagune & a Pholas-formed shell from within a large Corall. I believe a Porites.— Callao
3203	S	Shells, S. Lorenzo do. [Callao]
3204	B	Tyrannus. do. [listed as *Pyrocephalus obscurus* Gould in *Zoology* **3**:45, NHM 1855.12.19.195 type] *
3227	I	Buprestis: between Guasco & Coquimbo common

Septemb: Galapagos Is[ds] — Chatham Is[d].

3228	I	Acarus, from great black Sea Guana or Lizard
3229	I	Fly, from Caracara (of Galapagos)
3230:	I	31: 32. Three Coleopt. Heterom. under stones on a hill.— [for specimens 3229-3232 see *Insect Notes* pp. 90-1]
3233	P	Plant, on rocky most barren Volcanic Hill.— [see *Beagle Diary* pp. 351-2]
3240	I	Acarus same as (3228)
3241	I	Acarus, from Pudenda of common great land Tortoise.—

Charles Is[d]

3242	P X	Herbaceous Shrub, common in the [cont. opposite] higher & inland parts, smell something like the Geranium: [*Scalesia affinis*, see *Plant Notes* p. 182]
3243	P	Woody shrub: odor like Honeysuckle. [*Lantana peduncularis*, see *Plant Notes* p. 182]
3244	P	Parasite, growing on various kinds of trees.— These three came [cont. opposite] from Charles Is[d].— [*Phoradendron henslovii*, see *Plant Notes* p. 182]

<u>1835</u> Sept[r.] Galapagos Is[d.]

3245	I X	Scolytus. branches of dead Mimosa [note opposite] long cavities in whole length of bough. Very numerous:— [for specimens 3245-3246 see *Insect Notes* p. 91]
3246	I	Staphylinus, under dead bird
3251		Corallina (& encrusting Corallines Box 4) 12 Fathoms

3252 Muricæa (Gorgonia) V **324**

3253 P Common Spiny bush, small scarlet flowers
[for specimens 3253-3259 see *Plant Notes* pp. 182-3]

3254 P The commonest bush in the Island [cont. opposite] of Chatham (a dry Isl^{d}) grows straggling 6-12 ft.; leaves brownish green, very few in numbers;

[note opposite] NB All the following Plants from Chatham Is^{d}
[in pencil, circled] Ask me When writing plants

3255 P The Largest tree; low, thick, one to 2 ft in diameter, crooked branches, few leaves; Balsamic odor, trunk thick in proportion: common

3256 P Wild Cotton tree. one of the commonest shrubs

3257 P Green thickets, bright green generally common near sea-side

3258 P Convolvulus-like plant, on sea-sand; flower pink

3259 P One of the commonest low bushes, small yellow flower.—

3260 (Number lost)

3261 Caryophillia. reddish orange

3262 do. fine bright yellow, both at dead low Water.—

1835 Octob. Galapagos Is

3263 Various Cellariæ, encrusting Corallines &c from 40 Fathoms deep
[Busk Collection: *Mucronella ventricosa*]

3264 Gorgonia. V **326**

3284 P X Cactus. Flower yellow: leaves rounded oval attached to each other in same plane [cont. opposite] generally: branches in different planes: trunk cylindrical. tapers but little 6-10 ft high.— beset with strong spines, diverging from the points, hence hirsute with stars.— common on rocky ground.—
[*Opuntia galapageia*, see *Plant Notes* pp. 183-4 and sketch below]

3285 P Fungus on Mimosa tree.

3293 P Large, succulent, clinging plant, grows high up in damp plants
[*Peperomia galioides*, see *Plant Notes* p. 184]

3294 P X Syngynesia; the characteristic & abundant [cont. opposite] tree in the high ground: grows to a good size: foliage pale bright green. Trunk well formed cylindrical. branches regular.— [*Scalesia pedunculata*, see *Plant Notes* pp. 184-5]

3295 P X Common tree in the intermediate [cont. opposite] ground: the berrys are eaten by the inhabitants & form main food for Tortoise & yellow Lizard: called Guyavitas. taste acid little sweet, astringent & turpentinic [*Psidium galapageium*, see *Plant Notes* p. 185] NB. All the above 5 species of plants come from James Is^{d}

3296 B Heron female

3297 B Caracara. male [listed as *Craxirex Galapagoensis* Gould in *Zoology* **3**:23-5]

3298 B do young Female

3299 B Duck, salt water lagoons: bill lead coloured, base of upper mandible purple with black mark above.— Male [listed as *Pœcilonitta Bahamensis* Eyton in *Zoology* **3**:135]

3300 B Bittern F^{s}. [listed as *Nycticorax violaceus* in *Zoology* **3**:128]

3301 B do.

3302 B Tern F [listed as *Megalopterus stolidus* in *Zoology* **3**:145]

3303 B Owl. male [listed as *Otus Galapagoensis* Gould in *Zoology* **3**:32-3]

1835 Octob. Galapagos Is^{ds}.

3304 All Bs Gull. Male [listed as *Larus fuliginosus* Gould in *Zoology* **3**:141-2]

-305 Dove. do. excessively numerous [listed as *Zenaida Galapagoensis* in *Zoology* **3**:115-16]

-306 Thenca Male. Charles Is^{d} [listed as *Mimus trifasciatus* G.R.Gray in *Zoology* **3**:62]

-307 do do Chatham Is^{d}
[listed as *Mimus melanotis* in *Zoology* **3**:62, NHM 1855.12.19.223 type] *

-308 Yellow-breast Tyrannus. F Chatham Is^{d} [listed as *Myiobius magnirostris* in *Zoology* **3**:48, NHM 1856.3.15.10, labelled 3308D] *

309 Scarlet do. M [listed as male and female of *Pyrocephalus nanus* Gould in *Zoology* **3**:45-6, NHM 1855.12.19.198 type, labelled 3309D] *

310 Wren. F [listed as *Certhidea olivacea* Gould among the Geospinizinæ in *Zoology* **3**:106. Either NHM 1855.12.19.164 or .127] *

3311 A X Mouse; these were very numerous on [cont. opposite] Chatham Is^{d}, which is uninhabited. There is a skeleton head (3361) [listed as *Mus Galapagoensis* in *Zoology* **2**:65-6]

3312 Fringilla M

313 do. (sex not known)

314 do F

315 do F

316 do M

317 do M

318 do M

319 do M [in pencil] ? Chatham Is^{d}??

320: Icterus (??), jet black. M.

321: 22. do. both M^{s}.

323 do. F

324 Fringilla. Male. (young?)

1835 Octob. Galapagos Is^{d}

3325 All Bs Fringilla (F)

326 X do (F) (there were very many [cont. opposite] individuals with exactly same plumage).—

327 do (M)

328 do (F)

329 do (F)

330 Finch (with parrot beak) M. James Is^{d}

331 do (F)
[listed as *Camarhynchus psittaculus* Gould in *Zoology* **3**:103. NHM 1855.12.19.12, now missing, and NHM 1855.12.19.22] *

332 Finch (M)

3333 do (M)

334 do (M)

335 do (M)

336 do (M)

337 X do (F) [note opposite] The upper mandible is in in pill-Box (3361)

338 do (F)

339 do (F)

340 (M.) [listed as *Certhidea olivacea* Gould in *Zoology* **3**:106, NHM 1855.12.19.126, labelled 3340D] *

341 X Finch. (M) [note opposite] I saw specimen with precisely similar plumage a female
342 Tyrannus. (M) (young of 3309?)
343 —— M
3344 —— F [listed as *Pyrocephalus nanus* Gould in *Zoology* **3**:45-6, labelled 3344D] *

[specimens 3312-3341 were identified by Gould in *Zoology* **3**:98-105 as belonging to new genera or subgenera *Geospiza*, *Camarhynchus*, *Cactornis* and *Certhidea* containing the species *G. magnirostris*, *G. strenua*, *G. fortis*, *G. nebulosa*, *G. fuliginosa*, *G. dentirostris*, *G. parvula*, *G. dubia*, *Camarhynchus psittaculus*, *Camarhynchus crassirostris*, *Cactornis scandens*, *Cactornis assimilis* and *Certhidea olivacea*. For the reasons discussed by Frank J. Sulloway in his article 'Darwin and his finches: the evolution of a legend' (*Journal of the History of Biology* Vol. 15, pp. 1-53, 1982), it is not always possible to identify the surviving specimens with CD's original numbers, some of them may have been collected by FitzRoy or other members of the *Beagle*'s crew, and there are doubts in deciding on exactly which island some of them were collected. However, NHM 1855.12.19.80 and .113 are identified as *G. magnirostris*; NHM 1855.12.19.81, .83 and .114 are *G. strenua*; NHM 1855.12.19.44 and NHM 1857.11.28.247 are *G. fuliginosa*; NHM 1855.12.19.176 is *G. dentirostris*; NHM 1855.12.19.167 and .194 are *G. parvula*; NHM 1855.12.19.22 is *Camarhynchus psittaculus*; NHM 1855.12.19.15 type is *Cactornis assimilis*; NHM 1855.12.19.125 is *Cactornis scandens*; NHM 1855.12.19.164 and .127 are *Certhidea olivacea*.]

Mr Westwood
Tahiti. Insects

Down House Notebook 63.6

C.Darwin

<u>Printed numbers</u>
3345

Red = 1000 + &c
Green = 2000 + &c
Yellow = 3000 + &c &c

1835		Octob. Galapagos Is^{ds}
3345	B X	Tyrannus (M) [note opposite] I believe this species is certainly distinct from the Scarlet-breasted one & (its yellow breasted female?) (3309).— [listed as *Pyrocephalus dubius* Gould in *Zoology* 3:46]
-346	B X	Sylvia M. frequently near the Coast
-347	B	do. M [listed as *Sylvicola aureola* Gould in *Zoology* 3:86, NHM 1856.3.15.14, labelled 3347D] *
-348	B	do F
-349	B	Thenca F. Albermale Is^{d}. [listed as *Mimus parvulus* Gray in *Zoology* 3:63-4. NHM 1855.12.19.92 type] *
-350	B	do M. James Is^{d}. V **330b** [listed as *Mimus melanotis* Gray in *Zoology* 3:62, NHM 1855.12.19.223 type] *
-351	B	Water Hen. F. [listed as *Zapornia spilonota* Gould in *Zoology* 3:132-3]
-352	B	do F
-353	B	do M
-354	B	Charadrius F [? listed as *Hiaticula semipalmata* in *Zoology* 3:128]
-355	B	Tringa M [? listed as *Pelidna minutilla* in *Zoology* 3:131]
-356	B	Swallow M [listed as *Progne modesta* in *Zoology* 3:39-40]
-357	B	Charadrius F
-358	B	:359 Tringa F^{s}
3360	A	Rat James Is^{d}. These abound [cont. opposite] all over this Isd: they do not appear Carnivorous, like common rats.— Are they the same as domestic rat of S. America? Have they been brought by Ships?— [listed as *Mus Jacobiae* in *Zoology* 2:34-5] NB For ornithological Notes V **340**
3361	A	Head of Mouse (3311) & bill of finch (3337)
3362	B X	Contents of the stomach of a Flamingo [note opposite] These sphærical, *[illeg.]* globules of calcareous matter appeared to me worthy of being kept.— There is so little calcareous matter in this Archipelago. The Bird was shot in a shallow salt-water Lagoon.— There was a mucous matter with the Sand: but nothing else.
3363	I	:64. Small insects, sweeping; high up, central parts of Charles Isd [see *Insect Notes* p. 91]
<u>1835</u>		Octob. Galapagos Is^{ds}.
3365	I	:66. Small insects. do — do James Is^{d} [see *Insect Notes* pp. 94-6]
3367	S	Land shells, beneath stones. Charles & Chatham Is^{d}
3368	S	:69. Lands shells. do — do James Id.
3370	S	Sea shells, tidal rocks do
3371	S X	:72 Sea shells. do do — various islands in the Archipelago.— [note opposite] There is a large Buccinum, with large mouth. mamillated exterior. which emits much milky fluid,

which subsequently stains everything a most beautiful purple.— Is the common shell in Is^{ds}.—

3373	S	do Chatham Is^{d}.—
3374	B	Anthus James Is^{d}. V Ornithology [listed as *Dolichonyx oryzivorus* in *Zoology* **3**:106, labelled 3374D at NHM] *
3375	B	Sterna. Shot in ocean early in the night of the 3rd of November. Many 100 miles from the land in Pacifick [listed as *Megalopterus stolidus* Boiè in *Zoology* **3**:145]

November Tahiti

3388	S X	Land shells. interior dwarf mountain, under stones: the elongated yellow [cont. opposite] Succinea on the wild Bean
3389	S X	F. Water shells: although various [cont. opposite] genera, & from distinct streams, all have the Apices imperfect: I did not find the generality of any species, but what were thus injured & What is the cause? In the same box are four marine shells. I forget whether from the Galapagos or Lima.—
3390	I	Small insects — sweeping.—
3391	P	Lichen.—
3392		Corall.—
3393	I	: 94.— Insects

1835 Decemb^{r}: Bay of Islands. New Zealand

3413	B	Bird (common)
3414	S	Marine Shells
3415:	I	16 : 17 : 18. Insects, sweeping: [see *Insect Notes* p. 96]
3419	S	Land & fresh water Shells
3420	I	Cicindela in extraordinary numbers, in all parts of the Country
3421	I	: 22. Insects inhabiting rotten wood
3423	I	Bug caught at Iquique, Peru. [note opposite] Is called in the Mendoza country Benchuca: is mentioned by many travellers as so great a pest & bloodsucker: inhabits crevices in old walls.— This specimen, when caught, was very thin; even on showing it a finger would, when placed on a table, immediately run at it with protruded sucker. Being allowed, sucked for 10 minutes: became bloated & globular, 5 or 6 times original size; 18 days afterwards was again ready to suck: Being kept 4 & ½ months, became of proper proportions, as thin as at first; I then killed it.— A most bold & fearless insect.— [see *Insect Notes* pp. 96-7]

February Hobart town. Van Diemen's land

3445	I	Staphylinus; Carrion.
3446	I	Aphodius, Cow's dung
3503		Corallinas. V **279**(b)
3504	I	Aphodius. Horse's dung.
3505	I	Aphodius. Cowe's dung
3506	I	: 07: 08: 09: 10: 11: 12. I believe include 3 species of Onthophagus, 2 latter common in Cows dung [see *Insect Notes* pp. 97-8]
3513	I	Phalacrus, in rotten wood: Has a Phalacrus been taken before out of Europe? [see *Insect Notes* pp. 97-8]

1836		February	Hobart Town
3514	I	Larva. beneath stones fresh Water	
3515	C	Ligia. tidal rocks	
3516	C	: 17. Crust do.—	
3518		: 19: 20: 21. Planaria V **363**	
3522	S	Land shells	
3523	S	Fresh Water Shells	
3524	I	: 25: 26. Insects by sweeping [seè *Insect Notes* pp. 98-9]	
3527	I	do : Alpine : Mount Wellington Elevation 3000 ft.—	
3528	I	Insects, sweeping near Sydney [see *Insect Notes* pp. 99-100]	
3529	S	:30 Land shells — do —	
3531	S X	Shells, living in a muddy salt water pool almost separate from the sea.— Sydney [note opposite] Same locality as the Oyster in Spirits (1356)	
3532	S	Littoral shells. Hobart Town	

			K. George's Sound
3550	I	Beetle, inhabiting in numbers a large flower [see *Insect Notes* p. 100]	
3556	I	Curculio, one of the most abundant insects here	
3557		Corallina (inarticulata V **161 & 2**)	
3558		Four species of Corallina from tidal rock, K George Sound. Color as general.—	
1836		March.	King George's Sound
3559	S	Bulimus. New Zealand	
3560	S	Bulimus, (2 species} from calcareous sand hills at Bold head: & a Physa, fresh water lake: K. George's Sound	
3561	I	Small insects, sweeping in coarse grass or brushwood.— do [see *Insect Notes* pp. 100-1]	
3562	S X	Natica, taken off tidal rocks. [note opposite] Being kept by accident in some dry paper in my cabin, I found to my astonishment that 12 days afterwards that the animal was quite alive:	
3563	S	Marine tidal shells	
3564	A	A mouse, caught amongst bushes by trap baited with cheese.— [a bush rat, listed as *Mus fuscipes* in *Zoology* **2**:66-7]	

		April	Keeling Isd
3580	X	Mass composed of layers of a pale red encrusting Corallina; from the extreme breakers.—	
3583	X	Branched stony Millepora V **356** [note opposite] Outer reefs in the most exposed places [probably *M. tenella/tenera*]	
3584		Madrepore, in the lagoon; rather strongly branched. pale brown. columnar V **355** [probably *Acropora*]	
3586		Foliaceous. Madrepore.— lagoon V **355** [probably *Turbinaria*]	
3587		Other species of do.— do. V **356**	
1836		April.	Keeling Is^{ds}.—
3588	I	Beetle, taken on board the Beagle.	
3590	A	Rat, excessively common on certain of the Islets: said to be brought from Mauritius in a ship which was wrecked.	

3591	B	Land rail: very common, excepting a snipe, the only bird which I saw without web-feet.— [listed as *Rallus phillipensis* Linn. in *Zoology* **3**:133]
3592	S	Marine shells.
3593	I	Insects sweeping; the small ant swarms in countless numbers.
3594	I	Hemirobius.— [see *Insect Notes* p. 101]
3595	P	Fungus. common on the decaying trunks of the Cocoa nut tree
3596	P	Fruit of a large tree; milky. green. grows by pairs or three: likewise, root of a small plant, which is sweet when cooked & is sometimes eaten. [*Nesiosperma oppositifolia*, see *Plant Notes* p. 185]
3597	S	Marine shells.—
3599		Seriatopora. common in the Lagoon [still a valid scleractinian genus]

1836 April Keeling Is^{ds}.—

3600		White branched Madrepore, exceedingly common in lagoon V **354** [branched, so probably *Acropora*]
3601		Also a common coral: Lagoon
3602		Common Madrepore: grows in crowns, especially abundant at Tahiti: lagoon [probably *Acropora*]
3603	X	One of the commonest Corals in the lagoon: when alive yellow [notes opposite] on being placed in fresh water & afterwards to dry; a jet black slimy substance, glossy, was emitted from whole surface: grows in circles, the parts near the circumference only living: [probably *Porites nigrescens*] NB. All the specimens 3599 to 3608 from lagoon: the most abundant kinds are the branched sorts. 3599:3600:3603:3601
3604		Lagoon coral.
3605		Meandrina. V **351** [probably *Leptoria phrygia*]
3606		:3607. 3 common in the lagoon.—
3608		Astreæ, encrusting the Chama gigas in the lagoon.— [possibly a *Favia*]
3609		Millepora, branching. same as (3583), grows on outer reefs & in 12 fathoms water V **356** [probably *M. tenella/tenera*]
3610		Millepora, growing in plates in the midst of the outer breakers. V **358** [probably *M. platyphylla*]
3611		do do do
3612		Coral. when alive the most beautiful pale lake red, common [cont. opposite] in holes on the outer reefs.— V **355** [possibly an outer reef species of *Pocillopora*]

1836 April Keeling Is^{ds}

3613		Strong branched Madrepora growing in holes, outer reef.—
3614		Astrea, forming the grand masses in the midst of the outer breakers.—
3615		Pale colored Corallina, growing in rounded processes: common in the outer breakers.
3616		Delicate yet inflexible Corallina.— Hab. do: not so common: color when alive (has now faded) most beautiful "Peach blossom with lake red"

Mauritius

3633		Seriatopora, in 20 Fathom water
3634		Branching Millepora, part of it encrusting a tubiform shell.
3635	I	Water beetle, mountains stream [see *Insect Notes* p. 101]
3636	A	Animal.—

1836

3637	P	Moss on dead cocoa nut trees in woods of Keeling Is^{ds}. [*Hypnum rufescens*, see *Plant Notes* p. 185]
3638		Halimeda & other sea weeds on reefs. Keeling Is^{ds}.— [*Halimeda macroloba*, see *Plant Notes* p. 199]

June Cape of Good Hope

3686		Corallina V **57**(3)
3687	S	Land shells
3688	I	:89:90:91. Small insects sweeping in valleys of mountains near Simon's Bay.
3692	I	Acarus from the common land Tortoise of the Cape.—
3693	I	:94.95.96.97.98: small Aphodii very numerous beneath dung.
3699	S	Shell from the Keeling Is^{ds}.

St Helena

3729	S	Land Shells. the Succinea like shell very common on the bare Volcanic Hills

1836 July St Helena

3730	I	Small insects, sweeping high central land
3731		Spider caught in the ship
3818		Case of nereidous animal see (1462) Spirits.—
3819	I X	:20 Very common beetle beneath dung on higher parts of St. Helena [note opposite] This is the most extraordinary instance yet met with of transportal, or change in habits of Stercovorous insects [see *Insect Notes* p. 103]
3821	I	:22 Aphodius. do. do. do.
3823	I	: 24: 25: 26: 27: 28: 29: Flys & other insects taken on the mountainous parts & far from houses in Ascension

Bahia Brazil

3854	X	:55 Corallina on round stones tidal pools [note opposite] In some of these specimens orifices for the gemmules may be seen.—
3855		I believe same species.
3856		A distinct & common species extensively coating smooth surfaces of tidal pools in granitic rocks. Colour much darker than in last.
3858:	I X	59: 60. Small insects sweeping in forest & open places. [note opposite] These insects products of two whole days sweeping.— After winter rainy season. Beginning of August.— [see *Insect Notes* p. 103]

1836.

3861:	I	62: 63 : 64. Insects. Bahia
3865:	I	66: 67. Insects. Ascension
3882	S	Balanus, growing in clusters on the points of sandstone on the reef at Pernambuco
3900	B X	Bird from summit of barren arid mountain of Ascension [note opposite] Female

3901 A Rat, in great numbers inhabiting the high central part of mountain of Ascension, separated from coast by broad, perfectly arid waste of lava: was found here when island first settled. live in burrows. feed in the day.; are all black & glossy fur. Ascension
[listed as *Mus (Rattus* var.? *insularis)* in *Zoology* **2**:35-7]

3902 A Rat killed near the houses at the beach, at a spot where the turtle are killed. [another variety of *Mus Rattus*]

3903 A X : 3904. Mice, appear to be abori= [cont. opposite] ginal, like the Rat (3901). Inhabit same parts in numbers. Ascension. [listed as *Mus musculus* in *Zoology* **2**:38]

1836 Port Praya. C. de Verd

3905 B X Bird. Inhabits most arid lava plains, runs, habits very like the La Plata small Furnarius.
Female [note opposite] The common sparrow, is now building; habits like the S. American species.— common.— end of August.— [further note added later] 3905 Melancora c. [listed as *Melanocorypha cinctura* Gould in *Zoology* **3**: 87, labelled 3905D] *

3906 B Bird, inhabiting do plains. Male [note added later] 3906 Pyrrhulauda nigriceps
[listed as *Pyrrhalauda nigriceps* Gould in *Zoology* **3**:87-8, labelled 3906D] *

3907 B Swallow.— Female [listed as *Cypselus unicolor* Vieill. in *Zoology* **3**:41, 3907 in database] *

[A Catalogue follows of 'Everything not in Spirits', listing the specimens in bottles A to Q]

[On the last page of the list of Shells in Spirits of Wine copied out by Covington in CUL MS DAR 29.1, some additional specimens have later been added in CD's hand]

3915 Oysters Keeling Isl[ds]
3916 Ovules of Shell, (common Trochus?), Falkland Isl[d]
3917 Tubinicella from Whale (spermacetti) Lat 44°:30′ Jan. 1835. Chonos Arch:
3918 Balanus. Wollaston Isl[d] (very abundant)
3919 Mya. dug out of mud bank 6 inches beneath surface. very abundant — Bahia Blanca
3920 Bulla. body yellowish. Callao Bay. Lima — Peru
3921 Balanus. 19 Fathoms. 5 miles from the shore. Lat 48°:56′ S. coast of Patagonia

[Written at right angles across righthand side of page]

[illegible word] catalogue of everything *[more illegible words]* Spirits. The letter refers to the Bottle from A to Z. Each specimen is marked with one of the letters in the 8. Red Catalogues, so that it can be found by turning out contents of one bottle.—

[These lists contain no information now of value, and have therefore been omitted]

Index of animals and plants

[This index includes the often obsolete and sometimes incorrect generic names used by CD in the text, but excludes those mentioned only in the Specimen Lists. Some of his spellings are shown in brackets, but other inconsistencies have been ignored. More up-to-date generic and specific names are to be found in the footnotes.]

Index of people, ships and places

[CD's aberrant and sometimes inconsistent spelling of place names is given in brackets]